Lecture Notes in Computer Scie

Commenced Publication in 1973
Founding and Former Series Editors:
Gerhard Goos, Juris Hartmanis, and Jan van Leeuwen

Editorial Board

Alexander Gelbukh (Ed.)

Computational Linguistics and Intelligent Text Processing

14th International Conference, CICLing 2013
Samos, Greece, March 24-30, 2013
Proceedings, Part I

 Springer

Volume Editor

Alexander Gelbukh
National Polytechnic Institute
Center for Computing Research
Av. Juan Dios Bátiz, Col. Nueva Industrial Vallejo
07738 Mexico D.F., Mexico

ISSN 0302-9743   e-ISSN 1611-3349
ISBN 978-3-642-37246-9   e-ISBN 978-3-642-37247-6
DOI 10.1007/978-3-642-37247-6
Springer Heidelberg Dordrecht London New York

Library of Congress Control Number: 2013933372

CR Subject Classification (1998): H.3, H.4, F.1, I.2, H.5, H.2.8, I.5

LNCS Sublibrary: SL 1 – Theoretical Computer Science and General Issues

Typesetting: Camera-ready by author, data conversion by Scientific Publishing Services, Chennai, India

Printed on acid-free paper

Springer is part of Springer Science+Business Media (www.springer.com)

Preface

CICLing 2013 was the 14th Annual Conference on Intelligent Text Processing and Computational Linguistics. The CICLing conferences provide a wide-scope forum for discussion of the art and craft of natural language processing research as well as the best practices in its applications.

This set of two books contains four invited papers and a selection of regular papers accepted for presentation at the conference. Since 2001, the proceedings of the CICLing conferences have been published in Springer's *Lecture Notes in Computer Science* series as volume numbers 2004, 2276, 2588, 2945, 3406, 3878, 4394, 4919, 5449, 6008, 6608, 6609, 7181, and 7182.

The set has been structured into 12 sections:

- General Techniques
- Lexical Resources
- Morphology and Tokenization
- Syntax and Named Entity Recognition
- Word Sense Disambiguation and Coreference Resolution
- Semantics and Discourse
- Sentiment, Polarity, Emotion, Subjectivity, and Opinion
- Machine Translation and Multilingualism
- Text Mining, Information Extraction, and Information Retrieval
- Text Summarization
- Stylometry and Text Simplification
- Applications

The 2013 event received a record high number of submissions in the 14-year history of the CICLing series. A total of 354 papers by 788 authors from 55 countries were submitted for evaluation by the International Program Committee; see Figure 1 and Tables 1 and 2. This two-volume set contains revised versions of 87 regular papers selected for presentation; thus the acceptance rate for this set was 24.6%.

The book features invited papers by

- Sophia Ananiadou, University of Manchester, UK
- Walter Daelemans, University of Antwerp, Belgium
- Roberto Navigli, Sapienza University of Rome, Italy
- Michael Thelwall, University of Wolverhampton, UK

who presented excellent keynote lectures at the conference. Publication of full-text invited papers in the proceedings is a distinctive feature of the CICLing conferences. Furthermore, in addition to presentation of their invited papers, the keynote speakers organized separate vivid informal events; this is also a distinctive feature of this conference series.

Table 1. Number of submissions and accepted papers by topic[1]

Accepted	Submitted	% accepted	Topic
18	75	24	Text mining
18	64	28	Semantics, pragmatics, discourse
17	80	21	Information extraction
17	67	25	Lexical resources
14	44	32	Other
14	35	40	Emotions, sentiment analysis, opinion mining
13	40	33	Practical applications
11	52	21	Information retrieval
11	51	22	Machine translation and multilingualism
8	30	27	Syntax and chunking
7	40	17	Underresourced languages
7	39	18	Clustering and categorization
6	23	26	Summarization
5	32	16	Morphology
5	24	21	Word sense disambiguation
5	19	26	Named entity recognition
4	20	20	Noisy text processing and cleaning
4	17	24	Social networks and microblogging
4	13	31	Natural language generation
3	11	27	Coreference resolution
3	9	33	Natural language interfaces
3	8	38	Question answering
2	23	9	Formalisms and knowledge representation
2	18	11	POS tagging
2	2	100	Computational humor
1	11	9	Speech processing
1	11	9	Computational terminology
1	8	12	Spelling and grammar checking
1	3	33	Textual entailment

[1] As indicated by the authors. A paper may belong to several topics.

With this event we continued with our policy of giving preference to papers with verifiable and reproducible results: in addition to the verbal description of their findings given in the paper, we encouraged the authors to provide a proof of their claims in electronic form. If the paper claimed experimental results, we asked the authors to make available to the community all the input data necessary to verify and reproduce these results; if it claimed to introduce an algorithm, we encouraged the authors to make the algorithm itself, in a programming language, available to the public. This additional electronic material will be permanently stored on the CICLing's server, www.CICLing.org, and will be available to the readers of the corresponding paper for download under a license that permits its free use for research purposes.

In the long run we expect that computational linguistics will have verifiability and clarity standards similar to those of mathematics: in mathematics, each

Table 2. Number of submitted and accepted papers by country or region

Country or region	Authors Subm.	Papers[2] Subm.	Accp.	Country or region	Authors Subm.	Papers[2] Subm.	Accp.
Algeria	4	4	–	Malaysia	7	1.67	1
Argentina	3	1	–	Malta	1	1	–
Australia	3	1	–	Mexico	14	6.25	3.25
Austria	1	1	–	Moldova	3	1	–
Belgium	3	1	1	Morocco	7	4	1
Brazil	13	6.83	2	Netherlands	8	4.50	1
Canada	11	4.53	1.2	New Zealand	5	1.67	–
China	57	21.72	3.55	Norway	6	2.92	0.92
Colombia	2	1	1	Pakistan	5	2	–
Croatia	5	2	2	Poland	8	3.75	0.75
Czech Rep.	10	5	2	Portugal	9	3	–
Egypt	22	11.67	1	Qatar	2	0.67	–
Finland	2	0.67	–	Romania	14	9.67	2
France	64	25.9	5.65	Russia	15	4.75	1
Georgia	1	1	0.5	Singapore	5	2.25	0.25
Germany	32	13.92	6.08	Slovakia	2	1	–
Greece	21	6.12	2.12	Spain	39	15.50	8.75
Hong Kong	9	2.53	0.2	Sweden	2	2	–
Hungary	12	6	–	Switzerland	8	3.83	1.33
India	98	49.2	5.6	Taiwan	1	1	–
Iran	14	11.33	–	Tunisia	24	11	2
Ireland	6	4.5	1.5	Turkey	11	6.25	3.25
Italy	22	11.37	4.5	Ukraine	2	1.25	0.50
Japan	48	20.5	5	UAE	1	0.33	–
Kazakhstan	10	3.75	–	UK	35	15.73	5.20
Korea, South	7	3	–	USA	54	18.98	8.90
Latvia	6	2	1	Viet Nam	8	3.50	–
Macao	6	2	–	*Total:*	788	354	87

[2] By the number of authors: e.g., a paper by two authors from the USA and one from UK is counted as 0.67 for the USA and 0.33 for UK.

claim is accompanied by a complete and verifiable proof (usually much longer than the claim itself); each theorem's complete and precise proof—and not just a vague description of its general idea—is made available to the reader. Electronic media allow computational linguists to provide material analogous to the proofs and formulas in mathematics in full length—which can amount to megabytes or gigabytes of data—separately from a 12-page description published in the book. More information can be found on www.CICLing.org/why_verify.htm.

To encourage providing algorithms and data along with the published papers, we selected a winner of our Verifiability, Reproducibility, and Working Description Award. The main factors in choosing the awarded submission were technical correctness and completeness, readability of the code and documentation, simplicity of installation and use, and exact correspondence to the claims of the

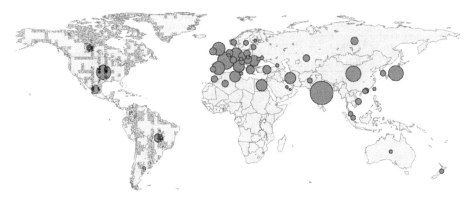

Fig. 1. Submissions by country or region. The area of a circle represents the number of submitted papers.

paper. Unnecessary sophistication of the user interface was discouraged; novelty and usefulness of the results were not evaluated—instead, they were evaluated for the paper itself and not for the data. This year's winning paper was published in a separate proceedings volume and is not included in this set.

The following papers received the Best Paper Awards, the Best Student Paper Award, as well as the Verifiability, Reproducibility, and Working Description Award, correspondingly (the best student paper was selected among papers of which the first author was a full-time student, excluding the papers that received a Best Paper Award):

1st Place: *Automatic Detection of Idiomatic Clauses*, by Anna Feldman and Jing Peng, USA;

2nd Place: *Topic-Oriented Words as Features for Named Entity Recognition*, by Ziqi Zhang, Trevor Cohn, and Fabio Ciravegna, UK;

3rd Place: *Five Languages are Better than One: An Attempt to Bypass the Data Acquisition Bottleneck for WSD*, by Els Lefever, Veronique Hoste, and Martine De Cock, Belgium;

Student: *Domain Adaptation in Statistical Machine Translation Using Comparable Corpora: Case Study for English-Latvian IT Localisation*, by Mārcis Pinnis, Inguna Skadiņa, and Andrejs Vasiļjevs, Latvia;

Verifiability: *Linguistically-Driven Selection of Correct Arcs for Dependency Parsing*, by Felice Dell'Orletta, Giulia Venturi, and Simonetta Montemagni, Italy.

The authors of the awarded papers (except for the Verifiability Award) were given extended time for their presentations. In addition, the Best Presentation Award and the Best Poster Award winners were selected by a ballot among the attendees of the conference.

Besides its high scientific level, one of the success factors of CICLing conferences is their excellent cultural program. The attendees of the conference had a chance to visit unique historical places: the Greek island of Samos, the birthplace of Pythagoras (Pythagorean theorem!), Aristarchus (who first realized that the Earth rotates around the Sun and not vice versa), and Epicurus (one of the founders of the scientific method); the Greek island of Patmos, where John the Apostle received his visions of the Apocalypse; and the huge and magnificent archeological site of Ephesus in Turkey, where stood the Temple of Artemis, one of the Seven Wonders of the World (destroyed by Herostratus), and where the Virgin Mary is believed to have spent the last years of her life.

I would like to thank all those involved in the organization of this conference. In the first place these are the authors of the papers that constitute this book: it is the excellence of their research work that gives value to the book and sense to the work of all other people. I thank all those who served on the Program Committee, Software Reviewing Committee, Award Selection Committee, as well as additional reviewers, for their hard and very professional work. Special thanks go to Ted Pedersen, Adam Kilgarriff, Viktor Pekar, Ken Church, Horacio Rodriguez, Grigori Sidorov, and Thamar Solorio for their invaluable support in the reviewing process.

I would like to thank the conference staff, volunteers, and the members of the local organization committee headed by Dr. Efstathios Stamatatos. In particular, we are grateful to Dr. Ergina Kavallieratou for her great effort in planning the cultural program and Mrs. Manto Katsiani for her invaluable secretarial and logistics support. We are deeply grateful to the Department of Information and Communication Systems Engineering of the University of the Aegean for its generous support and sponsorship. Special thanks go to the Union of Vinicultural Cooperatives of Samos (EOSS), A. Giannoulis Ltd., and the Municipality of Samos for their kind sponsorship. We also acknowledge the support received from the project WIQ-EI (FP7-PEOPLE-2010-IRSES: Web Information Quality Evaluation Initiative).

The entire submission and reviewing process was supported for free by the EasyChair system (www.EasyChair.org). Last but not least, I deeply appreciate the Springer staff's patience and help in editing these volumes and getting them printed in record short time—it is always a great pleasure to work with Springer.

February 2013   Alexander Gelbukh

Organization

CICLing 2013 is hosted by the University of the Aegean and is organized by the CICLing 2013 Organizing Committee in conjunction with the Natural Language and Text Processing Laboratory of the CIC (Centro de Investigación en Computación) of the IPN (Instituto Politécnico Nacional), Mexico.

Organizing Chair

Efstathios Stamatatos

Organizing Committee

Efstathios Stamatatos (chair)
Ergina Kavallieratou
Manolis Maragoudakis

Program Chair

Alexander Gelbukh

Program Committee

Ajith Abraham
Marianna Apidianaki
Bogdan Babych
Ricardo Baeza-Yates
Kalika Bali
Sivaji Bandyopadhyay
Srinivas Bangalore
Leslie Barrett
Roberto Basili
Anja Belz
Pushpak Bhattacharyya
Igor Boguslavsky
António Branco
Nicoletta Calzolari
Nick Campbell
Michael Carl
Ken Church

Dan Cristea
Walter Daelemans
Anna Feldman
Alexander Gelbukh (chair)
Gregory Grefenstette
Eva Hajicova
Yasunari Harada
Koiti Hasida
Iris Hendrickx
Ales Horak
Veronique Hoste
Nancy Ide
Diana Inkpen
Hitoshi Isahara
Sylvain Kahane
Alma Kharrat
Adam Kilgarriff

Philipp Koehn
Valia Kordoni
Leila Kosseim
Mathieu Lafourcade
Krister Lindén
Elena Lloret
Bente Maegaard
Bernardo Magnini
Cerstin Mahlow
Sun Maosong
Katja Markert
Diana Mccarthy
Rada Mihalcea
Jean-Luc Minel
Ruslan Mitkov
Dunja Mladenic
Marie-Francine Moens
Masaki Murata
Preslav Nakov
Vivi Nastase
Costanza Navarretta
Roberto Navigli
Vincent Ng
Kjetil Nørvåg
Constantin Orasan
Ekaterina Ovchinnikova
Ted Pedersen
Viktor Pekar
Anselmo Peñas
Maria Pinango
Octavian Popescu

Irina Prodanof
James Pustejovsky
German Rigau
Fabio Rinaldi
Horacio Rodriguez
Paolo Rosso
Vasile Rus
Horacio Saggion
Franco Salvetti
Roser Sauri
Hinrich Schütze
Satoshi Sekine
Serge Sharoff
Grigori Sidorov
Kiril Simov
Vaclav Snasel
Thamar Solorio
Lucia Specia
Efstathios Stamatatos
Josef Steinberger
Ralf Steinberger
Vera Lúcia Strube de Lima
Mike Thelwall
George Tsatsaronis
Dan Tufis
Olga Uryupina
Karin Verspoor
Manuel Vilares Ferro
Aline Villavicencio
Piotr W. Fuglewicz
Annie Zaenen

Software Reviewing Committee

Ted Pedersen
Florian Holz
Miloš Jakubíček

Sergio Jiménez Vargas
Miikka Silfverberg
Ronald Winnemöller

Award Committee

Alexander Gelbukh
Eduard Hovy
Rada Mihalcea

Ted Pedersen
Yorick Wiks

Additional Referees

Rodrigo Agerri
Katsiaryna Aharodnik
Ahmed Ali
Tanveer Ali
Alexandre Allauzen
Maya Ando
Javier Artiles
Wilker Aziz
Vt Baisa
Alexandra Balahur
Somnath Banerjee
Liliana Barrio-Alvers
Adrián Blanco
Francis Bond
Dave Carter
Chen Chen
Jae-Woong Choe
Simon Clematide
Geert Coorman
Victor Darriba
Dipankar Das
Orphee De Clercq
Ariani Di Felippo
Maud Ehrmann
Daniel Eisinger
Ismail El Maarouf
Tilia Ellendorff
Milagros Fernández Gavilanes
Santiago Fernández Lanza
Daniel Fernández-González
Karën Fort
Koldo Gojenola
Gintare Grigonyte
Masato Hagiwara
Kazi Saidul Hasan
Eva Hasler
Stefan Höfler
Chris Hokamp
Adrian Iftene
Iustina Ilisei
Leonid Iomdin
Milos Jakubicek
Francisco Javier Guzman

Nattiya Kanhabua
Aharodnik Katya
Kurt Keena
Natalia Konstantinova
Vojtech Kovar
Kow Kuroda
Gorka Labaka
Shibamouli Lahiri
Egoitz Laparra
Els Lefever
Lucelene Lopes
Oier López de La Calle
John Lowe
Shamima Mithun
Tapabrata Mondal
Silvia Moraes
Mihai Alex Moruz
Koji Murakami
Sofia N. Galicia-Haro
Vasek Nemcik
Zuzana Neverilova
Anthony Nguyen
Inna Novalija
Neil O'Hare
John Osborne
Santanu Pal
Feng Pan
Thiago Pardo
Veronica Perez Rosas
Michael Piotrowski
Ionut Cristian Pistol
Soujanya Poria
Luz Rello
Noushin Rezapour Asheghi
Francisco Ribadas-Pena
Alexandra Roshchina
Tobias Roth
Jan Rupnik
Upendra Sapkota
Gerold Schneider
Djamé Seddah
Keiji Shinzato
João Silva

Website and Contact

The webpage of the CICLing conference series is www.CICLing.org. It contains information about past CICLing conferences and their satellite events, including published papers or their abstracts, photos, video recordings of keynote talks, as well as information about the forthcoming CICLing conferences and contact options.

Table of Contents – Part I

Morphology and Tokenization

Syntax and Named Entity Recognition

Word Sense Disambiguation and Coreference Resolution

Semantics and Discourse

Table of Contents – Part II

Sentiment, Polarity, Emotion, Subjectivity, and Opinion

Machine Translation and Multilingualism

Text Mining, Information Extraction, and Information Retrieval

Text Summarization

Stylometry and Text Simplification

Applications

Unsupervised Feature Adaptation for Cross-Domain NLP with an Application to Compositionality Grading

Lukas Michelbacher, Qi Han, and Hinrich Schütze

Institute for Natural Language Processing, University of Stuttgart
{michells,hanqi}@ims.uni-stuttgart.de, hs999@ifnlp.org

Abstract. In this paper, we introduce *feature adaptation*, an unsupervised method for cross-domain natural language processing (NLP). Feature adaptation adapts a supervised NLP system to a new domain by recomputing feature values while retaining the model and the feature definitions used on the original domain. We demonstrate the effectiveness of feature adaptation through cross-domain experiments in compositionality grading and show that it rivals supervised target domain systems when moving from generic web text to a specialized physics text domain.

1 Introduction

Supervised machine learning has been successfully applied to many NLP problems. However, performance of machine-learned systems is often negatively impacted by the fact that the distributions of the labeled training set – the source domain – and that of the real-world data the system is applied to – the target domain – are different. To address this challenge, domain adaptation and cross-domain NLP have become a focus of NLP research [1,2]. Many successful adaptation approaches require some labeled target domain data (e.g. [1]). In this paper, we are interested in methods that do not make any such requirement. We will call these methods *cross-domain* and distinguish them from *domain adaptation*, a term we reserve for methods that use target labels.

In this paper, we propose a new method for cross-domain NLP: *feature adaptation*. The basic idea of feature adaptation is that the feature definitions and the model learned on the source domain are retained, but the feature values are recomputed in an unsupervised fashion on the target domain. As we will discuss below, feature adaptation can only be used for what we call *domain-independent features* whose values can be computed on target domain data without domain-specific knowledge. In contrast, a *domain-specific feature* either cannot be computed without domain-specific knowledge or does not make sense or changes its semantics when it is mechanically computed on the target domain according to its feature definition.

The problem to which we apply feature adaptation is *compositionality grading*, the task of assessing the level of semantic transparency of a word combination [3]. Compositionality grading is an important aspect of semantic processing and text

A. Gelbukh (Ed.): CICLing 2013, Part I, LNCS 7816, pp. 1–12, 2013.

processing in general because multi-word expressions such as lexicalized phrases and idiomatic expressions must often be treated differently from compositional combinations [4]. For example, a question answering system should know that Colombian coffee is a type of coffee, but Irish coffee is not.

We investigate two cross-domain settings for compositionality grading in this paper, where we use *domain* as a general umbrella term for *a type or genre of language with specific properties*. The first, more traditional cross-domain setting adapts a classifier trained on generic web text to a corpus of scientific physics text. The second setting is a cross-language setting: a system trained on English is applied to German and vice versa. We refer to the web-to-physics setting as *cross-topic* (as opposed to cross-domain) to be able to distinguish two different settings of cross-domain NLP in this paper: cross-topic and cross-language. To summarize our terminology: we use cross-domain as a generic term for unsupervised adaptation of a classifier to a target domain; and cross-topic and cross-language for the web/physics and English/German instances of cross-domain, respectively.

For evaluation, we use the public evaluation sets of DiSCo [3] and a new compositionality grading resource that we created for a scientific physics corpus and that we are publishing together with this paper.[1] We demonstrate experimentally that feature adaptation is effective for the two cross-domain scenarios, cross-topic and cross-language.

We make two main contributions in this paper. First, we introduce a new cross-domain NLP method, feature adaptation, and show its effectiveness for compositionality grading. Feature adaptation is simpler than other cross-domain methods. In contrast to domain adaptation, it requires no target labels. Second, we conduct the first cross-domain study on compositionality grading and show that – since it is a problem that can be addressed with domain-independent features – it is a task that is particularly well-suited for cross-domain approaches. We also provide a new resource for evaluation of compositionality grading in scientific text.

The paper is organized as follows. In Section 2 we describe and discuss feature adaptation, domain-independent and domain-specific features. Section 3 describes the experimental setup and the features and methods we use for compositionality grading. We discuss our results in Section 4. Section 5 discusses related work. Section 6 concludes.

2   Feature Adaptation and Feature Types

Feature Adaptation. The performance of supervised NLP systems trained on a source domain generally degrades when they are applied to a distinct target domain. This has been shown, for example, for parsing [5,6], sentiment classification [7] and machine translation [8]. We observe the same effect in our experiments on compositionality grading below.

---

[1] The dataset is available for download at `http://www.CICLing.org/2013/data/166`

The simplest cross-domain method is to use the source domain NLP system as is on the target domain without any adaptation. We refer to this baseline as *zero adaptation* in this paper.

We define *feature adaptation* as the process of taking a model trained on the source domain and using it unchanged on the target domain with the exception that the *feature values are recomputed on the target domain* according to the feature definitions. In other words, feature adaptation uses the instructions of how to compute a feature's value and carries them out to compute the value on the target domain. Feature adaptation is similar to zero adaptation in that the model and its parameters are not changed. It differs from zero adaptation in that the feature values are recomputed on the target domain.

Feature adaptation is not a method that can be applied in general to all problems because only particular types of features can be adapted by way of feature adaptation. We distinguish two types of features: domain-independent features, which can be recomputed on the target domain, and domain-specific features, which cannot be recomputed.

Domain-Independent Features are features that can be adapted using feature adaptation. The easiest way to explain domain-independence is to give as examples the two types of domain-independent features we will use below: association measures and vector similarity. They can be recomputed on the target domain without domain-specific knowledge and then have the same semantics for the target domain classifier as for the source domain classifier.

Consider feature adaptation for the feature "vector similarity between two words" from the domain mythology to the astrophysics domain. The vector similarity of "dwarf" and "star" will be low when computed for the source domain since mythological beings and celestial objects are semantically different. But when we recompute the similarity for the target domain – where both words refer to types of stars – the similarity will be high. After feature adaptation, the feature value reflects the distributional characteristics of the target domain. The feature is thus adapted by recomputing its value based on the target-domain distribution.

Note that because of the recalculation of the features on the target domain a phrase can in principle be correctly identified as compositional even though it was non-compositional (and identified as such) in the source domain.

Domain-Specific Features cannot be adapted by feature adaptation. There are a number of different subclasses of this case. (i) Computing the feature relies on an NLP tool whose behavior is domain-sensitive. Example: parsing-based features are domain-specific if a non-robust parser is used and the target domain (in contrast to the source domain) is user-generated content that is grammatically ill-formed and all lowercase. (ii) Computing the feature relies on a domain-dependent NLP resource. Example feature: the number of positive sentiment words occurring in a sentence, computed based on a lexicon of sentiment words. This feature can only be adapted by compiling a lexicon of sentiment words appropriate for the target domain. There is no automatic way of recomputing it that would not require some manual effort. (iii) The feature definition refers to a specific source domain datum whose behavior differs between source and target domains. This can be seen as a special case of (ii). The simplest example for this case are lexical features, a mainstay of many NLP classifiers; e.g. the word "like" is more likely to be a preposition in newspaper text. The feature [word="like"] changes its semantics for a part-of-speech tagger when moving to the target domain product reviews, where the frequency of "like" as a verb is higher. (iv) Even features that are domain-independent in some cross-domain problems may be domain-specific in others – e.g. the typical range of vector similarities may be different in a target domain with radically different distributional properties (e.g. large numbers of spelling errors).

We do not claim that feature adaptation is always applicable. However, some problems can be attacked using domain-independent features. Using only such features is beneficial because they suffer from little or no cross-domain degradation when adapted through feature adaptation. We will now describe our experiments that demonstrate this for compositionality grading.

3   Experimental Setup

Domains. We investigate two main settings of cross-domain NLP: a *cross-language* setting (German vs. English) and a *cross-topic* setting in which a classifier is adapted from generic web text to the physics domain. We use the ukWaC and deWaC corpora [9] as representatives for generic English ($\mathcal{D}_{en}$) and generic German ($\mathcal{D}_{de}$). The corpora are based on web crawls in the .de and .uk domains, respectively. We use the iSearch corpus [10] as a representative for the specialized domain physics, $\mathcal{D}_{ph}$. The iSearch corpus mainly consists of documents about astrophysics. All corpora are annotated with part-of-speech and lemma information using TreeTagger [11]. Table 1a summarizes the three domains.

Table 1. Domain summary and distribution of labeled data

(a) Summary of domain properties

domain	corpus	lang	topic	#tokens
$\mathcal{D}_{en}$	ukWaC	EN	generic	2 bn
$\mathcal{D}_{ph}$	iSearch	EN	specialized	1 bn
$\mathcal{D}_{de}$	deWaC	DE	generic	1.3 bn

(b) Train, dev and test sets

domain	train	dev	test	total
$\mathcal{D}_{en}$	58	10	77	145
$\mathcal{D}_{de}$	49	11	63	123
$\mathcal{D}_{ph}$	44	11	54	109

Compositionality Grading Task and Datasets. The objective of compositionality grading is to predict a numerical compositionality score for a given word pair. The gold standard datasets consist of word pairs and associated human compositionality judgments. To create annotated datasets, a list of adjective-noun pairs was extracted from each corpus. We focus on adjective-noun pairs because differences between domains are more likely to be visible in noun phrases

than in, for example, verb constructions. We split the annotated data into training, development and test sets. Table 1b shows the number of pairs in each
set.

The datasets for $\mathcal{D}_{en}$ and $\mathcal{D}_{de}$ were created by [3] using Amazon Mechani-
cal Turk (AMT) with four annotators per phrase. Each pair was presented in
five different contexts. For each context, annotators gave compositionality scores
ranging from 0 (fully non-compositional) to 10 (fully compositional) depending
on the usage. For example, *black box* is non-compositional in the context of avi-
ation (flight recorders are usually orange); it is compositional in the context of
storage (black containers). The scores were averaged over contexts and annotators and scaled to values between 0 and 100.

We know of no comparable dataset for $\mathcal{D}_{ph}$. Therefore, we created a dataset
for compositionality grading in the physics domain which we publish along with
this paper. We extracted adjective-noun pairs from the iSearch corpus. The
annotation procedure was similar to that used for DiSCo.[2] Three domain experts
annotated the pairs from $\mathcal{D}_{ph}$. We used experts because domain knowledge is
necessary for accurate and reliable annotation in the physics domain.

Scoring Method. The evaluation metric is the average point distance (APD)
between the gold standard scores $G = \{g_1 \ldots g_N\}$ and the compositionality
grading system scores $S = \{s_1 \ldots s_N\}$ where N is the size of the test set.
$APD(G, S)$ is defined as $\frac{1}{N} \sum_{i=1}^{N} |g_i - s_i|$. It measures how accurately the degree
of compositionality was predicted by the system. A lower APD is better; 0 is the
best possible APD. The baseline is a no-response baseline which assigns every
item a score of 50 [3].

Significance Test. We use a one-tailed binomial test to test for significance.
When comparing system S with predictions $s_1 \ldots s_N$ and system S' with predictions $s'_1 \ldots s'_N$, a success for system S is a score s_i that is closer to the gold
standard: $|g_i - s_i| < |g_i - s'_i|$. S is significantly better than S' if the number of
successes has probability smaller than p, for different values of p (see Section 4),
under the assumption that there is no difference between the systems.

3.1 Features and Methods

We choose a supervised learning approach for compositionality grading. The
classifiers are trained on the source domain and adapted to the target domain
using feature adaptation.

We first describe the domain-independent features we use: association measures and vector similarity features. These features are commonly used for compositionality grading [3].

Association Measures. Association measures are the standard approach to
measuring lexical association [12]. Lexical association can be an indicator for

---

[2] We used scores s between 1 and 5. Scores were scaled using $10 * (2s + 0.5)$ resulting
in scores between 15 and 95.

non-compositionality[3] in word pairs [13, Ch. 5]. We use all association measures used by [14] that can be computed from a pair's frequency data alone, i.e. pair frequency and marginal frequencies. These measures are: student's t-score, z-score, chi-square, pointwise mutual information, Dice coefficient, frequency, log-likelihood and symmetrical conditional probability. Each association measure is one feature in the feature vector of an adjective-noun pair. For example, the feature $am_{t.score}$ is the t-score of an adjective-noun pair.

Vector Similarity Features. Vector similarity features are a standard type of feature that is used to detect non-compositionality in word pairs. The features require computing distributional vectors for words and phrases in the word space model [15,16,17]. The word space model represents a word or phrase u as the semantic vector $v(u)$ that represents the contexts in which u appears. The semantic vector is the sum of all context vectors of u. Each context vector records $f_{u_i}(c_j)$, the number of occurrences for every item c_j in the context dictionary C in the context window of the ith occurrence of u in the corpus. The context window is a fixed-length window of tokens around the occurrence u_i.

These vectors are then used in accordance with the *distributional hypothesis* [18]: we can detect the non-compositionality of a word pair $p = w_1\ w_2$ by comparing the similarity of the contexts in which p appears and the contexts in which its parts w_1 and w_2 appear. For example, the non-compositional phrase *red tape* in the sense of excessive bureaucracy appears in different contexts than *red* and *tape*. In contrast, we expect the distribution of a compositional phrase like *young people* to be similar to those of *young* and *people*.

Different implementations of this idea exist in the literature. The most common are: (i) similarity between the pair $p = w_1\ w_2$ and a composition of its parts [14,19]; (ii) similarity between the pair and either the modifying adjective or the head [20]; (iii) similarity between the pair and the sum of its alternative pairs [21,22]. An alternative pair for $p = w_1\ w_2$ is a pair $w_1'\ w_2$ such that $w_1' \neq w_1$; for example, *sticky tape* is an alternative pair for *red tape*.

The most common composition function for semantic vectors is addition. Mitchell and Lapata [19] proposed element-wise multiplication as an alternative composition function and showed that it was superior to addition in a sentence similarity task. We use both addition and multiplication as composition functions in our experiments. We follow the notation of [23]: $v_1 + v_2$ denotes addition and $v_1 \times v_2$ element-by-element multiplication. Table 2 gives a summary of the vector similarity features we use and their definitions. In all cases, similarity is computed as cosine similarity.

4   Experiments, Results and Discussion

Experiments. For our three domains, we investigate the nine possible combinations of source and target domains: three in-domain setups and six cross-domain

---

[3] We now speak of non-compositionality rather than compositionality since this is the preferred terminology in the pertinent literature.

Table 2. Similarity features and formulas; cos: cosine similarity, $p = w_1 w_2$: adjective-noun phrase, v: semantic vector

$\textbf{sim}_{\textbf{add}}$ [14]	$\textbf{sim}_{\textbf{mult}}$ [19]	$\textbf{sim}_{\textbf{mod}}$ [20]
$\cos[v(p),\ v(w_1) + v(w_2)]$	$\cos[v(p),\ v(w_1) \times v(w_2)]$	$\cos[v(p),\ v(w_1)]$

$\textbf{sim}_{\textbf{head}}$ [20]	$\textbf{sim}_{\textbf{alt}}$ [21,22]
$\cos[v(p),\ v(w_2)]$	$\cos[v(p),\ \sum_{w_1' \neq w_1} v(w_1'\ w_2)]$

setups (see Table 3). The two main cross-domain setups are (i) cross-topic: English generic web text to English specialized-domain text ($\mathcal{D}_{en} \rightarrow \mathcal{D}_{ph}$) and vice versa ($\mathcal{D}_{ph} \rightarrow \mathcal{D}_{en}$) and (ii) cross-language: English generic web text to German generic web text ($\mathcal{D}_{en} \rightarrow \mathcal{D}_{de}$) and vice versa ($\mathcal{D}_{de} \rightarrow \mathcal{D}_{en}$). For completeness, we also run the experiments for the remaining two combinations: German generic web to English specialized-domain text ($\mathcal{D}_{de} \rightarrow \mathcal{D}_{ph}$) and vice versa ($\mathcal{D}_{ph} \rightarrow \mathcal{D}_{de}$).

We use random forest regression [24] for compositionality grading. Using feature selection, we identify the combination of association measure and vector similarity features that performs best on the validation data for a particular domain.[4] The best model for $\mathcal{D}_{en}$ uses two features: sim_{alt} and sim_{head}. The best model for $\mathcal{D}_{ph}$ uses four features: sim_{mult}, sim_{head}, sim_{alt} and $am_{t.score}$. The best model for $\mathcal{D}_{de}$ uses one feature: sim_{alt}.

The sim_{mod} feature is never used whereas sim_{head} occurs in two models which suggests that the similarity between a phrase and its head is a better predictor for non-compositionality than the similarity between the phrase and the modifier (cf. [20]). sim_{mult} is selected once whereas sim_{add} is not selected for any of the models. This can be viewed as weak evidence that multiplication is a better composition function than addition in accordance with the results of [19]. Finally, vector similarity features are selected more often than association measures: only one selected feature is an association measure, all others are vector similarity features.

In-Domain Results. The diagonal of Table 3 shows the results of the in-domain experiments. All our in-domain systems achieve results significantly better than the no-response baselines (see Section 3). The UoY system [25] won the English task with an APD of 14.62. Our system reaches 17.65 which means it would have ranked in the top half of the participating systems. On the German task, our system would have won with 25.58, outperforming the score of 27.09 achieved by the other contestant, UCPH [26]. Our system for $\mathcal{D}_{ph}$ achieves 13.36. Since this task is new, there is no reference system.

Cross-Domain Results. For cross-domain results, note that zero adaptation is equivalent to a no-response baseline if none of the test items occurred in the

---

[4] We omit the detailed description of the feature selection process. In this paper, we focus on proving the feasibility of feature adaptation for compositionality grading.

Table 3. The diagonal (1, I), (2, II), (3, III) contains in-domain (ID) results. Best in-domain results are bold. UoY and UCPH are the shared task winners. Column *sys* shows the system or method. FA and ZA are feature adaptation and zero adaptation, respectively. Column *APD* contains the distance from the gold standard (see Section 3). Column Δ indicates the difference from our best in-domain system. Column *sign.* shows if improvements over the baseline (b), feature adaptation (FA) and zero adaptation (ZA) are significant (at p=.1 (*), p=.05 (†) and p=.01 (‡)). e.g. an APD annotated with $b^{\ddagger}$ to the right is significantly lower than the baseline at p=.01. ZA* are no-response baselines (test items did not occur in the corpus).

target→ ↓ source	(I) $\mathcal{D}_{en}$				(II) $\mathcal{D}_{ph}$				(III) $\mathcal{D}_{de}$			
	sys	APD	Δ	sign.	sys	APD	Δ	sign.	sys	APD	Δ	sign.
(1) $\mathcal{D}_{en}$	base	24.67	+7.02		ZA	27.03	+13.67		ZA*	32.21	+6.93	
	ID	17.65		$b^{\ddagger} FA^{\ddagger\,de} FA^{*\,ph}$	FA	13.37	+0.01	$b^{\ddagger} ZA^{\ddagger}$	FA	28.06	+2.48	
	UoY	**14.62**	-3.02									
(2) $\mathcal{D}_{ph}$	ZA	22.65	+5.00		base	29.43	+16.07		ZA*	32.21	+6.93	
	FA	20.71	+3.06	$b^{\ddagger} ZA^{*}$	ID	**13.36**		$b^{\ddagger}$	FA	29.84	+4.26	$b^{\ddagger} ZA^{\ddagger}$
(3) $\mathcal{D}_{de}$	ZA	26.59	+8.94		ZA*	29.43	+16.07		base	32.21	+6.93	
	FA	22.04	+4.39	$b^{*} ZA^{\dagger}$	FA	15.89	+2.53	$b^{\ddagger} ZA^{\ddagger}$	ID	**25.58**	+4.26	b^{*}
									UCPH	27.09	+1.51	

respective corpus. These cases are marked with ZA* in Table 3 and we do not repeat the numbers below.

Row (1) of Table 3 shows the results for cross-domain experiments $\mathcal{D}_{en} \to \mathcal{D}_{ph}$ (column II) and $\mathcal{D}_{en} \to \mathcal{D}_{de}$ (column III). For $\mathcal{D}_{ph}$, the zero-adaptation system (27.03) reaches a lower APD than the $\mathcal{D}_{ph}$ baseline (29.43), but the improvement is not significant. With feature adaptation for $\mathcal{D}_{ph}$ (13.37), we achieve a result significantly better than both the baseline and zero adaptation. With feature adaptation for $\mathcal{D}_{de}$ (28.06), we outperform the baseline (32.21) but not significantly. For both $\mathcal{D}_{ph}$ and $\mathcal{D}_{de}$, the in-domain systems are not significantly better than feature adaptation.

Row (2) shows the results for the cross-domain experiments $\mathcal{D}_{ph} \to \mathcal{D}_{en}$ (I) and $\mathcal{D}_{ph} \to \mathcal{D}_{de}$ (III). For $\mathcal{D}_{en}$, feature adaptation (20.71) is significantly better than the baseline (24.67) and zero adaptation (22.65). For $\mathcal{D}_{de}$, feature adaptation (29.84) significantly outperforms the baseline (32.21). For $\mathcal{D}_{en}$, the in-domain system (17.65) significantly outperforms feature adaptation, but the in-domain system for $\mathcal{D}_{de}$ (25.58) does not.

Row (3) shows the results for the cross-domain experiments $\mathcal{D}_{de} \to \mathcal{D}_{en}$ (I) and $\mathcal{D}_{de} \to \mathcal{D}_{ph}$ (II). For $\mathcal{D}_{en}$, feature adaptation (22.04) is significantly better than the baseline (24.67) and zero adaptation (26.59). Zero adaptation even drops below the baseline. For $\mathcal{D}_{ph}$, feature adaptation (15.89) is significantly better than the baseline (29.43). For $\mathcal{D}_{en}$, the in-domain system (17.65) significantly outperforms feature adaptation but the in-domain system for $\mathcal{D}_{ph}$ (13.36) does not.

Discussion. Based on these results, we summarize the following findings for cross-domain compositionality grading:

(**A**) Zero adaptation is never significantly better than the baseline.

As we expected (see Section 2), performance degradation with zero adaptation is severe in 4 out of 6 experiments. In 3 experiments, none of the test items could be found in the source corpus. These are all cases where adaptation involves crossing into another language. For example, German words and phrases do not appear in the English corpus. When computing features for $\mathcal{D}_{en}$ test items from $\mathcal{D}_{de}$, there is even a performance drop compared to the baseline. Not surprisingly, the frequencies of English words in the German corpus are so low that it is impossible to obtain reliable vector similarities. In the two remaining cases ($\mathcal{D}_{en} \rightarrow \mathcal{D}_{ph}$ and $\mathcal{D}_{ph} \rightarrow \mathcal{D}_{en}$) – when adaptation happens only along the topic axis – zero adaptation brings minor improvements, but they are not significant.

(**B**) Feature adaptation is always better than the baseline, and significantly so in all cases except for $\mathcal{D}_{en} \rightarrow \mathcal{D}_{de}$[5].

We have shown the feasibility of feature adaptation and its benefits for cross-domain compositionality grading. In all experiments but one we see a significant improvement over the baselines without the need for labeling new data.

(**C**) One would expect in-domain performance to be better than cross-domain performance and this is the case for $\mathcal{D}_{en}$. However, for two of the domains, $\mathcal{D}_{de}$ and $\mathcal{D}_{ph}$, feature adaptation rivals in-domain performance.

For $\mathcal{D}_{ph}$ and $\mathcal{D}_{de}$ the in-domain systems do not bring significant improvements over feature adaptation. This is particularly interesting in the cross-topic setting $\mathcal{D}_{en} \rightarrow \mathcal{D}_{ph}$, where we apply the model learned on a more general domain (web corpus) to a specialized domain (physics) without a performance drop. This is good news because it is evidence that we can use feature adaptation instead of expensive labeling with trained domain experts in certain cases. We observe that the reverse is not the case. On $\mathcal{D}_{en}$, the in-domain system remains unrivaled and the models trained on $\mathcal{D}_{ph}$ and $\mathcal{D}_{de}$ are unable to exploit feature adaptation to the same extent.

5   Related Work

We use the term domain adaptation to refer to methods that use some labeled target-domain data. Such methods have been shown to achieve good results, but can only be applied when target labels are available [1,27]. For feature adaptation, we are interested in the scenario where no labeled target data is available. Typically, cross-domain methods exploit unlabeled target data [2,28,29,30]. However, during cross-domain application, these methods modify the original features and/or train a new model for the target domain. With feature adaptation, none of these steps are necessary because there is a clear separation of the

---

[5] There is a 4.15 APD improvement compared to baseline in the $\mathcal{D}_{en} \rightarrow \mathcal{D}_{de}$ experiment, but it is not significant.

model, the features and feature computation. Feature adaptation is simpler in that it only affects the third step, feature computation which is performed on unlabeled target domain data.

Lin [31] pioneered the investigation of non-compositionality in multi-word expressions. Schone and Jurafsky [14] combined statistical association measures and vector similarities for non-compositionality detection in multi-word expressions. The topic has since received steady attention in the literature (e.g. [32], [33], [20], [34], [35]). Compositionality grading has evolved from research on non-compositionality detection in multi-word expressions and has recently received increased attention in form of the DiSCo workshop at ACL 2011 and a corresponding shared task [3].

Task-unspecific and model-agnostic embeddings and other word representations can be computed in an unsupervised fashion and plugged into supervised NLP systems. One example of many is [36]: the authors report performance improvements for chunking and named-entity recognition. A direct comparison goes beyond the scope of this paper, but it could be argued that one would expect task-specific representations to perform better than task-unspecific representations.

Cross-language approaches to text classification based on a multilingual vector space model used as a joint representation of a pair of comparable corpora have also been proposed [37]. Models are trained on labeled documents from one language to classify documents in the other language with better accuracy than a simple baseline; however, performance falls short of a monolingual classifier. Feature adaptation also can be applied more generally, even if a – presumably domain-specific – parallel corpus is not available.

In general, the effort of porting a system to a target domain with feature adaptation is less than the effort for the source domain model because there is no training step for the target domain.

6   Conclusion

We proposed feature adaptation, a new unsupervised method for cross-domain NLP. Feature adaptation relies on domain-independent features which allow unsupervised recomputation of feature values in the target domain. Feature adaptation is simpler than other cross-domain methods and an effective way to tap target-domain knowledge without the need for target labels. In a case study we showed that for the example of compositionality grading the adapted systems achieve performance that is in many cases close to in-domain performance. We also provide a new resource for evaluation of compositionality grading in scientific text. Additionally, our results in this paper encourage research that uses only domain-independent features because they – as our experiments have shown – promise to be less susceptible to cross-domain degradation when adapted through feature adaptation.

Acknowledgments. This work was funded by DFG projects SFB 732 and WordGraph. We thank the anonymous reviewers for their comments.

References

1. Daumé III, H., Marcu, D.: Domain adaptation for statistical classifiers. Journal of Artificial Intelligence Research (JAIR) 26 (2006)
2. Blitzer, J., McDonald, R., Pereira, F.: Domain adaptation with structural correspondence learning. In: EMNLP, pp. 120–128 (2006)
3. Biemann, C., Giesbrecht, E.: Distributional semantics and compositionality 2011: Shared task description and results. In: ACL 2011 Workshop on Distributional Semantics and Compositionality, pp. 21–28 (2011)
4. Sag, I.A., Baldwin, T., Bond, F., Copestake, A., Flickinger, D.: Multiword Expressions: A Pain in the Neck for NLP. In: Gelbukh, A. (ed.) CICLing 2002. LNCS, vol. 2276, pp. 1–15. Springer, Heidelberg (2002)
5. Gildea, D.: Corpus variation and parser performance. In: EMNLP, pp. 167–202 (2001)
6. McClosky, D., Charniak, E., Johnson, M.: Reranking and self-training for parser adaptation. In: ACL/COLING, pp. 337–344 (2006)
7. Blitzer, J., Dredze, M., Pereira, F.: Biographies, bollywood, boom-boxes and blenders: Domain adaptation for sentiment classification. In: ACL, pp. 440–447 (2007)
8. Daume III, H., Jagarlamudi, J.: Domain adaptation for machine translation by mining unseen words. In: ACL/HLT, pp. 407–412 (2011)
9. Baroni, M., Bernardini, S., Ferraresi, A., Zanchetta, E.: The wacky wide web: a collection of very large linguistically processed web-crawled corpora. Language Resources and Evaluation 43(3) (2009)
10. Lykke, M., Larsen, B., Lund, H., Ingwersen, P.: Developing a Test Collection for the Evaluation of Integrated Search. In: Gurrin, C., He, Y., Kazai, G., Kruschwitz, U., Little, S., Roelleke, T., Rüger, S., van Rijsbergen, K. (eds.) ECIR 2010. LNCS, vol. 5993, pp. 627–630. Springer, Heidelberg (2010)
11. Schmid, H.: Probabilistic part-of-speech tagging using decision trees. In: International Conference on New Methods in Language Processing, pp. 44–49 (1994)
12. Evert, S.: The Statistics of Word Cooccurrences: Word Pairs and Collocations. PhD thesis, Institut für maschinelle Sprachverarbeitung (IMS), Universität Stuttgart (2004)
13. Manning, C.D., Schütze, H.: Foundations of Statistical Natural Language Processing. MIT Press (1999)
14. Schone, P., Jurafsky, D.: Is knowledge-free induction of multiword unit dictionary headwords a solved problem? In: EMNLP, pp. 100–108 (2001)
15. Schütze, H.: Dimensions of meaning. In: 1992 ACM/IEEE Conference on Supercomputing, Supercomputing 1992, pp. 787–796. IEEE (1992)
16. Sahlgren, M.: The Word-Space Model: Using distributional analysis to represent syntagmatic and paradigmatic relations between words in high-dimensional vector spaces. PhD thesis, Swedish Institute of Computer Science (2006)
17. Turney, P.D., Pantel, P.: From frequency to meaning: Vector space models of semantics. Journal of Artificial Intelligence Research 37(1) (2010)
18. Harris, Z.: Distributional structure. Word (1954)
19. Mitchell, J., Lapata, M.: Composition in distributional models of semantics. Cognitive Science 34(8) (2010)
20. Baldwin, T., Bannard, C., Tanaka, T., Widdows, D.: An empirical model of multiword expression decomposability. In: ACL 2003 Workshop on Multiword Expressions, pp. 89–96 (2003)

21. Michelbacher, L., Kothari, A., Forst, M., Lioma, C., Schütze, H.: A cascaded classification approach to semantic head recognition. In: EMNLP, pp. 793–803 (2011)
22. Garrido, G., Peñas, A.: Detecting compositionality using semantic vector space models based on syntactic context. shared task system description. In: ACL 2011 Workshop on Distributional Semantics and Compositionality, pp. 43–47 (2011)
23. Guevara, E.: A regression model of adjective-noun compositionality in distributional semantics. In: 2010 Workshop on Geometrical Models of Natural Language Semantics, pp. 33–37 (2010)
24. Breiman, L.: Random forests. Machine Learning 45(1) (2001)
25. Reddy, S., McCarthy, D., Manandhar, S., Gella, S.: Exemplar-based word-space model for compositionality detection: Shared task system description. In: ACL 2011 Workshop on Distributional Semantics and Compositionality, pp. 54–60 (2011)
26. Johannsen, A., Martinez, H., Rishøj, C., Søgaard, A.: Shared task system description: Frustratingly hard compositionality prediction. In: ACL 2011 Workshop on Distributional Semantics and Compositionality, pp. 29–32 (2011)
27. Chelba, C., Acero, A.: Adaptation of maximum entropy capitalizer: Little data can help a lot. Computer Speech & Language 20(4) (2006)
28. Huang, F., Yates, A.: Distributional representations for handling sparsity in supervised sequence-labeling. In: ACL/IJCNLP, pp. 495–503 (2009)
29. Bertoldi, N., Federico, M.: Domain adaptation for statistical machine translation with monolingual resources. In: Fourth Workshop on Statistical Machine Translation, pp. 167–174 (2009)
30. Pan, S.J., Ni, X., Sun, J.T., Yang, Q., Chen, Z.: Cross-domain sentiment classification via spectral feature alignment. In: WWW, pp. 751–760 (2010)
31. Lin, D.: Automatic identification of non-compositional phrases. In: ACL, pp. 317–324 (1999)
32. Bannard, C., Baldwin, T., Lascarides, A.: A statistical approach to the semantics of verb-particles. In: ACL 2003 Workshop on Multiword Expressions, pp. 65–72 (2003)
33. McCarthy, D., Keller, B., Carroll, J.: Detecting a continuum of compositionality in phrasal verbs. In: ACL 2003 Workshop on Multiword Expressions, pp. 73–80 (2003)
34. Katz, G., Giesbrecht, E.: Automatic identification of non-compositional multi-word expressions using latent semantic analysis. In: ACL 2006 Workshop on Multiword Expressions, pp. 12–19 (2006)
35. Sporleder, C., Li, L.: Unsupervised recognition of literal and non-literal use of idiomatic expressions. In: EACL, pp. 754–762 (2009)
36. Turian, J., Ratinov, L., Bengio, Y.: Word representations: A simple and general method for semi-supervised learning. In: ACL (2010)
37. Gliozzo, A., Strapparava, C.: Cross language text categorization by acquiring multilingual domain models from comparable corpora. In: Proceedings of the ACL Workshop on Building and Using Parallel Texts, pp. 9–16. Association for Computational Linguistics (2005)

Syntactic Dependency-Based N-grams:
More Evidence of Usefulness in Classification

Grigori Sidorov[1], Francisco Velasquez[1], Efstathios Stamatatos[2],
Alexander Gelbukh[1], and Liliana Chanona-Hernández[3]

[1] Center for Computing Research (CIC),
Instituto Politécnico Nacional (IPN), Mexico City,
Mexico
[2] University of the Aegean,
Greece
[3] ESIME, Instituto Politécnico Nacional (IPN), Mexico City,
Mexico
www.cic.ipn.mx/~sidorov

Abstract. The paper introduces and discusses a concept of syntactic n-grams (sn-grams) that can be applied instead of traditional n-grams in many NLP tasks. Sn-grams are constructed by following paths in syntactic trees, so sn-grams allow bringing syntactic knowledge into machine learning methods. Still, previous parsing is necessary for their construction. We applied sn-grams in the task of authorship attribution for corpora of three and seven authors with very promising results.

Keywords: Syntactic n-grams, sn-grams, syntactic paths, authorship attribution task, SVM classifier.

1   Introduction

First of all, let us clarify the term "syntactic n-grams". Syntactic n-grams are NOT n-grams constructed by using POS tags, as one can interpret in a naive fashion. In fact, this cannot be so strictly speaking, because POS tags represent morphological information, and not syntactic data.

We propose a definition of syntactic n-grams that it is based, as its name suggests, on syntactic information, i.e., information about word relations. The main idea of our proposal is related to the method of construction of these n-grams. We suggest obtaining them by following the paths in syntactic trees, as explained in detail below. Thus, we get rid of surface language-specific information in sentences so characteristic of traditional n-grams, and maintain only persistent and pertinent linguistic information that has very clear interpretation. In our opinion, this is the way how syntactic information can be introduced into machine learning methods. Note that syntactic n-grams, though they are obtained in a different manner, keep being n-grams and can be applied practically in any task when traditional n-grams are used.

A. Gelbukh (Ed.): CICLing 2013, Part I, LNCS 7816, pp. 13–24, 2013.

Obviously, there is a price to pay for using syntactic n-grams (sn-grams). Namely, parsing should be performed for getting the mentioned syntactic paths. There are many parsers available for many languages, but parsing takes time. Also, not for all languages there are parsers at one's disposal.

An interesting question for future work is to evaluate if shallow parsing —much faster than complete parsing— is enough for obtaining syntactic n-grams of good quality. Our intuition is that for many tasks it will be sufficient.

Another interesting question for future is if syntactic n-grams allow better comparison of results between languages. Obviously, in translation some syntactic relations are changes, but many of them are maintained. So, syntactic n-grams will be more "comparable" across the languages, since they smooth the influence of language-specific surface structures.

In this paper, we apply syntactic n-grams to authorship attribution problem using three popular classifiers and compare their performance with traditional n-grams.

The rest of the paper is organized as follows: the discussion and examples of syntactic n-grams are presented in Section 2. The problem of authorship attribution is briefly introduced in Section 3. Then experimental results for authorship attribution based on syntactic n-grams are presented and compared with baseline sets of features in Section 4. Finally, conclusions are drawn.

2   Construction of SN-grams Using Syntactic Paths

Modern natural language processing very widely uses the concept of n-grams. N-grams can be composed of various types of elements: words, POS tags, characters. Usually n-grams are constructed according to the appearance of its elements in a text (in a sequential order).

Syntactic n-grams (sn-grams) are n-grams that are obtained from texts by following paths in syntactic trees. We proposed this concept and discussed some properties of syntactic n-grams in [1]. In that work, we also compared syntactic n-grams with traditional n-grams, skip-grams and Maximal Frequent Sequences that are purely statistical techniques. Note that unlike n-grams obtained with statistical techniques, syntactic n-grams have clear linguistic interpretation, while keeping the property of being n-grams, i.e., they can be applied in the same tasks as traditional n-grams.

As traditional n-grams, syntactic n-grams can be composed by various types of elements like words/stems/lemmas or POS-tags. In [1] we mentioned that syntactic n-grams of characters are impossible. Now we change our mind: syntactic n-grams of character can be constructed from syntactic n-grams of words. A question for further research is if they are useful.

In case of syntactic n-grams, another type of elements can be used for their composition: tags of syntactic relations (SR tags), like *pobj, det, xcomp*, etc. These tags are similar to POS tags in the sense that they are morphosyntactic abstraction and are obtained during previous linguistic processing.

It is important to mention that also mixed syntactic n-grams can be used that represents an interesting direction of future research. For example, in a bigram, the first component can be a word, while the second one being a POS tag or SR tag, etc. This combination, for example, can be useful in a study of subcategorization frames.

Note that sn-grams are ordered according to the syntactic path. It means that the main word always is the first element and the dependent word is the second one (and the next dependent word will be the third element in case of trigrams, etc.). So, in case of syntactic n-grams, the mixed type can be especially useful.

Resuming, sn-grams can be composed of:

- Words/stems/lemmas,
- POS-tags,
- Characters,
- SR tags (tags of syntactic relations),
- Combination of the previous ones (mixed sn-grams).

Note that as in case of traditional n-grams, auxiliary words (stop words) can be either considered or ignored in sn-grams.

Let us consider an example, a phrase taken from Jules Verne novel:

A member of the Society then inquired of the president whether Dr. Ferguson was not to be officially introduced.

Stanford parser [12] returns the following syntactic information that corresponds to the tree in Fig. 1 and Fig. 2.

```
(ROOT
 (S
  (NP
   (NP (DT A) (NN member))
   (PP (IN of)
    (NP (DT the) (NNP Society))))
  (ADVP (RB then))
  (VP (VBD inquired)
   (PP (IN of)
    (NP (DT the) (NN president)))
   (SBAR (IN whether)
    (S
     (NP (NNP Dr.) (NNP Ferguson))
     (VP (VBD was) (RB not)
      (S
       (VP (TO to)
        (VP (VB be)
         (VP
          (ADVP (RB officially))
          (VBN introduced)))))))))
  (. .)))
```

The names of the syntactic relations that the parser returns as well are as following:. The format contains the relation name, the main word and its position in the sentence, the dependent word and its position in the sentence.

We do not represent graphically the relation for ROOT, but it indicates us where the syntactic tree starts.

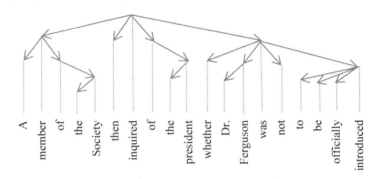

Fig. 1. Example of a syntactic tree

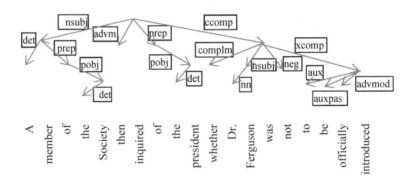

Fig. 2. Example of a syntactic tree with SR tags

det(member-2, A-1)	prep(inquired-7, of-8)	ccomp(inquired-7, was-14)
nsubj(inquired-7, member-2)	det(president-10, the-9)	neg(was-14, not-15)
prep(member-2, of-3)	pobj(of-8, president-10)	aux(introduced-19, to-16)
det(Society-5, the-4)	complm(was-14, whether-11)	auxpass(introduced-19, be-17)
pobj(of-3, Society-5)	nn(Ferguson-13, Dr.-12)	advmod(introduced-19, officially-18)
advmod(inquired-7, then-6)	nsubj(was-14, Ferguson-13)	xcomp(was-14, introduced-19)
root(ROOT-0, inquired-7)		

Table 1. Character-based n-grams, baseline (3 authors)

Profile size	Classifier	n-gram size			
		2	**3**	**4**	**5**
400	SVM	**90%**	**76%**	**81%**	**81%**
	NB	71%	62%	71%	67%
	J48	76%	62%	48%	76%
1,000	SVM	**95%**	**86%**	**86%**	76%
	NB	76%	76%	67%	**81%**
	J48	81%	67%	67%	71%
4,000	SVM	**90%**	**95%**	**90%**	**86%**
	NB	**90%**	71%	81%	71%
	J48	76%	76%	**90%**	81%
7,000	SVM	NA	**90%**	**86%**	86%
	NB	NA	62%	76%	71%
	J48	NA	76%	**86%**	**90%**
11,000	SVM	NA	<u>**100%**</u>	**90%**	**86%**
	NB	NA	67%	62%	71%
	J48	NA	71%	81%	**86%**

Fig. 1 and Fig. 2 show graphical representation of the corresponding syntactic tree. Note that the tree reflects depth levels of each word depending on how "far" it is from the root, i.e., what is the length of the corresponding path. The arrows are drawn from main words to dependent words. In Fig. 1, we add lines that show which word corresponds to tree nodes. In Fig. 2, the tags of syntactic relations (SR tags) are represented in the squares situated as close to the corresponding arrows as possible.

We use representation of a sentence as a dependency tree. As we discussed in [1], constituency representation and dependency representation are equal for our purposes, though dependency representation in case of sn-grams is more intuitive.

We hope that these figures will help in understanding of the method of construction of sn-grams. We just traverse the tree. The method is intuitively very clear: follow the paths represented by arrows and at each step take the words (or other elements) from the corresponding nodes.

More formal description of the method is as follows. We start from the root node R, choose the first arrow (we will pass through all arrows, so the order is not important) and take the node N on the other side of the arrow. Our first bigram (or part of a larger sn-gram) is R-N. Note that the order is important because R is the main word, and N is the dependent word. Now we move to the node N and repeat the operation, either for the next bigram, or for the larger sn-gram. The number of steps

Table 2. Word based n-grams, baseline (3 authors)

Profile size	Classifier	n-gram size			
		2	3	4	5
400	SVM	**86%**	**81%**	67%	45%
	NB	48%	67%	**81%**	**85%**
	J48	67%	76%	71%	60%
1,000	SVM	**86%**	71%	71%	48%
	NB	76%	**81%**	**95%**	**90%**
	J48	71%	67%	71%	67%
4,000	SVM	**86%**	<u>**95%**</u>	67%	48%
	NB	62%	65%	**81%**	**86%**
	J48	81%	70%	**81%**	57%
7,000	SVM	**86%**	**90%**	71%	45%
	NB	52%	48%	81%	**81%**
	J48	**86%**	71%	**86%**	51%
11,000	SVM	**89%**	**90%**	75%	33%
	NB	53%	52%	**90%**	**78%**
	J48	**89%**	81%	70%	44%

for construction of a sn-grams of a given size is equal to n-1, i.e., in case of bigrams we make only one step, in case of trigrams we make two steps, etc. In bifurcations, each possible direction corresponds to a new sn-gram or a set of new sn-grams (in case that there are more bifurcations at lower levels). When we finish with a sn-gram, we return to the nearest previous bifurcation and continue in the direction that was not explored yet.

Let us compare the results for extraction of traditional bigrams and syntactic bigrams.

Traditional bigrams of words for the example are: *a member, member of, of the, the Society, Society then, then inquired, inquired of, of the, the president, president whether, whether Dr., Dr. Ferguson, Ferguson was, was not, not to, to be, be officially, officially introduced.*

Syntactic bigrams of words for the example are: *inquired member, member a, member of, of Society, Society the, inquired then, inquired of, of president, president the, inquired was, was whether, was Ferguson, Ferguson Dr., was not, was introduced, introduced to, introduced be, introduced officially.*

In our opinion, syntactic bigrams are much more stable and less arbitrary, i.e., have more chances to be repeated in other sentences. A simple example: if we add an adjective for any noun. Traditional n-grams in the near context will be changed,

Table 3. POS n-grams, baseline (3 authors)

Profile size	Classifier	n-gram size			
		2	3	4	5
400	SVM	**90%**	**90%**	**76%**	62%
	NB	67%	62%	57%	52%
	J48	76%	57%	52%	71%
1,000	SVM	**95%**	**90%**	**86%**	**67%**
	NB	76%	57%	62%	52%
	J48	71%	62%	81%	57%
4,000	SVM	NA	<u>**100%**</u>	**86%**	**86%**
	NB	NA	57%	62%	57%
	J48	NA	62%	67%	76%
7,000	SVM	NA	<u>**100%**</u>	**90%**	**86%**
	NB	NA	38%	62%	57%
	J48	NA	38%	86%	**86%**
11,000	SVM	NA	**95%**	**90%**	86%
	NB	NA	43%	48%	57%
	J48	NA	57%	86%	**90%**

but syntactic n-grams will maintain stable, only one new sn-gram will be added: Noun-Adjective.

Note that while the number of syntactic bigrams is equal to the number of traditional bigrams, the number of sn-grams when n>2 can be less than in case of traditional n-grams. It is so because traditional n-grams consider just plain combinations, while for sn-grams there should exist "long" paths. It is very clear for greater values of n. Say, for n=5, there are many n-grams in the above mentioned example, while there is only one sn-gram: *inquired member of Society the.* It is obvious that the number of n-grams and sn-grams would be equal only if the whole phrase has only one path, i.e., there are no bifurcations.

Since we mentioned that there can be sn-grams of SR tags (tags of syntactic relations) and we will use them in experiments further in the paper, we would like to present the bigrams extracted from the same example sentence*: nsubj-det, nsubj-prep, prep-pobj (2), pobj-det (2), ccomp-complm, ccomp-nsubj, ccomp-neg, ccomp-xcomp, xcomp-aux, xcomp-auxpass, xcomp-advmod.* In this case we traverse the tree as well, but instead of nodes we take the names of arrows (arcs). In this work, we consider that SR tags are comparable with POS tags: they have similar nature and the quantity of both types of elements is similar: 36 and 53 elements correspondingly.

Table 4. Sn-grams of SR tags (3 authors)

Profile size	Classifier	n-gram size			
		2	3	4	5
400	SVM	**100%**	**100%**	87%	**93%**
	NB	**100%**	93%	73%	67%
	J48	87%	67%	**93%**	73%
1,000	SVM	**100%**	**100%**	87%	**93%**
	NB	80%	67%	80%	80%
	J48	87%	67%	**93%**	73%
4,000	SVM	**100%**	**100%**	**93%**	**73%**
	NB	40%	40%	53%	60%
	J48	67%	47%	73%	73%
7,000	SVM	**100%**	**100%**	87%	87%
	NB	53%	33%	33%	73%
	J48	67%	80%	67%	73%
11,000	SVM	**100%**	**100%**	**93%**	87%
	NB	40%	33%	33%	60%
	J48	67%	80%	53%	73%

3   Authorship Attribution Problem

We discussed the state of the art of the authorship attribution problem in our previous work on application of syntactic n-grams for authorship attribution [1]; also see various related works on authorship attribution [7, 8, 9, 10], among many others. Here we will just briefly state the problem.

In this work, we consider it as a supervised classification problem. In this case, the authors represent possible classes. Given a set of training data (texts) of known authors, the systems should learn from them the differences between classes. It is expected that the system would learn the author style, because the thematic differences between texts are arbitrary. It is an important point to choose texts of the same genre or at least on the same theme, otherwise the system could learn the thematic classification. After this, the system should classify the test data (other texts), according to the learned model. The classifiers can use as features the mentioned types of n-grams and sn-grams.

4   Experimental Results and Discussion

In our previous work [1], we conducted experiments for proving that the concept of sn-grams is useful using as an example the task of the authorship attribution for a

Table 5. Word based n-grams, baseline, SVM (7 authors)

Profile size	Calssifier	n-gram size			
		2	3	4	5
400	SVM	76%	44%	32%	27%
1,000	SVM	68%	51%	44%	22%
4,000	SVM	73%	61%	44%	29%
7,000	SVM	78%	76%	61%	NS
11,000	SVM	78%	78%	NS	NS

Table 6. Sn-grams of SR tags, SVM (7 authors)

Profile size	Classifier	n-gram size			
		2	3	4	5
400	SVM	86%	73%	62%	51%
1,000	SVM	84%	84%	59%	51%
4,000	SVM	86%	76%	59%	62%
7,000	SVM	86%	78%	68%	62%
11,000	SVM	86%	78%	62%	62%

corpus of three authors. There we used only one classifier, SVM, and obtained superior results for sn-grams than for traditional n-grams.

In this paper, we present also the results for two more popular classifiers: Naive Bayes and J48 (tree classifier). Besides, we obtained preliminary results for sn-grams for a corpus of seven authors.

The corpus for three authors used in both studies contains texts downloaded from the Project Gutenberg web site. We used novels of three English speaking authors: Booth Tarkington, George Vaizey, and Louis Tracy, who belong more or less to the same time period. There are 13 novels of each author. We used 8 novels (60%, totally about 3.7 MB for each author) for training and 5 novels for classification (40%, totally about 2 MB for each author).

We used WEKA software for classification [11]. In our previous work, we used only one classifier, SVM. In this work we present results for three classifiers: SVM (NormalizedPolyKernel of the SMO), Naive Bayes, and J48. Several baseline feature sets are analyzed that use traditional n-grams: words, POS tags and characters.

We use the term "profile size" for representing the first most frequent n-grams/sn-grams, e.g., for profile size of 1,000 only first 1,000 most frequent n-grams are used. We tested various thresholds for profile size and selected five of them as presented in all tables that contain results.

When the table cell contains the value *NA (not available)*, it means that it was impossible to obtain the corresponding number of n-grams for our corpus. This situation is presented only with bigrams, because in general there are less bigrams than trigrams etc. It means that the total number of all bigrams is less than the given profile size. It can happen for bigrams of POS tags or characters, but not for bigrams of words.

Results for various baseline sets of features are presented in Tables 1, 2, and 3. Table 1 represents character based feature set classification. The character based features show high results for the SVM classification method, reaching 100% for a profile of 11,000 trigrams.

In Table 2, results for word based feature set are presented. It can be noticed that the best results are obtained for SVM with bigrams and trigrams of words, reaching the maximum of 95% for a profile of 4,000 trigrams.

In Table 3, the use of the features based on POS tag is presented. It can be observed that the best results (100%) are obtained using SVM for profile sizes 4,000 and 7,000.

Table 4 presents the results obtained using sn-grams of SR tags, showing substantial improvement in comparison with traditionally obtained feature sets in case of the SVM classifier. The results confirm that syntactic n-grams outperform other features in the task of authorship attribution.

In the vast majority of cases the results of the SVM classification method are better than NB and J48. We obtained the best performance with SVM always getting a 100% of accuracy with bigrams and trigrams for any profile size for sn-grams.

The only case when NB gives the same result as SVM (of 100%) is for the profile size of 400 for bigrams. Still, we consider that it is due to the fact that our topline can be achieved relatively easy because of the corpus size and the number of authors (only three). Note that in all other cases SVM got better results.

On the other hand, the tendency that sn-grams outperform traditional n-grams is preserved and allows us to get rid of the necessity to choose the threshold values, like the profile size for this corpus, for example, traditional POS trigrams got 100% for profile sizes of 4,000 and 7,000 only, while sn-grams (bigrams and trigrams) give 100% for all considered profile sizes.

We also performed preliminary experiments for the corpus of the works of seven authors built in similar way as the corpus of three authors. This task is much more difficult than for three authors because there are more classes. Our results for one baseline feature set (words) and sn-grams are presented in Tables 5 and 6. We used only SVM since we got better results before using it, but the experiments for NB and J48 should be performed as well in future. Sn-grams obtained better results in all case, only once for case of 11,000 profile of trigrams the results are the same. The interpretation of the results related to the nature and complexity of the task is a matter of future work. We plan to perform the comparison of all techniques and classifiers in near future for the corpus of seven authors.

5   Conclusions and Future Work

This paper proposed a concept of syntactic n-grams (sn-grams), i.e., n-grams that are constructed using syntactic paths. For this we just traverse the syntactic tree. The concept of sn-grams allows bringing syntactic information into machine learning methods. Sn-grams can be applied in all tasks when traditional n-grams are used.

We analyzed several properties of syntactic n-grams and presented an example of a phrase with extracted sn-grams. A shortcoming of sn-grams is that syntactic parsing is necessary prior to their construction.

We tried the concept of sn-grams in the task of authorship attribution and obtained very promising results for a corpus of three and seven authors.

In our experiments, the best results were always achieved by SVM classifier, as compared with NB and J48.

We would like to mention the following directions of future work:

- Experiments with all feature sets on larger corpus (7 authors, or more).
- Analysis of the applicability of shallow parsing instead of full parsing.
- Analysis of usefulness of sn-grams of characters.
- Analysis of behavior of sn-grams between languages, e.g., in parallel texts or comparable texts.
- Application of sn-grams in other NLP tasks.
- Application of mixed sn-grams.
- Experiments that would consider combinations of the mentioned features in one feature vector.
- Evaluation of the optimal number and size of sn-grams for various tasks.
- Consideration of various profile sizes with more granularity.
- Application of sn-grams in other languages.

Acknowledgements. Work done under partial support of Mexican government (projects CONACYT 83270, SNI) and Instituto Politécnico Nacional, Mexico (projects SIP 20111146, 20113295, 20120418, COFAA, PIFI), Mexico City government (ICYT-DF project PICCO10-120) and FP7-PEOPLE-2010-IRSES: Web Information Quality - Evaluation Initiative (WIQ-EI) European Commission project 269180.

References

1. Sidorov, G., Velasquez, F., Stamatatos, E., Gelbukh, A., Chanona-Hernández, L.: Syntactic Dependency-Based N-grams as Classification Features. In: Mendoza, M.G. (ed.) MICAI 2012, Part II. LNCS (LNAI), vol. 7630, pp. 1–11. Springer, Heidelberg (2013)
2. Khalilov, M., Fonollosa, J.A.R.: N-gram-based Statistical Machine Translation versus Syntax Augmented Machine Translation: comparison and system combination. In: Proceedings of the 12th Conference of the European Chapter of the ACL, pp. 424–432 (2009)
3. Habash, N.: The Use of a Structural N-gram Language Model in Generation-Heavy Hybrid Machine Translation. In: Belz, A., Evans, R., Piwek, P. (eds.) INLG 2004. LNCS (LNAI), vol. 3123, pp. 61–69. Springer, Heidelberg (2004)
4. Agarwal, A., Biads, F., Mckeown, K.R.: Contextual Phrase-Level Polarity Analysis using Lexical Affect Scoring and Syntactic N-grams. In: Proceedings of the 12th Conference of the European Chapter of the ACL (EACL), pp. 24–32 (2009)

5. Cheng, W., Greaves, C., Warren, M.: From n-gram to skipgram to concgram. International Journal of Corpus Linguistics 11(4), 411–433 (2006)
6. Baayen, H., Tweedie, F., Halteren, H.: Outside The Cave of Shadows: Using Syntactic Annotation to Enhance Authorship Attribution. Literary and Linguistic Computing, pp. 121–131 (1996)
7. Stamatatos, E.: A survey of modern authorship attribution methods. Journal of the American Society for Information Science and Technology 60(3), 538–556 (2009)
8. Juola, P.: Authorship Attribution. Foundations and Trends in Information Retrieval 1(3), 233–334 (2006)
9. Argamon, S., Juola, P.: Overview of the international authorship identification competition at PAN-2011. In: 5th Int. Workshop on Uncovering Plagiarism, Authorship, and Social Software Misuse (2011)
10. Koppel, M., Schler, J., et al.: Authorship attribution in the wild. Language Resources and Evaluation 45(1), 83–94 (2011)
11. Hall, M., Frank, E., Holmes, G., Pfahringer, B., Reutemann, P., Witten, I.H.: The WEKA Data Mining Software: An Update. SIGKDD Explorations 11(1) (2009)
12. de Marneffe, M.C., MacCartney, B., Manning, C.D.: Generating Typed Dependency Parses from Phrase Structure Parses. In: Proc. of LREC (2006)

A Quick Tour of BabelNet 1.1

Roberto Navigli

Dipartimento di Informatica
Sapienza Università di Roma
Viale Regina Elena, 295 – Roma, Italy
navigli@di.uniroma1.it
http://lcl.uniroma1.it

Abstract. In this paper we present BabelNet 1.1, a brand-new release of the largest "encyclopedic dictionary", obtained from the automatic integration of the most popular computational lexicon of English, i.e. WordNet, and the largest multilingual Web encyclopedia, i.e. Wikipedia. BabelNet 1.1 covers 6 languages and comes with a renewed Web interface, graph explorer and programmatic API. BabelNet is available online at http://www.babelnet.org.

Keywords: BabelNet, knowledge acquisition, semantic networks, multilingual ontologies.

1 Introduction

In the information society, lexical knowledge is a key skill for understanding and decoding an ever-changing world. Indeed, lexical knowledge is not only an essential component for human understanding of text, it is also indispensable for Natural Language Processing tasks. Unfortunately, however, building such lexical knowledge resources manually is an onerous task requiring dozens of years – and what is more it has to be repeated from scratch for each new language.

The multilinguality aspect is key to this vision, in that it enables Natural Language Processing tasks which are not only cross-lingual, but are also independent of the language of the user input and of the other data utilized to perform the task.

In this paper we present BabelNet 1.1 (http://www.babelnet.org), a brand-new version of a very large multilingual ontology and semantic network, obtained as a result of a novel integration and enrichment methodology [11]. This resource is created by linking the largest multilingual Web encyclopedia – i.e., Wikipedia – to the most popular computational lexicon – i.e., WordNet [6]. The integration is performed via an automatic mapping and by filling in lexical gaps in resource-poor languages with the aid of Machine Translation (MT). The result is an "encyclopedic dictionary" that provides babel synsets, i.e., concepts and named entities lexicalized in many languages and connected with large amounts of semantic relations.

BabelNet complements existing resources, such as DBPedia [1], YAGO [4] or WikiNet [8]. In fact, these resources just provide coverage of Named Entities and

A. Gelbukh (Ed.): CICLing 2013, Part I, LNCS 7816, pp. 25–37, 2013.

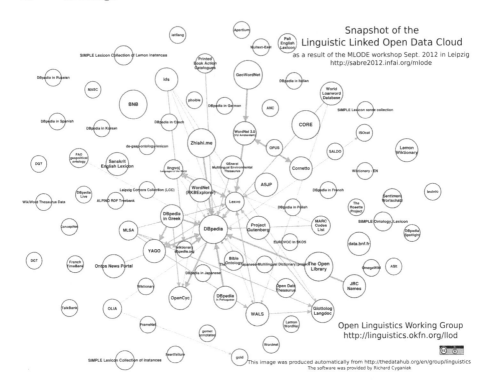

Fig. 1. Open Linguistics Working Group (2012), The Linguistic Linked Open Data cloud diagram (draft), version of September 2012, http://linguistics.okfn.org/llod.

more or less explicitly target the Linked Open Data (LOD) cloud, i.e. a vision of the (Semantic) Web in which related data that was not previously linked is connected. BabelNet, instead, focuses both on word senses and on Named Entities in many languages. Its aim, therefore, is to provide full lexicographic and encyclopedic coverage. BabelNet can be viewed both as a multilingual ontology, a large machine-readable encyclopedic dictionary and a multilingual semantic network. Compared to YAGO [16], which links Wikipedia categories to WordNet synsets using the first-sense heuristic, BabelNet integrates the two resources by means of a mapping strategy based on a Word Sense Disambiguation algorithm [9], and provides additional lexicalizations resulting from the application of Statistical Machine Translation. This is also in contrast to Wikipedia-based resources like WikiNet, which utilizes the lexicalizations available in Wikipedia articles and categories, or the Universal WordNet (UWN) [5], which is based on the lexicographic coverage of words.

Recently, BabelNet has also been linked to the LOD cloud [10], with the objective of contributing to the so-called Linguistic Linked Open Data (LLOD, see Figure 1), a vision fostered by the Open Linguistic Working Group (OWLG)[1]

---

[1] http://linguistics.okfn.org

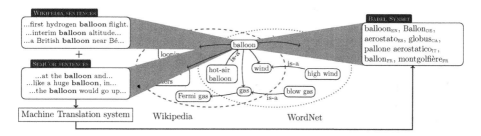

Fig. 2. An overview of BabelNet (nodes are labeled with English lexicalizations only): unlabeled edges are extracted from Wikipages (e.g., BALLOON (AIRCRAFT) links to MONTGOLFIER BROTHERS), labeled edges come from WordNet (e.g., balloon_n^1 *has-part* gasbag_n^1).

in which part of the Linked Open Data cloud is made up of interlinked linguistic resources [2], aimed at fostering integration, interoperability and reuse on the Semantic Web [3].

In the next Section we introduce BabelNet 1.1 and briefly illustrate its features. Then, in Section 3 we provide statistics, in Section 4 we take a tour of the new Web interface and in Section 5 we show how to get started with the BabelNet API. Finally, we give some conclusions in Section 6.

2   BabelNet

BabelNet [11] encodes knowledge as a labeled directed graph $G = (V, E)$ where V is the set of nodes – i.e., concepts such as balloon and named entities such as Montgolfier brothers – and $E \subseteq V \times R \times V$ is the set of edges connecting pairs of concepts (e.g., balloon *is-a* lighter-than-air craft). Each edge is labeled with a semantic relation from R, e.g., $\{is\text{-}a, \ part\text{-}of, \ \ldots, \ \epsilon\}$, where ϵ denotes an unspecified semantic relation. Each node $v \in V$ contains a set of lexicalizations of the concept for different languages, e.g., { $\text{balloon}_{\text{EN}}$, $\text{Ballon}_{\text{DE}}$, pallone $\text{aerostatico}_{\text{IT}}$, ..., $\text{montgolfière}_{\text{FR}}$ }. We call such multilingually lexicalized concepts *Babel synsets*. Concepts and relations in BabelNet are harvested from the largest available semantic lexicon of English, WordNet, and a wide-coverage collaboratively-edited encyclopedia, Wikipedia. In order to build the BabelNet graph, we collect at different stages:

a. from WordNet, all available word senses (as concepts) and all the lexical and semantic pointers between synsets (as relations);
b. from Wikipedia, all encyclopedic entries (i.e., Wikipages, as concepts) and semantically unspecified relations from hyperlinked text.

An overview of BabelNet is given in Figure 2. The excerpt highlights that Word-Net and Wikipedia can overlap both in terms of concepts and relations: accordingly, in order to provide a *unified resource*, we merge the intersection of these two knowledge sources. Next, to enable multilinguality, we collect the lexical realizations of the available concepts in different languages. Finally, we connect the multilingual Babel synsets by establishing semantic relations between them. Thus, our methodology consists of three main steps:

1. We **combine WordNet and Wikipedia** by automatically acquiring a mapping between WordNet senses and Wikipages. This avoids duplicate concepts and allows their inventories of concepts to complement each other.
2. We **harvest multilingual lexicalizations** of the available concepts (i.e., Babel synsets) by using (a) the human-generated translations provided by Wikipedia (the so-called *inter-language* links), as well as (b) a machine translation system to translate occurrences of the concepts within sense-tagged corpora.
3. We **establish relations between Babel synsets** by collecting all relations found in WordNet, as well as all wikipedias in the languages of interest: in order to encode the strength of association between synsets we compute their degree of correlation using a measure of relatedness based on the Dice coefficient.

3   Statistics

In this section we provide statistics for BabelNet 1.1, obtained by applying the construction methodology briefly described in the previous Section and detailed in [11].[2]

3.1   Lexicon

BabelNet currently covers 6 languages, namely: English, Catalan, French, German, Italian and Spanish. Its lexicon includes lemmas which denote both concepts (e.g., balloon) and named entities (e.g., Montgolfier brothers). The second column of Table 1 shows the number of lemmas for each language. The lexicons have the same order of magnitude for the 5 non-English languages, whereas English displays larger numbers due to the lack of inter-language links and annotated sentences for many terms, which prevents our construction approach from providing translations.

In Table 2 we report the number of monosemous and polysemous words divided by part of speech. Given that we work with nominal synsets only, the numbers for verbs, adjectives and adverbs are the same as in WordNet 3.0. As for nouns, we observe a very large number of monosemous words (almost 19 million),

---

[2] However, note that some additional heuristics, especially concerning Wikipedia redirections, have been applied in version 1.1 in order to improve the quality of the Wikipedia-WordNet mapping.

Table 1. Number of lemmas, synsets and word senses in the 6 languages currently covered by BabelNet

Language	Lemmas	Synsets	Senses
English	8,108,298	3,898,579	8,891,049
Catalan	1,319,089	754,187	1,483,131
French	2,901,230	1,530,017	3,222,211
German	2,846,656	1,591,536	3,100,059
Italian	1,989,959	1,264,333	2,212,186
Spanish	2,678,657	1,236,250	3,038,745
Total	19,843,889	5,581,954	21,947,381

Table 2. Number of monosemous and polysemous words by part of speech (verbs, adjectives and adverbs are the same as in WordNet 3.0)

POS	Monosemous words	Polysemous words
Noun	18,722,263	1,084,076
Verb	6,280	5,251
Adjective	16,591	4,947
Adverb	3,748	733
Total	18,748,882	1,095,007

but also a large number of polysemous words (more than 1 million). Both numbers are considerably larger than in WordNet, because – as remarked above – words here denote both concepts (mainly from WordNet) and named entities (mainly from Wikipedia).

3.2   Concepts

BabelNet 1.1 contains more than 5.5 million concepts, i.e., Babel synsets, and almost 22 million word senses (regardless of their language). In Table 1 we report the number of synsets covered for each language (third column) and the number of word senses lexicalized in each language (fourth column). The overall number of word senses in English is much higher than those in the other languages (owing to the high number of synonyms and redirections in English).

In Table 3 we show for each language the number of word senses obtained directly from WordNet, Wikipedia pages and redirections, as well as Wikipedia and WordNet translations.

3.3   Relations

We now turn to relations in BabelNet. Relations come either from Wikipedia hyperlinks (in any of the covered languages) or WordNet. All our relations are

Table 3. Composition of Babel synsets: number of synonyms from the English Word-Net, Wikipedia pages and translations, as well as translations of WordNet's monose-mous words and SemCor's sense annotations

		English	Catalan	French	German	Italian	Spanish	Total
English WordNet		206,978	-	-	-	-	-	206,978
Wikipedia	pages	3,829,110	375,046	1,231,374	1,287,134	932,536	915,046	8,570,246
	redirections	4,854,961	250,681	1,145,075	954,949	438,856	1,272,184	8,916,706
	translations	-	749,337	738,128	749,381	732,342	743,389	3,712,577
WordNet	translations	-	108,067	107,634	108,595	108,452	108,126	540,874
Total		8,891,049	1,483,131	3,222,211	3,100,059	2,212,186	3,038,745	21,947,381

Table 4. Number of lexico-semantic relations harvested from WordNet, WordNet glosses and the 6 wikipedias

	English	Catalan	French	German	Italian	Spanish	Total
WordNet	364,552	-	-	-	-	-	364,552
WordNet glosses	617,785	-	-	-	-	-	617,785
Wikipedia	68,626,079	4,553,922	18,390,507	21,942,058	14,443,937	12,758,598	140,715,101
Total	69,608,416	4,553,922	18,390,507	21,942,058	14,443,937	12,758,598	141,697,438

Table 5. Glosses for the Babel synset referring to the concept of balloon as aircraft

English	WordNet	Large tough nonrigid bag filled with gas or heated air.
	Wikipedia	A balloon is a type of aircraft that remains aloft due to its buoyancy.
German		Ein Ballon ist eine nicht selbsttragende, gasdichte Hülle, die mit Gas gefüllt ist und über keinen Eigenantrieb verfügt.
Italian		Un pallone aerostatico è un tipo di aeromobile, un aerostato che si solleva da terra grazie al principio di Archimede.
Spanish		Un aerostato, o globo aerostático, es una aeronave no propulsada que se sirve del principio de los fluidos de Arquímedes para volar, entendiendo el aire como un fluido.

semantic, in that they connect Babel synsets (rather than senses), however the re-lations obtained from Wikipedia are unlabeled.[3] In Table 4 we show the number of lexico-semantic relations from WordNet, WordNet glosses and the 6 wikipedias used in our work. We can see that the major contribution comes from the En-glish Wikipedia (68 million relations) and Wikipedias in other languages (a few million relations, depending on their size in terms of number of articles and links therein).

3.4   Glosses

Each Babel synset comes with one or more glosses (possibly available in many languages). In fact, WordNet provides a textual definition for each English synset, while in Wikipedia a textual definition can be reliably obtained from

---

[3] In a future release of the resource we plan to perform an automatic labeling based on work in the literature. See [7] for recent work on the topic.

the first sentence of each Wikipage[4]. Overall, BabelNet 1.1 includes 8,439,497 glosses (3,905,508 of which are in English). In Table 5 we show the glosses for the Babel synset which refer to the concept of balloon as 'aircraft'.

4   The Online Interface

The new online interface, shown in Figure 3 and available from http://www.babelnet.org, has 2 interactive options: search and explore. We describe the two options in the following subsections.

A very large multilingual ontology with millions of concepts • A wide-coverage "encyclopedic dictionary" • Obtained from the automatic integration of WordNet and Wikipedia • Enriched with automatic translations of its concepts • Connected to the Linguistic Linked Open Data cloud!

search explore

publications downloads

 BabelNet is an output of the MultiJEDI ERC Starting Grant No. 259234. Concept and application by Roberto Navigli. BabelNet and its API are licensed under a Creative Commons Attribution-Noncommercial-Share Alike 3 0 License

Fig. 3. The BabelNet homepage

4.1   Searching BabelNet

The search interface allows the user to look up a word in BabelNet by specifying its lemma and the language of interest (among the 6 available). In Figure 4 we

---

[4] "The article should begin with a short declarative sentence, answering two questions for the nonspecialist reader: *What (or who) is the subject?* and *Why is this subject notable?*", extracted from http://en.wikipedia.org/wiki/Wikipedia:Writing_better_articles. This simple, albeit powerful, heuristic has previously been used successfully to construct a corpus of definitional sentences [15] and learn a definition and hypernym extraction model [14].

show the first two results when *balloon* is searched in the English language. Each
entry in the search result page shows several kinds of information:

- **Basic information**, such as the main title (e.g. *balloon*[1] for the first nominal sense of balloon, following the notation of [9]), its unique ID (e.g.
  bn:00008187n) and its type (i.e. Concept vs. Named Entity).
- **the WordNet synset:** this includes all the synonyms in the synset which,
  if a mapping is available, denotes the target concept in WordNet.
- **Wikipedia titles:** i.e. the Wikipedia page titles in the various languages for
  the target concept. For instance, for the aircraft sense of balloon, we have
  Globus aerostàtic for Catalan, Balloon (aircraft) for English, Aérostation for
  French, etc.
- **Wikipedia redirections:** this includes the Wikipedia redirections to the
  corresponding Wikipage page titles. For instance, Hydrogen balloon redirects
  to Balloon (aircraft).
- **Automatic translations**, i.e. the majority translations obtained as a result of the application of Statistical Machine Translation to sense-tagged
  sentences. For instance, globo in Spanish was a majority translation of the
  aircraft sense of balloon.
- **Glosses:** we show the textual definitions available in the covered languages.
  For English, we might have two different definitions, one from WordNet, one
  from Wikipedia, if a link could be established.
- **Wikipedia categories:** the categories provided for the above Wikipedia
  page titles.
- **A link to DBPedia:** a hyperlink to the corresponding DBPedia entry is
  provided if available.

For instance, the lack of translations for the second sense of balloon (Toy balloon)
in Figure 4 implies either that there were not enough annotated sentences with
the sense of interest or that no majority translation could be found.

4.2   Exploring BabelNet

Beside each sense of a given lemma an "explore" link is available which allows
the user to move to the exploration interface. Alternatively, the user can click
on the link with the same name on top of the page and specify the lemma she
wants to explore. In either case, a graph representation of the selected Babel
synset is shown.

The graph shows up to 50 neighbours of the selected Babel synset. For example
in Figure 5 we show the graph obtained from the interface for the first sense of
balloon. The user can click on any node in the graph and follow that concept,
thus moving along the graph edges. Note that for presentation purposes only the
main title of each Babel synset is shown for each node.

BabelNet 1.1
A very large multilingual ontology

search · explore · publications · download

Q Type a **term:** balloon   [English ▼] [search]

(**examples:** plane, apple, star, Italian, bus driver, calcio, drive#n, bus#n#en, horse#en, mela#it)

Noun

Title: balloon¹ · **ID:** bn:00008187n · **Type:** Concept   ∴ explore

Senses: W ⬛ **balloon**¹

W ⬛ Globus aerostàtic, ⬛ **Balloon (aircraft)**, ▌▌Aérostation, ⬛ Ballon, ▌▌Pallone aerostatico, ⬛ Globo aerostático

W ⬛ Globus aerostàtics, ⬛ Balloonists, Charlière, Balloon flight, Hydrogen balloon, ▌▌Aérostation, ⬛ Ballongas, ⬛ Globo (aeronave), Globos aerostaticos, Globos aerostáticos, Balón aerostático, Globo aerostático, Balon aerostatico

⬛ globus, ▌▌ballon, ⬛ ballon, ⬛ globo

Glosses: W ⬛ Un globus aerostàtic és una aeronau aerostàtica no propulsada que es serveix del principi dels fluids d'Arquimedes per volar, entenent l'aire com un fluid.

W ⬛ A balloon is a type of aerostat that remains aloft due to its buoyancy.

W ⬛ large tough nonrigid bag filled with gas or heated air

W ▌▌L'aérostation est l'étude, la construction et la manœuvre des aérostats.

W ⬛ Das Wort Ballon leitet sich her von dem griechischen βάλλω: werfen.

W ▌▌Un pallone aerostatico è un tipo di aeromobile, un aerostato che si solleva da terra grazie al principio di Archimede.

W ⬛ Un globo aerostático es una aeronave aerostática no propulsada que se sirve del principio de los fluidos de Arquímedes para volar, entendiendo el aire como un fluido.

Categories: W ⬛ Globus aerostàtic, ⬛ Balloons (aircraft), Ballooning, Airship technology, Hydrogen technologies, Aeronautics, ▌▌Aérostation, Activité aérienne, ⬛ Ballon, ▌▌Aeromobili, Sport dell'aria, ⬛ Aeronaves por tipo

DBPedia: ⬛ Balloon (aircraft)

Title: balloon² · **ID:** bn:00008188n · **Type:** Concept   ∴ explore

Senses: W ⬛ **balloon**²

W ⬛ Toy balloon, ▌▌Ballon (jouet), ⬛ Luftballon, ▌▌Palloncino, ⬛ Globo (juguete)

W ⬛ Party balloon, Toy balloons, **Balloon (toy)**, Toy Balloon, Rubber balloon, ▌▌Ballon en mylar, Ballon en aluminium, Ballon (Jouet), Ballon en nylon métallisé, ▌▌Palloncini

Glosses: W ⬛ A toy balloon is an inflatable object which is often made of plastic or natural, biodegradable rubber.

W ⬛ small thin inflatable rubber bag with narrow neck

W ▌▌Un ballon est un récipient léger et relativement étanche destiné à être rempli de gaz, en général de l'air, parfois de l'hélium pour que le ballon vole.

W ⬛ Ein Luftballon ist ein elastischer Hohlkörper, der mit Gas befüllbar ist und sich dabei um ein Vielfaches seiner ursprünglichen Größe ausdehnt.

W ▌▌Un palloncino è un recipiente in lattice gonfiato a bocca o con l'aiuto di pompe con l'aria oppure usando le bombole di elio.

W ⬛ Un globo es un recipiente de material flexible relleno de gas, a menudo usado como juguete para los niños.

Categories: W ⬛ Balloons (entertainment), Traditional toys, Rubber toys, ▌▌Fête, Objet gonflable, ⬛ Spielzeug, ▌▌Giocattoli, ⬛ Juguetes, Inventos de Inglaterra

DBPedia: ⬛ Toy balloon

Fig. 4. The BabelNet search interface with the first two results for *balloon* (in English)

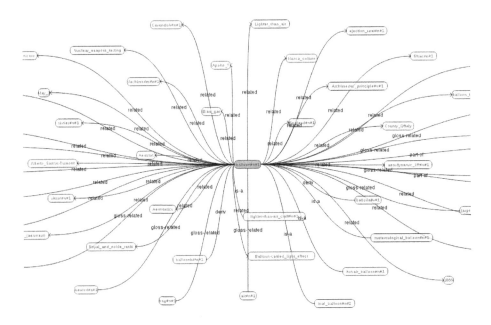

Fig. 5. The BabelNet explore interface with the first sense of *balloon* (in English)

5 The BabelNet API

Access to the BabelNet Java API can be obtained from the homepage (bottom-right icon of Figure 3). We will now take a quick tour of the main features of the API.

5.1 Creating a BabelNet Instance

An instance of BabelNet can be obtained with just one line of code:

```
BabelNet bn = BabelNet.getInstance();
```

5.2   Obtaining the Senses of a Lemma

We can obtain the senses of a given lemma as follows:

```
List<BabelSense> senses =
    bn.getSenses(languageToSearch, lemma, pos, includeRedirects);
```

where `languageToSearch` is any of the supported languages in BabelNet (i.e. `Language.CA`, `.EN`, `.ES`, `.FR`, `.DE`, `.IT`), `pos` is any of the four open-class part-of-speech tags (`POS.NOUN`, `.ADJECTIVE`, `.ADVERB`, `.VERB`) and `includeRedirects` is a boolean specifying whether we also want senses matching the lemma in the Wikipedia redirections. The `getSenses` method returns a list of `BabelSense` instances.

5.3   Obtaining the Babel Synsets Containing a Lemma

Similarly to the above method for obtaining senses, we can query BabelNet for synsets containing the input lemma:

```
List<BabelSynset> synsets =
    bn.getSynsets(languageToSearch, lemma, pos, includeRedirects);
```

5.4   Sorting Senses and Synsets

Senses can be sorted using the `BabelSenseComparator` according to the following criteria: WordNet senses first, sorted by sense number, next, Wikipedia senses in lexicographic order:

```
Collections.sort(senses, new BabelSenseComparator());
```

Similarly, one can sort synsets as follows:

```
Collections.sort(synsets, new BabelSynsetComparator());
```

using the `BabelSynsetComparator`.

5.5   Getting a String Representation for a Sense

To obtain a string representation for a given `BabelSense`:

```
System.out.println(sense.getSenseString());
```

where `sense` is a `BabelSense`. What is returned by the `getSenseString` method depends on the information available for the input sense:

- a lemma-pos-sense representation if it is a WordNet sense (e.g. balloon#n#1);
- the Wikipedia page title if it corresponds to a Wikipedia sense (e.g. Balloon (aircraft));
- the lemma if it is a translation (e.g. Ballon).

5.6 Important Information in a Babel Synset

Babel synsets contain many different kinds of information available by means of its methods, among which we have:

- `getId`, which returns the synset id;
- `getPOS`, which provides the synset's part-of-speech tag;
- `getSynsetSource`, which returns the synset source, i.e. whether it belongs to WordNet only, Wikipedia only or both (i.e. a mapping between a Wikipedia page and a WordNet sense could be established);
- `getSynsetType`: does the synset denote a `NAMED_ENTITY` or a `CONCEPT`?
- `getSenses`, which returns the list of senses in the synset;
- `getTranslations`, which returns a source-to-target map of the translations of Babel senses;
- `getRelatedMap`, which returns a map of all Babel synsets connected to the given synset by means of lexical-semantic relations;
- `getCategories`, which returns a list of Wikipedia categories for the given synset;
- `getImages`, which returns a list of image URIs for the given synset;
- `getWordNetOffsets`, which returns a list of WordNet offsets to which the synset is linked;
- `getMainSense`, which returns the main sense of the synset.

6 Conclusions

In this paper we have taken a quick tour of BabelNet 1.1, a very large multilingual ontology available at `http://www.babelnet.org`. We first introduced BabelNet, provided statistics for the new version 1.1, overviewed the new interface for search and graph exploration, and surveyed the Java programmatic API. This resource has already proven useful in tasks such as multilingual Word Sense Disambiguation [13] and semantic relatedness [12], but many horizons are still to be explored.

Acknowledgments.

 The author gratefully acknowledges the support of the ERC Starting Grant MultiJEDI No. 259234. Thanks go to Daniele Vannella for his help on calculating statistics.

References

1. Bizer, C., Lehmann, J., Kobilarov, G., Auer, S., Becker, C., Cyganiak, R., Hellmann, S.: Dbpedia - a crystallization point for the web of data. Journal of Web Semantics 7(3), 154–165 (2009)
2. Chiarcos, C., Hellmann, S., Nordhoff, S.: Towards a linguistic linked open data cloud: The Open Linguistics Working Group. TAL 52(3), 245–275 (2011)

3. Gracia, J., Montiel-Ponsoda, E., Cimiano, P., Gómez-Pérez, A., Buitelaar, P., McCrae, J.: Challenges for the multilingual web of data. J. Web Sem. 11, 63–71 (2012)
4. Hoffart, J., Suchanek, F.M., Berberich, K., Weikum, G.: Yago2: A spatially and temporally enhanced knowledge base from wikipedia. Artificial Intelligence 194, 28–61 (2013)
5. de Melo, G., Weikum, G.: Towards a universal wordnet by learning from combined evidence. In: Proceedings of the Eighteenth ACM Conference on Information and Knowledge Management, Hong Kong, China, pp. 513–522 (2009)
6. Miller, G.A., Beckwith, R., Fellbaum, C.D., Gross, D., Miller, K.: WordNet: an online lexical database. International Journal of Lexicography 3(4), 235–244 (1990)
7. Moro, A., Navigli, R.: WiSeNet: Building a Wikipedia-based semantic network with ontologized relations. In: Proceedings of the 21st ACM Conference on Information and Knowledge Management (CIKM 2012), Maui, HI, USA (2012)
8. Nastase, V., Strube, M.: Transforming wikipedia into a large scale multilingual concept network. Artificial Intelligence 194, 62–85 (2013)
9. Navigli, R.: Word Sense Disambiguation: A survey. ACM Computing Surveys 41(2), 1–69 (2009)
10. Navigli, R.: BabelNet goes to the (Multilingual) Semantic Web. In: ISWC 2012 Workshop on Multilingual Semantic Web (2012)
11. Navigli, R., Ponzetto, S.P.: BabelNet: The automatic construction, evaluation and application of a wide-coverage multilingual semantic network. Artificial Intelligence 193, 217–250 (2012)
12. Navigli, R., Ponzetto, S.P.: BabelRelate! a joint multilingual approach to computing semantic relatedness. In: Proceedings of the Twenty-Sixth AAAI Conference on Artificial Intelligence, AAAI (2012)
13. Navigli, R., Ponzetto, S.P.: Joining forces pays off: Multilingual joint word sense disambiguation. In: Proceedings of the 2012 Joint Conference on Empirical Methods in Natural Language Processing and Computational Natural Language Learning (EMNLP-CoNLL), pp. 1399–1410 (2012)
14. Navigli, R., Velardi, P.: Learning Word-Class Lattices for definition and hypernym extraction. In: Proceedings of the 48th Annual Meeting of the Association for Computational Linguistics (ACL), Uppsala, Sweden, pp. 1318–1327 (2010)
15. Navigli, R., Velardi, P., Ruiz-Martínez, J.M.: An annotated dataset for extracting definitions and hypernyms from the web. In: Proceedings of the 7th International Conference on Language Resources and Evaluation, Valletta, Malta, May 19-21 (2010)
16. Suchanek, F.M., Kasneci, G., Weikum, G.: Yago: A large ontology from Wikipedia and WordNet. Journal of Web Semantics 6(3), 203–217 (2008)

Automatic Pipeline Construction
for Real-Time Annotation

Henning Wachsmuth, Mirko Rose, and Gregor Engels

Universität Paderborn
s-lab – Software Quality Lab
Paderborn, Germany
hwachsmuth@s-lab.upb.de, {mrose,engels}@upb.de

Abstract. Many annotation tasks in computational linguistics are tackled with manually constructed pipelines of algorithms. In real-time tasks where information needs are stated and addressed ad-hoc, however, manual construction is infeasible. This paper presents an artificial intelligence approach to *automatically* construct annotation pipelines for given information needs and quality prioritizations. Based on an abstract ontological model, we use partial order planning to select a pipeline's algorithms and informed search to obtain an efficient pipeline schedule. We realized the approach as an expert system on top of Apache UIMA, which offers evidence that pipelines can be constructed ad-hoc in near-zero time.

1 Introduction

Information extraction and other applications of computational linguistics deal with the annotation of text, which often takes several interdependent steps. A typical annotation task is to relate different entity types to event anchors in a text. E.g., the BioNLP shared task GENIA [13] included event types like *PositiveRegulation(Theme, Cause, Site, CSite)* whose instances are contained in sentences like: "Eo-VP16, but not the empty GFP retrovirus, increased perforin expression in both WT and T-bet-deficient CD8+ T cells." Before entities and events can be related, the respective types must have been recognized, which normally requires linguistic annotations, e.g. part-of-speech tags. These in turn can only be added to a text that has been segmented into lexical units, e.g. into sentences.

Because of the interdependencies of the steps, the standard way to address such an information need is with an annotation pipeline $\Pi = \langle \mathbf{A}, \pi \rangle$, where $\mathbf{A}$ is a set of algorithms and π is the schedule of these algorithms. Each algorithm in $\mathbf{A}$ takes on one analysis by annotating certain types of output information, and it requires certain types of information as input. The schedule π has to ensure that all input requirements are fulfilled. Different algorithms for an analysis may vary in their requirements, thereby placing different constraints on π. Moreover, Π may have to meet efficiency and effectiveness criteria, e.g. measured as run-time or F_1-score. Often, an increase in effectiveness is paid with a decrease in efficiency and vice versa. The set $\mathbf{A}$ must therefore be composed in respect of the quality criteria at hand. When Π incorporates filtering steps, π affects the efficiency of Π

A. Gelbukh (Ed.): CICLing 2013, Part I, LNCS 7816, pp. 38–49, 2013.

as well [21]. As a result, pipeline construction faces two involved challenges: (1) The selection of a set of algorithms that addresses a given information need and that complies with the given quality criteria. (2) The determination of an efficient schedule that fulfills the input requirements of all algorithms.

Traditionally, the construction of efficient and effective annotation pipelines is performed manually by experts with domain knowledge. However, recent developments suggest that the future of annotation tasks will be real-time search applications, where regular internet users state information needs ad-hoc [7,15]. In such scenarios, the only way to be able to directly respond to a user is to construct and execute annotation pipelines in an automatic manner.

In this paper, we present an approach to automatically construct annotation pipelines based on techniques from artificial intelligence [18]. To this end, we formalize expert knowledge on algorithms and information types within an abstract ontological model. Given an information need, we rely on *partial order planning* to select a set of algorithms with a defined partial order. Where appropriate, filtering steps are integrated, while a prioritization of quality criteria allows to influence the trade-off between a pipeline's presumed efficiency and effectiveness. To obtain a correct and efficient schedule, we apply *informed search* using estimations of the algorithms' run-times. In large-scale scenarios, A* search could find a near-optimal schedule on a sample of texts. For ad-hoc pipeline construction, however, we argue that a greedy best-first search strategy is more reasonable.

We realized our approach as an open-source *expert system* on top of Apache UIMA [2]. Experiments with this system suggest that automatic pipeline construction can be performed in near-zero time for realistic numbers of algorithms. Altogether, our main contributions are three-fold:

1. We formalize the expert knowledge that is needed to automatically address annotation tasks in an abstract ontological model (Section 3).
2. We approach the automatic construction of efficient and effective annotation pipelines with partial order planning and informed search (Section 4).
3. We provide an expert system that constructs and executes annotation pipelines ad-hoc, thereby qualifying for real-time applications (Section 5).

2 Related Work

Planning and informed search represent algorithmic foundations of artificial intelligence [18]. The latter has been used to speed up several tasks in natural language processing, e.g. parsing [16]. In [20], we solve optimal scheduling theoretically, but we suggest to apply informed search in practice. In case of planning, our work resembles [6] where partial order planning is proposed to compose information extraction algorithms. The authors do not consider quality criteria, though, nor do they realize and evaluate their approach. This also holds for [22], in which knowledge discovery workflows of minimum length are planned for an ontology of data mining algorithms. A planning approach that sums up the costs of steps is given in [17] for data stream processing. In annotation tasks, however, the values of criteria such as precision cannot simply be aggregated.

Annotation tasks are often tackled with pipelines of algorithms [3]. For manually constructing such pipelines, software frameworks like Apache UIMA [2] and GATE [10] provide tool support. Still, manual pipeline construction tends to be error-prone and cost-intensive [5]. Recently, a UIMA-based system that automatically composes and executes a defined set of algorithms was presented in [11], but it is not yet available. The author previously worked on U-COMPARE [12], which allows a fast but manual construction and evaluation of pipelines. In contrast, we perform construction fully automatically. We realized our approach as an expert system for regular users. Expert systems have been used for scheduling and construction since the early times of artificial intelligence [9].

Automatic pipeline construction is important when different tasks require different algorithms. Major open information extraction systems such as REVERB currently avoid this scenario by analyzing task-independent syntactical structures rather than semantic concepts [8]. However, they are restricted to binary relations. More complex tasks are e.g. addressed by declarative approaches like SYSTEMT [4]: users specify needs by defining logical constraints of analysis steps, while the system manages the workflow. SYSTEMT optimizes schedules, but it targets at expert users and works only with rule-based algorithms.

Efficient schedules benefit from the filtering of relevant information, which has a long tradition in information extraction [1]. In [21], we show how to construct efficient pipelines for any set of extraction algorithms. We adopt parts of this approach here, but we address arbitrary annotation tasks.

3   An Ontological Model of Annotation Tasks

In this section, we model the expert knowledge needed to automatically address annotation tasks. Many annotation tasks target at information of domain- and application-specific *type systems*, as e.g. in the GENIA example from Section 1. Still, these type systems instantiate the same abstract structures. In particular, most works distinguish between *primitive types*, such as integers or strings, and hierarchically organized *annotation types*, which assign syntactic or semantic concepts to spans of text. Annotation types can have *features* whose values are either primitives or annotations. Among others, features thereby allow to model concepts like relations and events as annotations. Not all features of an annotation are always set. We call a feature *active* if it has a value assigned.

Now, an information need refers to a set of annotation types and features. Besides, it may place *value constraints* on the text spans to be annotated. Within GENIA, a constraint could be to keep only positive regulation events whose cause is "Eo-VP16", i.e., *PositiveRegulation(_, "Eo-VP16", _, _)*. In general, we define the abstract *information type* to be found in annotation tasks as follows.

Information Type. A set of annotations of an arbitrary but fixed type denotes an information type C if it contains all annotations that meet two conditions:

1. **Active Feature.** The annotations in C either have no active feature or they have the same single active feature.
2. **Constraints.** The annotations in C fulfill the same set of value constraints.

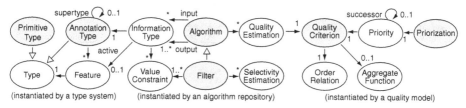

Fig. 1. Abstract ontological model of the expert knowledge for addressing annotation tasks. White and black arrowheads denote "subclass" and "has" relations, respectively. Grey-colored concepts are instantiated by concrete classes within an application.

By defining C to have at most one active feature, we obtain a normalized unit of information in annotation tasks. A single information need can be stated as a set of information types $\mathbf{C} = \{C_1, \ldots, C_{|\mathbf{C}|}\}$, meaning a conjunction $C_1 \wedge \ldots \wedge C_{|\mathbf{C}|}$. Different information needs result in disjunctions of such conjunctions.

When fulfilling a need like *PositiveRegulation(_, "Eo-VP16", _, _)*, a certain effectiveness and efficiency is achieved. Such *quality criteria* are used to evaluate whether a solution is good or is better than another one. Conceptually, a quality criterion Q defines an *order relation* for a set of values. Q may have an *aggregate function* that maps any two values $q_1, q_2 \in Q$ (say, run-times) to a value $q \in Q$ (say, the total run-time). An aggregate function allows to infer the quality of a solution from the quality of partial solutions. However, far from all criteria entail such a function. For instance, there is no general way of inferring an overall precision from the precision of two algorithms. Similarly, functions that aggregate values of *different* quality criteria rarely make sense. In contrast to other multi-criteria optimization problems [14], weighting different Pareto-optimal solutions (where any improvement in one criterion worsens others) hence does not seem reasonable in annotation tasks. Instead, we propose *quality prioritizations*:

Quality Prioritization. A prioritization $P = (Q'_1, \ldots, Q'_{|\mathbf{Q}|})$ is a permutation of a set of quality criteria $\mathbf{Q} = \{Q_1, \ldots, Q_{|\mathbf{Q}|}\}$ that defines an order of importance.

E.g., *(run-time, precision, recall)* targets at the solution with highest recall under all solutions with highest precision under all solutions with lowest run-time. Together, $\mathbf{Q}$ and a set of prioritizations $\mathbf{P} = \{P_1, \ldots, P_{|\mathbf{P}|}\}$ define a *quality model*.

To select a set of algorithms $\mathbf{A}$ for an information need $\mathbf{C}$ that complies with a prioritization P, internal operations of the algorithms do not matter, but only their *input* and *output* behavior. The actual efficiency and effectiveness of an algorithm on a collection or a stream of texts is unknown beforehand. For many algorithms, typical *quality estimations* are known from evaluations, though.

Algorithm. Let $\mathbf{C}$ be a set of information types and $\mathbf{Q}$ a set of quality criteria. Then an algorithm A is a 3-tuple $\langle \mathbf{C}_{in}, \mathbf{C}_{out}, \vec{q} \rangle$ such that $\mathbf{C}_{in} \neq \mathbf{C}_{out}$ and

- **Input.** $\mathbf{C}_{in} \subseteq \mathbf{C}$ is the set of input information types required by A,
- **Output.** $\mathbf{C}_{out} \subseteq \mathbf{C}$ is the set of output information types A produces, and
- **Estimations.** $\vec{q} \in (Q_1 \cup \{\bot\}) \times \ldots \times (Q_{|\mathbf{Q}|} \cup \{\bot\})$ contains one value q_i for each $Q_i \in \mathbf{Q}$. q_i defines a quality estimation or it is unknown, denoted as $\bot$.

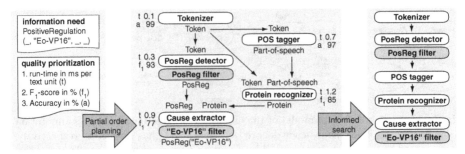

Fig. 2. Pipeline construction for an information need and a quality prioritization. Seven algorithms are selected with partial order planning and scheduled with informed search.

Assume that an algorithm has produced a type $C \in \mathbf{C}$, e.g. *PositiveRegulation*. Then a means to improve efficiency is to further analyze only text units (say, sentences) that contain positive regulation events and that, hence, may be relevant for $\mathbf{C}$ [21]. We call an algorithm a *filter* if it discards text units that do not meet some defined value constraints. In practice, algorithms may combine annotation and filtering operations. Here, we separate filters as they can be created on-the-fly for given information types. Such a filter has a single input type $\mathbf{C}_{in} = \{C_{in}\}$ that equals its output type $\mathbf{C}_{out} = \{C_{out}\}$ except that C_{out} additionally meets the filter's value constraints.[1] While a filter implies a certain selectivity (i.e., a fraction of filtered text units), selectivities strongly depend on the input [20]. So, reasonable *selectivity estimations* can only be obtained during analysis.

Altogether, Figure 1 associates the described concepts in an ontological model. The helper concepts *Type* and *Priority* realize the diversity of features and the permutations of prioritizations. As depicted, the modeled knowledge can be partitioned into three parts that are instiantiated within a concrete ontology:

Annotation Task Ontology. An annotation task ontology Ω denotes a 3-tuple $\langle \mathbf{C}_\Omega, \mathbf{P}_\Omega, \mathbf{A}_\Omega \rangle$ such that $\mathbf{C}_\Omega$ is a set of available information types, $\mathbf{P}_\Omega$ is a set of available quality prioritizations, and $\mathbf{A}_\Omega$ is a set of available algorithms.

4   Automatic Pipeline Construction

We now introduce an artificial intelligence approach to automatically construct annotation pipelines. In particular, we use *partial order planning* to select a set of algorithms and a greedy *informed search* strategy to find an efficient schedule. Figure 2 exemplarily illustrates the application of our approach.

We consider pipeline construction as a planning problem. In artificial intelligence, the term "planning" denotes the process of generating a sequence of actions that transforms an initial state of the world into a specified goal state [18]. A planning problem is defined by its domain and by the task to be addressed. For pipeline construction, we describe the planning problem as follows:

---

[1] For filters that check the presence of an information type in a text unit, we define that all information types have an implicit constraint "is contained in text unit".

Algorithm PIPELINEPARTIALORDERPLANNING($\mathbf{C}_0, \mathbf{C}_\phi, P_\phi, \mathbf{A}_\Omega$)

1: Algorithm set $\mathbf{A}_\phi \leftarrow \{A_\phi\}$
2: Partial schedule $\pi_\phi \leftarrow \emptyset$
3: Input requirements $\mathbf{\Lambda} \leftarrow \{\langle C_\phi, A_\phi\rangle \mid C_\phi \in \mathbf{C}_\phi \backslash \mathbf{C}_0\}$
4: **while** $\mathbf{\Lambda} \neq \emptyset$ **do**
5:     Input requirement $\langle C_\mathbf{\Lambda}, A_\mathbf{\Lambda}\rangle \leftarrow \mathbf{\Lambda}.\text{poll}()$
6:     **if** $C_\mathbf{\Lambda} \in \mathbf{C}_\phi$ **then**
7:         Filter $A_F \leftarrow$ CREATEFILTER($C_\mathbf{\Lambda}$)
8:         $\mathbf{A}_\phi \leftarrow \mathbf{A}_\phi \cup \{A_F\}$
9:         $\pi_\phi \leftarrow \pi_\phi \cup \{(A_F < A_\mathbf{\Lambda})\}$
10:         $\langle C_\mathbf{\Lambda}, A_\mathbf{\Lambda}\rangle \leftarrow \langle A_F.C_{in}, A_F\rangle$
11:     Algorithm $A \leftarrow$ SELECTBESTALGORITHM($C_\mathbf{\Lambda}, P_\phi, \mathbf{A}_\Omega$)
12:     **if** $A = \perp$ **then return** $\perp$
13:     $\mathbf{A}_\phi \leftarrow \mathbf{A}_\phi \cup \{A\}$
14:     $\pi_\phi \leftarrow \pi_\phi \cup \{(A < A_\mathbf{\Lambda})\}$
15:     $\mathbf{\Lambda} \leftarrow \mathbf{\Lambda} \cup \{\langle C, A\rangle \mid C \in A.\mathbf{C}_{in} \backslash \mathbf{C}_0\}$
16: **return** $\langle \mathbf{A}_\phi, \pi_\phi\rangle$

Fig. 3. Pseudocode of partial order planning for selecting a set of algorithms $\mathbf{A}_\phi$ (with a partial schedule π_ϕ) that addresses a planning problem $\phi^{(\Omega)} = \langle \mathbf{C}_0, \mathbf{C}_\phi, P_\phi, \mathbf{A}_\Omega\rangle$

Planning Problem. Let $\Omega = \langle \mathbf{C}_\Omega, \mathbf{P}_\Omega, \mathbf{A}_\Omega\rangle$ be an annotation task ontology. Then a planning problem $\phi^{(\Omega)}$ denotes a 4-tuple $\langle \mathbf{C}_0, \mathbf{C}_\phi, P_\phi, \mathbf{A}_\Omega\rangle$ such that

- **Initial State.** $\mathbf{C}_0 \subseteq \mathbf{C}_\Omega$ is the initially given information,
- **Goal.** $\mathbf{C}_\phi \subseteq \mathbf{C}_\Omega$ is the information need to be fulfilled,
- **Quality.** $P_\phi \in \mathbf{P}_\Omega$ is the quality prioritization to be met, and
- **Actions.** $\mathbf{A}_\Omega$ is the set of algorithms available to fulfill $\mathbf{C}_\phi$.

We implicitly model states of a planning domain as sets of information types, thereby reflecting the states of analysis of an input text.[2] So, all states $\mathbf{C}$ with $\mathbf{C}_\phi \subseteq \mathbf{C}$ are goal states. Algorithms represent actions in that they modify states by adding new information types. To solve a problem $\phi^{(\Omega)}$, we hence need a pipeline of algorithms that produces $\mathbf{C}_\phi \backslash \mathbf{C}_0$ while complying with P_ϕ.

4.1 Algorithm Selection Based on Partial Order Planning

We chose partial order planning to select a set of algorithms $\mathbf{A}_\phi$ and to define a partial schedule π_ϕ. In general, this backward approach recursively generates and combines subplans for all preconditions of those actions that achieve a planning goal [18]. Actions may conflict, namely, if an effect of one action violates a precondition of another one. In annotation tasks, however, algorithms only produce information. While filters reduce the input to be processed, they never prevent subsequent algorithms from being applicable [6].

In Figure 3, we adapt partial order planning for algorithm selection. For planning purposes only, a helper "finish algorithm" A_ϕ is initially added to $\mathbf{A}_\phi$. Also, a set of input requirements $\mathbf{\Lambda}$ (the "agenda") is derived from $\mathbf{C}_\phi \backslash \mathbf{C}_0$ (lines 1–3).

---

[2] In practice, $\mathbf{C}_0$ will often be the empty set, meaning that an analysis starts on plain text. Accordingly, a non-empty set $\mathbf{C}_0$ indicates texts that already have annotations.

Each requirement specifies an information type, e.g. *PositiveRegulation*, and an algorithm that needs this type as input. While Λ is not empty, lines 4–15 insert algorithms into $\mathbf{A}_\phi$ and update both π_ϕ and Λ. In particular, line 5 retrieves an input requirement $\langle C_\Lambda, A_\Lambda \rangle$ with the deterministic method *poll()*. If C_Λ belongs to $\mathbf{C}_\phi$, a filter A_F is integrated on-the-fly, whereas $\langle C_\Lambda, A_\Lambda \rangle$ is replaced with the input requirement of A_F (lines 6–10). Then, line 11 selects an algorithm A that produces C_Λ. If any input requirement cannot be fulfilled, planning fails (line 12) and does not reach line 16 to return a partially ordered pipeline $\langle \mathbf{A}_\phi, \pi_\phi \rangle$.

For space reasons, we only sketch SELECTBESTALGORITHM: First, the set $\mathbf{A}_\Lambda$ of algorithms that produce C_Λ is determined. These algorithms are compared iteratively for each quality criterion Q of the prioritization P_ϕ. If Q has no aggregate function, $\mathbf{A}_\Lambda$ is reduced to the algorithms with the best estimation for Q. Else, the estimations of possible predecessor algorithms of $\mathbf{A}_\Lambda$ are aggregated. In the worst case, this requires to recursively create plans for all input requirements of $\mathbf{A}_\Lambda$. However, predecessors are stopped taken into account as soon as a filter is encountered: filtering changes the input to be processed, hence it does not make sense to aggregate estimations of algorithms before and after filtering. In case, only one algorithm remains for any $Q \in P_\phi$, it constitutes the single best algorithm. Otherwise, any best algorithm is returned.

The filtering view conveys a benefit of partial order planning: As we discussed in [21], *early filtering* of information and *lazy evaluation* improve the efficiency of pipelines while leaving their effectiveness unaffected. Since our planner proceeds backwards, the constraints in π_ϕ prescribe only to execute algorithms right before needed, which implies lazy evaluation. Also, π_ϕ allows to execute a filter directly after its respective annotation algorithm, thereby enabling early filtering.

We defined an information need as one set $\mathbf{C}_\phi$, but many tasks address $k > 1$ needs concurrently. Aside from *PositiveRegulation*, for instance, GENIA faced eight other event types, e.g. *Binding* [13]. The principle generalization for k problems $\phi_1, \ldots, \phi_k$ is simple: We apply partial order planning to each ϕ_i, which results in k partially ordered pipelines $\langle \mathbf{A}_{\phi_1}, \pi_{\phi_1} \rangle, \ldots, \langle \mathbf{A}_{\phi_k}, \pi_{\phi_k} \rangle$. Then, we unify all these pipelines as $\langle \mathbf{A}_\phi, \pi_\phi \rangle = \langle \bigcup_{i=1}^{k} \mathbf{A}_{\phi_i}, \bigcup_{i=1}^{k} \pi_{\phi_i} \rangle$. However, attention must be paid to filters, e.g. a text unit without positive regulations still may yield a binding event. To handle such cases, a set of relevant text units should be maintained independently for each ϕ_i, which is beyond the scope of this paper. Below, we assume that an according maintenance system is given.

4.2 Scheduling with Informed Best-First Search

Informed search aims at efficiently finding solutions by exploiting problem-specific knowledge [18]. During search, a directed acyclic graph is generated stepwise, in which nodes correspond to partial solutions and edges to solving subproblems. For scheduling the set of algorithms $\mathbf{A}_\phi$, we let a node with depth d in the graph denote a pipeline $\langle \mathbf{A}, \pi \rangle$ with d algorithms. The graph's root node is the empty pipeline, and each leaf a pipeline $\langle \mathbf{A}_\phi, \pi \rangle$ with a correct schedule π. An edge represents the execution of an applicable *filter stage*. Here, a filter stage $\langle \mathbf{A}_F, \pi_F \rangle$ is a pipeline where $\mathbf{A}_F$ consists of a filter A_F and all algorithms

---

Algorithm GREEDYBESTFIRSTSCHEDULING($\mathbf{A}_\phi, \pi_\phi$)

1: Algorithm set $\mathbf{A} \leftarrow \emptyset$
2: Schedule $\pi \leftarrow \emptyset$
3: **while** $\mathbf{A} \neq \mathbf{A}_\phi$ **do**
4:     Filter stages $\mathbf{\Pi} \leftarrow \emptyset$
5:     **for each** Filter $A_F \in \{A \in \mathbf{A}_\phi \backslash \mathbf{A} \mid A$ is a filter$\}$ **do**
6:         Algorithm set $\mathbf{A}_F \leftarrow \{A_F\} \cup$ GETALLPREDECESSORS($\mathbf{A}_\phi \backslash \mathbf{A}, \pi_\phi, A_F$)
7:         Schedule $\pi_F \leftarrow$ GETANYTOTALORDERING($\mathbf{A}_F, \pi_\phi$)
8:         Estimated cost $h[\langle \mathbf{A}_F, \pi_F \rangle] \leftarrow$ GETAGGREGATEESTIMATION($\mathbf{A}_F$)
9:         $\mathbf{\Pi} \leftarrow \mathbf{\Pi} \cup \{\langle \mathbf{A}_F, \pi_F \rangle\}$
10:     Filter stage $\langle \mathbf{A}_t, \pi_t \rangle \leftarrow \underset{\langle \mathbf{A}_F, \pi_F \rangle \in \mathbf{\Pi}}{\arg\min} \ h[\langle \mathbf{A}_F, \pi_F \rangle]$
11:     $\pi \leftarrow \pi \cup \pi_t \cup \{(A < A_t) \mid A \in \mathbf{A} \wedge A_t \in \mathbf{A}_t\}$
12:     $\mathbf{A} \leftarrow \mathbf{A} \cup \mathbf{A}_t$
13: **return** $\langle \mathbf{A}_\phi, \pi \rangle$

Fig. 4. Pseudocode of greedy best-first search for scheduling the filter stages $\langle \mathbf{A}_t, \pi_t \rangle$ of a partially ordered pipeline $\langle \mathbf{A}_\phi, \pi_\phi \rangle$ according to increasing estimated run-time

in $\mathbf{A}_\phi \backslash \mathbf{A}$ that precede A_F within π_ϕ. $\langle \mathbf{A}_F, \pi_F \rangle$ is *applicable* at node $\langle \mathbf{A}, \pi \rangle$ if for all ordering constraints $(A' < A) \in \pi_\phi$ with $A \in \mathbf{A}_F$, we have $A' \in \mathbf{A}$. Given A_F is scheduled last, all schedules of the algorithms in a filter stage entail the same run-time. Thus, it suffices to schedule all filter stages instead of all algorithms.

To efficiently find solutions, a common informed search strategy, called "best-first search", is to generate successor nodes of the node with the lowest *estimated solution cost* first. For this purpose, a *heuristic function* h provides an estimated cost of a path from a node to a leaf. The widely used best-first approach A* then obtains the estimated solution cost of a path through a node by aggregating the cost of reaching the node with the value of h. If h is optimistic (i.e., h never overestimates costs), the first solution found by A* is optimal [18].

Now, let $R(\langle \mathbf{A}, \pi \rangle)$ be the units of an input text filtered by a pipeline $\langle \mathbf{A}, \pi \rangle$. Further, let $t(\langle \mathbf{A}_F, \pi_F \rangle)$ be the estimation of the aggregate run-time per text unit of each filter stage $\langle \mathbf{A}_F, \pi_F \rangle$, and $\mathbf{\Pi}$ the set of all applicable filter stages at node $\langle \mathbf{A}, \pi \rangle$.[3] Then we estimate the costs of reaching a leaf from $\langle \mathbf{A}, \pi \rangle$ as:

$$h(\langle \mathbf{A}, \pi \rangle) = |R(\langle \mathbf{A}, \pi \rangle)| \cdot \min \{t(\langle \mathbf{A}_F, \pi_F \rangle) \mid \langle \mathbf{A}_F, \pi_F \rangle \in \mathbf{\Pi}\}$$

In case $t(\langle \mathbf{A}_F, \pi_F \rangle)$ is optimistic, $h(\langle \mathbf{A}, \pi \rangle)$ is also optimistic, since at least one stage must process $R(\langle \mathbf{A}, \pi \rangle)$. In large-scale scenarios, A* can use h to find an efficient schedule on a sample of texts. To compute $R(\langle \mathbf{A}, \pi \rangle)$, however, the filtered text units must be processed by each successor of the current best node. For ad-hoc scenarios, A* thus imposes much computational overhead, as the information contained in the sample may already suffice to directly return first results.

Instead, we propose greedy best-first search, using the algorithms' estimated run-times only. As sketched in Figure 2, we always apply the filter stage $\langle \mathbf{A}_t, \pi_t \rangle$ with lowest $t(\langle \mathbf{A}_t, \pi_t \rangle)$ first. Since no text unit is taken into account, scheduling can be performed without a sample of texts. Figure 4 shows the greedy search

---

[3] Here, we assume that run-time estimations of all algorithms in $\mathbf{A}_\phi$ are given. For algorithms without run-time estimations, at least default values can be used.

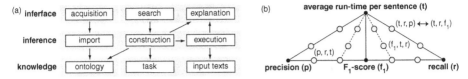

Fig. 5. (a) Architecture of the expert system. (b) Visualization of the system's quality model. The grey partially labeled circles represent the set of quality prioritizations $\mathbf{P}_\Omega$.

for $\langle \mathbf{A}_\phi, \pi \rangle$. Lines 3–12 subsequently add filter stages to $\langle \mathbf{A}, \pi \rangle$. Here, the set $\mathbf{\Pi}$ is built with one stage for each filter A_F in $\mathbf{A}_\phi \backslash \mathbf{A}$. Line 6 identifies all remaining algorithms that must precede A_F, and lines 7–8 compute an ordering and a run-time estimation of the filter stage. Then, line 10 determines $\langle \mathbf{A}_t, \pi_t \rangle$. Before $\mathbf{A}$ and $\mathbf{A}_t$ are merged in line 12, π_t is inserted into π as well as additional ordering constraints to schedule the algorithms in $\mathbf{A}_t$ after those in $\mathbf{A}$ (line 11).

GREEDYBESTFIRSTSCHEDULING may fail when rather slow filter stages filter much less text. However, real-time applications will often not allow to preprocess a statistically significant sample of texts. Moreover, the first applied stage always processes the whole input, which makes a greedy strategy seem reasonable.

5   An Expert System for Real-Time Annotation

We implemented our approach as an expert system on top of Apache UIMA in order to evaluate whether it qualifies for real-time annotation. The system can be accessed at http://www.CICLing.org/2013/data/122 together with its Java source code, usage instructions, and the algorithms used below.

Apache UIMA is a software framework for applications that annotate natural language text [2]. It allows to compose *primitive analysis engines* (say, annotation algorithms) as *aggregate analysis engines* (say, pipelines). To this end, analysis engines are accompanied by a descriptor file with metadata, such as its input and output annotation types. The respective type system itself is also specified in a descriptor file. These files are all that we need for pipeline construction.[4]

Figure 5(a) shows the architecture of our expert system. All permanent knowledge is stored in an OWL *ontology* according to the model from Section 3. Via an *acquisition* module, the automatic *import* of an annotation task ontology $\langle \mathbf{C}_\Omega, \mathbf{P}_\Omega, \mathbf{A}_\Omega \rangle$ from a set of descriptor files can be triggered, except for $\mathbf{P}_\Omega$: For convenience, we built in the quality model in Figure 5(b) and defined additional "implies" relations between prioritizations. In this manner, the expert system can compare algorithms whose effectiveness is e.g. given as accuracy, when e.g. F_1-score is prioritized. The system's prototypical *search* interface enables users to specify a collection of *input texts* as well as a *task* consisting of a prioritization, a set of information types, and different kinds of value constraints. The system then starts the ontology-based ad-hoc *construction* and the *execution* of an aggregate analysis engine. Analysis results are presented by an *explanation* module.

---

[4] By default, quality estimations are not specified in descriptor files. We integrated them in the files' informal description field via a fixed notation, e.g. "@Recall 0.7".

Table 1. The time in ms for algorithm selection (t_A), scheduling (t_π), and automatic pipeline construction in total (t_{apc}) as well as the number of employed algorithms $|A|$ for each C_ϕ and the prioritization (p, r, t) based on the 38 / 76 algorithms of A_{Ω_1} / A_{Ω_2}.

Information need C_ϕ	t_A	t_π	t_{apc}	$|A|$
PositiveRegulation(_, _, _, _)	5.0 / 5.4	1.0 / 1.4	20.2 / 22.9	6
PositiveRegulation(_, "Eo-VP16", _, _)	5.6 / 12.1	2.9 / 3.3	26.5 / 37.7	13
PositiveReg.(Theme, "Eo-VP16", Site, CSite)	14.6 / 17.9	5.0 / 5.7	39.5 / 46.9	20

Table 2. Properties, run-time t in seconds averaged over 10 runs, precision p, and recall r of the pipelines constructed for the information need *StatementOnRevenue(Time, Money)* and each quality prioritization P on the test set of the Revenue corpus [19].

Priorization P	Properties of constructed pipeline	t	p	r
$(t, p, r), (t, r, p)$	fully rule-based; fast preprocessing	**3.5**	0.6	0.48
(r, t, p)	mostly rule-based; exact preprocessing	21.3	0.67	0.58
$(p, r, t), (p, t, r), (r, p, t)$	mostly statistical; exact preprocessing	125.4	**0.76**	**0.66**

5.1   Experimental Analysis of Automatic Pipeline Construction

We experimented with two ontologies $\langle C_\Omega, P_\Omega, A_{\Omega_1} \rangle$ and $\langle C_\Omega, P_\Omega, A_{\Omega_2} \rangle$ based on P_Ω and a set C_Ω that refers to 40 concrete annotation types of GENIA [13] and the statement on revenue task [19]. A_{Ω_2} consists of 76 preprocessing and extraction algorithms, while A_{Ω_1} contains only half of them. Up to three algorithms exist for an information type in both cases. All experiments were conducted on a 2 GHz Intel Core 2 Duo MacBook with 4 GB memory.

We evaluated the expert system for information needs of different complexity. In particular, we measured the pipeline construction time for both ontologies and three needs related to *PositiveRegulation*, as detailed in Table 1.[5] Irrespective of the underlying ontology, algorithm selection and scheduling take only a couple of milliseconds in all cases. The remaining part of t_{apc} refers to operations such as creating descriptor files. Altogether, the measured run-times seem to grow linear in the number of employed algorithms and even sublinear in the number of available algorithms. Hence, we argue that our approach is suitable for real-time annotation. In contrast, manual construction would take at least minutes.

In a second experiment with A_{Ω_2}, we ran the expert system for the information need *StatementOnRevenue(Time, Money)* and each possible prioritization of *run-time*, *precision*, and *recall*. This resulted in the three different pipelines listed in Table 2, which we then executed on the test set of the *Revenue corpus* [19]. The pipeline for (t, p, r) and (t, r, p) relies only on fast rule-based algorithms. As expected, this pipeline was one to two orders of magnitude faster than the other ones, while achieving much less precision and recall. The low recall of 0.58 under (r, t, p) may seem surprising. However, it indicates the restricted feasibility of predicting quality in annotation tasks: favoring high quality algorithms does not

---

[5] We averaged the run-times t_A, t_π, and t_{apc} over 25 runs, as their standard deviations were partly over half as high as the run-times themselves due to I/O operations.

guarantee a high overall quality, since the latter is also influenced by the interactions of the algorithms. In the end, high quality can never be ensured, though,
as it depends on the domain of application and the processed input texts.

Finally, our realization revealed additional challenges of automatic pipeline
construction, which we summarize under three distinct aspects:

Joint Annotation. Algorithms that produce more than one information type
can compromise a quality prioritization. For instance, a constructed pipeline may
schedule a tagger A_t before a chunker A_{tc} that also performs tagging but less
accurate. In this case, the tags of A_t are overwritten by A_{tc}. On the contrary, our
expert system recognizes "dominating" algorithms. E.g., if A_{tc} precedes A_t and
efficiency is of upmost priority, then A_t is omitted. Still, automatic pipeline construction benefits from a maximum decomposition of the analysis steps.

Inheritance. By concept, information types can inherit features from super-
types. In an information need *PositiveRegulation(Theme, _, _, _)*, for instance,
Theme might be inherited from a general type *Event*. To handle such cases, we
normalize the need into *PositiveRegulation $\land$ Event(Theme)*, while ensuring that
only positive regulation events are kept. However, *Theme* also exemplifies a more
complex problem: In GENIA, different event types can be themes of regulations.
But, for scheduling, it suffices to detect one event type before theme extraction,
so the expert system does not select further algorithms. A solution would be to
require all subtypes for types like *Event*, but this is left to future work.

Limitations of Automation. Some limitations result from performing pipeline
construction fully automatically. E.g., our approach does not allow to force certain algorithms to be employed (such as a sentence splitter tuned for biomedical
texts), except for cases that can be realized based on quality criteria. Also, algorithms that target at a middle ground between efficiency and effectiveness tend
not to be chosen because of the prioritization concept. Similarly, it is not possible to prioritize efficiency in one stage (say, preprocessing) and effectiveness in
another (say, relation extraction). While such possibilities could be integrated in
our approach, they require more user interaction and thereby reflect the inherent
trade-off of pipeline construction between automation and manual tuning.

6   Conclusion

Annotation tasks like information extraction are often tackled with pipelines of
algorithms manually constructed by experts. In contrast, we provide an artificial
intelligence approach to automatically construct pipelines, which we realized as
an open-source expert system. Experiments with this system suggest that our approach renders it possible to efficiently and effectively tackle ad-hoc annotation
tasks in real-time applications. Currently, the approach relies on abstract expert
knowledge and techniques like relevance filtering. However, it uses *general* efficiency and effectiveness estimations only. In the future, we will investigate how
to exploit samples of input texts and knowledge about the domain of application in order to optimize pipeline construction. Also, we will work on a relevance
maintenance system to filter information for different tasks at the same time.

Acknowledgments. This work was partly funded by the German Federal Ministry of Education and Research (BMBF) under contract number 01IS11016A.

References

1. Agichtein, E.: Scaling Information Extraction to Large Document Collections. Bulletin of the IEEE Computer Society TCDE 28, 3–10 (2005)
2. Apache UIMA, http://uima.apache.org
3. Bangalore, S.: Thinking Outside the Box for Natural Language Processing. In: Gelbukh, A. (ed.) CICLing 2012, Part I. LNCS, vol. 7181, pp. 1–16. Springer, Heidelberg (2012)
4. Chiticariu, L., Krishnamurthy, R., Li, Y., Raghavan, S., Reiss, F.R., Vaithyanathan, S.: SystemT: An Algebraic Approach to Declarative Information Extraction. In: Proc. of the 48th ACL, pp. 128–137 (2010)
5. Das Sarma, A., Jain, A., Bohannon, P.: Building a Generic Debugger for Information Extraction Pipelines. In: Proc. of the 20th CIKM, pp. 2229–2232 (2011)
6. Dezsényi, C., Dobrowiecki, T.P., Mészáros, T.: Adaptive Document Analysis with Planning. In: Pěchouček, M., Petta, P., Varga, L.Z. (eds.) CEEMAS 2005. LNCS (LNAI), vol. 3690, pp. 620–623. Springer, Heidelberg (2005)
7. Etzioni, O.: Search Needs a Shake-up. Nature 476, 25–26 (2011)
8. Fader, A., Soderland, S., Etzioni, O.: Identifying Relations for Open Information Extraction. In: Proc. of the EMNLP, pp. 1535–1545 (2011)
9. Fox, M.S., Smith, S.F.: ISIS: A Knowledge-based System for Factory Scheduling. Expert Systems 1, 25–49 (1984)
10. GATE, http://gate.ac.uk
11. Kano, Y.: Kachako: Towards a Data-centric Platform for Full Automation of Service Selection, Composition, Scalable Deployment and Evaluation. In: Proc. of the 19th IEEE ICWS, pp. 642–643 (2012)
12. Kano, Y., Dorado, R., McCrohon, L., Ananiadou, S., Tsujii, J.: U-Compare: An Integrated Language Resource Evaluation Platform Including a Comprehensive UIMA Resource Library. In: Proc. of the Seventh LREC, pp. 428–434 (2010)
13. Kim, J.D., Wang, Y., Takagi, T., Yonezawa, A.: Overview of Genia Event Task in BioNLP Shared Task 2011. In: BioNLP Shared Task Workshop, pp. 7–15 (2011)
14. Marler, R.T., Arora, J.S.: Survey of Multi-Objective Optimization Methods for Engineering. Structural and Multidisciplinary Optimization 26(6), 369–395 (2004)
15. Pasca, M.: Web-based Open-Domain Information Extraction. In: Proc. of the 20th CIKM, pp. 2605–2606 (2011)
16. Pauls, A., Klein, D.: k-best A^* Parsing. In: Proc. of the Joint Conference of the 47th ACL and the 4th IJCNLP, pp. 958–966 (2009)
17. Riabov, A., Liu, Z.: Scalable Planning for Distributed Stream Processing Systems. In: Proc. of the 16th ICAPS, pp. 31–41 (2006)
18. Russell, S.J., Norvig, P.: Artificial Intelligence: A Modern Approach, 3rd edn. Prentice-Hall (2009)
19. Wachsmuth, H., Prettenhofer, P., Stein, B.: Efficient Statement Identification for Automatic Market Forecasting. In: Proc. of the 23rd COLING, pp. 1128–1136 (2010)
20. Wachsmuth, H., Stein, B.: Optimal Scheduling of Information Extraction Algorithms. In: Proc. of the 24th COLING: Posters, pp. 1281–1290 (2012)
21. Wachsmuth, H., Stein, B., Engels, G.: Constructing Efficient Information Extraction Pipelines. In: Proc. of the 20th CIKM, pp. 2237–2240 (2011)
22. Žáková, M., Křemen, P., Železný, F., Lavrač, N.: Automating Knowledge Discovery Workflow Composition through Ontology-based Planning. IEEE Transactions on Automation Science and Engineering 8(2), 253–264 (2011)

A Multilingual GRUG Treebank for Underresourced Languages

Oleg Kapanadze[1] and Alla Mishchenko[2]

[1] Ilia State University, Tbilisi, Georgia
ok@caucasus.net
[2] Ukrainian Lingua-Information Fund, Kiev, Ukraine
alla.mishchenko@gmail.com

Abstract. In this paper, we describe outcomes of an undertaking on building Treebanks for underresourced languages Georgian, Russian, Ukrainian, and German - one of the "major" languages in the NLT world. The monolingual parallel sentences in four languages were syntactically annotated manually using *the Synpathy* tool. The tagsets follow an adapted version of the German TIGER guidelines with necessary changes relevant for the Georgian, the Russian and the Ukrainian languages grammar formal description. An output of the monolingual syntactic annotation is in the TIGER-XML format. Alignment of monolingual repository into the bilingual Treebanks was done by *the Stockholm TreeAligner* software. A demo of the GRUG treebank resources will be held during a poster session.

Keywords: underresourced languages, Treebanks, annotation, alignment, Georgian, Russian, Ukrainian, German.

1 Introduction

Parallel corpora are language resources that contain texts and their translations, where the texts, paragraphs, sentences, and words are linked to each other. In the past decades they became useful not only for NLP applications, such as machine translation and multilingual lexicography, but are considered indispensable for empirical language research in contrastive and translation studies.

Naturally-occurring text in many languages are annotated for linguistic structure. A Treebank is a text corpus in which each sentence has been annotated with syntactic structure. Treebanks are often created on top of a corpus that has already been annotated with part-of-speech tags. The annotation can vary from constituent to dependency or tecto-grammatical structures. In turn, Treebanks are sometimes enhanced with semantic or other linguistic information and are skeletal parses of sentences showing rough syntactic and semantic information.

In this paper we describe an experimental undertaking on building parallel Treebanks for German-Georgian, German-Russian, German-Ukrainian and Georgian-Ukrainian language pairs. The languages (except German) involved in the project from the computational viewpoint are considered underresourced, for which paral-

A. Gelbukh (Ed.): CICLing 2013, Part I, LNCS 7816, pp. 50–59, 2013.

lel texts are very rare, if at all existent. In the presented study we used a multilingual parallel corpus appended to a German-Russian-English-Georgian (GREG) valency lexicon for Natural Language Processing [4], [5].

The GREG lexicon itself contains a manually aligned German, Russian, English and Georgian valency data supplied with syntactic subcategorization frames and are saturated with semantic role labels. The multilingual verb lexicon is expended with examples of sentences in 4 languages involved. They unfold lexical entries' meaning and are considered as mutual translation equivalents. The size of bilingual sublexicons, depending to a specific language pair, varies between 1200-1300 entries and the number of example sentences appended to the lexicons are different. For example, a German-Georgian subcorpus, used for this study, has a size of roughly 2600 sentence pairs that correspond to different syntactic subcategorization frames. For the German-Russian language pair had been extracted more fine grained subcorpus with about 4000 sentences as translation equivalents. A German-Ukrainian subcorpus, created for the GRUG initiative support, is relatively small.

Starting the mentioned experiment, on the first phase of the project we intended

- to tag and lemmatize terminal nodes in four languages.
- to produce syntactic parses for multilingual parallel text.
- to determine "good" and "fuzzy" matches between phrases across the tree structures in the languages involved.

Drawing on the created monolingual resources the further objective of the initiative anticipated:

- Production of the parallel trees using the developed monolingual treebanks.
- Alignment of the German-Georgian, German-Russian, German-Ukrainian and Georgian-Ukrainian parallel trees.
- Making general conclusions concerning feasibility of development a full-scale parallel treebanks for the mentioned language pairs.

2   Morphological and Syntactic Annotation of a Multilingual GRUG Text

Initially emphasis was made on development of a parallel treebank for a typologically dissimilar language pair - German and Georgian, since the later is an agglutinative language using both suffixing and prefixing. The Georgian text morphological annotation, tagging and lemmatizing procedures were done with a finite-state morphological transducer which is based on the XEROX FST tools [6], [7].

Before starting syntactic annotation procedures for the Georgian text, we made an overview of experience in building parallel treebanks for languages with different structures [8], [9], [3], [10]. In a Quechua-Spanish parallel treebank, due to strong agglutinative features of the Quechua language, the monolingual Quechua treebank was annotated on morphemes rather than words. This allowed to link morpho-syntactic information precisely to its source. Besides, according to the authors, building phrase structure trees over Quechua sentences does not capture the characteristics of the language. Therefore, for its description a Role and Reference Grammar has been opted that allowed by using nodes, edges and secondary edges to represent the most important aspects of Role and Reference syntax for Quechua sentences [10].

Although Georgian is also an agglutinative language, there is no need to annotate the Georgian Treebank on morphemes. The Georgian syntax can be sufficiently well represent by dependency relations. Therefore, the Georgian Treebank was annotated according to an adapted version of the German TIGER guidelines. To the rest two languages, Russian and Ukrainian, involved in the experiment, morphological features, including POS tags, were assigned manually in script encoding process pursuant to the NEGRA-Treebank (Stuttgart-Tübinger Tagset, STTS) guidelines with the necessary changes relevant to the Russian and Ukrainian grammar formal description. The German Treebank annotation follows the TIGER general annotation scheme [1], [13].

The POS-tagged and lemmatized monolingual sentences in 4 languages were syntactically annotated by means of *the Synpathy* tool developed at Max Plank Institute for Psycholinguistics, Nijmegen [14]. It employs a SyntaxViewer developed for the TIGER-Research project (Institut für Maschinelle Sprachverarbeitung, Universität Stuttgart).

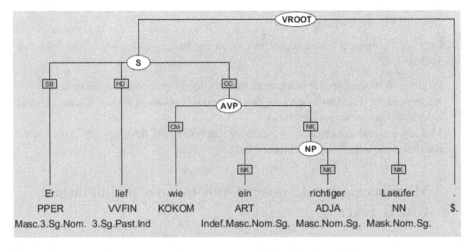

Fig. 1. A syntactically annotated German sentence (English gloss: "He ran as a proper runner") in a TIGER-XML format

A German syntactically analyzed sentence in the Figure 1 visualizes a hybrid approach to the syntactic annotation procedure as tree-like graph structures and integrates annotation according to the constituency and functional relations. Consequently, in a tree structure the node labels are phrasal categories, whereas the parental and secondary edge labels correspond to syntactic functions.

An annotated German sentence from Figure 1 may have two alternative translation equivalents in Georgian depicted in Figure 2 and 3.

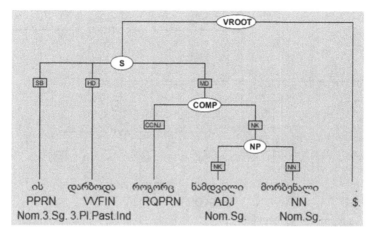

Fig. 2. A translation equivalent in the Georgian language of an annotated German sentence from Figure 1. (lit. "he [used to] ran as-like proper runner").

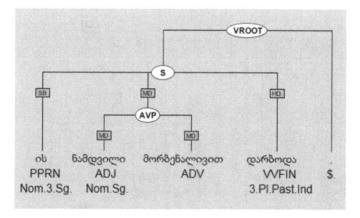

Fig. 3. An alternative translation equivalent in Georgian (lit. "he proper runner-like ran")

Syntactic annotations for a Russian and an Ukrainian translation equivalents of the source sentence from Figure 1 are very close to each-other due to the structural similarity of those languages.

The Russian and the Ukrainian language, as all the Eastern Slavic languages, also posses rich inflectional morphology and fairly free word order.

The Russian and the Ukrainian languages typologically are more closely related languages also to German than Georgian is. Consequently, the TIGER tagsets for these two languages underwent minor changes by incorporated additional POS and CAT features. The changes for the Georgian language tagsets and CAT values are more significant, but in general they conform to the TIGER annotation scheme used as a source in compiling the feature sets and their values for three new languages involved.

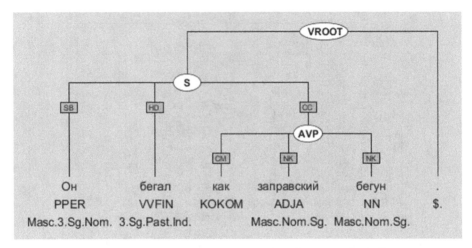

Fig. 4. A translation equivalent in the Russian language of an annotated German sentence from Figure 1. ("He ran as a proper runner").

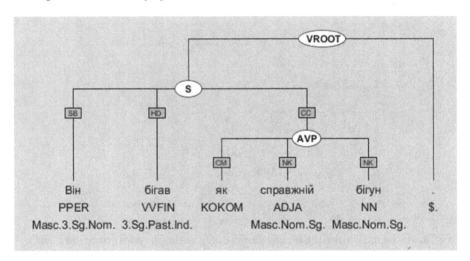

Fig. 5. A translation equivalent in the Ukrainian language of an annotated German sentence from Figure 1 ("He ran as a proper runner")

3   Building Bilingual Treebanks. Alignment of the GRUG Monolingual Resources into Parallel Treebanks

The monolingual treebanks converted into TIGER-XML are a powerful database-oriented representation for graph structures.   In a TIGER-XML graph each leaf (= token) and each node (= linguistic constituent) has a unique identifier [11]. We use these unique identifiers for the phrase and word alignment across trees in translation equivalents.

Alignment of the German, Georgian, Russian and Ukrainian Treebank resources into parallel treebanks is done with help of *the Stockholm TreeAligner*, a tool which inserts alignments between pairs of syntax trees [11], [12]. It handles in parallel treebanks alignment of tree structures, in addition to word alignment, which – according to its developers - is unique and makes difference between two types of alignment: Nodes and words representing exactly the same meaning are aligned as exact translation correspondences using the green color. If nodes and words represent just approximately the same meaning, they are aligned as "fuzzy" translations by means of lines in the red color.

Phrase alignment can be regarded as an additional layer of information on top of the syntax structure. It shows which part of a sentence in L1 language is equivalent to the part of a corresponding sentence in L2 language. This is provided by a graphical user interface of *the Stockholm TreeAligner*. We drew alignment lines manually between pairs of sentences, phrases and words over parallel syntax trees. We intended to align as many phrases as possible. The goal is to show translation equivalence. Phrases shall be aligned only if the tokens, that they span, represent the same meaning and if they could be recognized as translation equivalents outside the current sentence context. The grammatical forms of the phrases need not fit into other contexts, but the meaning has to fit. However, for syntactic annotation in the Georgian language a precise description of a specific mechanism of its clause is necessary. "The Georgian clause is a word collocation which draws on coordination and government of the linked verb and noun sequence" [2]. The most significant divergence in syntactic relations model is, that in the Georgian clause we observe a mutual government and agreement relations or a bilateral coordination phenomenon between verb-predicate and noun-actants which number may reach up to three in a single clause. It anticipates control of the noun case forms by verbs, whereas the verbs, in their turn, are governed by nouns with respect to a grammatical person. Nevertheless, constituency and functional relations employed in the TIGER scheme is sufficiently powerful also for the Georgian syntax description.

Despite existing typological divergences between the GRUG languages, an annotation scheme according to the constituency structures and functional dependency adopted in the TIGER format, can sufficiently well visualize alignment outcomes between bilingual parallel trees. The first significant structural divergence from German is absence of articles as grammatical category in the rest three GRUG languages. Another major difference is word order freedom. For the German language there is an assumed basic word order, which is postulated to be either SOV in dependent clauses and SVO in main clauses. Among the three languages, Georgian has the most free word order due to its rich morphology. Russian and Ukrainian also posses relatively free word order compared to German, but to a lesser extent than the Georgian language. *The Stockholm TreeAligner* guidelines allow phrase alignments within m : n sentence alignments and 1 : n phrase alignments. Even though m : n phrase alignments are technically possible, we have only used 1 : n phrase alignments, for simplicity and clarity reasons.

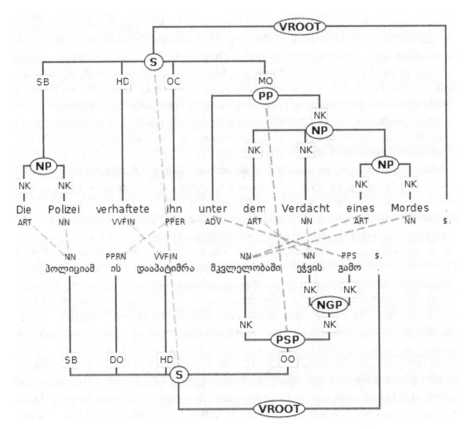

Fig. 6. A German-Georgian parallel tree with "good" and "fuzzy" alignments. English gloss: "The police arrested him under the suspicion of a murder". Georgian lit. "police[*Nar*] he[*Nom*] arrested under[*Pstf_In*] suspicion[*Gen*] for". ([*Nar*] – Narrative case marker, [*Nom*] – Nominative case marker, [*Pstf_In*] – Postfix "in", [*Gen*]-Genitive case marker).

Nevertheless, a 1:1 alignment on word, phrase and sentence level can be often viewed in the German-Ukrainian and the German-Russian parallel trees.

Despite the depicted divergences in syntactic structures of the Georgian Russian, the Ukrainian and the German languages, development of a parallel Treebank by means of two software we have opted, can sufficiently well cope with differences in linguistic structures of the languages involved in the study.

4   Conclusion and Future Plans

In the presented paper we gave an outline of an initiative for building multilingual parallel treebanks for three underesourced languages - Georgian, Russian and Ukrainian and one of the technologically most advanced language – German. This research has been curried out in the framework of the CLARIN-D project (http://fedora.clarin-d.uni-saarland.de/grug/).

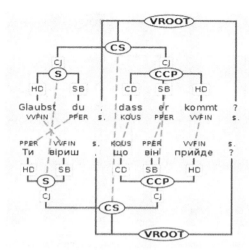

Fig. 7. A German-Ukrainian parallel tree with "good" alignments on all levels. (English gloss: "Do you believe that he will come?") (German Lit. "Believe you, that he comes[*Present_Tense*]?") (Ukrainian Lit. "you believe, that/what he [will] come?").

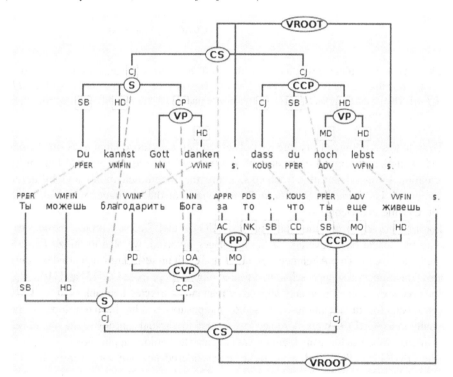

Fig. 8. A German-Russian parallel tree with "good" and "fuzzy" alignments (English gloss: "You can thank the God that you are still alive"). (German lit. "You can God thank, that you [are] still living") (Russian lit. "You can/may thank God for that, what/that you [are] still living").

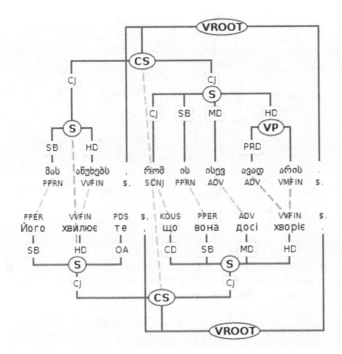

Fig. 9. A Georgin-Ukrainian parallel tree with "good" and "fuzzy" alignments. English gloss: "He is worried that he/she is still ill". Georgian lit. "He[*Dat*] worries, that he/she still ill[*Abl*] is". ([*Dat*]- Dative case marker, [*Abl*] -Ablative case marker) Ukrainian lit. "Him worries that, that/what she hitherto ill [is]".

Despite the typological divergences observed between those languages, in the depicted examples of aligned bilingual parallel trees we tried to demonstrate feasibility of creation bilingual parallel treebanks, using the mentioned mono and bilingual alignment tools (*Synpathy, Stockholm TreeAligner*) for the languages with non Latin script, Cyrillic and Georgian.

For a further advance in the multilingual GRUG parallel Treebank project, besides extending the lexical base coverage of the languages involved, we plan to utilize *Lingua-Align*, a toolbox for Tree Alignment (cl. http://stp.lingfil.uu.se/~joerg/Lingua/index.html). It draws on a supervised approach to tree-to-tree alignment proposed in [15] and [16]. This toolbox requires a small manually aligned or manually corrected treebank of at least 100 sentence pairs for training purposes. Based on a repository that has been developed so far manually in the GRUG Treebank project, we intend to expend capacity of the mentioned multilingual resources for using them in real language technology applications.

The GRUG parallel Treebanks could be considered not only as a repository for empirical language research in contrastive and translation studies, but also as the evaluation corpora for NLT systems for training and testing parsers and as a database for Translation Memory systems.

References

1. Brants, S., Hansen, S.: Developments in the TIGER Annotation Scheme and their Realization in the Corpus. In: Proceedings of the Third Conference on Language Resources and Evaluation (LREC 2002), Las Palmas, pp. 1643–1649 (2002)
2. ჩიქობავა, არ.: მარტივი წინადადების პრობლემა ქართულში, თბილისი (1928); Chikobava, A.: The Problem of the Simple Sentence in Georgian, Tbilisi (1928)
3. Grimes, S., Li, X., Bies, A., Kulick, S., Ma, X., Strassel, S.: Creating Arabic-English Parallel Word-Aligned Treebank Corpora at LDC. In: Proceedings of the Second Workshop on Annotation and Exploitation of Parallel Corpora, The 8th International Conference on Recent Advances in Natural Language Processing (RANLP 2011), Hissar, Bulgaria (2011)
4. Kapanadze, O., Kapanadze, N., Wanner, L., Klatt, S.: Towards A Semantically Motivated Organization of A Valency Lexicon for Natural Language Processing: A GREG Proposal. In: Proceedings of the EURALEX Conference, Copenhagen (2002)
5. Kapanadze, O.: Verbal Valency in Multilingual Lexica. In: Workshop Abstracts of the 7th Language Resources and Evaluation Conference, LREC 2010, Valletta, Malta (2010)
6. Kapanadze, O.: Describing Georgian Morphology with a Finite-State System. In: Yli-Jyrä, A., Kornai, A., Sakarovitch, J., Watson, B. (eds.) FSMNLP 2009. LNCS, vol. 6062, pp. 114–122. Springer, Heidelberg (2010)
7. Kapanadze, O.: Finite State Morphology for the Low-Density Georgian Language. In: FSMNLP 2009 Pre-proceedings of the Eighth International Workshop on Finite-State Methods and Natural Language Processing, Pretoria, South Africa (2009)
8. Megyesi, B., Dahlqvist, B.: A Turkish-Swedish Parallel Corpus and Tools for its Creation. In: Proceedings of Nordiska Datalingvistdagarna, NoDaL- iDa 2007 (2007)
9. Megyesi, B., Hein Sågvall, A., Csató, E.A., Johanson, E.: Building a Swedish-Turkish Parallel Corpus. In: Proceedings of the Fifth International Conference on Language Resources and Evaluation, LREC 2006 (2006)
10. Rios, A., Göhring, A., Volk, M.: Quechua-Spanish Parallel Treebank. In: 7th Conference on Treebanks and Linguistic Theories, Groningen (2009)
11. Samuelsson, Y., Volk, M.: Presentation and Representation of Parallel Treebanks. In: Proceedings of the Treebank-Workshop at Nodalida, Joensuu, Finland (2005)
12. Samuelsson, Y., Volk, M.: Phrase Alignment in Parallel Treebanks. In: Proceedings of 5th Workshop on Treebanks and Linguistic Theories, Prague, Czech Republic (2006)
13. Smith, G.: A Brief Introduction to the TIGER Treebank, Version 1. Potsdam Universität (2003)
14. Synphaty: Syntax Editor – Manual – Nijmegen: Max Planck Institute for Psycholinguistics (2006)
15. Tiedemann, J., Kotzé, G.: Building a Large Machine-Aligned Parallel Tree- bank. In: Proceedings of the Eighth International Workshop on Treebanks and Linguistic Theories (TLT 2008), pp. 197–208. EDUCatt, Milano (2009)
16. Tiedemann, J.: Lingua-Align: An Experimental Toolbox for Automatic Tree- to-Tree Alignment. In: Proceedings of the 7th International Conference on Language Resources and Evaluation (LREC 2010), Valetta, Malta (2010)

Creating an Annotated Corpus
for Extracting Canonical Citations
from Classics-Related Texts
by Using Active Annotation

Matteo Romanello

King's College London
Department of Digital Humanities
26-29 Drury Lane, London WC2B 5RL
matteo.romanello@kcl.ac.uk

Abstract. This paper describes the creation of an annotated corpus supporting the task of extracting information–particularly canonical citations, that are references to the ancient sources–from Classics-related texts. The corpus is multilingual and contains approximately 30,000 tokens of POS-tagged, cleanly transcribed text drawn from the *L'Année Philologique*. In the corpus the named entities that are needed to capture such citations were annotated by using an annotation scheme devised specifically for this task.

The contribution of the paper is two-fold: firstly, it describes how the corpus was created using Active Annotation, an approach which combines automatic and manual annotation to optimize the human resources required to create any corpus. Secondly, the performances of an NER classifier, based on Conditional Random Fields, are evaluated using the created corpus as training and test set: the results obtained by using three different feature sets are compared and discussed.

1 Introduction

Modern publications in Classics–such as commentaries, monographs and journal papers–contain a wealth of information that is essential to scholars. The references to primary sources therein such as inscriptions, papyri, manuscripts, archaeological findings and ancient texts, that is scholars' primary sources, all constitute meaningful entry points to information. Despite the attention that other disciplines, such as Bioinformatics, have paid to investigating the automatic extraction of information from texts from a discipline-specific perspective very little research has been done to date in the field of Classics on this topic.

This paper describes the creation of a multilingual annotated corpus to support the task of Named Entity Recognition (NER) and information extraction from Classics-related secondary sources. The corpus covers the main European languages used in Classics, namely English, French, German, Italian and Spanish: this aspect is important if one considers that Classics, like (perhaps) other

A. Gelbukh (Ed.): CICLing 2013, Part I, LNCS 7816, pp. 60–76, 2013.

Humanities disciplines, showed the tendency to preserve the use of national languages within scholarly communications rather than adopting English as *lingua franca* [1]. The named entities that were annotated in the corpus are mentions of ancient authors, titles of ancient works and references to specific texts, or parts of them–which from now on will be referred to as *canonical references*. The corpus has been released under an open license, together with the software that was used to create it, with the hope that the increased availability of tools and data will foster new research on information extraction in the field of Classics[1].

Although the characteristics of canonical citations are thoroughly examined in section 4.2, let us consider for the sake of clarity just an example: 'Homer *Iliad* 1.1' is a reference to the very first line of Homer's *Iliad* and can be alternatively expressed as 'Hom. *Il.* I 1'–note the use of abbreviations and Roman numerals– or, in an even more concise fashion, as 'A 1'. This last variant form, which uses the uppercase letters of the Greek alphabet to identify the books of the *Iliad* and the lowercase ones for the *Odyssey*, is used especially in publications with a very strong focus on the study of epics, where the number of references to the homeric poems tends to be in the order of hundreds.

The main goal in creating such a corpus is to lay the foundations for the creation of an expert system that allows one to explore any corpus of Classics-related publications by using the citations of ancient texts as search key. The first step in doing so is to devise an NER system to extract citations and other named entities of interest from texts: to train such a system is, in fact, the main purpose of this corpus. Once such canonical citations and their meanings have been extracted from texts the system uses them to create a semantic index to the texts in the corpus as well as a citation graph that can be used for information retrieval or network analysis purposes. The possible applications of such a system to Classical scholarship–and its consequent impact–become evident as soon as one thinks, for example, about intertextuality. Intertextuality research is all about studying a text in relation to a system of texts and canonical references are the way in which such texts are cited in modern publications.

The paper is organized as follows: section 2 presents a brief overview of studies that have dealt with canonical citations from a computation perspective and work done in other disciplines in relation to the extraction of discipline-specific information from texts. In section 3 the statistical model of choice to build an NER classifier and the annotation approach that was used, are introduced. Section 4 provides information about the publication from which the corpus texts were drawn, presents the challenges and limits of extracting canonical citations from texts and describes the annotation scheme that was devised to annotate the corpus. In section 5 the details of the corpus annotation process are given and finally in section 6 the evaluation of the performances of an NER system, trained with this corpus using different feature sets, are presented and discussed.

---

[1] Software and data that were used to produce the results described in this paper can be found on the conference website at http://www.cicling.org/2013/data/216

2   Related Work

Canonical citations–that is references to ancient texts expressed in a concise fashion–are the standard citations used by scholars in Classics to refer to primary sources (i.e. ancient texts): as such, they have never really been an object of research in this field.

However, some attention to this practice of referencing started being paid as Humanities Computing–later called Digital Humanities–began emerging and being theorized [2]. Early publications in this field immediately identified such references as a suitable target for possible applications of hypertext [3,4,5]. Indeed, the hyperlink seemed the natural way to translate citations into machine actionable links between the citing and the cited texts, a practice that would later be explored in its hermeneutical implications by McCarty [2, pp. 101-103].

More recent studies in the Digital Classics community have focussed on possible ways of representing and exploiting canonical references in a digital environment. [6] and [7] have tackled the issue of devising network protocols that allow us to share and retrieve passages of ancient texts, and related metadata, by leveraging this traditional way of referencing texts: their research lead to two different yet complementary standards, respectively the Canonical Text Services protocol[2] and a metadata format based on the OpenURL protocol[3]. Moreover, Romanello [8,9] explored new value added services for electronic publications that can be offered once the semantics of canonical citations have been properly encoded within web documents.

As large scale digitization initiatives started producing the first tangible results, the opportunity and necessity of extracting automatically named entities from texts appeared clearly. Services for the automatic extraction of named entities were deemed to be a crucial part of the emerging cyberinfrastructure for research in Classics [10]. Such named entities include not only names of people and geographical places but also some specialized entities such as canonical citations, thus confirming a tendency–observed in the field of Natural Language Processing and Information Extraction–to extend the hierarchy and number of named entities in order to capture discipline-specific information, such as proteins or genomes in bioinformatics [11, pp. 2-4]. If on this research topic there have been to date only a few studies and some preliminary results [12,13], more has been done on improving the accuracy of named entity classification and disambiguation from historical documents [14,15].

However, looking outside the domain of Classics, the automatic extraction of citations and bibliographic references from modern journal papers is a relatively well explored topic [16,17,18]. Although research on this area focussed mainly on references to modern publications the methodology they employ, such as machine-learning techniques and the feature sets used for training, can be largely applied to extraction of references to ancient texts.

---

[2] Homer Multitext project, http://www.homermultitext.org/hmt-doc/

[3] Canonical Citation Metadata Format, http://cwkb.org/matrix/20100922/

3   Methods

3.1   Conditional Random Fields

At first sight the extraction of canonical citations may seem a problem that can easily be solved by using a rule-based approach, such as that proposed by Galibert et al. [19] to extract from patents citations to other patents. The existence of standard abbreviations to refer to ancient authors and works, together with the fact that the body of classical literature is a finite set, seem to indicate that a set of hand-crafted rules can be used to extract such citations.

However, given the number of factors that can cause variations in the way canonical citations are expressed (see section 4.2 for a more detailed analysis), the compilation of a comprehensive set of rules to capture them can become a very time-consuming task. A machine-learning based system, instead, seems to offer a more scalable approach, particularly when combined with an annotation method, such as Active Annotation, which seeks to reduce the effort of producing new training data.

The supervised training method used here is a probabilistic undirected graphical model called Conditional Random Fields (CRFs). CRFs were theorized by Lafferty et al. [20] and, although they have been applied to a wider range of classification problems including computer vision and bioinformatics, they became the state-of-the-art method in sequence labeling tasks, such as Named Entity Recognition (see [21] for an introduction to CRFs and its possible applications).

The main benefit of using a model based on conditional probability, as opposed one based on joint probability, is that they can account for multiple and/or conditionally dependent features. In addition to CRFs, other supervised training methods that are well established in relation to NER tasks are Support Vector Machines (SVM) and Maximum Entropy. However, the comparison of the performances that can be achieved using different training methods, although it is undoubtedly of some interest from a technical point of view, goes beyond the scope of this paper.

Although here a C++ implementation of CRFs called CRF++[4] has been employed, implementations written in other languages are available such as CRFsuite[5] and Wapiti[6] in Python or a Java implementation that is distributed as part of the MAchine Learning for LanguagE Toolkit (Mallet)[7].

3.2   Active Annotation

Active Annotation indicates the use of the Active Learning paradigm in the specific context of creating a corpus for NLP tasks. The main idea underlying Active Learning is that the accuracy of a classifier is higher if the training examples are selected from the most informative. The more informative the

---

[4] CRF++, http://crfpp.googlecode.com/

[5] CRFsuite, http://www.chokkan.org/software/crfsuite/

[6] Wapiti, http://wapiti.limsi.fr/

[7] Mallet, http://mallet.cs.umass.edu/

training examples, the higher will be the performances of a model trained with them. The situations where it makes sense to use this paradigm are those where much unlabelled data can easily be obtained but labelled instances are time-consuming and thus expensive to produce. [22] contains an in-depth discussion of the differences and similarities between Active Learning and Active Annotation. To identify informative instances an uncertainty sampling method based on the least confident strategy is used [23], but other methods use entropy as uncertainty measure [24, pp. 12-26].

In order to optimize the effort of manually annotating the data the Active Annotation algorithm described in [23] was applied. This method proved to be more effective than random selection of candidates when performing an NER task on data drawn from the Humanities domain. The rationale behind it is that training data is more effective when it is drawn from those instances that are most difficult to classify for a given statistical model.

Active Annotation is an iterative process which stops when there is no further improvement in the performances between two consequent iterations. During each iteration a set of candidates for annotation is selected from the development set and, after manual correction, is added to the training set. The improvement is assessed by comparing the system performances–typically the F1 measure–before and after adding this set of candidates to the training set. An instance is added to the candidate set when the Confidence Interval (CI) for one or more of its tokens is above a given threshold.

The CI value is calculated by computing the difference between the probability of the two best labels predicted by the statistical model for a given token. Let us consider an example of how the CI is calculated. For example, the CI for the token "Philon" is 0.161877 as the two best labels outputted by the classifier have a probability respectively of 0.578489 and 0.416612.

4   Annotating and Extracting Canonical Citations

4.1   APh as Corpus

The data used to create our corpus was drawn from the *L'Année Philologique* (APh), a critical and analytical bibliographic index of publications in the field of Classics that has been published annually since 1924. Its thorough coverage, guaranteed also by a structure based on national offices, makes the APh an essential resource for everyone who studies classical texts. To give an idea of the scale of data, the work presented in this paper uses approximately 7.5-8% of a single volume (APh vol. 75) out of the 80 volumes already published at the time of writing.

Such a huge and constantly growing body of data calls for an automatic, and thus scalable, way of extracting information from it and therefore makes it suitable for our purposes. In addition to this, what makes an annotated corpus of APh records extremely valuable is its information density. The analytical reviews in the APh are rather concise (see fig. 1) and contain a variety of references not only to canonical texts, papyri, manuscripts and inscriptions, but also to archaeological objects, such as coins or pottery. Although for the time being our work is limited to canonical references, additional annotations for the other entity types could be added to the corpus in the future.

Example of APh abstract

(a) In **Statius'** « **Achilleid** » **(2, 96-102)** Achilles describes his diet of wild animals in infancy, which rendered him fearless and may indicate another aspect of his character - a tendency toward aggression and anger.

(b) The portrayal of angry warriors in Roman epic is effected for the most part not by direct descriptions but indirectly, by similes of wild beasts (e.g. **Vergil, Aen. 12, 101-109 ; Lucan 1, 204-212 ; Statius, Th. 12, 736-740 ; Silius 5, 306-315**).

(c) These similes may be compared to two passages from **Statius (Th. 1, 395-433 and 8, 383-394)** that portray the onset of anger in direct narrative.

Fig. 1. APh vol. 75 n. 06697: analytical review of Susanna Braund and Giles Gilbert, *An ABC of epic ira: anger, beasts, and cannibalism*

4.2   Challenges of Extracting and Resolving Citations

The extraction of citations is modelled as a typical NER problem consisting of two sub-tasks: NE classification and NE resolutions. After having defined the basic components of a citation (see 4.3), those elements are extracted from text: this is the classification task and consists of identifying the correct entity type for each token in the text.

Once named entities and relations between them have been extracted, the entity referred to needs to be determined. For example, once the citation "Lucan 1, 204-212" has been captured one needs to determine its content, that is a reference to the text span of Lucan's *Bellum Civile* going from line 204 to line 212 of the first book.

Although the use of Latin abbreviations to refer to authors and works makes canonical citations quite regular, the language in which a given text is written introduces another cause of variation in the way they can be expressed. In a text written in German, for example, author names and work titles that can be given either in their full or abbreviated form, are likely to be in German, especially when expressed in a discursive rather than concise form.

Before describing in detail the annotation scheme that was used, let us consider the main challenges of extracting and resolving such citations. The first challenge is the **ambiguity** of citations which is often caused by the concise abbreviations that are used to refer to a given author or work. A citation such as "Th." can stand for Thucydides, *Theogonia* or *Thebaid* if considered out of its context. If, instead, the preceding mention of Statius is taken into account, 'Th." can only refer to Statius' *Thebaid*.

A second challenge is posed by the fact that canonical citations often imply **implicit domain knowledge**. In the citations "Lucan 1, 204-212" or "Silius

5, 306-315", for example, the author name is given whilst the title of the work is left implicit. The reason for this–as any scholar or student of classical texts knows–is that the work indication can be implied when one refers to the *opus maximum* of an author, which is the case for Lucan's *Bellum Civile*, or when only one work is survived or ascribable with some certainty to that author, as for Silius Italicus' *Punica*.

Another challenge is related to citations expressed in a **discursive form** (e.g. "In Statius' « Achilleid » (2, 96-102)") as opposed to a more formalized and structured one (e.g. "Statius, Th. 12, 736-740") as they require a deeper parsing of the natural language. Finally, in some cases the information is not just hard to extract because of ambiguity or discursiveness but is not even recoverable: in such case we speak of **underspecification** or underspecified references. An example of this is the use of the abbreviation "ff." in the context of a reference to mean "and the following sections/lines/verses" (e.g. "Hom. *Il.* 1,1 ff.").

4.3   Annotation Scheme

The entity types that have been annotated in the corpus are: ancient author names, ancient work titles and canonical citations. The first two entity types are marked using respectively the tags AAUTHOR and AWORK. To annotate the canonical citations two different tags were used to distinguish between their different components: REFAUWORK denotes the part of the citation string which contains information about the cited author and/or work, whereas REFSCOPE was used to capture information about the specific text passage which is being cited.

This scheme differs from the one devised by Romanello et al. [12] which had one single tag, namely REF, to capture the entire citation string. The reason for using different tags for different parts of the citation string lies mainly in its characteristics. The part captured by REFAUWORK normally contains abbreviations of author names or work title, therefore consists of alphabetic characters and punctuation, whereas the one which is captured by the tag REFSCOPE contains references to specific sections of a work expressed mainly by a combination of numeric characters and punctuation.

Citations, author names and work titles, however, are only some of the named entities that one can identify in the APh data, and more generally in modern publications in Classics. For example, references to papyri (e.g. "P. Hamb. 312") proved to be, during the creation of the corpus, among the entities with which canonical citations are most likely to be confused because of their very similar surface appearance. Another class of references which was not annotated in the APh corpus–mainly because of resource and time constraints–were references to literary fragments (e.g. "frr. 331-358 Kassel-Austin").

5   Active Annotation Details

In the Active Annotation phase a CRF-based classifier is used to predict the label to be assigned to each token: the probability values of the two most likely labels are needed in order to compute the Confidence Interval (CI) which, in turn, determines whether an instance should be added to the list of effective candidates or not. The CRF model was trained using the full feature set described in section 6.1 with the only exception being the Part of Speech (POS) information: it was not possible to extract the POS tags during the corpus creation phase, as explained at the end of this section, because of the tokenization method that was initially used.

Let us see now in detail the decisions that were made in applying the Active Annotation method to creating the APh corpus, given that our situation differed in some respects from that of the experiments described by Ekbal et al. [23].

A first difference is that this corpus was created from scratch and therefore some seed instances had to be selected in order to create an initial dataset for training and testing. Given the large size of the development set (7k records, ~480k tokens) and the fact that many of its instances are *negative*–that is they do not contain the NEs in the annotation scheme–the seeds were selected partly randomly and partly manually in order to keep a reasonable balance between positive and negative instances. Therefore, some 100 instances were selected, containing approximately 6.4k tokens, out of the ~330k tokens in the training set. For the sake of clarity, an *instance* is each of the sentences an APh abstract is made of, whereas an APh abstract is considered as a record.

The CI threshold was set to 0.2 as this was the value that lead to the best results in the experiments described by [23]. In practice, this means that for an instance to be considered an effective candidate it needs to contain one or more tokens with CI value lower or equal to 0.2.

For each round of Active Annotation, all tokens with CI over this threshold were added to the candidate set, which was then pruned in order to avoid having duplicate records–each record in the list may contain several multiple tokens that are considered effective. At this point the 30 highest scoring records in the effective candidate are sent to the annotators for manual correction of the results obtained by automatic annotation and are then added to the training set. Due to constraints, the two domain-experts that annotated the corpus worked on separate datasets and therefore the inter-annotator agreement could not be reported.

Another issue we faced relates to the size of the training and test set, and specifically to the proportion between them. The test set, obtained during the seed selection phase, had an initial size of about 2k tokens. Keeping its size fixed throughout the Active Annotation process would have lead to a disproportion between training and test set, with a consequent impact on F-score, precision and recall. The main consequence of such disproportion is the risk of overfitting the statistical model, that is training a model that will perform well on a dataset similar to the training set but will not be general enough to perform as well on a dataset with different characteristics.

This problem was solved by increasing at regular intervals the size of the test set by adding a certain number of tokens, selected using the same method. The size of the test set was increased to 4146 tokens at the beginning of round 4, and then to 6594 tokens at the start of round 7, as reported in Table 1. The table gives some details for each round of Active Annotation: what was the initial size of the training set, how many tokens were added from effective candidates, what was the initial performance of the classifier (F-score) and what the improvement (F-score gain) after the manually corrected effective candidates were added to the training set.

Table 1. Details of the Active Annotation process

#	r	p	F	F gain	train.	test	added
1	45.45	80.65	58.14	1.51	4233	2178	2032
2	51.82	79.17	62.64	4.50	6265	2178	1968
3	55.45	75.31	63.87	1.24	8233	2178	1762
4	71.18	77.16	74.05	1.53	8027	4146	1688
5	72.06	76.47	74.20	0.15	9715	4146	2100
6	73.17	77.28	75.17	0.97	11815	4146	1433
7	71.82	70.58	71.19	1.11	13248	6594	1813
8	71.66	72.00	71.83	0.64	15061	6594	1593
9	73.73	70.69	72.17	0.35	16654	6594	1856

Time and resourcing meant we were only able to complete 9 rounds, however, the corpus had by then reached a size which made it comparable to other datasets used for similar tasks.

After round 9, the size of the corpus was ~23k tokens, which became slightly less than ~26k after re-tokenizing the text. The re-tokenization was needed in order to be able POS-tag the text–and then include this information in the feature set–as the whitespace-based tokenization that was initially applied proved not to be suitable for this purpose. The reason for doing this at two separate stages was the poor performances of the tokenizer when an additional list of abbreviations was not provided: this is due to the high number of abbreviations that are present in our texts and lead to a very high number of wrongly tokenized words (e.g. "Hom." being split into "Hom" and "."). This problem was solved by providing the tokenizer with a list of abbreviations that had been extracted from the corpus.

6   Evaluation of the NER System

6.1   Named Entity Features

Linguistics Features. Since the system was designed to be language-independent, the number of linguistic features was kept to a minimum. The neighbouring words of each token w_i in the range $w_{i-3} \ldots w_{i+3}$ were considered as features, whereas experiments with using word suffixes and prefixes of length up to 4 characters showed a degradation of the performance. This may be due to the fact that this feature was extracted also for tokens containing digits and/or numbers.

The POS information of the current token is extracted automatically for all languages using TreeTagger[8] and included in the feature set without any manual correction.

Orthographic Features

- **punctuation**: this feature takes value `no_punctuation` when the token does not contain any punctuation sign at all. Otherwise it takes one of the following values: `continuing_punctuation`, `stopping_punctuation`, `final_dot`, `quotation_mark` or `has_hyphen` which is particularly important in range-like notations.
- **brackets**: when a token contains both an open and a closed parenthesis, e.g. "(10)" or "[Xen.]", the feature value is set either to `paired_round_brackets` or `paired_squared_brackets` depending on the kind of parenthesis. Similarly, when it contains either an open or a closed one, possible values are `unpaired_round_brackets` or `unpaired_square_brackets`.
- **case**: this feature is set to `all_lower` or `all_caps` when the token contains all lowercase or all uppercase characters. Other possibilities are that the token contains a mix of lower and uppercase characters (`mixed_caps`) or that only the first letter is uppercase (`init_caps`).
- **number**: three possible values of this feature are determined by the presence (`number` or `mixed_alphanum`) or absence of numeric characters (`no_digits`). The values `dot_separated_number` and `dot_separated_plus_range`, are used to identify known sequences of numbers and punctuation signs, such as "1.1.1" or "1.1.1-1.1.3", that are often found in canonical citations and particularly in the part of a citation indicating the scope of the reference.
- **pattern**: the surface similarity between tokens is captured by means of two features: a compressed pattern and an extended pattern. The former is computed by replacing lowercase characters with "a", uppercase ones with "A", numbers with "0" and punctuation signs with "-", whereas in the latter sequences of similar characters are replaced by one single pattern character.

Semantic Features. Since the corpus of classical texts is a finite one, the use of semantic features should improve quite significantly the performances of our system, at least as far as the entities `AAUTHOR`, `AWORK` and `REFAUWORK` are concerned. Such features are extracted by matching each token against dictionaries of author names, work titles and their abbreviations. For the sake of performance, the dictionaries that are used for look-up are converted into a suffix array, a highly efficient data structure for indexing strings. This was implemented by using Pysuffix[9], a Python implementation of Kärkkäinen's algorithm for constructing suffix arrays [25].

---

[8] TreeTagger, http://www.ims.uni-stuttgart.de/projekte/corplex/TreeTagger/
[9] Pysuffix http://code.google.com/p/pysuffix/

Data from the Classical Works Knowledge Base (CWKB) project [10], together with a list of canonical abbreviations that is distributed as part of the Perseus digital library[11], was used in order to create such dictionaries. CWKB in particular proved to be an essential source as it contains the canonical abbreviations and the name variants in the main European languages of 1,559 authors and 5,209 works.

Four separate features are extracted from each token to capture the fact that it is successfully matched against the author dictionary or the work dictionary: the feature takes value `match_authors_dict` or `match_works_dict` if the matching is total, whereas if the matching is partial–meaning that the token is contained in a dictionary entry, but the token length is smaller than the length of the matching entry–it is set either to `contained_authors_dict` or `contained_works_dict`.

6.2   Error Analysis

Examining the tokens that were added at each round to the list of effective candidates turned out to be extremely instructive as one can see which tokens are most problematic for the classifier, that is those with lowest CI score, for instance distinguishing canonical references from papyri references given the high surface similarity between the two.

Let us now look more closely at the performances of the classifier obtained by taking into consideration the training and test set as in the last round (n=9) of Active Annotation. The results obtained in this section were obtained by using the full feature set now including also the POS tags, as opposed to the feature set used during the candidate selection phase.

Table 2. Evaluation results aggregated by class for the last round (9) of active learning. Precision, recall and F-score are based on the number of absolute correct tags: values for the same measures, but limited to entirely correct entities are given in round brackets.

Class	p	r	F
AAUTHOR	57.89 (62.75)	38.60 (40)	46.32 (48.85)
AWORK	68.11 (62.20)	78.85 (72.86)	73.09 (67.11)
REFAUWORK	71.58 (71.43)	78.16 (75)	74.73 (73.17)
REFSCOPE	72.37 (66.34)	86.14 (67.68)	78.66 (67)
Overall	69.64 (65.22)	79.73 (62.28)	74.34 (63.72)

The entities on which the classifier records the worst performances are the names of ancient authors (`AAUTHOR`), where precision, recall and F-score are respectively 57.89%, 38.60% and 46.32%. There are two facts that emerge when looking more closely at the errors. On the one hand, the recall appears to be very low when compared to the results achieved on all other entities. On the other

---

[10] Classical Works Knowledge Base `http://cwkb.org/`
[11] Perseus Digital Library `http://www.perseus.tufts.edu/hopper/`

hand, approximately 40% of all the AAUTHOR entities retrieved by the classifier are in fact named entities, just not ancient authors. Among the errors one can find names of historical figures who in some case were also authors, such as Caesar, with the resulting issue of distinguishing contexts where someone is mentioned particularly in relation to his role of author from other contexts. Interestingly, errors in classifying such entities are less frequent when the mention of an ancient author immediately follows a canonical reference, as in the example of fig. 1, and more common when such mentions appear within a discursive context. This issue may be related to how the corpus was annotated and specifically to the fact that it is lacking a generic named entity tag for people (e.g. PERSON) who are not specifically ancient authors and also to fact that no features related to the global context are extracted.

With the exception of the performances on AAUTHOR entities, which are particularly poor as we have just seen, those on the remaining entities are pretty much in line, with F-score values respectively of 73.09%, 74.73% and 78.66%. The identification and classification of titles of ancient works mentioned in the text (AWORK) presented at least two issues. The first one concerns the number of false positives and explains the relatively low precision: in the APh data, title of works are normally given between French quotation marks (i.e. « and »), also known as Guillemets, but they are also used for honorific titles, concepts, and the like, all cases where in other styles scare quotes are used instead. This is due to the style which is adopted throughout the publications and leads to some ambiguity and relative problems of classifications. A second issue is the number of errors in identifying the boundaries of such entities, which is evident when comparing the F-score on absolute correct tags with the F-score calculated on correct entities: this may be related to feature set, and specifically to the fact that no specific features were used to capture multiword names.

6.3   Evaluation

As has been stated above, there were two main motivations for creating such an annotated corpus: the lack of both datasets and software for this specific kind of NER and the intuition that the bigger the corpus the more representative the results of evaluation achieved when using such a corpus as training and test set. Therefore, the evaluation was carried out by using different feature sets, so to have at least a baseline to be used for comparison. Moreover, the 10-fold cross-validation performed on chunks of the corpus of varying size largely confirmed by empirical evidence the initial intuition as explained below.

Firstly, we performed the cross-validation on the whole corpus (size = 28,893 tokens) by using three different feature sets: the results are given in table 4. With the first set, considered as the baseline and consisting solely of POS tags as features, precision, recall and F-score of respectively 66.34%, 42.22% and 51.12% were achieved. The second feature set used includes POS tag information as well as the wide range of orthographic features: with this feature set an improvement of respectively +11.75%, +21.61% and +18.37% on precision, recall and F-score was registered. Yet the highest scores were obtained when using the full feature

Table 3. Break-down of the evaluation results by class. The results are relative to the last round (n=9) of active learning.

Class	TP	FP	FN	Tot. retr.	Total	p	r	F
B-AAUTHOR	29	18	50	47	79	0.62	0.37	0.46
I-AAUTHOR	15	14	20	29	35	0.52	0.43	0.47
B-AWORK	49	31	20	80	69	0.61	0.71	0.66
I-AWORK	171	72	39	243	210	0.70	0.81	0.75
B-REFAUWORK	29	12	11	41	40	0.71	0.73	0.72
I-REFAUWORK	39	15	8	54	47	0.72	0.83	0.77
B-REFSCOPE	78	23	21	101	99	0.77	0.79	0.78
I-REFSCOPE	239	98	30	337	1269	0.71	0.89	0.79
O	6389	165	249	6554	6638	0.97	0.96	0.97

set, which consists of the previous two plus the semantic features. The precision, recall and F-score obtained when using this third feature set were respectively 79.85%, 69.07% and 73.44%. It is noteworthy that the use of semantic features–typically indicating whether a given token matches successfully against one or more dictionaries or lexica–does not always lead to an improvement of the overall performances, as observed by [26].

Table 4. Results of the 10-fold cross-validation using the whole corpus (25104 tokens)

Feature Set	r	p	F
POS	42.22	66.34	51.12
POS+ortho	63.83	78.09	69.49
POS+ortho+sem	69.07	79.85	73.44

Secondly, an analogous 10-fold cross-validation was performed but on chunks of the corpus of varying size: the purpose of this evaluation experiment was to verify the correlation between the results and the size of training and test set. In total 10 iterations were performed: in the first one only 10% of the corpus was considered, then for each new iteration another 10% was added until in the 10th and last iteration the entire corpus was used. For training the full feature set was used, that is POS tag information, orthographic and semantic features.

The results of this experiment are plotted in fig. 2. The first pattern it is possible to observe is the gradual convergence of precision and recall as the training and test sets increase in size. A similar pattern can also be found in the accuracy scores that were registered during the corpus creation as reported in Table 1. Another, and more interesting, phenomenon that was observed is the spike in performances when using less than 50% of the whole corpus for both training and testing. A similar, although not identical, pattern can also be found in round n=6 of Active Annotation, where the highest F-score value was measured. What this shows is that the reason for relatively high measure of

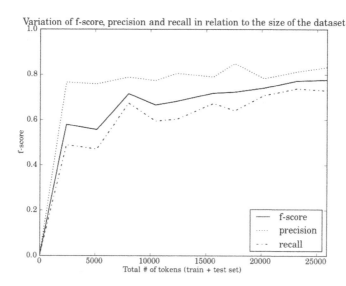

Fig. 2. This diagram shows how accuracy measures vary in relation to the size of the dataset. F-score, precision and recall were calculated using a 10-fold cross-validation on data chunks of regularly increasing size.

precision, recall and F-score sometimes is to be found in the size of the dataset used.

7   Conclusions and Further Work

The significance of the work presented in this paper lies mainly in that it starts to fill the lack of datasets and software for discipline-specific NER on modern texts related to Classics. The evaluation results themselves–although encouraging, particularly when considering that were obtained on a multilingual corpus–were not entirely satisfactory. However, our main hope in releasing both the dataset and the software under an open license is to fuel new research on this topic, so that better results can eventually be achieved.

In addition to the gap that this resource fills, it makes possible to perform with greater accuracy basic yet essential steps of text processing such as tokenization, POS tagging and sentence segmentation. The main reason for this is that texts in this field contain a high number of abbreviations that are not common in other fields. Canonical references, in fact, are essentially an abbreviation of the title of the cited work followed by the indication of the citation scope. As a result, a list of abbreviations can easily be extracted from the corpus and then supplied to the tokenizer or POS tagger. Similarly, sentence breaks were manually identified in the corpus texts, thus making possible the use of this corpus to train more accurate sentence tokenizers.

Furthermore, the corpus annotation could be improved and extended in several ways. A first basic enhancement would be to manually correct the POS tags since they were assigned in an unsupervised fashion. Second, the layer of named entity annotations could be extended in order to improve the overall performances by introducing a generic class for those entities that do not fall into any specific category, as suggested by the evaluation results presented in section 6.3. Other named entities that could be annotated in the corpus include references to papyri, manuscripts and fragmentary texts. Third, the layers of syntactic annotation and anaphoric annotation could be added to the existing ones: this would allow for a deeper language parsing and therefore it would make possible to capture those citations that are expressed in a more discursive fashion.

Finally, further work is currently being carried out in the following main directions: 1) the use of other supervised learning methods, in addition to CRF, in order to compare the results obtained; 2) the disambiguation and co-reference resolution of the automatically extracted named entities; 3) the comparison and evaluation of the results that are obtained by applying a model trained with cleanly transcribed texts on a corpus of potentially noisy OCRed documents.

Acknowledgments. The author wants to gratefully thank Prof. Eric Rebillard for giving him access to the APh data and for his precious help in annotating the corpus, as well as the anonymous reviewers of the paper for their very helpful comments.

References

1. Mimno, D.: Computational Historiography: Data Mining in a Century of Classics Journals. ACM Transactions on Computational Logic, 1–19 (2005)
2. McCarty, W.: Humanities Computing. Palgrave Macmillan (2005)
3. Crane, G.: From the old to the new: intergrating hypertext into traditional scholarship. In: Proceedings of the ACM Conference on Hypertext, Chapel Hill, North Carolina, United States, pp. 51–55. ACM (1987)
4. Bolter, J.D.: The Computer, Hypertext, and Classical Studies. The American Journal of Philology 112, 541–545 (1991)
5. Bolter, J.D.: Hypertext and the Classical Commentary. In: Accessing Antiquity: The Computerization of Classical Studies, pp. 157–171. University of Arizona Press, Tucson (1993)
6. Ruddy, D., Rebillard, E.: Text Linking in the Humanities: Citing Canonical Works Using OpenURL (2009)
7. Smith, N.: Digital Infrastructure and the Homer Multitext Project. In: Bodard, G., Mahony, S. (eds.) Digital Research in the Study of Classical Antiquity, pp. 121–137. Ashgate Publishing, Burlington (2010)
8. Romanello, M.: New Value-Added Services for Electronic Journals in Classics. JLIS.it 2 (2011)
9. Romanello, M.: A semantic linking framework to provide critical value-added services for E-journals on classics. In: Mornati, S., Chan, L. (eds.) ELPUB 2008. Open Scholarship: Authority, Community, and Sustainability in the Age of Web 2.0 - Proceedings of the 12th International Conference on Electronic Publishing held in Toronto, Canada, June 25-27, pp. 401–414 (2008)

10. Crane, G., Seales, B., Terras, M.: Cyberinfrastructure for Classical Philology. Digital Humanities Quarterly 3 (2009)
11. Nadeau, D., Sekine, S.: A survey of named entity recognition and classification. Lingvisticae Investigationes 30, 3–26 (2007)
12. Romanello, M., Boschetti, F., Crane, G.: Citations in the digital library of classics: extracting canonical references by using conditional random fields. In: Proceedings of the 2009 Workshop on Text and Citation Analysis for Scholarly Digital Libraries. NLPIR4DL 2009, Morristown, NJ, USA, pp. 80–87. Association for Computational Linguistics (2009)
13. Romanello, M., Thomas, A.: The World of Thucydides: From Texts to Artefacts and Back. In: Zhou, M., Romanowska, I., Zhongke, W., Pengfei, X., Verhagen, P. (eds.) Revive the Past. Proceeding of the 39th Conference on Computer Applications and Quantitative Methods in Archaeology, Beijing, April 12-16, pp. 276–284. Amsterdam University Press (2012)
14. Smith, D.A., Crane, G.: Disambiguating Geographic Names in a Historical Digital Library. LNCS, pp. 127–136 (2001)
15. Babeu, A., Bamman, D., Crane, G., Kummer, R., Weaver, G.: Named Entity Identification and Cyberinfrastructure. In: Kovács, L., Fuhr, N., Meghini, C. (eds.) ECDL 2007. LNCS, vol. 4675, pp. 259–270. Springer, Heidelberg (2007)
16. Kramer, M., Kaprykowsky, H., Keysers, D., Breuel, T.: Bibliographic Meta-Data Extraction Using Probabilistic Finite State Transducers (2007)
17. Councill, I.G., Giles, C.L., Kan, M.Y.: ParsCit: An open-source CRF Reference String Parsing Package. In: Calzolari, N., Choukri, K., Maegaard, B., Mariani, J., Odjik, J., Piperidis, S., Tapias, D. (eds.) Proceedings of LREC, vol. (3), pp. 661–667. Citeseer, European Language Resources Association, ELRA (2008)
18. Kim, Y.M., Bellot, P., Faath, E., Dacos, M.: Automatic annotation of incomplete and scattered bibliographical references in Digital Humanities papers. In: Beigbeder, M., Eglin, V., Ragot, N., Géry, M. (eds.) CORIA, pp. 329–340 (2012)
19. Galibert, O., Rosset, S., Tannier, X., Grandry, F.: Hybrid Citation Extraction from Patents. In: Chair, N.C.C., Choukri, K., Maegaard, B., Mariani, J., Odijk, J., Piperidis, S., Rosner, M., Tapias, D. (eds.) Proceedings of the Seventh Conference on International Language Resources and Evaluation, LREC 2010. European Language Resources Association, ELRA (2010)
20. Lafferty, J., McCallum, A., Pereira, F.: Conditional random fields: Probabilistic models for segmenting and labeling sequence data. In: Brodley, C.E., Danyluk, A.P. (eds.) Machine Learning International Workshop then Conference, ICML 2001, pp. 282–289. Citeseer (2001)
21. Sutton, C., McCallum, A.: An Introduction to Conditional Random Fields for Relational Learning. In: Getoor, L., Taskar, B. (eds.) Introduction to Statistical Relational Learning. MIT Press (2006)
22. Vlachos, A.: Active annotation. In: Proceedings of the Workshop on Adaptive Text Extraction and Mining (ATEM 2006), pp. 64–71 (2006)
23. Ekbal, A., Bonin, F., Saha, S., Stemle, E., Barbu, E., Cavulli, F., Girardi, C., Poesio, M.: Rapid Adaptation of NE Resolvers for Humanities Domains using Active Annotation. Journal for Language Technology and Computational Linguistics 26, 39–51 (2011)
24. Settles, B.: Active Learning Literature Survey. Computer Sciences Technical Report 1648, University of Wisconsin- Madison (2009)

25. Kärkkäinen, J., Sanders, P., Burkhardt, S.: Linear work suffix array construction. Journal of the ACM 53, 918–936 (2006)
26. Settles, B.: Biomedical named entity recognition using conditional random fields and rich feature sets. In: JNLPBA 2004: Proceedings of the International Joint Workshop on Natural Language Processing in Biomedicine and its Applications, Morristown, NJ, USA, pp. 104–107. Association for Computational Linguistics (2004)

Approaches of Anonymisation of an SMS Corpus

Namrata Patel[1], Pierre Accorsi[1], Diana Inkpen[2],
Cédric Lopez[3], and Mathieu Roche[1]

[1] LIRMM – CNRS, Univ. Montpellier 2, France
[2] Univ. of Ottawa, Canada
[3] Objet Direct – VISEO, France

Abstract. This paper presents two anonymisation methods to process an SMS corpus. The first one is based on an unsupervised approach called *Seek&Hide*. The implemented system uses several dictionaries and rules in order to predict if a SMS needs anonymisation process. The second method is based on a supervised approach using machine learning techniques. We evaluate the two approaches and we propose a way to use them together. Only when the two methods do not agree on their prediction, will the SMS be checked by a human expert. This greatly reduces the cost of anonymising the corpus.

1 Introduction

In the past few years, SMS (Short Message Service) communication has become a veritable social phenomenon. Although numerous scientific studies (namely in the fields of linguistics, sociology, psychology, mass communication, etc.) have been conducted on this recent form of communication, there remains a general gap in our accumulated knowledge of the subject. This is mainly due to the fact that researchers have limited access to suitable data for their studies. Typically, they require large volumes of authentic data for their work to be significant.

The international project *sms4science* (http://www.sms4science.org/) aims at building and studying precisely such a body of data by collecting authentic text messages from different parts of the world. In the context of the sud4science project (http://www.sud4science.org/), over 90,000 authentic text messages in French have been collected. But the publication of these resources requires to meticulously remove all traces of identification from each SMS. In order to perform this anonymisation task, we have developed the *Seek&Hide*[1] software [1]. After the summarisation of the principle of this system, this paper focuses on machine learning approaches in order to predict SMS to anonymise.

In this paper, we begin by introducing the distinctive aspects of our work by looking at pre-existing anonymisation techniques (section 2). We present two solutions: the first one based on rules (section 3) and the second one based on machine learning (section 4). A combined approach is finally proposed in section 5. To conclude our study, we present and discuss the obtained results (section 6).

---

[1] Not "Hide and Seek", but "Seek and Hide": with this tool, we seek to hide words that are to be anonymised.

A. Gelbukh (Ed.): CICLing 2013, Part I, LNCS 7816, pp. 77–88, 2013.

2 Related Work

Anonymisation is indispensable when one seeks to mask an individual's identity. For example, this is essential before the distribution of court orders pertaining to children, juvenile delinquents, victims of sexual harassment, etc. [2], or when one needs to put together a medical corpus [3,4]. In the medical field, it is customary to resort to automatic anonymisation techniques using rules and medical dictionaries in order to process the most common cases [2,5,6,7,8]. These systems primarily aim at the automatic recognition of names, dates, places, and other elements which could lead to the identification of people covered by publication restrictions. Generaly the used methods to recognize Named Entity are based on specific rules and dictionaries. Moreover, supervised methods can be applied. For instance [9] have trained several classifiers, and they have combined decision functions for an anonymisation task.

We agree with [10] that the process of anonymisation cannot be entirely automated. Their work focuses on the creation of an interface by which the researcher can identify personal data and decide whether or not to render it anonymous. Given the size of our *sud4science LR* corpus (over 90,000 SMS), an automated procedure considerably benefits the annotator, as shall be evidenced in our paper. The distinguishing feature in our approach, as put forth by [11], is that we take into consideration the numerous linguistic particularities of the forms used in SMS writing.

[12] present the first freely available corpus of Dutch SMS, where anonymisation was performed automatically by replacing sensitive data, including dates, times, decimal amounts, and numbers with more than one digit (telephone numbers, bank accounts, street numbers, etc.), e-mail addresses, URLs, and IP addresses. [13] have collected about 60,000 SMS, focusing on English and Mandarin Chinese. Previous works consider the same sensistive data: It seems that no names (or nicknames) were automatically anonymised. We call attention to the fact that our paper focuses on the most complex part of the anonymisation process: that of the processing of first names.

Sometimes, the identity of the markers that need to be anonymised are trivial names made up by the senders themselves, are subject to syntactic variations (often significant) and become cultural footprints (nicknames, diminutives, repetition of letters into the name) [14], i.e., in the following text message, "cece" requires anonymisation.

Coucou mon cece *! J'espere [. . .]*

Other anonymisation tasks are accomplished using regular expressions to identify the appropriate words; for example, e-mail addresses, telephone numbers, and URLs.

In the following section we propose two anonymisation techniques which are adapted to the demands of text messages.

3   Seek&Hide and Anonymisation of SMS Data

3.1   Principle

As stated in the previous section, our approach to anonymise/de-identify corpus of French text messages is to adopt a two-phase procedure.

This two-way procedure ensures the dependability of the system: the combined use of Natural Language Processing (NLP) techniques and human evaluation helps minimise computer as well as human errors, greatly improving the overall result.

Let us now take a deeper look into the workings of the system by considering each of its processes individually.

The main purpose of the system is to process a corpus using (a) dictionaries as reference material and (b) word-processing techniques so as to identify and eventually hide words that have to be anonymised. This constitutes the preliminary treatment of the corpus. Text messages that are processed by this phase undergo the following transformation, the details of which are given in the next paragraph:

"Coucou Patrice, ça va?" ↦ *"Coucou <PRE_7_17316>, ça va?"*

When a word is anonymised, it is replaced by a code conforming to the following format: <[Tag]_[#characters]_[cross-reference]> where "Tag" indicates the type of the word (e.g. First name, Last name). Thus, "Patrice" is replaced by < *PRE_7_17316* > where:

- *PRE* → First name (prénom)
- *7* → number of characters in "Patrice"
- *17316* → "Patrice"'s ID in the dictionary of first names

3.2   Global Process

Having seen what *Seek&Hide* does in its automatic phase, let us now find out how this is done. *Seek&Hide* operates in a three-part procedure:

(1) Pre-processing. Each SMS in the corpus, in its raw state, is basically just a string of characters. In order for *Seek&Hide* to make any sense out of this data, it needs to break the string of characters into words. This is the pre-processing phase, called "Tokenisation". Once tokenised, the SMS becomes a coherent sentence: a series of identifiable words. The SMS tokenisation is a complex processing [15]. For an anonymisation task that does not need a precise analysis of message content, such as ours, we consider that the simple use of a "space" as separator for tokenisation is satisfactory.

(2) Identification. In this phase of the automatic process, *Seek&Hide* uses a technique of identification which uses specific kinds of dictionaries to analyse each word of an SMS. The idea behind is simple: each word of an SMS can either be classified as "To anonymise" or as "Nothing to anonymise". We thus use two kinds of dictionaries corresponding to this classification, distinguishing them by "Dictionary" and "Anti-dictionary" on the basis of their content (pertaining to the task of anonymisation):

- The "Dictionary" contains words that need to be anonymised.
- The "Anti-dictionary" contains words that do not require anonymisation.

The following list shows the different resources used as reference material to identify the words in the text messages:

Dictionary: Dictionary of first names (21,921 first names)
Anti-dictionaries:

- Dictionary of inflected forms of the French language (LExique des Formes Fléchies du Français, LEFFF)[2] (105,595 lemmas),
- Dictionary of some forms used in SMS writing (739 words),
- Dictionary of places (9,463 cities and 194 countries).

Each word is then labelled according to its presence, or the lack of it, in the dictionaries used by *Seek&Hide* (*cf.* Table 1).

(3) Treatment. The words of the corpus are processed according to their labels and are thus (a) anonymised, (b) ignored, (c) highlighted. Words that could not be identified in a dictionary, and words that were identified in both types of dictionaries, are highlighted. These will be processed by users via a web-interface of the system [1].

Table 1 summarises this treatment by giving the range of possible cases encountered. As can be seen, "Cédric" is anonymised because it is identified only in the dictionary (of first and last names). Similarly, "crayon" is ignored as it is identified only in the anti-dictionary (LEFFF). "Pierre" and "Namrata" are problematic: "Pierre" is ambiguous as it belongs to both, the dictionary and the anti-dictionary. "Namrata" is unknown as it belongs to neither of the dictionaries. These two words are consequently highlighted for further processing. This one is based on a semi-automatic system based on human-machine interactions.

Moreover, in order to take into account the specificities of the SMS data, we added different heuristics to solve these cases:

Misspelled words: ex.: *surment* (instead of *sûrement*)
Words written without their accents: ex.: *desole* (instead of *désolé*)
Words with misplaced accents: ex.: *dèsolè* (instead of *désolé*)
Letter repetitions: ex.: *nicoooolllaassss* (instead of *nicolas*)
Onomatopoeias: ex.: *mouhahaha*
Omission of the apostrophe: ex.: *jexplique* (instead of *j'explique*)
Concatenation: ex.: *jtaime* (instead of *je t'aime*)

---

[2] http://www.labri.fr/perso/clement/lefff/

These heuristics work particularly well for the SMS data because there are a number of cases in text messages in which words are not always written using their correct spellings. Performing an "accent-insensitive" word-search in such cases, for example, is one of the heuristics employed by *Seek&Hide*. This and other heuristic solutions are further discussed in [1].

Table 1. Different cases to take into account

Word	Dico	Anti-dico	Label	Treatment
Cédric	yes	no	Dictionary	Automatically anoymised
crayon	no	yes	Anti-dictionary	Ignored (not to be anonymised)
Pierre	yes	yes	Ambiguous	Candidate to anonymise
Namrata	no	no	Unknown	Candidate to anonymise

In order to improve the system and to reduce the workload in the manual validation phase, we decided to implement a second technique, based on supervised machine learning. These methods learn from annotated training data, and are able to make predictions on new test data.

4   Machine Learning Approach to SMS Anonymisation

We wanted to see if it is possible to train a classifier to decide which messages need to be anonymised and which do not need. The classifier works at the SMS level not at word level. That is, if there are no words to be anonymise, the classifier will signal that there is nothing to anonymise; but if there is one or more words to anonymise, the classifier will signal that this SMS needs anonymisation. The reason we trained the classifier this way is that the data available for training a classifier were labeled at message level (not at word level).

The features used by the classifier are inspired from the linguistics analysis from the previous section. The values of the features are calculated by using the lexical resources mentioned above.

Here is the list of features extracted from each SMS:

- The number of words from the SMS that are in the dictionary of abbreviated forms specific to SMS texts (anti-dictionary).
- The number of words that are in the LEFFF dictionary of French (anti-dictionary).
- The number of words that are in the dictionary of first names. We expect this to be particularly useful for the class of messages to be anonymised.
- The number of words that are in the dictionary of country names.
- The length of the SMS.
- The number of words in upper case in the text.
- The average words length in the SMS.

- The number of pronouns in the text.
- The number of numbers.
- The number of punctuation tokens.
- Elongation: the number of words with elongated / repeated vowels.

For the features that count numbers of various elements, we experimented with the counts and we also normalized by the length of each SMS, but the results were similar, because most of the messages have similar lengths (usually short texts).

The algorithm that we selected for the classification is the Decision Trees algorithm (DT). In fact we tested several algorithms from Weka [16], but the DT algorithm worked better than other classifiers that we tried and it has the advantage that we can see what the classifier learnt by examining the learnt decision tree. Other classifiers, such as SVM and Naive Bayes, learn separation planes or probabilities, and these numbers are not understandable for a human examiner.

We also experimented with several meta-classifiers, and the Bagging algorithm based on Decision Trees was successful in improving the results with another 2 percentage points (as it will be shown in Section 6.3). Bagging is a form of voting with several classifiers trained on various parts of the training data, in order to obtain a more generic classifier.

The following section describes how we can combine the learnt model (section 4) with *Seek&Hide* system based on the use of rules (section 3).

5   Discussion: How to Combine Both Approaches?

Seek&Hide system predicts if it is necessary

- to anonymise the SMS (TA)
- to anonymise nothing (NTA)
- to give the SMS to the expert because the prediction is impossible (Untagged).

The learnt model obtained by machine learning methods proposes two classes (TA and NTA). So, we can combine both approaches to propose a general prediction. This general prediction is based on the

- agreement/disagreement between *Seek&Hide* and the learnt model,
- class found with the learnt model if *Seek&Hide* can not predict.

Then the general prediction is based on the situation presented in Table 2. The principle is to minimize the intervention of the expert. With the use of both methods (i.e. *Seek&Hide* and the learnt model), the manual analysis by an expert is useful only if there exists a disagreement between the two automatic methods.

Table 2. Different possible actions regarding the predictions of both systems (*Seek&Hide* and learnt models)

Seek&Hide	Learnt model (Machine leaning)	Action
TA	TA	**TA**
TA	NTA	expert
NTA	TA	expert
NTA	NTA	**NTA**
Untagged	TA	**TA**
Untagged	NTA	**NTA**

6   Experiments

6.1   Experimental Protocol

Seek&Hide was tested on a sample of our SMS corpus containing 23,055 SMS that were manually tagged as "To anonymise (TA)" or "Nothing to anonymise (NTA)" by a student-annotator. During the acquisition of the corpus, a fourth year student was employed for a three-month internship, in order to read the incoming messages and make sure they respected certain rules and regulations. He thus labelled those text messages that needed to be anonymised as "To anonymise" and those that were to be left as-is as "Nothing to anonymise (NTA)". Out of the 23,055 SMS in our sample, 90.7% (i.e. 20,913 SMS) were noted by him as NTA and 9.3% (i.e. 2,142 SMS) as TA.

In the following section, we propose a method of evaluation whereby *Seek&Hide*'s results on the sample are compared with those of the student-annotator.

6.2   Global Analysis of *Seek&Hide* Results

Table 3 presents the SMS distribution in the sample according to 3 categories: Those tagged TA, those tagged NTA, and those left untagged. The untagged label corresponds to the text messages that our automatic system cannot tag (TA or NTA) because they contain ambiguous and/or unknown words. These will be processed via the web-interface of the semi-automatic phase of our tool.

We note that our system returns results for 65.3% of the sample corpus (i.e. 15,052 SMS with TA and NTA tags). The other part of the corpus (34.7 % of the corpus left untagged) has to be processed by the semi-automatic system. In this section we focus our analysis on the evaluation of the 15,052 SMS automatically processed by *Seek&Hide* as NTA and TA. In this context, the confusion matrix (see Table 4) shows a more detailed analysis of the results obtained.

The top left box indicates the number of true positives - TP (i.e. the 13,904 SMS correctly classified by the application as 'NTA'), the top right, the false negatives - FN (i.e. the 59 SMS classified as 'NTA' by *Seek&Hide* and 'TA' by the student-annotator), the bottom left, the false positives - FP (i.e. the 413 SMS classified as 'TA' by Seek&Hide and 'NTA' by the student-annotator), and the bottom right, the true negatives TN (i.e. the 676 SMS correctly classified as 'TA'). Note that *correct* and *incorrect* terms are based on the tags specified by the student-annotator. A deeper analysis of Table 4 shows that *Seek&Hide* predicts 13,963 SMS (i.e. first line of Table 4: $13,904 + 59$) that do not need anonymisation (tagged NTA). Of these, only 59 SMS are irrelevant, when compared with the manual evaluation done by the student-annotator. This shows that the NTA-tagging performed by *Seek&Hide* is very efficient. However, the *Seek&Hide* prediction of text messages that require anonymisation (i.e. second line of Table 4) is not as good, as only 676 of 1,089 SMS are relevant.

In our case we obtain a value of accuracy at 0.96. This score validates the relevance of our methods in predicting text messages that may or may not require anonymisation.

Table 3. Results of the *Seek&Hide* System on the Sample Analysed by the Student-Annotator

Sample Processed	Tag: TA	Tag: NTA	Untagged	Total
By Seek&Hide	1,089	13,963	8,003	23,055
	4.72%	**60.57%**	**34.71%**	

Table 4. Confusion Matrix

Confusion Matrix	Student-Annotator: NTA	Student-Annotator: TA
Seek&Hide: NTA	13,904	59
Seek&Hide: TA	413	676

The following section presents results obained with other methods based on machine learning approaches.

6.3   Global Analysis of Machine Learning Results

The first 4000 SMS texts from our corpus were selected as training data for the classifier. The labels (to anonymise or not) were available from the student annotator, as mentioned above. The reason to limit the size of the training data is that we do not want to burden the human expert with a lot of manual annotation, since the goal of the system is to save expert's time for the SMS anonymisation task. In fact, when we experimented with half the amount of the training data, we obtained similar results.

In the 4000 training messages, there were 529 that were positive examples (labelled with the class TA, to anonymise) and the rest of 3471 were negative examples. With this high imbalance, the classifier learns mostly the characteristics of the negative class. Such a classifier is not useful, since it would anonymise only very few examples.

To deal with the high data imbalance, we use a simple undersampling technique. We balanced the training data by keeping only 529 of the negative examples. We successfully trained a classifier on the 1058 examples (529 positive examples, 529 negative examples).

An example of learnt decision tree is presented in figure 1. Only a part of the tree is presented, because it has 127 nodes, from which 64 are leaf nodes. The leaf nodes contain the predicted class, followed in brackets by the number of correct / incorrect instances from the training data classified into the node. By examining the learnt decision tree, we note that the best feature, used in the root of the tree is the number of words that start with capital letters. The length of the words also seems to be important. We expected the occurrences in the dictionary of First names to be higher up in the tree. Table 5 shows the InfoGain values for each feature. The table ranks the features by their ability to discriminate between the two classes, since the InfoGain measures the entropy of classification when one feature at a time is used. We can see that the FristName feature is the third most important.

```
UpperCase <= 1
|    Numbers <= 2: NTA (273.0/8.0)
|    Numbers > 2
|    |    WordLength <= 5: NTA (5.0/1.0)
|    |    WordLength > 5: TA (18.0/3.0)
UpperCase > 1
|    Numbers <= 3
|    |    UpperCase <= 2
|    |    |    Punctuation <= 1
|    |    |    |    SMSLength <= 27: NTA (5.0)
|    |    |    |    SMSLenght > 27
|    |    |    |    |    WordLength <= 7: TA (48.0/6.0)
|    |    |    |    |    WordLength > 7: NTA (2.0)
......
|    Numbers > 3
|    |    WordLength <= 5
|    |    |    FirstName <= 7
|    |    |    |    Pronouns <= 0
|    |    |    |    |    Elongations <= 0: TA (5.0)
|    |    |    |    |    Elongations > 0
|    |    |    |    |    |    SMSLength <= 78: NTA (2.0)
|    |    |    |    |    |    SMSLength > 78: TA (4.0)
|    |    |    |    Pronouns > 0
|    |    |    |    |    Punctuation <= 6: NTA (6.0)
|    |    |    |    |    Punctuation > 6: TA (2.0)
|    |    |    FirstName > 7: TA (35.0/3.0)
|    |    WordLength > 5: TA (94.0/4.0)
```

Fig. 1. Part of the learnt decision tree

Table 6 presents some results of the Bagging DT classifier, by 10-fold cross-validation. This standard evaluation technique in machine learning uses 9 parts of the data for training and tests on the remaining part, then it repeats this for other 9 parts. At the end, it averages over the 10 folds. The baseline in the table is 50%, for a random classifier that would choose any of the two classes. The DT algorithm achieves an accuracy of 77.8%, therefore it is much better than the baseline. The meta-classifier (Bagging DT) reaches 79.4%. The performance for

Table 5. Importance of the features in the classification

Rank	Feature	InfoGain Value
1	UpperCase	0.2754
2	SMSLength	0.2121
3	FirstName	0.1366
4	Countries	0.0986
5	Cities	0.0986
6	WordLenght	0.0776
7	Numbers	0.0367
8	Elongations	0.0353
9	Abbreviations	0.034
10	Punctuation	0.0269
11	LEFFF	0.0262
12	Pronouns	0.000

each of the two classes is presented in terms of Precision (how many examples classified into that class are correct), Recall (how many correct examples of that class are retrieved) and F-measure (the harmonic mean of Precision and Recall). The values in Table 6 show that the classifier is doing equally well for the two classes.

We did additional testing for the machine learning method. We trained the classifier on the entire training data, and we tested it on the next 2000 messages from the corpus (from the 4000th SMS to 5999th). This is a realistic test set, because the test data comes later in the time line, and it might have differences compared to the training data. For this test, the accuracy of the Decision Tree was 74.6% DT, while Bagging DT achieved an accuracy of 76.9%. The accuracy is slightly lower than the one obtained by 10-fold cross validation. This shows that the classifier is general enough to obtain similar results on new test data.

Table 6. Machine learning results for each class (i.e., confusion matrix and precision/recall/F-measure)

Classes	Real: NTA	Real: TA	Class	Precision	Recall	F-Measure
Model: NTA	383	72	NTA	0.842	0.724	0.778
Model: TA	146	457	TA	0.758	0.864	0.807

7   Conclusion and Future Work

The system proposed in this article performs the anonymisation/de-identification of a corpus. To this end, it uses (a) a dictionary of first names and (b) anti-dictionaries (of ordinary language and of some forms of SMS writing) to identify the words that require anonymisation. Note that the adopted principle is sufficiently generic for it to be adapted to various types of corpus, irrespective of their language.

In its automatic phase, the system processes over 70% of the corpus. This corresponds to the unambiguous text messages present in it: Those that contain words that are neither unknown, nor ambiguous (found in both the dictionary as well as the anti-dictionaries). A comparative analysis of its performance, based on the manual evaluation of a significant albeit small portion of the corpus (i.e. 23,055 SMS), yielded positive results on 96% of the text messages processed (whether considered to be anonymised or not to be anonymised).

As future work, two students will perform the task of anonymisation. A thorough analysis on their part will allow us to improve our techniques and enrich our dictionaries. Another immediate direction of future work is to design more features for the supervised machine learning algorithms, in order to increase their accuracy for this task.

We would also like to apply the tool and its associated algorithms to other types of data (e.g., medical data) that require anonymisation/de-identification.

Acknowledgements. We thank **Rachel Panckhurst**, leader of the sud4science LR project. This project is part of a vast international SMS data collection project, entitled sms4science (`http://www.sms4science.org`), and started at the CENTAL (Centre for Natural Language Processing, Université Catholique de Louvain, Belgium) in 2004. Our work is supported by the MSH-M (Maison des Sciences de l'Homme de Montpellier – France) and the DGLFLF (Délégation générale à la langue française et aux langues de France).

References

1. Accorsi, P., Patel, N., Lopez, C., Panckhurst, R., Roche, M.: Seek&Hide: Anonymising a French SMS Corpus Using Natural Language Processing Techniques. Lingvisticæ Investigationes 35(2) (2012)
2. Plamondon, L., Lapalme, G., Pelletier, F.: Anonymisation de décisions de justice. In: TALN 2004 (2004)
3. Grouin, C., Rosier, A., Dameron, O., Zweigenbaum, P.: Une procédure d'anonymisation à deux niveaux pour créer un corpus de comptes rendus hospitaliers. Risques, technologies de l'information pour les pratiques médicales, 23–34 (2009)
4. Tamersoy, A., Loukides, G., Nergiz, M., Saygin, Y., Malin, B.: Anonymization of longitudinal electronic medical records. IEEE Transactions on Information Technology in Biomedicine 16(3), 413–423 (2012)
5. Sweeney, L.: Replacing personally-identifying information in medical records, the scrub system. In: Proceedings of the AMIA Annual Fall Symposium, American Medical Informatics Association, p. 333 (1996)
6. Aramaki, E., Imai, T., Miyo, K., Ohe, K.: Automatic deidentification by using sentence features and label consistency. In: i2b2 Workshop on Challenges in Natural Language Processing for Clinical Data, pp. 10–11 (2006)
7. Gardner, J., Xiong, L., Wang, F., Post, A., Saltz, J., Grandison, T.: An evaluation of feature sets and sampling techniques for de-identification of medical records. In: Proceedings of the 1st ACM International Health Informatics Symposium, pp. 183–190. ACM (2010)

8. Gicquel, Q., Proux, D., Marchal, P., Hagége, C., Berrouane, Y., Darmoni, S., Pereira, S., Segond, F., Metzger, M.: Évaluation d'un outil d'aide á l'anonymisation des documents médicaux basé sur le traitement automatique du langage naturel. Systèmes d'information pour l'amélioration de la qualité en santé, 165–176 (2012)

9. Szarvas, G., Farkas, R., Busa-Fekete, R.: State-of-the-art anonymization of medical records using an iterative machine learning framework. JAMIA 14(5), 574–580 (2007)

10. Reffay, C., Blondel, F., Giguet, E., et al.: Stratégies pour l'anonymisation systématique d'un corpus d'interactions plurilingues. In: Proceedings of IC 2012, pp. 1–21 (2012)

11. Fairon, C., Klein, J.: Les écritures et graphies inventives des sms face aux graphies normées. Le Français aujourd'hui (3), 113–122 (2010)

12. Treurniet, M., De Clercq, O., van den Heuvel, H., Oostdijk, N.: Collection of a corpus of dutch sms (2012)

13. Chen, T., Kan, M.: Creating a live, public short message service corpus: The nus sms corpus. Language Resources and Evaluation, 1–37

14. Reffay, C., Teutsch, P.: Anonymisation de corpus réutilisables Masquer l'identité sans altérer l'analyse des interactions. In: Proceedings of EIAH 2007: Environnements Informatiques pour l'Apprentissage Humain (2007)

15. Beaufort, R., Roekhaut, S., Cougnon, L.A., Fairon, C.: A hybrid rule/model-based finite-state framework for normalizing sms messages. In: Proceedings of ACL, pp. 770–779 (2010)

16. Hall, M., Frank, E., Holmes, G., Pfahringer, B., Reutemann, P., Witten, I.: The weka data mining software: An update. SIGKDD Explorations 11(1) (2009)

A Corpus Based Approach for the Automatic Creation of Arabic Broken Plural Dictionaries

Samhaa R. El-Beltagy[1] and Ahmed Rafea[2]

[1] Nile University, Center for Informatics Science, Giza, Egypt
selbeltagy@nileuniversity.edu.eg
[2] The American University in Cairo, Computer Science Department, Cairo, Egypt
rafea@aucegypt.edu

Abstract. Research has shown that Arabic broken plurals constitute approximately 10% of the content of Arabic texts. Detecting Arabic broken plurals and mapping them to their singular forms is a task that can greatly affect the performance of information retrieval, annotation or tagging tasks, and many other text mining applications. It has been reported that the most effective way of detecting broken plurals is through the use of dictionaries. However, if the target domain is a specialized one, or one for which there are no such dictionaries, building those manually becomes a tiresome, not to mention expensive task. This paper presents a corpus based approach for automatically building broken plural dictionaries. The approach utilizes a set of rules for mapping broken plural patterns to their candidate singular forms, and a corpus based co-occurrence statistic to determine when an entry should be added to the broken plural dictionary. Evaluation of the approach has shown that it is capable of creating dictionaries with high levels of precision and recall.

1   Introduction

Broken plurals (BP) are an important part of any Arabic text, as they form around 10% of any Arabic content and 40% of encountered plurals [1]. This makes broken plurals much more common than irregular nouns in English. Unlike regular or sound plurals, which simply result in the addition of suffixes, broken plurals change the entire structure of a word's singular form in order to convert it to a plural. This structural change usually involves the addition of infixes as well as re-ordering or deletion of letters that existed in the original word. While a language like English has irregular plurals, it does not really have an equivalent of a broken plural. Gowder el al [2] have shown that identifying broken plurals and handling them results in improved information retrieval results, while El-Beltagy and Rafea [3] have also shown that handling broken plurals can significantly improve the results of semantic tagging systems. Yet the identification of broken plurals has proven to be a difficult task. While broken plurals exhibit well defined and known patterns, simply relying on these patterns for their identification results in considerable errors. In [4], it was shown that applying simple BP pattern matching on a test corpus for identifying broken plurals resulted in a precision value of 13.7% and a recall value of 99.7%.

A. Gelbukh (Ed.): CICLing 2013, Part I, LNCS 7816, pp. 89–97, 2013.

The extremely low precision value is indicative of the uselessness of relying on BP pattern matching alone for accurate identification of broken plurals. Work presented in [4] also concluded that a dictionary based approach yields the best results in identifying broken plurals. Building dictionaries, especially for specific domains, is however a tiresome, expensive and a time consuming process.

This work proposes a framework for automatically building broken plural dictionaries from a large input corpus. Experimental results show that the proposed method is capable of achieving its task with a high level of accuracy.

The rest of this paper is organized as follows: Section 2 briefly reviews related work, section 3 provides an overview of the proposed system, section 4 presents the experiment carried out to evaluate the work and its results, and finally section 5 concludes this paper.

2   Related Work

Highly relevant to this work is that of Goweder et al[4] in which they attempt the difficult problem of identifying broken plurals and reducing them to their singular forms. In their work, Goweder et al carry out a series of experiments for detecting broken plurals using a test corpus. In all experiments, input words in the corpus are firstly lightly stemmed using a modified version of the aggressive Khoja stemmer [5]. In the first and least accurate experiment, all forms that fit the pattern of a broken plural were detected and analyzed to see whether or not words that fit these patterns are in fact broken plurals. Having found that this technique results in very low precision, an alternative method which adds further restrictions to existing patterns based on the authors' observations, was adopted. Using this method precision increased significantly to 53.92% from 13.7% and recall reached 100% after it was 99.7%. To improve the results further, a third variation that employs a machine learning approach was followed to automatically add restriction rules based on a training dataset. Using this approach the results were further improved as precision reached 75.1% and recall 95.9%. The best results however, were obtained using a dictionary based approach. The dictionary in this instance was built semi-automatically and resulted in a precision value of 81.2% and a   recall of 100%.

Adopting the premise that broken plural identification is important for conducting accurate stemming, the work of (El-Beltagy and Rafea)[3]   employed a corpus based approach for identifying broken plurals. Like in [4], the authors of [3] identified several target broken plural patterns. In their work, they try to match words that fit identified broken plural patterns to their candidate singular forms which they generate using a set of rules (one for each pattern). A match is said to be found, if the candidate transformation appears in any document in their input corpus. While the underlying premise adopted in the work of [3] is quite   similar to the one proposed in this paper, it differs from it in four   significant ways. 1. In [3] , the initial process of identifying broken plurals and building a BP list accordingly is semi-supervised, whereas in this work it is fully automated. 2. In [3], when a match is found as described above,   only the singular form is stored in what the authors call a stem list. During their stemming stage, words to be stemmed are again matched against BP patterns. If a match is found, a possible transformation is generated; if this transformation is found in the stem list, then conflation takes place. This process however, is highly error prone because a word matching a BP pattern may incorrectly map to an entry in the stem list. For example, if the word كلب (dog), the broken plural of which is كلاب (dogs) is stored in the stem list, and if the word كلوب (lantern) is encountered, كلوب will match with one of the BP patterns even though it is not a broken plural and it will be incorrectly conflated to the word كلب . Our proposed model avoids this problem by storing both the broken plural word, and its singular form in a dictionary. 3. The model proposed in [3] attempts to find a match between a candidate transformation and any word in the input corpus, again an error prone process. The work presented herein, proposes a set of restrictions when matching between a word and its possible candidate form to reduce errors. 4. The set of broken plural patterns covered by this work is wider, and cases where a single BP pattern may have multiple possible transformations, are handled as detailed in section 3.

The work presented in [6] proposes a model based on machine translation to detect and convert broken plurals to their singular forms. In this model, words that match BP patterns are translated to the English language. If the resulting English word is found to end with an "s" or if the word exists in a list of irregular English nouns, then it is identified as a broken plural. The English term is then stemmed and the stem is retranslated to the Arabic language to obtain the Arabic singular form. [7] and [8] discuss the complexity of Arabic broken plurals from a linguistic perspective.

Using a corpus for transformation validation purposes is not a novel idea. In fact Xu and Croft made use of this idea to build a stemmer and demonstrated that for the English language this is a very effective approach [9]. In their work, Xu and Croft define corpus based stemming as the process of automatically modifying "equivalence classes to suit the characteristics of a given text corpus" their assumption being that a stemmer that can adapt to a certain domain using the characteristics of its corpus, should perform better than one that cannot. Another assumption underlying their work is that words and their stems are likely to occur in the same document or even more specifically, in the same text window. This is a similar assumption to what the proposed work takes with respect to broken plurals and their singular forms. Rather than use any linguistic knowledge to generate equivalence classes, an n-gram model is employed to carry out that task, which is different than what we propose in this work as candidate singular forms are generated based on linguistic knowledge. After experimenting with Xu and Croft's approach on Arabic, Larkey et al [10] have shown that stemming using an n-gram approach is not the most appropriate approach to use for a language such as Arabic.

3   Overview of the Proposed Approach

In order to detect and create a dictionary of broken plurals where an entry in the dictionary is a broken plural term mapped to its singular form, we follow a corpus based approach. The main difficulty in detecting Arabic broken plurals can be attributed to two main factors: 1. A word that matches a BP pattern may not be a broken plural at all, and 2. A single BP pattern may have more than one possible transformation pattern. For example, the terms ملايين (millions), براميل (barrels), and اعاصير(hurricanes) all match the BP pattern فعاليل (f3Alyl). However, to transform each of these words to its singular form, each requires a different rule. Please refer to table 1. for a list of BP patterns covered by this work, examples of each pattern, and the possible transformation patterns for each. The "ه]" that appear at the end of some transformation patterns, indicates that the singular form may or may not have a "ه" at the last letter. The basic premise on which this work builds is that in any given corpus, BPs and their singular forms are likely to co-occur within documents of this corpus.

Table 1. List of BP patterns covered by this work

#	Pattern	Plural Example	Trans. Form	Singular Transformation	Transformation Pattern
			Word length = 4		
P1	فعول (f3Wl)	حقوق (rights)	C_1	حق (rights)	فع (f3)
		قروض (loans)	C_2	قرض (loans)	فعل (f3l)
			Word length = 5		
P2	فوائل (fWA2l)	سوائل (liquids)	C_1	سائل (liquid)	فائل[ه] (fA2l[h])
P3	فعائل (f3A2l)	رهائن (hostages)	C_1	رهينه (hostage)	فعيله (f3ylh)
		خسائر (losses)	C_2	خساره (loss)	فعاله (f3Alh)
		بدائل (alternatives)	C_3	بديل (alternative)	فعيل (f3yl)
P4	فعايا (f3AyA)	خلايا (cells)	C_1	خليه (cell)	فعيه (f3yh)
P5	افعال (Af3Al)	اصوات (sounds)	C_1	صوت (sound)	فعل (f3l)
P6	فواعل (fWA3l)	شوارع (streets)	C_1	شارع (street)	فاعل (fA3l)
		مواسم (seasons)	C_2	موسم (season)	فوعل (fW3l)
P7	فعلاء (f3lA2)	خبراء (experts)	C_1	خبير (expert)	فعيل(f3yl)
			Word length = 6		
P8	فواعيل(fWA3yl)	صواريخ (rockets)	C_1	صاروخ (rocket)	فاعول(fA3Wl)
P9	فعاليل (f3Alyl)	ملايين (millions)	C_1	مليون (million)	فعويل(f3wyl)
		براميل (barrels)	C_2	برميل (barrel)	فعليل(f3lyl)
		اعاصير(hurricanes)	C_3	اعصار (hurricane)	فعلال(f3lAl)
P10	افعياء (Af3yA2)	اذكياء (Stupid plural)	C_1	ذكي (Stupid singular)	فعي (f3y)

The first step in building a dictionary using the proposed approach, is to index all documents of the input corpus via search engine. When experimenting with this approach, the Apache Lucune search engine was used [11]. Some of the steps taken in preprocessing are dependent on the nature of the input documents. For example, html documents would require the removal of html tags. Same thing with XML or SGML documents, but in general the only preprocessing done on input documents is normalization and stop word removal. Most of the approaches that addressed broken plural detection, carried out light stemming on words in a corpus during preprocessing. The reason we did not follow a similar approach is that doing so would result in missing broken plurals that have leading or trailing letters that can be mistakenly eliminated using a light stemmer, as an integral part of the word. For example the word الوان (colors), which matches with broken plural pattern P5 has "ال" as its leading two letters. Since "ال" as a prefix often denotes "the", it is always removed by a light stemmer. However, in the above given example, it does not stand for "the", but is simply part of the word and thus it will be incorrectly removed by a light stemmer preventing its detection and the detection of it singular form لون (color). Another example is that of the word قوانين (laws) (broken plural pattern 9) in which the trailing letters "ين" would also be mistaken as a suffix by a light stemmer and removed.

To build the dictionary, words in the corpus are scanned for matches with any of the BP patterns presented in Table 1. The followed approach relies on the generation of candidate transformations for each of the encountered patterns and on attempting to measure the extent to which the original word that matched a BP pattern collocates with its candidate transformation. The collocation metric employed by this work is that of normalized point wise mutual information [12] which is calculated using equation (1) and where x represents the original BP word and y its candidate transformation.

$$i(x,y) = \left(\ln \frac{p(x,y)}{p(x)p(y)} \right) \Big/ {-\ln p(x,y)} \tag{1}$$

A simplified algorithm for building the BP dictionary is as follows:

```
seen = {}
for each document d_i ∈ collection c
  For each term ti ∈ d_i and t_i ∉ seen
    seen = seen ∪ t_i
    if((t_i.length >= 4) and  (t_i.length <=6))
    Pattern P = getBPPatternFor(t_i)
    If (P!=null)
        Candidates cans = P.generateCandidatesFor(t_i)
        ScoreEntry e = getMaxScoreFrom(cans)
        if(e.maxScore > Ω)       // Ω is a threshold
          dictionary.add(new entry(t_i,e.term)
        end
    end
  end
end
```

Actual implementation is more involved. For broken plurals with transformation patterns that have a possible trailing [ه], a candidate term without the "ه" is generated first and it's only if the this term fails to match the threshold Ω that the "ه" is added and the search and scoring repeated. Matching each of the patterns has some additional constraints, such as letters it should not start with, end with or have as a second letter. These are summarized in table 2.

Table 2. Constraints placed on various patterns

Pattern	Does not start with	Does not end with	Second letter is Not
P1	No constraints	No constraints	No constraints
P2	و ا ي	ي ه ا	No constraints
P3	و ا ي	ي ه ا	No constraints
P4	و ا ي	No constraints	No constraints
P5	No constraints	ي ه ء و	ي ت
P6	ت و ا ي	ي ه ا	No constraints
P7	و ا ي	No constraints	No constraints
P8	No constraints	ي ه و ا	No constraints
P9	و	ي ه ا	No constraints
P10	No constraints	No constraints	ل

Also, when searching for any term in the corpus, whether the term is the original BP or the proposed singular form, the term in addition to the term + prefix"ال" are both entered and Or(ed) in the search query.

Before evaluating the system using a large dataset, it was first applied on a small document collection (39 documents). This collection is part of a larger dataset which is described in the next section and which was fully indexed before experimenting with the smaller subset. The purpose of experimentation was to determine the best threshold Ω for each of the presented patterns and for detecting any problems with the algorithm. One of the problems that emerged during this experimentation was related to the detection and conversion of pattern P1. When applying the transformation rules of this BP pattern, many nouns which happened to match with this pattern and which should have been left as is, were mapped to closely related verbs that co-occurred with the nouns, which is of course incorrect. An example of such a faulty mapping is that between the word ذهول (amazement) to the verb ذهل (amaze). This happened at a frequency that threatened to considerably affect the precision of this pattern mapping. To avoid such a faulty mapping, the knowledge that the prefix "ال" never attaches to verbs was employed. So basically another factor called alScore was introduced based on this observation. The alScore factor is calculated using equation 2.

$$\text{alScore} = \frac{Total \ \# \ of \ documents \ in \ which \ BP \ term \ and \ (ال + candidate \ term) \ appear \ together}{Total \ \# \ of \ documents \ in \ which \ BP \ term \ appears \ in \ corpus} \quad (2)$$

For any valid mapping, the alScore is calculated. If this score falls below a threshold α, the mapping is not added to the dictionary, otherwise it is. This restriction improved the precision of this rule considerably.

Since the documents used were small ones (average number of words = 220), colocation counts were carried out on the level of an entire document. However, in larger documents, it might be preferable to carry out search using a proximity window.

4   Experiment and Results

In order to experiment with the proposed approach the Arabic Newswire A Corpus [13] which consists of approximately 383,872 articles collected from the Agence France Press (AFP) Arabic Newswire, was used. As stated in the previous section, documents in the corpus were indexed using Lucune [11]. SGML tags were removed before indexing.

The evaluation metrics used were precision, recall, f-score, accuracy, and specificity (true negative rate).   Each of these metrics is defined as follows:

Let

TP	=	number of correctly extracted dictionary entries
TN	=	number of correctly rejected dictionary entries. These represent words that fit a broken plural pattern, but which were correctly rejected due to constraints introduced by the proposed model
FP	=	number of incorrectly extracted dictionary entries
FN	=	number of broken plural words that match with one of the target BP patterns and should have been added as an entry in the dictionary, but was ignored.

Then

$$\text{Precision} = \frac{TP}{TP+FP} \tag{3}$$

$$\text{Recall} = \frac{TP}{TP+FN} \tag{4}$$

$$\text{F-score} = \frac{2 \ X \ Precsion \ X \ Recall}{Precsion+ \ Recall} \tag{5}$$

$$\text{Accuracy} \ = \frac{TP+TN}{TP+TN+FP+FN} \tag{6}$$

$$\text{Specificity} = \frac{TN}{TN+FP} \tag{7}$$

To determine the values of the above metrics, an Arabic native speaker was asked to go over all instances where a term matched a BP pattern and to indicate whether or not this term should be mapped to a singular term, and to select the value of the singular form from suggested alternatives where appropriate. When in doubt, the native speaker referred to the on-line dictionary Almaany [14]. The system generated dictionary was then compared to manually created one.

Table 3 shows the pattern transformation   statistics, while table 4 displays the overall results of the performance of the system based on the above metrics.

The used Ω thresholds   were empirically set using a small subset of documents as described in the previous section. By looking at table 3, it can be seen that the pattern with the most word matches is that of P1, followed by P5 and P9, but despite the fact these patterns' match rate is quite high, the system manages to successfully avoid making in-correct mappings. Patterns P10 and P4, seem to be the rarest of all patterns, while pattern P5 seems to account for the highest number of correct entries in the dictionary.

Table 3. The different patterns and their matcing results

Pattern	TP	FP	FN	TN	Used Ω Threshold
P1	99	19	54	2693	0.1
P2	14	1	2	10	0.1
P3	72	1	26	104	0.001
P4	9	0	3	46	0.001
P5	171	30	42	915	0.15
P6	63	3	20	223	0.15
P7	33	4	9	294	0.1
P8	12	0	7	183	0.2
P9	87	9	23	952	0.15
P10	6	0	7	5	0.1
Overall	566	67	193	5425	

Table 4. Overall system performance

Pattern	Precision %	Recall %	F-score %	Accuracy %	Specificity %
P1	83.9	64.7	73.1	97.5	99.3
P2	93.3	87.5	90.3	88.9	90.9
P3	98.6	73.5	84.2	86.7	99.1
P4	100	75	85.7	94.8	100
P5	85.1	80.3	82.6	93.8	96.8
P6	95.5	75.9	84.6	92.6	98.7
P7	89.2	78.6	83.5	96.2	98.7
P8	100	63.2	77.4	96.5	100
P9	90.6	79.1	84.5	97.0	99
P10	100	46.2	63.2	61.1	100
Overall	89.5	74.5	81.3	95.8	98.8

From table 4, it can be concluded that the overall precision and accuracy of the system are quite high for a fully automated dictionary builder. However, recall while not low, could certainly be higher. When analyzing the reason for the low recall, it was found that the majority of BP words that should have been included in the dictionary, but were not, did not collocate with their singular forms within the corpus. However, experimenting with light stemming of words that do not match with any of the BP patterns may improve this recall value.

5   Conclusion

This paper presented a corpus based approach for automatically building dictionaries that map broken plurals to their singular forms. Evaluation of the system has shown that it can carry out this task with a high level of accuracy and precision. The main advantage of following the proposed approach is that it can be applied on any corpora, whether domain specific or even colloquial text, as colloquial Arabic broken plurals follow the same transformation patterns as the more formal modern standard Arabic. The generated dictionaries can be easily integrated into any Arabic stemmer or text mining application.

References

1. Goweder, A., De Roeck, A.: Assessment of a Significant Arabic Corpus. In: Proceedings of the Arabic NLP Workshop at ACL/EACL, Toulouse, France, pp. 73–79 (2001)
2. Goweder, A., Poesio, M., De Roeck, A.: Broken Plural Detection for Arabic Information Retrieval. In: SIGIR 2004. pp. 566–567 (2004)
3. El-Beltagy, S.R., Rafea, A.: An Accuracy Enhanced Light Stemmer for Arabic Text. ACM Transactions on Speech and Language Processing 7, 2–23 (2011)
4. Goweder, A., Poesio, M., De Roeck, A., Reynolds, J.: Identifying broken plurals in unvowelised Arabic text. In: EMNLP 2004, Barcelona, Spain (2004)
5. Khoja, S., Garside, R.: Stemming Arabic text. Computing Department, Lancaster University, Lancaster, UK (1999)
6. Goweder, A.M., Almerhag, I.A., Ennakoa, A.A.: Arabic Broken Plural Recognition Using a Machine Translation Technique. In: ACIT 2008, Hammamet, Tunisia (2008)
7. McCarthy, J.J.: A prosodic account of Arabic broken plurals (1983)
8. Kiraz, G.A.: Analysis of the Arabic Broken Plural and Diminutive. In: The 5th International Conference and Exhibition on Multi-Lingual Computing, Cambridge, UK (1996)
9. Xu, J., Croft, W.B.: Corpus, based stemming using co-occurrence of word variants. ACM Transactions on Information Systems 16, 61–81 (1998)
10. Larkey, L.S., Ballesteros, L., Connell, M.E.: Improving Stemming for Arabic Information Retrieval: Light Stemming and Co-occurrence Analysis. In: Proceeedings of SIGIR 2002, Tampere, Finland (2002)
11. Apache: Lucene, http://lucene.apache.org/
12. Bouma, G.: Normalized (Pointwise) Mutual Information in Collocation Extraction. In: The Biennial GSCL Conference, pp. 31–40 (2009)
13. LDC: Arabic Newswire A Corpus (1994), http://www.ldc.upenn.edu/Catalog/CatalogEntry.jsp?catalogId=LDC2001T55
14. Almaany: Almaany on-line dictionary, http://www.almaany.com/

Temporal Classifiers for Predicting the Expansion of Medical Subject Headings

George Tsatsaronis[1], Iraklis Varlamis[2], Nattiya Kanhabua[3], and Kjetil Nørvåg[4]

[1] Biotechnology Center,
Technische Universität Dresden
george.tsatsaronis@biotec.tu-dresden.de
[2] Department of Informatics and Telematics,
Harokopio University of Athens
varlamis@hua.gr
[3] L3S Research Center,
Leibniz Universität Hannover
kanhabua@L3S.de
[4] Department of Computer and Information Science,
Norwegian University of Science and Technology
Kjetil.Norvag@idi.ntnu.no

Abstract. Ontologies such as the *Medical Subject Headings* (*MeSH*) and the *Gene Ontology* (*GO*) play a major role in biology and medicine since they facilitate data integration and the consistent exchange of information between different entities. They can also be used to index and annotate data and literature, thus enabling efficient search and analysis. Unfortunately, maintaining the ontologies manually is a complex, error-prone, and time and personnel-consuming effort. One major problem is the continuous growth of the biomedical literature, which expands by almost 1 million new scientific papers per year, indexed by *Medline*. The enormous annual increase of scientific publications constitutes the task of monitoring and following the changes and trends in the biomedical domain extremely difficult. For this purpose, approaches that try to learn and maintain ontologies automatically from text and data have been developed in the past. The goal of this paper is to develop temporal classifiers in order to create, for the first time to the best of our knowledge, an automated method that may predict which regions of the *MeSH* ontology will expand in the near future.

1 Introduction and Motivation

The biomedical domain is characterized by an exponential growth in the produced data volumes, primarily scientific published articles, knowledge and databases, nucleotide sequences and protein structures [6]. For instance, the number of the scientific articles that are published and indexed by *PubMed*[1] is nowadays approximately close to 23 million, with an average of almost 15,000 new articles being added each week.

---

[1] The main search engine for the life sciences developed by the US National Library of Medicine. http://www.ncbi.nlm.nih.gov/pubmed

A. Gelbukh (Ed.): CICLing 2013, Part I, LNCS 7816, pp. 98–113, 2013.

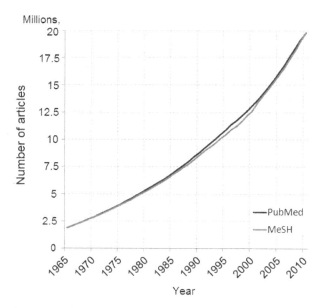

Fig. 1. Growth of the biomedical scientific literature (indexed by *Medline*) in absolute number of articles over the past 45 years

This overwhelming amount of information constitutes the task of monitoring and following the trends in the biomedical domain almost impossible for researchers without the aid of automated tools and efficient search engines. In this direction, we have experienced in the last decade large activity in the area of semantic-enabled technologies. Semantic search technologies, e.g., ontology-based search of articles, achieve the goal of analyzing efficiently the textual information which lies inside the scientific publications, and index the articles using domain ontology concepts so that researchers can browse thematically the new citations, and filter out easily the irrelevant results. Two such popular search engines for the life sciences are *PubMed* and *GoPubMed*[2][4]. Both of these engines are based on annotating the scientific publications indexed by *Medline*[3] with ontology concepts. The former engine uses the *Medical Subject Headings*[4], while the latter uses in addition the *Gene Ontology*[5] and the *Universal Protein Resource*[6].

Within the process pipeline of semantic search engines for the biomedical domain, such as the aforementioned examples, there are two major issues that need to be addressed efficiently. The first issue pertains to the ability of the engine to annotate timely and accurately the new scientific articles using concepts of the underlying ontologies. The latest advances in the field of text classification and text alignment have provided the respective research communities of semantic search with novel methodologies which

---

[2] A semantic search engine for the life sciences.
  http://www.gopubmed.org/web/gopubmed/
[3] The U.S. National Library of Medicine's (NLM) premier bibliographic database.
[4] http://www.nlm.nih.gov/mesh/
[5] http://www.geneontology.org/
[6] http://www.uniprot.org/

can address efficiently this task. For example, in Figure 1 we show how the *PubMed* engine is able to follow the exponential increase of published research articles (blue line) by annotating almost all of the new articles (red line) with *MeSH* concepts in a timely fashion. In addition, there have been several recently published methodologies, e.g., [13], which claim that can address efficiently the specific task of annotating scientific literature text with *MeSH* concepts using machine learning techniques, which in this case were shown to be robust to the ambiguity of *MeSH* terms. Thus, maintaining the pace of the annotations in a level that can follow the increasing amount of newly published articles is an issue that has been sufficiently addressed in the bibliography and can be conducted in a satisfactory manner with automated methods.

The second problem faced by the biomedical semantic search engines is the maintenance of the underlying ontologies, so that the changes and advances of the biomedical domain are depicted in the used conceptualizations. Though the problem, also known in the literature as *ontology evolution*[8], has been studied for a long time, in the biomedical domain it is far from being solved. The intrinsic difference of the biomedical domain compared to other disciplines is the exponential pace with which new facts and findings are communicated via newly published articles. Thus, the cost of maintaining manually the underlying ontologies is extremely large, given the tens of thousands of new articles indexed weekly by *PubMed*.

Motivated by this problem, in this paper we present a new methodology that may aid in an automated manner the maintenance of *MeSH*. We approach the problem, for the first time to the best of our knowledge, using temporal classifiers. More precisely, we construct classifiers that learn to predict which headings may be expanded in the near future, based on a feature set that contains both static and temporal features pertaining to the structure of *MeSH*, the appearance of the headings in the *PubMed* indexed articles' major and minor annotations, and the *PubMed* results retrieved when querying with the specific heading. As a result of this process, we may then use the learned classifiers to predict in an unseen instance of *MeSH* which headings will be expanded, and evaluate the performance of the classifiers based on the success of the predictions.

The rest of the paper is organized as follows. Section 2 presents some background knowledge regarding the *Medical Subject Headings*, as well as related work and methodological background in techniques for automated biomedical ontology extension. Section 3 presents in detail the suggested methodology. Section 4 analyzes the experimental evaluation and the produced results, and, Section 5 concludes and provides pointers to future work.

2   Background and Related Work

In the following we provide background information on the structure and properties of *MeSH*, as well as discussion of the related work with regards to automated methods that suggest biomedical ontologies' extensions.

2.1   The Medical Subject Headings Hierarchy

Medical Subject Headings (*MeSH*) is a hierarchy of terms maintained by the *United States National Library of Medicine* and its purpose is to provide headings (terms)

Table 1. Changes in the *MeSH* hierarchy from 1999 until 2012. Difference from the previous year, and the average minimum and maximum depths of the changes are presented.

Year	Number of *MeSH* Headings	Difference from Previous Year	Average Min. Depth	Average Max. Depth
1999	19, 354	-	-	-
2000	19, 537	183	4.61	5.26
2001	20, 374	837	6.12	6.61
2002	21, 624	1, 250	6.35	6.76
2003	22, 281	657	5.00	5.69
2004	22, 767	486	5.18	5.94
2005	23, 709	942	5.38	6.02
2006	24, 200	491	4.96	6.02
2007	24, 656	456	5.05	6.16
2008	25, 102	446	4.07	4.68
2009	25, 523	421	4.21	4.95
2010	26, 095	572	4.44	5.17
2011	26, 549	454	4.10	4.85
2012	26, 850	301	4.33	5.24

which can be used to index scientific publications in the life sciences, e.g., journal articles, books, and articles in conference proceedings. The indexed publications may be then searched through popular search engines, such as *PubMed* or *GoPubMed*, using the *MeSH* headings to filter semantically the results. It has been reported in the past that such a retrieval methodology is beneficial, especially for the precision of the retrieved results [4].

MeSH includes three types of data: (i) *descriptors*, also known as *subject headings*, (ii) *qualifiers*, and, (iii) *supplementary concept records*. *Descriptors* are the main terms that are used to index scientific publications. They are organized into 16 trees, and as of 2013 they number $26, 853 MeSH$[7]. They include a short description or definition of the term, and they frequently have synonyms, known as *entry terms*. *Qualifiers*, also known as *subheadings*, may be used additionally to the *descriptors* to narrow down the topic of each of the *descriptors*. In total there are approximately 80 *qualifiers* in *MeSH*. *Supplementary concept records*, approximately 214, 000 in the most recent *MeSH* release, describe mainly chemical substances and are linked to respective *descriptors* in order to enlarge the thesaurus with information for specific substances.

In this work we focus only on *MeSH descriptors*[8], and more precisely we aim at learning classifiers, using both static and temporal features, which may predict the *MeSH* headings that will be expanded in the next *MeSH* releases. In Tables 1 and 2 we present some statistics for the *MeSH* hierarchy over the past years that showcase the prediction problem we are addressing. Table 1 shows the number of *MeSH* headings per year from 1999[9] until 2012. It shows additionally the difference in number of headings

---

[7] The most recent *MeSH* version is released as *MeSH* 2013.

[8] For the remaining of the paper, *MeSH headings*, *MeSH descriptors* and *MeSH terms* may be used interchangeably, referring always to the *descriptors* of *MeSH*.

[9] *MeSH* exists since 1963. However, is it only since 1999 that it has been systematically maintained and the changes are thoroughly tracked.

Table 2. Changes in the *MeSH* hierarchy per *MeSH* tree, from 1999 until 2012. Difference from the previous year, and the total number of heading additions is presented. Also the depth of each tree in the latest release is presented in the last line.

Year	Tree A	Tree B	Tree C	Tree D	Tree E	Tree F	Tree G	Tree H	Tree I	Tree J	Tree K	Tree L	Tree M	Tree N	Tree V	Tree Z
2000	32	18	14	70	26	2	20	2	2	1	0	2	0	3	7	0
2001	11	448	44	222	40	10	30	3	9	8	20	8	2	15	1	0
2002	46	716	41	250	58	28	52	16	26	10	17	17	11	36	0	0
2003	28	242	45	266	24	4	39	3	3	9	0	1	7	3	0	2
2004	12	169	53	139	39	4	38	2	4	1	24	11	0	8	0	1
2005	35	93	58	684	24	2	42	1	2	5	0	4	0	3	6	0
2006	27	60	70	255	35	6	43	8	4	7	0	1	1	3	0	0
2007	39	15	73	218	31	2	25	5	27	6	1	1	5	36	3	1
2008	28	57	57	101	51	13	95	15	7	8	4	3	3	28	3	4
2009	23	45	90	101	66	20	35	13	11	12	2	11	3	30	4	4
2010	32	62	90	167	110	21	70	8	10	23	0	11	3	40	4	2
2011	30	25	83	112	82	21	75	3	4	13	2	6	8	30	0	2
2012	8	4	24	107	74	18	28	5	11	16	1	8	4	46	1	2
Total	351	1,954	742	2,692	660	151	592	84	120	119	71	84	47	281	29	18
Depth	11	12	10	11	10	7	10	7	9	10	6	9	7	9	4	7

compared to the previous year, and the average minimum and maximum depths that the changes (insertions) of *MeSH* headings occurred. Each of the *MeSH* headings may appear in different locations in the *MeSH* hierarchy of the 16 trees, and this is why in Table 1 we show both the average minimum and the average maximum of the changes occurred. A first view of the prediction problem we address can be formed by examining thoroughly Table 1. Annually, the past 12 years there is an average addition of 576 new *MeSH* headings. The changes occur on average between the levels of depth 4 and 6 in the *MeSH* trees. Thus, taking structural features into account, such as depth of *MeSH* headings and number of children and siblings of each heading, can be easily motivated by the statistics shown in Table 1.

Table 2 shows the added *MeSH* headings per year and per tree. Also the total number of additions is presented from 1999 till date. The last line of the table presents the maximum depth of each tree, at the latest *MeSH* release. The depth is computed by finding the shortest path to the root, as there might be several paths leading to the root of a *MeSH* tree, starting from any given node, due to the fact that each heading may appear in several different positions in the *MeSH* tree. The letters of the *MeSH* trees correspond to thematic categories as follows: *(A) Anatomy, (B) Organisms, (C) Diseases, (D) Chemicals and Drugs, (E) Analytical, Diagnostic and Therapeutic Techniques and Equipment, (F) Psychiatry and Psychology, (G) Phenomena and Processes, (H) Disciplines and Occupations, (I) Anthropology, Education, Sociology and Social Phenomena, (J) Technology, Industry, Agriculture, (K) Humanities, (L) Information Science, (M) Named Groups, (N) Health Care, (V) Publication Characteristics,* and *(Z) Geographicals.*

Several of the aforementioned *MeSH* trees neither evolve fast, nor frequently. For example the *MeSH* trees V and Z have been expanded by only 29 and 18 new *MeSH* headings respectively in the past 13 years. In addition, there are trees that evolve constantly, but extremely slowly, like for example the *MeSH* trees *K* and *M*, for which there are even years that no new *MeSH* headings are added. We believe that the approach in this paper may be beneficial and meaningful for the *MeSH* trees that have a large number of new *MeSH* headings, are updated yearly, and the changes are usually large, at least enough to provide sufficient number of positive training examples for the temporal classifiers. For this purpose, in this study we focus only in the three largest *MeSH* trees, namely trees *B*, *C* and *D*, which change annually and usually there are minimum tens of new *MeSH* headings added in them each new *MeSH* release. These three trees number approximately 17, 000 *MeSH* headings, constitute almost 64% of the *MeSH* hierarchy, and contain approximately the 68% of all new *MeSH* headings additions since 1999.

2.2 Biomedical Ontology Evolution and Extension

Ontologies in the biomedical domain are the main tools for information and data integration. Gene product annotations[1], analysis of high-throughput data[16] and search[14] are just three examples of processes in which biomedical ontologies are used. However, maintaining the high degree of coverage of biomedical ontologies constitutes a major problem in the biomedical domain, since to keep up with new information, ontologies must be revised and newly added terms need to be enriched with definitions, cross-references and additional properties.

In the biomedical domain, ontologies are manually curated, and, thus, developing and maintaining them is often a slow, tedious and error-prone process. Alleviating the bottleneck of automated maintenance requires the application of advanced text mining and related techniques. In the past, in the majority of the cases the developed techniques utilized term recognition and pattern-based relationship extraction [9] algorithms. Perhaps closer to our work are the approaches that either apply analysis of Web search results and *PubMed* articles [5], or machine learning to predict the regions that may be expanded [12].

Fabian et al. [5] developed an approach for automated sibling generation for *MeSH*. The approach has two main directions. First, it examines the possibility to extract *MeSH* term siblings by analyzing the structure of Web documents, and more precisely the *HTML* divisions or paragraphs that the terms are mentioned. The motivation behind this direction lies in the hypothesis that terms which are in a sibling relationship to each other are often located together in tables, lists, or headings in Web documents. Second, they examine a text-based approach by analyzing enumerations in sentences using natural language processing. For example, in the following sentence which is part of a *PubMed* article, there are enumerations of *endocrine cells*: *'...several adenohypophysial endocrine cells such as somatotrophs, thyrotrophs, and gonadotrophs'*.[10] Finally, they also examined the benefits of an approach that combines the two aforementioned approaches for sibling generation. In experimental evaluation that they conducted using 1, 000 *MeSH* terms, the authors report that they are able to recover 79.3% of the terms' siblings, when the system is given as input an initial seed of example siblings, e.g., three seed siblings. With regards to precision, using only the top-10 returned suggestions as siblings, e.g., using 10 as a cut-off threshold in the returned sibling list, the combined approach achieves 60.8%, the structure-based 53% and the text-based 48%.

---

[10] http://www.ncbi.nlm.nih.gov/pubmed/11478270

The aforedescribed sibling generation approach exists as part of the *DOG4DAG* system [15] which supports the semi-automated creation and extension of *OBO* ontologies.

Pesquita and Couto [12] very recently published a new methodology for predicting the extension of the *Gene Ontology*. The idea is to use supervised learning in order to predict areas of an ontology that will undergo extension in a future version. The basic sets of features that are used for the learning process are: *structural*, e.g., number of all descendants of a term, *annotation*, e.g., all manual annotations for the term as they are given by the *Gene Ontology*, *citation*, e.g., number of *PubMed* articles mentioning the term, and *hybrid*, e.g., ratios of features that belong to the rest of the sets. In their experimental evaluation, after tuning of the parameters they report an F-Measure of approximately 79% in predicting the regions of the *Gene Ontology* that will undergo refinement in future versions.

In this paper we suggest a method that is absolutely complementary to the two aforedescribed approaches. With regards to the method in [5], our approach may constitute a first step for the selection of the terms to be expanded. For instance, our predicting methodology can suggest which *MeSH* terms may be expanded in future versions of *MeSH*, and for the descendants of these terms, the approach in [5] can be used to extract sibling terms. With regards to the approach in [12], our method adds a complementary insight to the role of temporal features in the learned classifiers.

Thus, a major difference of the current approach with the approach in [12] is that the authors in the latter work do not examine the potentiality of adding temporal features in the classifiers that are learned, e.g., acceleration in their appearance in the *PubMed* corpus. In this direction there are several approaches that may be considered in order to include time-based features in the learning of language models and classifiers, e.g., the reader may wish to consult the work in [7]. Furthermore, the target ontology in the work of Pesquita and Couto is the *Gene Ontology*, while in our case the feature set and the tuning is optimized for the *MeSH* hierarchy. These two knowledge-bases have significant differences, with the main difference being that *Gene Ontology* offers manual annotations of gene products, with respective evidence codes that support this annotation. In contrast, *MeSH* is basically a hierarchy rather than an ontology, and the only annotations for the terms that exist refer to annotations that may be used for the indexing of *PubMed* documents. Finally, the two knowledge-bases are designed for totally different purposes, with the former being an ontology to describe functional aspects of gene products and the latter being a hierarchy to provide headings that can be used as indexing indicators for scientific citations with application to retrieval of documents.

3   Temporal Classifier Models for *MeSH* Terms

In this section we provide in detail the components and the steps of our methodology to construct temporal classifiers in order to predict *MeSH* headings that are going to be expanded in future *MeSH* releases. More precisely, we give all the details with regards to the three aspects of our approach, namely the time parameters and class labels, the feature engineering of the classifiers, and the actual methodology of building the temporal classifiers.

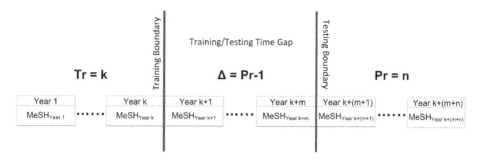

Fig. 2. Illustration of the introduced time parameters. *Tr* is the number of years used for training, *Pr* the prediction time frame, and Δ the training/testing time gap to avoid bias and overlap between the training and the testing process.

3.1 Formulation of the Classification Problem and Time Parameters

The basic time unit in our approach is the notion of a *year*, since every new release of *MeSH* is distributed annually. However, the suggested approach may be generalized to cover time units with different granularity in other applications. Based on the *year* unit, we can formally define our classification problem as follows; let C be the class label based on which the training of the classifiers takes place. C in our case is a binary value, with the value of 1 denoting a positive example and the value of 0 a negative example. Each example (instance) in our case is a *MeSH* heading, denoted with I, at a given year t, and for which a set of features X has been computed, which are explained in the next subsection, thus $I = [X_1, ..., X_N]$. If I is a *MeSH* heading that is going to be expanded in the next n years, e.g., a heading for which direct children will be added in the *MeSH* hierarchy, then I is a positive example, and, hence $C = 1$ in this case, where n is a time parameter that we will denote as *Pr* (*Prediction Time Window*). In the opposite case, I is a negative example, and, hence, $C = 0$ in this case. Given $Pr = n$, the classification problem we are solving is thus formulated as shown in Equation 1.

$$M : I \times C \longrightarrow 0, 1 \qquad (1)$$

where $M(I, C) = 1$ if $I \in C$, and $M(I, C) = 0$ otherwise. The class label is formulated parametrically in our case based on *Pr*, as shown in Equation 2.

$$C(Pr, I) = \begin{cases} 1, & \text{if } I \text{ obtains new direct children in the next } Pr \text{ years} \\ 0, & \text{otherwise} \end{cases} \qquad (2)$$

Thus, we are learning a classifier which attempts to capture based on the feature values, the pattern based on which the *MeSH* headings are expanded in future *MeSH* releases. Besides the parameter *Pr*, which defines the time window of the prediction in years, we are also introducing one more parameter, namely *Tr* (*Training Time Window*). *Tr* represents the number of years that are going to be used for producing the training examples of the classifier. Thus, a *Tr* value of 1 denotes that the *MeSH* version of only one year is used for the training of the classifier. In theory, *Tr* and *Pr* can have any positive integer values; however, to avoid the mixing of training and testing examples during the training, and, thus, securing that there is no bias inserted in the training, towards test examples that have actually been seen, there has to be a time gap between the years of *MeSH* selected for training, and the years of *MeSH* selected for testing and evaluation. We define this gap as Δ (*Training/Testing Time Gap*), and it is defined as a function of *Pr* as follows: $\Delta = Pr - 1$. The rationale behind this definition is straightforward, and we will explain it through the following example; let $Tr = 1$, and $Pr = 2$, which means that we use a single year of *MeSH* to produce training examples, let it be y_1. Then, the class labels are 1 if the respective headings are going to be expanded in the next two years (y_2 or y_3), and zero otherwise. If we did not leave any time gap between training and testing, we could use y_2 for testing. Assume in y_1 there is an instance I that has $C = 1$, because it is expanded in y_3. The classifier has been trained also from this instance's feature values, and, thus, using y_2 for testing where I would still have $C = 1$, is repetition of the positive training example from the training set to the test set, and, hence, it introduces bias. However, adding a time gap $\Delta = Pr - 1 = 1$ in this case, does not allow us to select y_2 as testing. In fact in this case the testing can start from y_3 and on. It may be the case that I has again $C = 1$ in y_3, but this is something the classifier has not learned from y_1, because in this case $C = 1$ means that I can only be expanded in y_4 or y_5. Thus, introducing Δ as explained, ensures that there is no bias in the training of the classifier. The three parameters (Tr, Δ, Pr) and their relation are also illustrated in Figure 2.

3.2   Feature Engineering

The feature set that we have designed for learning the classifier explained in Equations 1 and 2 are of five different categories, namely *structural, citation, annotation, hybrid* and *temporal*. In the following we explain the rationale behind the engineering of these features. Each of the following features may be computed at a specific time point (year), let it be y_n, and, thus, the values from year to year may change. This latter behavior of the feature values is captured by the *temporal* features. For the shake of description we assume that each of these features is, thus, computed at year y_n for *MeSH* heading I.

Structural Features. Motivated by the findings shown in Table 1, we introduce five structural features, namely *minDepth, maxDepth, siblings, direct children*, and *all children*. The first two features refer to the minimum and the maximum depth of I in year y_n as this is computed from the *MeSH* hierarchy in the *MeSH* version of year y_n. The *siblings* is the number of *MeSH* headings that share at least one common parent with I. The *direct children* is the number of all *MeSH* headings that have I as a parent, and *all children* is the number of *MeSH* headings that have I as an ancestor.

Citation Features. One of the basic hypothesis behind the introduction of citation features lies from the need of the ontologies to cover as widely as possible the topics that are heavily discussed in the literature of the domain, e.g., in scientific publications. Based on this, we introduce three *citation* features that are computed using the *PubMed* corpus (cf. Section 4.1 for more details on the used data sets). *PubMed results* computes

Table 3. Summary of the used features. Features are computed for *MeSH* heading I in time point y_n.

Category	Feature	Name	Description
Structural	X_1	*minDepth*	Minimum depth I appears
	X_2	*maxDepth*	Maximum depth I appears
	X_3	*siblings*	# *MeSH* heading siblings to I
	X_4	*direct children*	# *MeSH* heading direct children of I
	X_5	*all children*	# *MeSH* heading descendants of I
Citation	X_6	*PubMed results*	# *PubMed* results with I as query
	X_7	*direct children results*	# *PubMed* results with I's children as query
	X_8	*all children results*	# *PubMed* results with I's descendants as query
Annotation	X_9	*PubMed annotations*	# major/minor *PubMed* annotations with I
	X_{10}	*direct children annotations*	# major/minor *PubMed* annotations with I's children
	X_{11}	*all children annotations*	# major/minor *PubMed* annotations with I's descendants
Hybrid	X_{12}	*annRatioAll*	$\frac{all\ children\ annotations}{all\ children}$
	X_{13}	*annRatioDir*	$\frac{direct\ children\ annotations}{direct\ children}$
	X_{14}	*resRatioAll*	$\frac{all\ children\ results}{all\ children}$
	X_{15}	*resRatioDir*	$\frac{direct\ children\ results}{direct\ children}$
	X_{16}	*annRatioResults*	$\frac{PubMed\ annotations}{PubMed\ Results}$
Temporal	$X_i{}'$	*temporal* X_i	$\frac{X_{i,y_n} - X_{i,y_{n-1}}}{X_{i,y_n}}$

the number of scientific citations returned using the heading I and its synonyms as query and the *PubMed* indexed documents as a corpus. For the computation of *PubMed results* only the indexed documents up to year y_n are considered. Respectively, *direct children results* and *all children results* are the sum of the number of the *PubMed results* returned for all direct children of I, or for all descendants of I.

Annotation Features. *MeSH* is used as a controlled vocabulary to index *Medline* documents. The annotation *Medline* documents with *MeSH* terms is done by curators, and the product is high quality annotations that may be used for text mining. In order to embed this knowledge in our features, we introduce three *annotation* features, similar to the *citation* features, namely *PubMed annotations*, *direct children annotations* and *all children annotations*, which compute respectively the number of times I was used as a major or minor heading in *PubMed* documents, the sum of the number of times any of I's direct children was used, and the sum of the number of times any I's descendant was used.

Hybrid Features. Several of the aforementioned features may receive very high integer values, like for example the *all children results* feature, given the volume of the *MeSH* headings. For this purpose we introduce a set of *hybrid* features that normalize the value of some of the aforementioned features, using their ratios. These features are: *annRatioAll*, the ratio between *all children annotations* and *all children*, *annRatioDir*, the ratio between *direct children annotations* and *direct children*, *resRatioAll*, the ratio between *all children results* and *all children*, *resRatioDir*, the ratio between *direct children results* and *direct children*, *annRatioResults*, the ratio between

PubMed annotations and *PubMed results*, and, *dirRatioAll*, the ratio between *direct children* and *all children*.

Temporal Features. All the categories of the aforedescribed features are static, in the sense that they do not take into account the evolution of the feature values over a certain period of time, e.g., the changes from the last year. For this purpose, we introduce a series of temporal features, based on all of the aforementioned feature categories, to capture the change of the feature values from year to year. Formally, let X_i be any of the introduced features, and year y_n the examined time point. For each X_i we introduced $X_i\prime$ defined in Equation 3.

$$X_i\prime = \frac{X_{i,y_n} - X_{i,y_{n-1}}}{X_{i,y_n}} \tag{3}$$

where X_{i,y_n} is the feature value for feature X_i at time point y_n, and $X_{i,y_{n-1}}$ at the previous time point, e.g., y_{n-1}, respectively. The aim of each feature $X_i\prime$ is, thus, to quantify the feature value changes, compared to the previous time unit.

The summary of all of the aforedescribed features can be found in Table 3.

3.3 Implementing Classifiers to Predict Expansions of *MeSH* Terms

Given the asymmetrical expansion of the *MeSH* hierarchy per tree, and also the large imbalance between positive (headings that will be expanded) and negative examples, as it was shown in Table 2, here we explain how we implement the classification that was described in the previous sections. The basic idea is that a classifier is constructed per tree. The goal is to train an expert for each of the *MeSH* trees, since every tree expands with a totally different pattern, and, thus, mixing training examples from different trees might lead to inconsistencies, e.g., training examples that in one tree might have $C = 1$, in another tree with similar feature values might have $C = 0$, because of the different patterns that the the *MeSH* trees expand. Thus, to avoid such inconsistencies, a classifier C_i is trained for each of the trees T_i.

More formally, for a given set of parameters Tr, Pr, T, where Tr is the number of *MeSH* versions (one per year) to use in order to draw the training examples, Pr determines the time window of the prediction, and T is the letter of a specific *MeSH* tree, a training instance is built from each $I \in T$ that was present in *MeSH* in the respective year, with a class label that is determined by Equation 2 and feature values computed as summarized in Table 3. Hence, if $Tr = 2$, there will be two instances in the training examples for any I that was present in both examined years. The years in our case start from 1999 and run until 2011[11]. Finally, given the Δ time gap, a test set can be built for purposes of evaluation, where for all headings I in the respective *MeSH* version, features can be computed again as described in Table3. The reader may wish to note, however, that a test year cannot be larger than the maximum allowed year (2011) in our case. Thus, if $Pr = k$, and y_n is the examined test year, evaluation cannot take place if $y_n + k > 2011$.

---

[11] The 2013 release of *MeSH* contains the full 2012 *MeSH*, however, the *PubMed* files for the whole of 2012 are not yet available. Thus, we stop in year 2011.

Table 4. Experimental setup of the parameters *Tr* and *Pr*. Each cell shows the number of test years in which we can evaluate the classifier, for the respective *Tr* as training years and the respective *Pr* as the prediction years.

Tr	PR= 1($\Delta = 0$)	PR= 2($\Delta = 1$)	PR= 3($\Delta = 2$)	PR= 4($\Delta = 3$)	PR= 5($\Delta = 4$)	PR= 6($\Delta = 5$)
1	11	9	7	5	3	1
2	10	8	6	4	2	–
3	9	7	5	3	1	–
4	8	6	4	2	–	–
5	7	5	3	1	–	–
6	6	4	2	–	–	–
7	5	3	1	–	–	–
8	4	2	–	–	–	–
9	3	1	–	–	–	–
10	2	–	–	–	–	–

With regards to the used learner, we are using *Random Forests*, which have been shown to be among the state of the art machine learners available [2]. As our data set is highly imbalanced, we are using on top of the *Random Forests*, cost sensitive classification, using the *MetaCost* method [3]. In practice, every classifier may be transformed into a cost-sensitive classifier. In our case, we use in every training set, a series of pre-constructed cost matrices to be given as input into the *MetaCost* approach. The used cost matrices start with the two class assignment $[0.0\ 1.0;\ 5.0\ 0.0]$ of costs, e.g., the penalty for for false negatives is twice as double as for false positives in order to handle the imbalance of the data sets, and increase the penalty of false negatives with a step of 5.0 units until $[0.0\ 1.0;\ 30.0\ 0.0]$. With 10-fold cross validation on the training, we identify the best cost matrix set up, and use this for the testing, e.g., we tune the cost matrix values on the training set. Eventually, for each parameter *Pr*, and for each tree, there may be a different cost matrix used.

4   Experimental Evaluation

In this section we present the results of our experimental evaluation. We describe in detail the data sets used and the setup of our experimental evaluation, and we discuss analytically the importance of the findings.

4.1   Data Sets and Overall Experimental Setup

For our evaluation we are using all the *MeSH* releases from 1999 until 2011 (inclusive). As our corpus, we are using all of the *PubMed* indexed articles that were indexed until 31/12/2011, which number approximately 22 million scientific publications. The titles, abstracts, years, *MeSH* major and minor annotations of the articles are indexed locally in a *Lucene* index[12]. The versions of the *MeSH* hierarchy are indexed separately, also in

---

[12] http://lucene.apache.org/core/

Table 5. Experimental results for all *Tr* and *Pr* parameters for *MeSH* Tree B (*Organisms*)

Tr	PR= 1($\Delta = 0$)			PR= 2($\Delta = 1$)			PR= 3($\Delta = 2$)			PR= 4($\Delta = 3$)			PR= 5($\Delta = 4$)			PR= 6($\Delta = 5$)		
-	P	R	F	P	R	F	P	R	F	P	R	F	P	R	F	P	R	F
1	16.7	0.5	1.0	11.5	29.1	16.5	38.1	27.1	31.7	14.9	48.8	22.8	11.9	62.3	19.9	**12.1**	**62.8**	**20.3**
2	4.8	12.0	5.6	10.8	37.6	16.8	20.8	37.6	26.7	32.9	33.1	33.0	12.9	65.0	21.5	-	-	-
3	3.2	13.5	4.7	8.7	41.0	14.3	30.8	30.7	30.7	36.7	33.6	35.1	**16.4**	**35.0**	**22.3**	-	-	-
4	6.5	13.2	7.2	12.9	30.2	18.1	29.1	28.6	28.8	**43.5**	**40.4**	**41.9**	-	-	-	-	-	-
5	11.4	10.3	9.3	13.1	27.7	17.7	28.1	33.5	30.6	**64.9**	**53.6**	**58.7**	-	-	-	-	-	-
6	8.0	9.8	8.5	20.1	30.6	24.3	**59.3**	**40.0**	**47.8**	-	-	-	-	-	-	-	-	-
7	12.5	5.6	7.7	16.1	20.7	18.1	**88.2**	**50.0**	**63.8**	-	-	-	-	-	-	-	-	-
8	14.3	4.8	7.1	**35.3**	**17.9**	**23.8**	-	-	-	-	-	-	-	-	-	-	-	-
9	35.0	5.4	9.0	**28.4**	**28.1**	**28.3**	-	-	-	-	-	-	-	-	-	-	-	-
10	**42.9**	**14.3**	**21.4**	-	-	-	-	-	-	-	-	-	-	-	-	-	-	-

a *Lucene* index. Table 4 shows the experimental setup regarding the parameters *Tr* and *Pr*. More precisely, in each cell we show the number of test years we can evaluate the classifiers, if we use *Tr* years of *MeSH* as training, and aim to capture *MeSH* expansions in the following *Pr* years. In parenthesis, the respective time gap between training and testing sets is also presented (Δ), which is computed as described in Section 3.1. All the features are computed as described in Table 3, and the resulting instances are stored per year and per tree separately in *Weka*[13] file format, which is the data mining tool we use to perform our evaluation. Regarding the learners used, as discussed, we are using the cost sensitive classification approach on-top of *Random Forests*, which constitute our base learners. In practice, for each *Pr* value, as shown in Table 4, we generate respective instance files per year and per tree, e.g., for *PR*= 1 we have 12 *Weka* files for each of the *MeSH* trees *B, C,* and *D* (total of 36 files), which we then use according to the setup of Table 4 to conduct the evaluation[14]. The training files always start from year 1999.

4.2   Evaluation and Analysis of Results

Tables 5, 6, and 7 show respectively the results of the testing for the *MeSH* tress *B, C,* and *D,* following the experimental set-up shown in Table 4. The tables present the *micro-averaged Precision* (*P*), *Recall* (*P*) and *F-Measure* (*F*) for the positive class ($C = 1$)[15]. *Micro-averaging* is more appropriate in our case compared to *Macro-averaging* due to the large discrepancies among the number of test instances present in each year. The best results for each Pr value are reported in bold.

A first immediate conclusion from the reported results is that using more years for training (higher *Tr*), is definitely beneficial for the construction of the classifiers and always produced the best results. The reported results also show that the suggested methodology is able to predict with precision reaching up to 88.9% and respective recall

---

[13] http://www.cs.waikato.ac.nz/ml/weka/

[14] The files corresponding to *PR*= 3 and *PR*= 4 are publicly available at the following URL: http://www.CICLing.org/2013/data/180

[15] In the majority of the cases, the results for the negative class obtained almost always an *F-Measure* larger than 90%.

Table 6. Experimental results for all *Tr* and *Pr* parameters for *MeSH* Tree C (*Diseases*)

Tr	PR= 1($\Delta = 0$)			PR= 2($\Delta = 1$)			PR= 3($\Delta = 2$)			PR= 4($\Delta = 3$)			PR= 5($\Delta = 4$)			PR= 6($\Delta = 5$)		
-	P	R	F	P	R	F	P	R	F	P	R	F	P	R	F	P	R	F
1	16.4	1.2	2.3	37.8	7.6	12.7	40.4	16.3	23.2	44.4	30.2	35.9	36.3	22.9	28.1	37	36.9	36.9
2	31.3	1.4	2.8	39.7	6.2	10.7	45.6	16.5	24.2	45.1	36.4	40.3	41.3	30.3	35	-	-	-
3	26.5	2	3.8	49.1	7.8	13.5	45.4	21.7	29.4	46.2	48.5	47.3	40.4	39.8	40.1	-	-	-
4	36	1.5	2.9	34.7	6.9	11.5	42.7	28.3	34	65.8	43.8	52.6	-	-	-	-	-	-
5	25	1.9	3.5	38.3	12.8	19.3	51.1	34.2	41	79.6	57	66.4	-	-	-	-	-	-
6	26.3	3.2	5.7	35.7	12.9	18.9	67.7	34.7	45.9	-	-	-	-	-	-	-	-	-
7	32.4	6.1	10.2	40.5	16.7	23.7	81.9	45.4	58.4	-	-	-	-	-	-	-	-	-
8	36.1	4.1	7.3	48.4	23.7	31.8	-	-	-	-	-	-	-	-	-	-	-	-
9	37.5	5.8	10.1	53.1	41	46.2	-	-	-	-	-	-	-	-	-	-	-	-
10	47.4	9.3	10.9	-	-	-	-	-	-	-	-	-	-	-	-	-	-	-

of 50% (*F-Measure*= 63.8%) the headings that are going to be expanded in the Tree B (in this case) of the *MeSH* hierarchy in maximum three years from the testing year. Overall, for all three trees the results show that the prediction of expansions using *Pr* in the range [2, 4] is possible with satisfactory results, if sufficient number of training *MeSH* years is used. In all other cases, the predictions are poor, e.g., predicting if in the immediate year a *MeSH* heading will be expanded. Interpreting the results, the top *F-Measures*, e.g., in the range 59% − 63%, and the respective precision and recall scores, mean that our methodology may suggest a list of typically few *headings* for the respective cases, in which more than 80% of the listed *headings* will be expanded in the predicted period, and in which almost half of all of the *MeSH* headings that will be expanded in that period may be found. As far as the difficulty of each tree is concerned, results show that the suggested methodology has higher success with the *Diseases* tree (*F-Measure* 66.4% for *Pr*= 4). Finally, with regards to the importance of features, we conducted 10-fold cross validation on almost all of the used training sets, and analyzed the importance of features using the information gain score. The top 5 features proved to be: *temporal siblings, temporal all children, temporal direct children, annRatioAll*, and, *all children results*, which shows that the notion of temporal features aids significantly the prediction, and also that the use of the offered *PubMed* annotations, and the wider use of the *PubMed* corpus is extremely beneficial.

Analyzing the reported results from a broader view, we have shown that under conditions it is possible for the suggested methodology to predict the *MeSH* regions that will be expanded with a relatively high *Precision*, if sufficient number of training years is provided, and a lengthier prediction span is given as a parameter. This work alone may constitute a first step for automated ontology evolution, provided that it is augmented with a second step which may also suggest specific new terms to be added below the *MeSH* headings that are predicted as positive, i.e., that should be expanded in the next few years. In this direction, a possible expansion of the approach is to apply the extraction of temporal text rules [10,11], annotate the rules with existing biomedical ontology concepts (e.g., *UMLS*), and analyze the rules the contain in the *antecendents* the *MeSH* headings that should be expanded. In these later rules, we expect that the

Table 7. Experimental results for all *Tr* and *Pr* parameters for *MeSH* Tree D (*Chemicals and Drugs*)

Tr	*PR*= 1($\Delta = 0$)			*PR*= 2($\Delta = 1$)			*PR*= 3($\Delta = 2$)			*PR*= 4($\Delta = 3$)			*PR*= 5($\Delta = 4$)			*PR*= 6($\Delta = 5$)		
-	*P*	*R*	*F*	*P*	*R*	*F*	*P*	*R*	*F*	*P*	*R*	*F*	*P*	*R*	*F*	*P*	*R*	*F*
1	54.5	5.2	9.5	46.4	11.8	18.8	36.5	26.8	30.9	44.1	39.1	41.4	32.4	35.6	33.9	**30.4**	**34.1**	**32.1**
2	53.9	5.8	10.5	48.2	12.6	20	33.5	33.9	33.7	50.1	36.7	42.3	27	38.4	31.7	-	-	-
3	33.3	5.9	10	36.7	15.5	21.8	47.1	28.4	35.5	59	38.5	46.6	**41.4**	**30.9**	**35.4**	-	-	-
4	51.3	8.5	14.5	43.8	17.9	25.5	39	37.8	38.4	52.9	45.9	49.2	-	-	-	-	-	-
5	22.3	17.3	19.5	37.4	21.7	27.5	53.8	32.7	40.7	**63.2**	**56.5**	**59.7**	-	-	-	-	-	-
6	32	**15.7**	**21.1**	41.6	25.9	31.9	49.1	36	41.6	-	-	-	-	-	-	-	-	-
7	23.9	16.3	19.4	30.5	31.5	31	**64.5**	**51.7**	**57.4**	-	-	-	-	-	-	-	-	-
8	31.2	14.1	19.4	57.7	22	31.8	-	-	-	-	-	-	-	-	-	-	-	-
9	53.6	8.3	14.4	**75.3**	**41.7**	**53.6**	-	-	-	-	-	-	-	-	-	-	-	-
10	34.8	11	16.6	-	-	-	-	-	-	-	-	-	-	-	-	-	-	-

consequents may contain important terms which could be added under the *MeSH* heading that requires expansion.

5   Conclusions and Future Work

In this paper we presented a novel methodology for constructing temporal classifiers in order to predict the evolution of the *Medical Subject Headings* hierarchy. We engineered a set of of features that utilize the *MeSH structure*, *PubMed citations*, *PubMed results*, combinations of the aforementioned (*hybrid*), and *temporal* changes of the feature values, and applied temporal classification to make the predictions. Our results show that predicting the *MeSH* headings to be expanded is feasible, if a prediction window of at least 2 years is used and sufficient *MeSH* versions of previous years are employed for training. To the best of our knowledge, this is the first approach in the bibliography to address the issue of predicting *MesH* evolution, and we hope that our results motivate future work towards the use of *temporal cost-sensitive* classifiers for predicting ontology evolution in the biomedical domain for other ontologies as well.

References

1. Ashburner, M., Ball, C.A., Blake, J.A., Botstein, D., Butler, H., Cherry, J.M., Davis, A.P., Dolinski, K., Dwight, S.S., Eppig, J.T., Harris, M.A., Hill, D.P., Issel-Tarver, L., Kasarskis, A., Lewis, S., Matese, J.C., Richardson, J.E., Ringwald, M., Rubin, G.M., Sherlock, G.: Gene Ontology: tool for the unification of biology. Nature Genetics 25, 25–29 (2000)
2. Breiman, L.: Random forests. Machine Learning 45(1), 5–32 (2001)
3. Domingos, P.: Metacost: A general method for making classifiers cost-sensitive. In: KDD, pp. 155–164 (1999)
4. Doms, A., Schroeder, M.: GoPubMed: exploring PubMed with the Gene Ontology. Nucleic Acids Research 33, 783–786 (2005)
5. Fabian, G., Wächter, T., Schroeder, M.: Extending ontologies by finding siblings using set expansion techniques. Bioinformatics 28(12), 292–300 (2012)

6. Howe, D., Costanzo, M., Fey, P., Gojobori, T., Hannick, L., Hide, W., Hill, D.P., Kania, R., Schaeffer, M., Pierre, S.S., Twigger, S., White, O., Rhee, S.Y., Rhee, S.Y.: Big data: The future of biocuration. Nature, 47–50 (2008)
7. Kanhabua, N., Nørvåg, K.: Improving Temporal Language Models for Determining Time of Non-timestamped Documents. In: Christensen-Dalsgaard, B., Castelli, D., Ammitzbøll Jurik, B., Lippincott, J. (eds.) ECDL 2008. LNCS, vol. 5173, pp. 358–370. Springer, Heidelberg (2008)
8. Leenheer, P.D., Mens, T.: Ontology evolution. In: Ontology Management, pp. 131–176 (2008)
9. Liu, K., Hogan, W.R., Crowley, R.S.: Natural language processing methods and systems for biomedical ontology learning. Journal of Biomedical Informatics 44(1), 163–179 (2011)
10. Neumayer, R., Tsatsaronis, G., Nørvåg, K.: TRUMIT: A Tool to Support Large-Scale Mining of Text Association Rules. In: Gunopulos, D., Hofmann, T., Malerba, D., Vazirgiannis, M. (eds.) ECML PKDD 2011, Part III. LNCS (LNAI), vol. 6913, pp. 646–649. Springer, Heidelberg (2011)
11. Nørvåg, K., Eriksen, T.Ø., Skogstad, K.-I.: Mining Association Rules in Temporal Document Collections. In: Esposito, F., Raś, Z.W., Malerba, D., Semeraro, G. (eds.) ISMIS 2006. LNCS (LNAI), vol. 4203, pp. 745–754. Springer, Heidelberg (2006)
12. Pesquita, C., Couto, F.M.: Predicting the extension of biomedical ontologies. PLoS Computational Biology 8(9) (2012)
13. Tsatsaronis, G., Macari, N., Torge, S., Dietze, H., Schroeder, M.: A maximum-entropy approach for accurate document annotation in the biomedical domain. BMC Journal of Biomedical Semantics 3(suppl. 1), S2 (2012)
14. Tsuruoka, Y., Ichi Tsujii, J., Ananiadou, S.: FACTA: a text search engine for finding associated biomedical concepts. Bioinformatics 24(21), 2559–2560 (2008)
15. Wächter, T., Fabian, G., Schroeder, M.: DOG4DAG: semi-automated ontology generation in obo-edit and protégé. In: SWAT4LS, pp. 119–120 (2011)
16. Whetzel, P.L., Parkinson, H.E., Causton, H.C., Fan, L., Fostel, J., Fragoso, G., Game, L., Heiskanen, M., Morrison, N., Rocca-Serra, P., Sansone, S.-A., Taylor, C.J., White, J., Stoeckert Jr., C.J.: The MGED Ontology: a resource for semantics-based description of microarray experiments. Bioinformatics 22(7), 866–873 (2006)

Knowledge Discovery
on Incompatibility of Medical Concepts

Adam Grycner, Patrick Ernst, Amy Siu, and Gerhard Weikum

Max-Planck Institute for Informatics, Saarbrücken, Germany
{agrycner,pernst,siu,weikum}@mpi-inf.mpg.de

Abstract. This work proposes a method for automatically discovering incompatible medical concepts in text corpora. The approach is distantly supervised based on a seed set of incompatible concept pairs like symptoms or conditions that rule each other out. Two concepts are considered incompatible if their definitions match a template, and contain an antonym pair derived from WordNet, VerbOcean, or a hand-crafted lexicon. Our method creates templates from dependency parse trees of definitional texts, using seed pairs. The templates are applied to a text corpus, and the resulting candidate pairs are categorized and ranked by statistical measures. Since experiments show that the results face semantic ambiguity problems, we further cluster the results into different categories. We applied this approach to the concepts in Unified Medical Language System, Human Phenotype Ontology, and Mammalian Phenotype Ontology. Out of 77,496 definitions, 1,958 concept pairs were detected as incompatible with an average precision of 0.80.

1 Introduction

1.1 Motivation

In recent years, the biomedical research community developed many ontologies and knowledge bases. They contain information about diverse concepts, such as diseases, symptoms, and medications. The goal is to structure knowledge in order to facilitate machine processing. Most ontologies form extensive class hierarchies. However, incompatibility relationships are very sparse. If one could enrich ontologies with such knowledge, one would enable applications such as consistency checking and automatic evaluation of ontologies [1]. In the literature, there are multiple possible definitions of incompatibility [1,2], ranging from crisp ontological disjointness to fuzzy contradictions in natural language. In this work, concepts are defined as incompatible based on the information found in the concepts' dictionary definitions. As dictionary we use the Unified Medical Language System (UMLS) [3], which is one of the largest collections of entities in the biomedical domain. Additionally, we use two phenotype ontologies, the Human Phenotype Ontology [4] and the Mammalian Phenotype Ontology [5], since they are rich in definitions. We conclude that two concepts are incompatible if their definitions contradict. For determining if two definitions are contradictory,

A. Gelbukh (Ed.): CICLing 2013, Part I, LNCS 7816, pp. 114–125, 2013.

we adopt the definition provided in [2] – "two pieces of text are contradictory if they are extremely unlikely to be considered true simultaneously." For example, the concepts *hypertension* and *hypotension* are incompatible because their definitions, "high blood pressure" and "low blood pressure" respectively, contradict. Later in this paper, we show that we can distinguish between several categories of incompatibility.

1.2 Contribution

The manual curation of ontologies is a labor-intensive and time-consuming process. However, Biomedical Natural Language Processing can expedite this process. One approach is to exploit the rich amount of knowledge buried in text sources. This work presents an automatic, distantly supervised method that extracts incompatible pairs of concepts based on their definitions with high-precision. Additionally, the method clusters incompatible pairs according to their categories and creates rankings representing a measure of reliability.

Specifically, we make the following contributions:

1. Candidate selection for incompatible pairs
   We present an approach for ranking medical definitions based on inverse document frequencies. As a result, we reduce the search space for incompatibility detection.
2. Incompatibility detection
   We present a linguistic machinery to decide whether two arbitrary concepts are incompatible. Techniques are introduced which rely on dependency parsing, regular expressions on sentence graphs, and antonym lexica such as WordNet [6], VerbOcean [7] and a hand-crafted biomedical antonym lexicon.
3. Category clustering
   We propose a clustering approach that refines the results by ranking and assigning incompatibility categories.

2 Related Work

This work considers incompatibility as textual contradiction between medical definitional texts. Such definitions are very short pieces of biomedical texts, and finding contradictions amongst them is a special case of performing the same task over longer texts in the general domain. The works presented in [2,8] addresses the phenomenon of contradiction in this general setting. They explain situations in which a contradiction may occur, and what the indicating factors are. To extract contradictions, then, they propose a system based on sentence alignment and a suite of linguistic features such as antonymy, modality, and polarity.

This work is also closely related to the task of finding antonyms in text. Two other works are particularly of interest, as they both utilize pattern-based approaches. [9] proposes to find antonyms and other information via textual patterns. Using a supervised Machine Learning technique, related words are searched for and then divided into categories – antonyms, synonyms, analogies, and associations. The second pattern-based approach, proposed in [10], places much emphasis on antonyms. Here, instead of using textual patterns, dependency patterns are compared to the structure of a sentence in the form of its dependency graph. For example, in the following pattern:

$$ANT1 : conj \leftarrow neither\ nor \rightarrow conj : ANT2$$

$conj$ describes the dependency type. Apart from leveraging patterns, the approach of [11] leverages the structure of a thesaurus to group contrasting categories of words. Antonymy checking is performed depending on the categories to which words belong. To further estimate the strength of opposition in constrasting words, distributional and co-occurrence statistics are applied.

The approach presented in [12] uses, like this work, textual definitions as the corpus. Using a modified Vector Space Model, every word is described by a context vector created from the word's definition. Weights in a context vector are adjusted according to the sense of the word. The closer two context vectors are to each other, the more antonymous the two corresponding words are.

To our knowledge, detecting incompatibility is not a well explored area in the biomedical domain. There are only a few works investigating this problem. One such work [13] presents a system for detecting contradictory protein-protein interactions in text. Each protein-protein interaction has a context formed by a combination of semantic components such as interaction polarity, manner polarity, and verb direction. This context in turn determines whether certain interactions contradict.

Besides leveraging text to extract incompatible concepts, [1] leverages information captured by ontologies. Their approach automatically enriches learned or manually engineered ontologies with disjointness axioms. Relying on statistics-based lexico-syntatical and ontological features, a confidence value is calculated to indicate if two concepts are disjoint.

3   Data Preparation

Our method operates on definitions of medical concepts. We assemble a corpus of 77,496 definitions using those available in UMLS, the Human Phenotype Ontology, and the Mammalian Phenotype Ontology. We apply to this corpus various linguistic analyses, which will become ingredients to the method later. Each definition is processed with the Stanford CoreNLP toolkit[1] to produce a typed dependency parse tree. In addition, each definition is processed with MetaMap [14], an entity recognition tool specifically developed for the biomedical domain.

Since the basis of this work is a distantly supervised approach, positive examples, namely pairs of incompatible concepts, are required. We therefore manually select from the corpus 200 such pairs as *seed pairs*.

---

[1] http://nlp.stanford.edu/software/corenlp.shtml

Another critical ingredient is antonyms, derived from three sources. First, we use JWNL[2] to access WordNet. When a word $\mathcal{A}$ or any of its synonyms are in an antonym relation with another word $\mathcal{B}$ or any of its synonyms, we say $\mathcal{A}$ and $\mathcal{B}$ form an antonym pair. Second, 1,973 antonym pairs are extracted from VerbOcean. Based on our observations over the aforementioned corpus, however, common medical antonyms such as "protrusion" and "retrusion," and "acidosis" and "alkalosis" are not found in WordNet and VerbOcean. Therefore, we augment the antonyms collection with 122 such hand-crafted pairs.

4   Method

Our method is a three-step process, as shown in Figure 1. Starting from a corpus of medical definitions, we end up with pairs of incompatible concepts, clustered by categories and ranked by reliability. We elaborate each step with a subsection below.

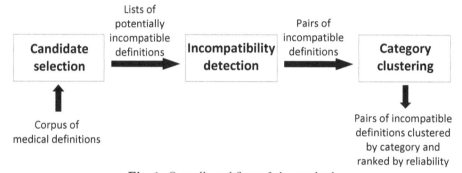

Fig. 1. Overall workflow of the method

4.1   Candidate Selection

We observe that incompatible concepts very frequently have similar definitions, in that they share common biomedical entities. With this insight, it is inefficient and unnecessary to consider every pair of definitions. Therefore, as the first step, we reduce the search space by finding and ranking similar definitions. Preliminary experiments showed that incompatible concepts are found within the top ranked definitions.

In order to measure similarity between two definitions, we introduce a cosine similarity metric based on recognized entities and their inverse document frequency (idf) scores. Consider a running example illustrated in Figure 2, starting with the definition of *Rhabdomyosarcoma*. This definition contains five biomedical entities. Within the corpus, 6,278 other definitions contain at least one of these five entities. However, notice that the entities *tumor* and *muscle* have low idf scores. This implies that these two "stop" entities are highly prevalent in

---

[2] http://jwordnet.sourceforge.net/

Concept: Rhabdomyosarcoma
Definition: It is a malignant tumor derived from skeletal (striated) muscle

Entity	malignant tumor	malignant	tumor	derived	skeletal system	striated	muscle
CUI	C0006826	C0205282	C0027651	C1441547	C0037253	C0205364	C0026845
idf	4.25	5.01	3.92	5.80	6.09	7.98	4.62

Concept: Rhabdomyoma
Definition: It is a benign tumor of striated muscle

Entity	benign tumor	benign	tumor	striated	muscle
CUI	C0086692	C0205183	C0027651	C0205364	C0026845
idf	6.64	5.40	3.92	7.98	4.62

(a)

Vector representing *Rhabdomyosarcoma*, $\vec{x} = [\,0 \quad 0 \quad 4.25 \quad 5.01 \quad 3.92 \quad 5.80 \quad 6.09 \quad 7.98 \quad 4.62\,]$
Vector representing *Rhabdomyoma*, $\vec{y} = [\,6.64 \quad 5.40 \quad 0 \quad 0 \quad 3.92 \quad 0 \quad 0 \quad 7.98 \quad 4.62\,]$

$$\cos\theta = \frac{\vec{x} \cdot \vec{y}}{\|\vec{x}\|\,\|\vec{y}\|} = \frac{3.92^2 + 7.98^2 + 4.62^2}{13.18 * 14.64} \simeq 0.5204$$

(b)

Fig. 2. In (a), the entities and their CUIs found in two medical definitions are tabulated. Three entities occur in both definitions. In (b), each definition's *idf* scores form a vector. The cosine similarity between the two vectors, 0.5204, is the similarity score between the two definitions.

the corpus, and that they lead to too many definitions, similar or otherwise. We drop stop entities below a threshold *idf* score of 5, as determined by preliminary experiments. Now with the remaining three entities, only 1,240 other definitions contain at least one of these three entities.

Next, between the starting definition of *Rhabdomyosarcoma*, and each of the 1,240 definitions, calculate their cosine similarity scores as follows. Formulate the *idf* values of a definition's entities as a vector. Apply the standard cosine similarity function to pairs of vectors. In the running example, *Rhabdomyosarcoma* and *Rhabdomyoma* thus yield a similarity score of 0.5204. Sorting the 1,240 definitions by their similarity scores in descending order, we arrive at a ranked list. Table 1 shows the top definitions thus ranked for *Rhabdomyosarcoma*. Notice that two concepts, *Rhabdomyoma* and *Benign skeletal muscle neoplasm*, which will eventually be determined to be incompatible to *Rhabdomyosarcoma*, are ranked high in this list.

Table 1. The top 20 medical concepts whose definitions have the highest similarity scores when compared to *Rhabdomyosarcoma*'s definition. Notice that the concepts incompatible to *Rhabdomyosarcoma* (marked in bold) are amongst the highest ranked.

Similarity	Concept
0.5204	**Rhabdomyoma**
0.4086	Extracardiac rhabdomyoma
0.3503	Neoplasm of skeletal muscle
0.3384	Malignant skeletal muscle neoplasm
0.3340	Embryonal rhabdomyosarcoma of prostate
0.3263	Adult rhabdomyosarcoma
0.3256	Rhabdomyosarcoma of liver
0.3251	Embryonal rhabdomyosarcoma of orbit
0.3247	Breast rhabdomyosarcoma
0.3227	**Benign skeletal muscle neoplasm**
0.3204	Rhabdomyosarcoma of prostate
0.3198	Anal rhabdomyosarcoma
0.3189	Rectal rhabdomyosarcoma
0.3178	Round cell liposarcoma
0.3176	Gallbladder rhabdomyosarcoma
0.3176	Rhabdomyosarcoma of mediastinum
0.3163	Rhabdomyosarcoma of the orbit
0.3118	Rhabdomyosarcoma of testis
0.2978	Odontoma
0.2914	Bile duct rhabdomyosarcomas

4.2   Incompatibility Detection

So far, we have reduced the search space, and have arrived at rankings of concepts with similar definitions. The next step is to detect, for a given concept, if there are any incompatible concepts with its corresponding ranking. We perform this incompatibility detection via *templates*, the core of our method. Intuitively, a template is a graphical regular expression operating over typed dependency parse trees. We claim that two concepts are incompatible if their corresponding definitions match a template. Specifically, we start with a seed pair, and locate the antonyms in their definitions. Mapping these antonyms to the definitions' typed dependency parse trees, mark the corresponding nodes as *anchors*, and take the common subtree attached to these anchors. Further, we transform this subtree by replacing nodes with placeholders. The edges remain unchanged. They represent the dependency types and hence the template's semantics. This approach generates 67 templates from 200 seed pairs.

Part (a) of Figure 3 shows a sample template, where each rectangle represents a placeholder. Parts (b) and (c) illustrate how the template is applied to the pair of definitions in our running example. Using the Semgrex tool from the Stanford CoreNLP toolkit, typed dependency parse trees of *Rhabdomyosarcoma* and *Rhabdomyoma* are compared to the template. After verifying that there is a match between the template and the trees, we attempt to fill in all the

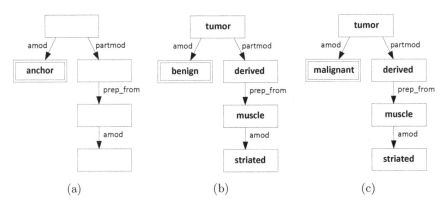

Fig. 3. Part (a) depicts a sample template, derived from the common subtree in a seed pair's typed dependency parse trees. Each rectangle represents a placeholder. Applying this template to the definitions of *Rhabdomyosarcoma* and *Rhabdomyoma* yields (b) and (c) respectively. The anchor placeholders (in double border) contain an antonym pair. The rest of the placeholders contain exact word matches.

placeholders. The anchor placeholders must contain antonyms. All other placeholders must either share the same words, lemma, or part-of-speech tag, or be synonyms. With all placeholders successfully filled in, *Rhabdomyosarcoma* and *Rhabdomyoma* are declared incompatible.

4.3   Category Clustering

As mentioned in the Introduction, there is no clear-cut, authoritative interpretation of incompatibility. In this work, we propose the following five categories characterizing incompatibility: biomedical type, level, spatial, temporal, and miscellaneous. These categories are not only salient to the biomedical domain, but are also amenable to applications such as spatio-anatomical reasoning.

In order to cluster the incompatible concept pairs, we start by assigning antonym pairs to these five categories. The antonyms occurring within the filled-in templates determine the cluster to which the concept pair belongs. Where an antonym pair appears in two categories, we perform further disambiguation by leveraging context. For instance, "higher" and "lower" appear in both level and spatial categories. In the phrase "lower chamber of heart," typed dependencies state that the *heart* anatomical entity is modified by "lower." Whenever an anatomical entity is modified by a word belonging to two categories, the spatial category is preferred. "Higher concentration of calcium" does not satisfy this condition, and hence falls into the level category.

Within a cluster, we further sort the concept pairs in order to estimate the reliability of the pairing. We use the following features in descending order of significance for this sorting: number of nodes and edges in the most specific matching template; cosine similarity of detected entities and words in the definitions; size of the largest common subtree in the typed dependency parse trees; number of words in the definitions; and number of matching templates.

5   Evaluation

5.1   User Study

Although comparing our results to a gold standard would conform to best practices, to the best of our knowledge there is no such gold standard for pairs of incompatible medical concept. Therefore, we decided to manually evaluate a subset of the concept pairs extracted by the aforementioned method. This subset consisted of 20 randomly selected concept pairs from every incompatibility category; in total, we evaluated 100 pairs for their correctness.

Eight annotators assessed if a pair is incompatible or not for a given category. Among the annotators, five have medical experience and three work in biomedical informatics. To facilitate the assessment, we generated natural language statements out of the incompatible pairs. The templates of these statements differ according to their categories, as follows:

Biomedical: Something cannot be both $\mathcal{X}$ and $\mathcal{Y}$.
Level: $\mathcal{X}$ and $\mathcal{Y}$ cannot occur in the same body at the same time.
Miscellaneous: $\mathcal{X}$ and $\mathcal{Y}$ are contradictory conditions.
Spatial: $\mathcal{X}$ and $\mathcal{Y}$ cannot occur in the same location.
Temporal: $\mathcal{X}$ and $\mathcal{Y}$ cannot occur at the same time.

where $\mathcal{X}$ and $\mathcal{Y}$ were replaced with a particular concept pair. For example, for the Miscellaneous category, we generated the statement "*increased coping response* and *decreased coping response* are contradictory conditions." Another example for the Temporal category reads, "*increased tumor latency* and *decreased tumor latency* cannot occur at the same time." Additionally, we provide the concepts' definitions, since it may be very difficult to rate if two concepts are incompatible based solely on their names. Every annotator had to assess 40 statements, and thus not every statement was assessed by every annotator. Each annotator assigned one of three labels: "Correct," "Incorrect," and "No answer" if the annotator could not understand a definition or was unsure about the assessment. On average, we received 2.64 "Correct" or "Incorrect" assessments per statement. To present the questions, we generated web surveys using LimeSurvey[3], an open source online survey application. Figure 4 shows a sample item from the surveys.

5.2   Results

Using the data and method detailed in Section 4, we extracted $1,958$ incompatible concept pairs out of $77,496$ definitions in either UMLS, the Human Phenotype Ontology, or the Mammalian Phenotype Ontology, with an overall precision of 0.80.

Table 2 shows the distributions of incompatible pairs for the different categories as well as the corresponding precision values. Table 3 shows some

---

[3] http://www.limesurvey.org/

non-secretory adrenocortical adenoma
and
secretory adrenocortical adenoma
are contradictory conditions.
Choose one of the following answers

⦿ Correct

○ Incorrect

○ No answer

non-secretory adrenocortical adenoma
It is an hormonally inactive adrenocortical adenoma , that is , an adenoma that does not secrete excessive amounts of adrenal hormones .

secretory adrenocortical adenoma
It is an hormonally active adrenocortical adenoma , that is , an adenoma that secretes excessive amounts of adrenal hormones .

Fig. 4. Sample item from the evaluation surveys. An item is structured as follows. The upper part displays the generated natural language statement involving two incompatible concepts. The middle part displays the label choices. The lower part displays the definitions of the two concepts.

exemplary results for each category. Besides precision, we also computed an inter-annotator agreement value. As for the metric, we chose Krippendorff's alpha, a coefficient based on disagreement [15]. Recall that, in the user study, there were more than two annotators, and not every statement was assessed by every annotator. Krippendorff's alpha takes these factors into account, and is therefore particularly appropriate for our evaluation setting. The inner-annotator agreement was thus 0.38. This low value was caused by the dominant number of assessments labelled as "Correct," leading to a low expected disagreement. This skews the Krippendorff's alpha calculation.

Table 2. Result overview. For every incompatibility category, the table shows the number of extracted incompatible pairs and the precision.

Category	Pairs	Precision
Biomedical	272	0.78
Level	568	0.89
Miscellaneous	613	0.75
Spatial	457	0.92
Temporal	54	0.67
Overall	1,958	0.80

As can be seen from Table 2, Level and Spatial categories achieved the best precisions. For the Level category, the biggest contributing factor was that most of the concept pairs involve an increase or decrease of a specific substance. This information is reflected in the structures of definitions, which in turn is matched reliably by the templates. As for the Spatial category, spatial or locational information often occurs in the same textual patterns. For instance, a spatial qualifier frequently precedes an anatomical concept.

Table 3. Exemplary results for each category

Category	Concept 1	Concept 2
Biomedical	Lipoma of skin	Cutaneous liposarcoma
	Benign fibroma of liver	Hepatic fibrosarcoma
Level	White blood cell count increased	Leukopenia
	Increased circulating low-density lipoprotein cholesterol	Hypobetalipoproteinemia
Miscellaneous	Struma ovarii	Malignant teratoma, undifferentiated
	Genu varum	Knee joint valgus deformity
Spatial	Takayasu arteritis	Leukocytoclastic vasculitis
	Disorder of the optic nerve	Olfactory nerve diseases
Temporal	Delayed puberty	Precocious puberty
	Premature aging	Slow aging

The precision for the Temporal category is the lowest, at 0.67. The main contributing factor was that we defined "primary," "secondary," and other terms as temporal antonyms. In "delayed eruption of primary teeth" and "delayed eruption of permanent teeth," "primary" and "permanent" are indeed temporal antonyms. Unfortunately, this is not true for all occurrences of "primary." For instance, in *primary hyperparathyroidism* and *secondary hyperparathyroidism*, "primary" and "secondary" are used to distinguish two types of the same disease, and are not in a temporal relationship.

Another source of error was the presence of multiple antonyms modifying a single concept, such that the correct antonym pair could not be identified. For example, when the definition "abnormal increase or decrease in the level of lipoprotein cholesterol" is matched against "abnormal decrease in the level of lipoprotein cholesterol," our method only picks up "abnormal increase" in the first definition and compares it with "abnormal decrease" in the second definition. This problem arises because disjunction of modifiers is not considered. Moreover, when a modifier has an additional negation marker, antonym detection fails. In "large cell lung carcinoma" and "non-small cell lung carcinoma," "large" and "non-small" are incorrectly matched as antonyms.

6   Conclusion and Future Work

In this paper, we presented a distantly supervised approach for automatically finding incompatible medical concepts. To determine if two medical concepts are incompatible, our method finds contradictions between their definitions. The core of the method consists of building templates out of typed dependency parse trees, matching them to definitions, and detecting antonymy between matches. After applying our method to 77, 496 definitions in our corpus, 1, 958 pairs were determined to be incompatible. In the evaluation, we showed that the method achieved an overall precision of 0.80. In addition, we also outlined the major sources of error.

In the future, we plan to extend the antonym checking module beyond a simple dictionary-based approach, to a more elaborate one using other natural language processing techniques. We would also like to enhance the clustering module so as to improve the precision for ambiguous categories, especially the temporal and spatial ones.

References

1. Völker, J., Vrandečić, D., Sure, Y., Hotho, A.: Learning Disjointness. In: Franconi, E., Kifer, M., May, W. (eds.) ESWC 2007. LNCS, vol. 4519, pp. 175–189. Springer, Heidelberg (2007)
2. de Marneffe, M.C., Rafferty, A.N., Manning, C.D.: Finding contradictions in text. In: McKeown, K., Moore, J.D., Teufel, S., Allan, J., Furui, S. (eds.) ACL, The Association for Computer Linguistics, pp. 1039–1047 (2008)
3. Lindberg, D., Humphreys, B., McCray, A.: The unified medical language system. Methods of Information in Medicine 32(4), 281–291 (1993)
4. Robinson, P.N., Mundlos, S.: The Human Phenotype Ontology. Clinical Genetics 77(6), 525–534 (2010)
5. Smith, C., Goldsmith, C.A., Eppig, J.: The Mammalian Phenotype Ontology as a tool for annotating, analyzing and comparing phenotypic information. Genome Biology 6(1), R7+(2004)
6. Fellbaum, C.: WordNet: An Electronic Lexical Database (Language, Speech, and Communication). The MIT Press (May 1998)
7. Chklovski, T., Pantel, P.: VerbOcean: Mining the Web for Fine-Grained Semantic Verb Relations. In: Lin, D., Wu, D. (eds.) Proceedings of EMNLP, pp. 33–40 (2004)
8. Pado, S., de Marneffe, M.C., MacCartney, B., Rafferty, A.N., Yeh, E., Manning, C.D.: Deciding entailment and contradiction with stochastic and edit distance-based alignment. In: Proceedings of the TAC 2008 (2009)
9. Turney, P.D.: A uniform approach to analogies, synonyms, antonyms, and associations. In: Scott, D., Uszkoreit, H. (eds.) COLING, pp. 905–912 (2008)
10. Lobanova, A., Bouma, G., Sang, E.T.K.: Using a treebank for finding opposites. In: Dickinson, M., Mrisep, K., Passarotti, M. (eds.) Proceedings of TLT9, pp. 139–150 (2010)
11. Mohammad, S., Dorr, B.J., Hirst, G.: Computing word-pair antonymy. In: Tseng, S., Chen, T., Liu, Y. (eds.) EMNLP, pp. 982–991. ACL (2008)
12. Schwab, D., Lafourcade, M., Prince, V.: Antonymy and conceptual vectors. In: Tseng, S., Chen, T., Liu, Y. (eds.) Proceedings of COLING, pp. 1–7 (2002)

13. Sanchez-Graillet, O., Poesio, M.: Discovering contradicting protein-protein interactions in text. In: Proceedings of the Workshop on BioNLP (2007)
14. Aronson, A.R., Lang, F.M.: An overview of metamap: historical perspective and recent advances. JAMIA 17(3), 229–236 (2010)
15. Krippendorff, K.: Content Analysis: An Introduction to Its Methodology. SAGE Publications (September 1980)

Extraction of Part-Whole Relations from Turkish Corpora

Tuğba Yıldız, Savaş Yıldırım, and Banu Diri

Department of Computer Engineering, İstanbul Bilgi University
Dolapdere, 34440 İstanbul, Turkey
Department of Computer Engineering, Yıldız Technical University
Davutpaşa, 34349 İstanbul, Turkey
{tdalyan,savasy}@bilgi.edu.tr,
banu@ce.yildiz.edu.tr

Abstract. In this work, we present a model for semi-automatically extracting part-whole relations from a Turkish raw text. The model takes a list of manually prepared seeds to induce syntactic patterns and estimates their reliabilities. It then captures the variations of part-whole candidates from the corpus. To get precise meronymic relationships, the candidates are ranked and selected according to their reliability scores. We use and compare some metrics to evaluate the strength of association between a pattern and matched pairs. We conclude with a discussion of the result and show that the model presented here gives promising results for Turkish text.

Keywords: Meronym, Part-Whole, Semantic Lexicon, Semantic Similarity.

1 Introduction

The meronym has been referred to as a part-whole relation that represents the relationship between a part and its corresponding whole. The discovery of meronym relations plays an important role in many NLP applications, such as question answering, information extraction [1–3], query expansion [4] and formal ontology [5, 6].

Different types of part-whole relations have been proposed in the literature [7–9]. One of the most important and well-known taxonomies, designed by Winston [8] identified part-whole relations as falling into six types: Component-Integral, Member-Collection, Portion-Mass, Stuff-Object, Feature-Activity and Place-Area. On the other hand, the most popular and useful ontologies such as WordNet have also classified meronyms into three types: component-of (HAS-PART), member-of (HAS-MEMBER) and stuff-of (HAS-SUBSTANCE)[10].

A variety of methods have been proposed to identify part-whole relations from a text source. Some studies employed lexico-syntactic patterns for indicating part-whole relations. There have also been other approaches such as statistical, supervised, semi-supervised or WordNet corporation[1, 14, 16–18].

A. Gelbukh (Ed.): CICLing 2013, Part I, LNCS 7816, pp. 126–138, 2013.

This study is a major attempt to semi-automatically extract part-whole relations from a Turkish corpus. Other recent studies to harvest meronym relations and types of meronym relations for Turkish are based on dictionary definition (TDK) and WikiDictionary [11–13].

The rest of this paper is organized as follows: Section 2 presents and compares related works. We explain our methodology in Section 3. Implementation details are explained in Section 4. Experimental results and their evaluation are reported in Section 5.

2   Related Works

Many studies for automatically discovering part-whole relations from text have been based on Hearst's [14] pattern-based approach. Hearst developed a method to identify hyponym (is-a) relation from raw text with using lexico-syntactic patterns. Although the same technique was applied to extract meronym relations in [14], it was reported that efforts concluded without great success.

In [16], it was proposed a statistical methods in very large corpus to find parts. Using Hearst's methods, five lexical patterns and six seeds (book, building, car, hospital, plant, school) for wholes were identified. Part-whole relations extracted by using patterns were ranked according to some statistical criteria with an accuracy of 55% for the top 50 words and an accuracy of 70% for the top 20 words.

A semi-automatic method was presented in [1] for learning semantic constraints to detect part-whole relations. The method picks up pairs from WordNet and searches them on text collection: SemCor and LA Times from TREC-9. Sentences that containing pairs were extracted and manually inspected to obtain list of lexico-syntactic patterns. Training corpus was generated by manually annotating positive and negative examples. C4.5 decision tree was used as learning procedure. The model's accuracy was 83%. The extended version of this study was proposed in [17]. An average precision of 80.95% was obtained.

Hage [3] developed a method to discover part-whole relations from vocabularies and text. The method followed two main phases: learning part-whole patterns and learning wholes by applying the patterns. An average precision of 74% was achieved.

A weakly-supervised algorithm, Espresso [18] used patterns to find several semantic relations besides meronymic relations. The method automatically detected generic patterns to choose correct and incorrect ones and to filter with the reliability scoring of patterns and instances. System performance for part-of relations on TREC was 80% precision.

Another attempt at automatic extraction of part-whole relation was for a Chinese Corpus [19]. The sentence containing part-whole relations was manually picked and then annotated to get lexico-syntactic patterns. Patterns were employed on training corpus to find pairs of concepts. A set of heuristic rules were proposed to confirm part-whole relations. The model performance was evaluated with a precision of 86%.

Another important studies were proposed in [2, 20]. In [2], a set of seeds for each type of part-whole relations was defined. The minimally-supervised information extraction algorithm, Espresso [18] successfully retrieved part-whole relations from corpus. For English corpora, the precision was 80% for general seeds and 82% for structural part-of seeds. In [20], an approach extracted meronym relation from domain-specific text for product development and customer services.

3   Methodology

The pattern-based approach proposed here is implemented in two phases: Pattern identification and part-whole pair detection. Fig. 1 represents how the system is split up into its components and shows data flow among these components. The system takes a huge corpus and a set of unambiguous part-whole pairs. It then proposes a list of parts for a given whole.

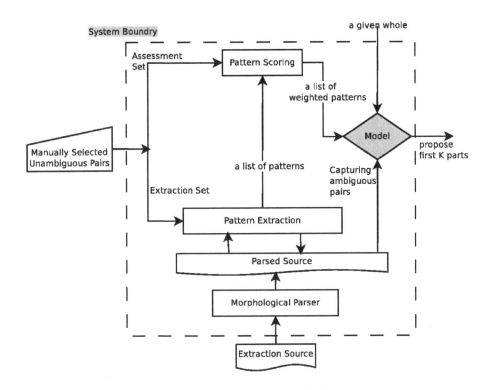

Fig. 1. High-level representation of the system

3.1   Pattern Identification

We begin by manually preparing a set of unambiguous seed pairs that convey a part-whole relation. For instance, the pair (engine, car) would be member of that set. The seed set is further divided into two subsets: an *extraction set* and an *assessment set*.

Each pair in the extraction set is used as query for retrieving sentences containing that pair. Then we generalize many lexico-syntactic expressions by replacing part and whole token with a wildcard or any meta character. The second set, the assessment set, is then used to compute the usefulness or reliability scores of all the generalized patterns. Those patterns whose reliability scores, $rel(p)$, are very low are eliminated. The remaining patterns are kept, along with their reliability scores. A classic way to estimate $rel(p)$ of an extraction pattern is to measure how it correctly identifies the parts of a given whole. The success rate is obtained by dividing the number of correctly extracted pairs by the number of all extracted pairs. The outcome of entire phase is a list of reliable lexico-syntactic expressions along with their reliability scores.

3.2   Part-Whole Pair Detection

In order to extract the pairs among which there is a part-whole relation, the previously generated patterns are applied to an extraction source that is a Turkish raw text. The instantiated instances (part-whole pairs) are assessed and ranked according to their reliability scores, where reliability score of a pair is described below.

There are several ways to compute a reliability score for both pattern and instance. In [18], the reliability score of a pattern, $rel(p)$, is proposed as shown in equation (1) and that of an instance, $rel(i)$, is formulated as in equation (2).

$$rel(p) = \frac{\sum_{p \in P} \left(\frac{pmi(i,p)}{max_{pmi}} \times rel(i) \right)}{|P|} \qquad (1)$$

$$rel(i) = \frac{\sum_{i \in I} \left(\frac{pmi(i,p)}{max_{pmi}} \times rel(p) \right)}{|I|} \qquad (2)$$

pmi is the pointwise mutual information that is one of the commonly used metrics for the strength of association between two variables, where max_{pmi} is the maximum pmi value between all pairs and all patterns and where $rel(i)$ is the reliability of instance i. Initially, all reliability scores of instances in set of unambiguous pairs are set to 1. Then, reliability score of a pattern is calculated based on these $rel(i)$ scores.

In [18], the pmi score between an instance $i(x,y)$ and pattern p was formulated as in following equation (3).

$$pmi(i,p) = log\frac{|x,p,y|}{|x,*,y||*,p,*|} \qquad (3)$$

where $|x, p, y|$ is the number of times instance $i(x, y)$ is instantiated with pattern p, $|x, *, y|$, $|*, p, *|$ are the individual distributions of instance and pattern respectively. However, the defect in the formula is that the pmi score always takes negative values. This leads a ranking the reverse of the expected. It must be multiplied by the numbers of all pairs matched by all patterns, $|*, *, *|$. Thus, we redefined the formula as shown in equation (4).

$$pmi(i, p) = log \frac{|x, p, y||*, *, *|}{|x, *, y||*, p, *|} \qquad (4)$$

A frequent pair in a particular pattern does not necessarily convey a part-whole relation. Thus, to calculate reliability of a pair, all patterns are taken into consideration as shown in equation (2).

In our research, we experiment with three different measures of association (pmi, dice, tscore) to evaluate their performance. We also utilized *inverse document frequency (idf)* to cover more specific parts. The motivation for use of *idf* is to differentiate distinctive features from other common ones.

We categorized our parts into two groups; *distinctive* and *general* parts. If a part of a given whole is inheritable from hypernyms of that whole, we call this kind of part general or inheritable. If, not, we call such part specific or distinctive part. E.g. a desk has has-part relationship with drawer and segment as in WordNet. While drawer is distinctive part of desk, segment is a general part that inherits from its hypernym *artifact*. Indeed, it is really difficult to apply this chaining approach to all nouns. Instead of using all hypernym chain, the parts come from *physical entity* or some particular hypernyms and the distinctive ones can be merged.

Thus, to distinguish the distinctiveness, we utilized *idf* that is obtained by dividing the number of times a part occurs in part position by how many pairs retrieved by all the patterns. We observed that the most frequent part instances are top, inside, segment, side, bottom, back, front and state, head etc. All of these resemble general features.

3.3   Baseline Algorithm

Pointwise mutual information and other measures can be alternatively used between part and whole rather than instance pair and pattern. To designate a baseline algorithm, for a given whole, its possible parts are retrieved from a list ranked by association measure between whole and part that are instantiated by a reliable pattern as formulated in equation (5).

$$assoc(whole, part) = \frac{|whole, pattern, part|}{|*, pattern, part||whole, pattern, *|} \qquad (5)$$

We intuitively designated a baseline algorithm to compare the results and the expectation is that a proposed model should outperform the baseline algorithm. The baseline function is based on most reliable and productive pattern, the genitive pattern. As Table 1 suggests, the rel(genitive-pattern) has the best score in accordance with average of all three measures (pmi, dice and tscore) and the capacity is about 2M part-whole pairs.

For a given whole, all parts that co-occur with that whole in the genitive pattern are extracted. Taking co-occurrence frequency between the whole and part could be misleading due to some nouns frequently placed in part/head position such as side, front, behind, outside. To overcome the problem, the co-occurrence, the individual distributions of both whole and part must be taken into account as shown in equation (5). These final scores are ranked and their first K parts are selected as the output of baseline algorithm.

4   Experimental Design

In our experiment, we used a set of natural language resources for Turkish; a huge corpus of 500M tokens and a morphological parser provided by [15]. The morphological parser based on a two-level morphology has an accuracy of %98. The web corpus contains four sub-corpora. Three of them are from major Turkish news portals and another corpus is a general sampling of web pages in the Turkish Language. The corpus is tokenized and encoded at paragraph and sentence level.

The morphological parser splits a surface token into its morphemes in system architecture as shown in Figure 1. The representation of a parsed token is in the form of **surface/root/pos/[and all other markers]**. When the genitive phrase "arabanın kapısı (door of the car)" is given, the parser split it into the parts as below.

```
English: (door of the car)
Turkish: arabanın kapısı
Parsed : araba+noun+a3sg+pnon+gen kapı+a3sg+pnon+p3sg
```

In order to identify lexical forms that express part-whole relations, we manually selected 200 seed pairs. Out of 200 pairs, 50 are used as pattern extraction set to extract the lexico-syntactic patterns and 150 are used as assessment set to compute the reliability scores of each pattern, *rel(p)*. All sentences containing *part* and corresponding *whole* token in extraction set are retrieved. Replacing part/whole token with a meta character, e.g. wildcard, we extracted many patterns.

However, due to the noisy nature of the web corpus and the difficulties of an agglutinative language, many patterns have poor extraction capacity. Turkish is a relatively free word order language with agglutinating word structures. The noun phrases can easily change their position in a sentence without changing the meaning of the sentence, and only affecting its emphasis. This is a big challenge for syntactic pattern extraction. Based on reliability scores, we decided to filter out some generated patterns and finally obtained six different significant patterns. Here is the list of the patterns, their examples and related regular expression formula:

1. **Genitive Pattern: NP+gen NP+pos**

    In Turkish, there is only one genitive form: The modifier morphologically takes a genitive case, *Gen (nHn)* and the head takes possessive agreement *pos(sH)* as shown before ("arabanın kapısı/ door of the car"). The morphological feature of genitive is a good indicator to disclose a semantic relation between a head and its modifier. In this case, we found that the genitive has a good indicative capacity, although it can encode various semantic interpretations. Taking the example, *Ali's team*, the first interpretation could be that *the team belongs to Ali*, the second interpretation is that *Ali's favorite team* or *the team he supports*. To overcome such problem, researcher have done many studies based on statistical evidence, some well-known semantic similarity measurements and semantic constraints based on world knowledge resources.

    The regular expression of genitive pattern for "'arabanın kapısı' is as follows:

    ```
    Regex   : \w+\+noun[\w\+]+gen +\w+\+noun[\w\+]+p3sg
    ```

2. **NP+nom NP+pos**

    ```
    English: (car door)
    Turkish: araba kapısı
    Parsed : araba+noun+a3sg+pnon+nom  kapı+a3sg+pnon+p3sg
    Regex  : \w+\+noun\+a3sg\+pnon\+nom \w+\+noun\+[\w\+]+p3sg
    ```

3. **NP+Gen (N—ADJ)+ NP+Pos**

    ```
    English: (back garden gate of the house)
    Turkish: Evin arka bahçe kapısı
    Parsed : Evin+ev+noun+a3sg+pnon+gen
             arka+arka+noun+a3sg+pnon+nom
             bahçe+bahçe+noun+a3sg+pnon+nom
             kapısı+kapı+noun+a3sg+p3sg+nom

    Regex: \w+\+noun[\w\+]+gen
           (\w+\+noun\+a3sg\+pnon\+nom |\w+\+adj[\w\+]+ )
           \w+\+noun[\w\+]+p3sg
    ```

4. **NP of one-of NPs**

    ```
    English: (the door of one of the houses)
    Turkish: Evlerden birinin kapısı
    Parsed : Evlerden+ev+noun+a3pl+pnon+abl
             birinin+biri+pron+quant+a3sg+p3sg+gen
             kapısı+kapı+noun+a3sg+p3sg+nom
    Regex  : \w+\+noun\+a3pl\+pnon\+abl
             birinin\+biri\+pron\+quant\+a3sg\+p3sg\+gen
             \w+\+noun\+\w+\+p3sg
    ```

5. **NP whose NP**

English: The house whose door is locked
Turkish: Kapısı kilitli olan ev
Parsed: Kapısı+kapı+noun+a3sg+p3sg+nom
         kilitli+kilit+noun+a3sg+pnon+nom-adj*with
         olan+ol+verb+pos-adj*prespart
         ev+ev+noun+a3sg+pnon+nom
Regex : \w+\+noun[\w\+]+p3sg\+\w+
         (\w+\+noun\+a3sg\+pnon\+nom|
         \w+\+noun\+a3sg\+pnon\+nom\-adj*with)
         (\w+\+verb\+pos\-adj*prespart|
         \w+\+verb\+pos\+narr\+a3sg) \w+\+noun\+a3sg

6. **NP with NPs**

English: the house with garden and pool
Turkish: bahçeli ve havuzlu ev
Parsed : bahçeli+bahçeli+adj ve+ve+conj
         havuzlu+havuz+noun+a3sg+pnon+nom-adj*with
         ev+ev+noun+a3sg+pnon+nom
Regex: \w+\+noun\+a3sg\+pnon\+nom\-adj*with \w+\+noun\+

All patterns are evaluated according to their usefulness. To assess them, output of each pattern is checked against a given assessment set. Setting instance reliability of all pairs in the set to 1, reliability score of the patterns are computed as shown before. For a assessment set size of 150 pairs, all pattern and their *rel(p)* are given in Table 1

When comparing the patterns, P1 is the most reliable pattern with respect to all measures. P1 is based on genitive case which many studies utilized it for the problem. We roughly order the pattern as P1, P2, P3, P6, P4, P5 by their normalized average scores in the Table 1.

Table 1. Reliability of Patterns

	rel(P1)	rel(P2)	rel(P3)	rel(P4)	rel(P5)	rel(P6)
pmi	1.58	1.53	0.45	0.04	0.07	0.57
dice	0.01	0.003	0.01	0.004	0.001	0.003
tscore	0.11	0.12	0.022	0.0004	0.001	0.03

To calculate reliability of instances, we utilize not only pmi measure, but also dice, t-score and idf measures. In equation(1), *rel(p)* and equation (2), *rel(i)*, association measure can be **pmi, pmi-idf, dice, dice-idf, tscore,** and **tscore-idf.** For a particular whole noun, all possible parts instantiated by patterns are selected as a candidate set. For each association measure, their *rel(p)* and *rel(i)* scores are calculated and further sorted. The first K candidate parts are checked against the expected parts.

5   Evaluation

For the evaluation phase, we manually and randomly selected five whole words: *book, computer, ship, gun* and *building*. For each whole noun, the experimental results are given in Table 2.

Table 2. The results of the scores for five wholes

whole	pmi	pmi-idf	dice	dice-idf	tscore	tscore-idf	baseline	average
gun-10	2	**4**	1	1	0	1	2	1.57
gun-20	4	**5**	3	2	1	1	4	2.86
gun-30	**6**	**6**	**6**	**6**	2	3	**6**	5
book-10	9	3	**10**	**10**	8	7	8	7.86
book-20	**18**	9	**18**	**18**	16	12	13	14.86
book-30	22	14	22	**23**	21	20	17	19.86
building-10	4	2	5	**7**	**7**	6	**7**	5.43
building-20	11	8	**15**	14	**15**	13	**15**	13
building-30	17	13	22	**23**	20	19	18	18.86
ship-10	**9**	7	**9**	**9**	6	5	**9**	7.71
ship-20	14	13	**18**	**18**	9	10	15	13.86
ship-30	18	17	**26**	24	13	14	21	19
computer-10	8	**9**	**9**	**9**	6	7	8	8
computer-20	**16**	15	13	15	8	11	10	12.57
computer-30	**21**	16	20	20	10	15	14	16.57
average prec.								
precision10	64%	50%	68%	**72%**	54%	52%	68%	61.14%
precision20	63%	50%	**67%**	**67%**	49%	47%	57%	57.14%
precision30	56%	44%	**64%**	**64%**	44%	47.3%	50.6%	52.86%

Where gun-10 means that we evaluated first 10 selection of all measures for whole gun. For a better evaluation, we selected first 10, 20 and 30 candidates ranked by the association measure defined above. The proposed parts were manually evaluated by looking at their semantic role. We needed to differentiate part-whole relations from other possible meanings. Indeed, all the proposed parts are somehow strongly associated with corresponding whole. However, our specific goal here is to discover meronymic relationship and, thus we tested our results with respect to the component-integral meronymic relationship as defined in [8] or HAS-PART in WordNet. For first 10 output, dice-idf with precision of 72% performs better than others on average. For first 20 selection, dice and dice-idf share the highest scores of 67%. For first 30 selection, dice, dice-idf with precision of 64% outperforms other measures.

Looking at the Table 2, for the first 10 selection, all measures perform well against all wholes but gun. This is simply because *gun* gives less corpus evidence to discover parts of it. With a deeper observation, we have manually captured only 9 distinctive parts and 10 general parts, whereas whole *building* has 51 parts, out of which 13 are general parts.

We conducted another experiment to distinguish distinctive parts from general ones. Excluding general parts from the expected list, we re-evaluated the result of the experiments. The results were, of course, less successful but a better fine-grained model was obtained. The result are shown in Table 3. The table showed that all idf weighted measures are better than others. For the first 30 selection, when idf is applied, pmi, dice measures are increased by 2% and tscore measure is increased by 7.3% on average as expected.

Table 3. The results for distinctive parts

precision	pmi	pmi-idf	dice	dice-idf	tscore	tscore-idf	baseline	average
precision10	50	50	58	**64**	34	44	60	51.43
precision20	48	50	48	**53**	34	40	51	46.29
precision30	40.67	42.67	47.33	**49.33**	31.33	38.67	40.67	41.52

General parts can easily be captured when running the system for *entity* or any hypernym. To do so, we checked noun *"şey"'* (thing) to cover more general parts or features. We retrieved some meaningful nouns such as *top, end, side, base, front, inside, back, out* as well as other meaningless parts.

Additionally, we can easily apply is-a relation, whereas we cannot always apply the same principle to part-whole hierarchy. For instance if tail is a meronym of cat and tiger is a hyponym of cat, by inheritance, tail must be a meronym of tiger then. However, transitivity could be limited in the part-whole relation. Handle is meronym of door, door is a meronym of house. It can incorrectly implied that the house has a handle. On the other hand, finger-hand-body hierarchy is a workable example to say that a body has a finger.

As our another result, Table 4 partly confirms our expectation that the success rate from a larger training seed set is slightly better than those from a smaller one. As we increase the seed size from 50 to 150, only pmi measure clearly improves and the other measures did now show significant improvements.

Table 4. The precision (prec) results for training set (TS) size of 50,100 and 150

#of_TS	results	pmi	pmi-idf	dice	dice-idf	tscore	tscore-idf	baseline	avg.
train50	prec10	64	52	**70**	68	52	44	68	59.71
	prec20	59	50	70	**68**	48	45	57	56.71
	prec30	51.33	43.33	62.67	**64**	42	46	50.67	51.43
train100	prec10	68	50	**72**	70	52	48	66	60.86
	prec20	63	49	**68**	66	48	44	58	56.57
	prec30	56	44.67	63.33	**64**	42.67	45.33	50.67	52.38
train150	prec10	64	50	68	**72**	54	52	68	61.14
	prec20	63	50	**67**	**67**	49	47	57	57.14
	prec30	56	44	**64**	**64**	44	47.33	50.67	52.86

The goal of the study is to retrieve meronymic relation, more specifically component integral (CI) relation. Looking at the result in Table 5, almost those candidates that are incorrectly proposed in terms of component integral relations, however, fall into other semantic relations such as property, cause etc. When evaluating the results with respect to these semantic relations SR that includes CI as well, we obtain better precision but coarser relations. The average score for first 10 selection is 68 and 90 with respect to CI and SR respectively. For first 20 and 30 selection, we can conclude that system successfully disclose semantic relation in coarser manner.

The last remark is that only dice and dice-idf significantly outperformed the baseline algorithm. Looking at Table 2 and Table 3, all other measures did not show significant performance with respect to the baseline algorithm.

Table 5. The results for Component-Integral (CI) relation and other Semantic Relations(SRs)

whole	#CI	#Other SRs
gun-10	1	5
gun-20	3	11
gun-30	6	14
building-10	5	10
building-20	15	19
building-30	22	28
computer-10	9	10
computer-20	13	16
computer-30	20	24
ship-10	9	10
ship-20	18	20
ship-30	26	28
book-10	10	10
book-20	18	18
book-30	22	25
average-10	68	90
average-20	67	84
average-30	64	79.3

6   Conclusion

In this study, we proposed a model that semi-automatically acquires meronymy relations from a Turkish raw text. The study is a major corpus driven attempt to extract part-whole relations from a Turkish corpus. The raw text has been tokenized and morphologically parsed. Some manually prepared seeds were used to induce lexico-syntactic patterns and determine their usefulness. Six reliable patterns were extracted and scored. Based on these patterns, the model captures a variety of part-whole candidates from the parsed corpus.

We conducted several experiments on a huge corpus of size of 500M tokens. All experiments indicate that proposed model has promising results for Turkish Language. According to experimental results and observation, we conclude some points as follows.

We compared the strength of some association measure with respect to their precisions. Looking at the first 10 selection of each measure, all measures are equally same in terms of precision, whereas for first 30 selection dice and dice-idf outperformed other ones with precision of 64%. Looking at the all resulting tables, we can say that only dice and dice-idf significantly outperformed the baseline algorithm. The performance of other measures were not confirmed our expectation.

For distinctive part retrieval rather than general parts, we expected that idf weighted measures are significantly better than others. For the first 30 selection, when idf is applied, pmi, dice measures are increased by 2% and tscore measure is increased by 7.3% on average as expected. To get general part for a given whole, hypernymy and meronymy relation can be intertwined. While we cannot always apply part-whole hierarchy due to lack of transitivity, hypernymy relation can be easily and safely applied. We showed that some common parts can be hierarchically inherited from artifact or entity synset.

We conducted an experiment to see the importance of larger seed set. It partly confirms our expectation that the success rate from a larger training seed set is slightly better than those from a smaller one. As we increase the seed size from 50 to 150, only pmi measure clearly improves and the other measures did now show significant improvements.

Initial goal of the study is to retrieve meronymic relation, more specifically component-integral relation. Most of the incorrectly retrieved candidates actually fall into other kind of semantic relations such as has-a, property, cause etc. When evaluating the results with respect to these semantic relations, we obtained better precision but coarser relations as expected.

References

1. Girju, R., Badulescu, A., Moldovan, D.I.: Learning Semantic Constraints for the Automatic Discovery of Part-Whole Relations. In: Proceedings of the Human Language Technology Conference of the North American Chapter of the Association for Computational Linguistics (HLT-NAACL 2003), pp. 1–8. Association for Computational Linguistics, Morristown (2003)
2. Ittoo, A., Bouma, G.: On Learning Subtypes of the Part-Whole Relation: Do Not Mix Your Seeds. In: Proceedings of the 48th Annual Meeting of the Association for Computational Linguistics, pp. 1328–1336 (2010)
3. Van Hage, W.R., Kolb, H., Schreiber, G.: A Method for Learning Part-Whole Relations. In: Cruz, I., Decker, S., Allemang, D., Preist, C., Schwabe, D., Mika, P., Uschold, M., Aroyo, L.M. (eds.) ISWC 2006. LNCS, vol. 4273, pp. 723–735. Springer, Heidelberg (2006)

4. Buscaldi, D., Rosso, P., Arnal, E.S.: Using the WordNet Ontology in the GeoCLEF Geographical Information Retrieval Task. In: Peters, C., Gey, F.C., Gonzalo, J., Müller, H., Jones, G.J.F., Kluck, M., Magnini, B., de Rijke, M., Giampiccolo, D. (eds.) CLEF 2005. LNCS, vol. 4022, pp. 939–946. Springer, Heidelberg (2006)

5. Artale, A., Franconi, E., Guarino, N., Pazzi, L.: Part-Whole Relations in Object-Centered Systems: An Overview. Data Knowledge Engineering 20(3), 347–383 (1996)

6. Schulz, S., Hahn, U.: Part-whole representation and reasoning in formal biomedical ontologies. Artificial Intelligence in Medicine 34(3), 179–200 (2005)

7. Madelyn, A.I.: Problems of the part-whole relation. In: Evens, M. (ed.) Relational Models of the Lexicon, pp. 261–288. Cambridge University Press, New York (1989)

8. Winston, M.E., Chaffin, R., Herrmann, D.: A Taxonomy of Part-Whole Relations. Cognitive Science 11(4), 417–444 (1987)

9. Keet, C.M., Artale, A.: Representing and reasoning over a taxonomy of part-whole relations. Applied Ontology 3(1-2), 91–110 (2008)

10. Miller, G.A., Beckwith, R., Fellbaum, C., Gross, D., Miller, K.: WordNet: An on-line lexical database. International Journal of Lexicography 3, 235–244 (1990)

11. Yazıcı, E., Amasyalı, M.F.: Automatic Extraction of Semantic Relationships Using Turkish Dictionary Definitions. EMO Bilimsel Dergi 1(1), 1–13 (2011)

12. Ayşe Şerbetçi, A., Orhan, Z., Pehlivan, İ.: Extraction of Semantic Word Relations in Turkish from Dictionary Definitions. In: Proceedings of the ACL 2011 Workshop on Relational Models of Semantics (RELMS 2011), Portland, Oregon, USA, pp. 11–18. Association for Computational Linguistics (2011)

13. Orhan, Z., Pehlivan, İ., Uslan, V., Önder, P.: Automated Extraction of Semantic Word Relations in Turkish Lexicon. Mathematical and Computational Applications 16(1), 13–22 (2011)

14. Hearst, M.A.: Automatic Acquisition of Hyponyms from Large Text Corpora. In: COLING 1992 Proceedings of the 14th conference on Computational linguistics, vol. 2, pp. 539–545. Association for Computational Linguistics (1992)

15. Sak, H., Güngör, T., Saraçlar, M.: Turkish Language Resources: Morphological Parser, Morphological Disambiguator and Web Corpus. In: Nordström, B., Ranta, A. (eds.) GoTAL 2008. LNCS (LNAI), vol. 5221, pp. 417–427. Springer, Heidelberg (2008)

16. Berland, M., Charniak, E.: Finding Parts in Very Large Corpora. In: Proceedings of the 37th Annual meeting of the Association for Computational Linguistics on Computational Linguistics, pp. 57–64 (1999)

17. Girju, R., Badulescu, A., Moldovan, D.I.: Automatic Discovery of Part-Whole Relations. Computational Linguistics 32(1), 83–135 (2006)

18. Pantel, P., Pennacchiotti, M.: Espresso: Leveraging Generic Patterns for Automatically Harvesting Semantic Relations. In: Proceedings of the 21st International Conference on Computational Linguistics and the 44th Annual Meeting of the ACL, pp. 113–120. ACL Press, Sydney (2006)

19. Cao, X., Cao, C., Wang, S., Lu, H.: Extracting Part-Whole Relations from Unstructured Chinese Corpus. In: Proceedings of the 2008 Fifth International Conference on Fuzzy Systems and Knowledge Discovery, pp. 175–179. IEEE Computer Society (2008)

20. Ittoo, A., Bouma, G., Maruster, L., Wortmann, H.: Extracting Meronymy Relationships from Domain-Specific, Textual Corporate Databases. In: Hopfe, C.J., Rezgui, Y., Métais, E., Preece, A., Li, H. (eds.) NLDB 2010. LNCS, vol. 6177, pp. 48–59. Springer, Heidelberg (2010)

Chinese Terminology Extraction Using EM-Based Transfer Learning Method

Yanxia Qin[1,2], Dequan Zheng[1], Tiejun Zhao[1], and Min Zhang[1,2]

[1] School of Computer Science and Technology,
Harbin Institute of Technology, Harbin 150001, China
{yxqin,dqzheng,tjzhao}@mtlab.hit.edu.cn
[2] Human Language Technology, Institute for Infocomm Research,
Singapore 138632
mzhang@i2r.a-star.edu.sg

Abstract. As an important part of information extraction, terminology extraction attracts more attention. Currently, statistical and rule-based methods are used to extract terminologies in a specific domain. However, cross-domain terminology extraction task has not been well addressed yet. In this paper we propose using EM-based transfer learning method for cross-domain Chinese terminology extraction. Firstly, a naive bayes model is learned from source domain. Then EM-based transfer learning algorithm is used to adapt the classifier learnt from source domain to target domain, which is in different data distribution and domain from source domain. The advantage of our proposed method is to enable the target domain to utilize the knowledge from the source domain. Experimental results between computer domain and environment domain show the proposed Chinese terminology extraction with EM-based transfer learning method outperforms traditional statistical terminology extraction method significantly.

Keywords: Chinese, Terminology Extraction, EM, Transfer Learning.

1 Introduction

Information extraction aims to extract useful structural information from unstructured free texts to facilitate people reading and learning. Terminologies, as the basic semantic unit of domain information, are very useful to other downstream applications [19]. Terminology extraction technologies could recognize and extract terminologies of a specific domain automatically. Thus could both relieve experts from laborious work and assist achieving natural language processing tasks like text summarization, machine translation and so on. The state-of-the-art terminology extraction technology consists of two parts: 1) candidate terminology extraction, during which candidate terms would be extracted from original texts according to their unihood; 2) terminology verification, in which candidate terms are evaluated and ranked by their termhood. Unihood measures the probability of a string to be a valid phrase while termhood represents the probability of a candidate term being a terminology [23].

Currently, terminology extraction technology aims at one domain task since terminologies are highly domain-dependent. Statistical and rule-based methods are commonly applied to conduct terminology extraction task in one domain, most of which are

A. Gelbukh (Ed.): CICLing 2013, Part I, LNCS 7816, pp. 139–152, 2013.

domain-independent. Statistical methods utilize statistic values to measure termhood of candidate terms. Rule-based methods mainly extract rules according to terminologies' grammatical features and domain-related information. When there's no enough labelled data to train an effective terminology extraction method, traditional terminology extraction methods are not suitable any more. Then we come up with transfer learning methods to deal with terminology extraction task instead of use traditional methods. That's the purpose of our work.

In most circumstances, researchers are faced with some problems, such as lacking enough labelled training data and testing data having different data distribution with training data. And these problems are believed to decrease the performance of traditional machine learning methods. Transfer learning methods are proposed to address these problems faced with traditional machine learning methods. Transfer learning methods could transfer knowledge obtained from a source domain to assist learning model in a target domain [18]. The theoretical basis behind transfer learning is that, people could learn a new task easily when they have already gained knowledge of a related task. Like people who are familiar to "Perl" can learn to program with "Python" easily. Similarly, transfer learning methods work on target domain's task by getting help from source domain.

In this paper, we focus on cross-domain Chinese terminology extraction tasks. First, we utilize a naive bayes classifier trained on source domain data to initially estimate the degree of candidate terms to be a terminology in target domains. Second, an EM-based transfer learning method is proposed to adapt the initial model to target domain iteratively. During the process, some detail techniques are designed to fit Chinese.

We would introduce our work as follows. Section 2 gives a brief review of related work. In section 3, Chinese terminology extraction with naive bayes classifier is introduced, including candidate terminology extraction, naive bayes terminology verification and post-processing is introduced. Then, a Chinese terminology extraction using EM-based transfer learning method is discussed in section 4. In section 5, corpus construction, evaluation metrics, experimental setting and result analysis are presented. At last, Conclusion and future work are given in the last section.

2   Related Work

Currently, statistical methods and rule-based methods are commonly applied to extract terminologies in one domain. Statistics are used to measure whether a string is terminology or not in statistical terminology extraction methods. Term frequency (TF), mutual information (MI) and C-value and so on are the widely used statistics. TF and MI statistics compute candidate terms' termhood by measuring their inter bond strength [24]. C-vlaue statistic evaluates candidate terms' termhood based on the term frequency value and a special part sensitive to nested terms [9]. Zhang evaluated five statistical methods and experiments showed that hybrid methods outperformed methods with single termhood metric [25]. While rule-based methods rely on syntactic rules and domain related rules to extract terminologies, in which big amount of human effort may be required to achieve high precision and recall [11,15].

As for Chinese terminology extraction task, besides terminology verification methods, researchers focus on segmentation and candidate terminology extraction task, during which internal and contextual information are used to compute unihood of a string [10]. Bilingual terminology extraction with internal and external information attracts more interests. Internal information includes statistical, lexical, contextual information in comparable corpora [13]. External resources consist of wikipedia, UMLS/MeSH and so on [7,8,4]. Kea, a keyphrase extraction tool was developed based on machine learning method, which uses naive bayes classifier to classify candidate terms into terminology category or non-terminology category [21]. Yang achieved cross-domain terminology extraction task with candidate terms' contextual information. Domain independent term delimiters are extracted from corpus and used to segment sentences to get candidate terms [23]. However, it's better that two domains share same delimiters.

Transfer learning methods are used in many tasks like text classification, sentiment classification, clustering and regression. In text classification transfer task, EM algorithm and naive bayes classifier are commonly combined [2,20,17]. TrAdaBoost algorithm was proposed to achieve transfer task by reweighting source domain data [3]. A structural correspondence learning algorithm makes use of pivot features to reduce difference between source domain and target domain resolving the transfer task [1]. Traditional probabilistic latent semantic analysis algorithm (PLSA) is extended to form Topic-bridged PLSA algorithm (TPLSA) transferring knowledge through shared topic [22].

Inspired by Kea [21] and EM-based text classification transfer learning method [2], an EM-based Chinese terminology extraction transfer learning method was proposed to deal with Chinese terminology extraction transfer task in this paper.

3   Chinese Terminology Extraction with Naive Bayes Classifier

In this section, we would present the naive bayes-based Chinese terminology extraction method.The terminology extraction process consists of three parts: Chinese candidate terminology extraction, terminology verification and post-processing.

3.1   Chinese Candidate Terminology Extraction

The Chinese candidate terminology extraction part takes raw texts written in Chinese as input and outputs a list of candidate terms. This task includes three parts: 1) Noise Removal, 2) Segmentation and 3) Repartition. Segmentation and Repartition parts are designed especially for Chinese.

Noisy data in the texts like figures, tables, xml, html tags and so on are removed firstly. Segmentation on Chinese seems to be more difficult as Chinese has not word delimiters like spaces in English. Stanford Chinese word segmenter[1] is applied to do word segmentation in this paper.

We obtain an sequence of words which contains one to five characters after Segmentation. While a Chinese terminology contains two to eight words, we are required to

---

[1] Stanford Chinese word segmenter :
http://nlp.stanford.edu/software/segmenter.shtml

repartition the word sequence with some segmentation tags. In this paper, punctuation marks, Chinese and Arabic numbers, auxiliary words and so on are used to construct a segmentation tag list. A list of candidate terms will be yielded after the repartition stage. Obviously, terms composed of verbs, adjectives and adverbs are unlikely to be terminologies. A non-Noun filter is used to filter all these terms out. Stanford log-linear part-of-speech tagger[2] is utilized in our filter.

3.2   Naive Bayes Terminology Verification

After getting candidate terms, we should proceed to verify whether these terms are terminologies or not. Especially, this paper verifies terminology with a machine learning method. Since a candidate term appeared in a text is either a terminology or a non-terminology, a binary classifier could be applied to verify its termhood. Appropriate terminologies may be obtained after classifying phrases in texts into terminology or non-terminology categories. Naive bayes classifier is utilized as it is straightforward and effective [16]. The naive bayes terminology verification mainly builds a naive bayes classifier to estimate the probability of a candidate term being a terminology. Building a classifier needs to choose some features and train a classification model.

Features. Candidate terms' features like Term Frequency - Inverse Document Frequency (TF-IDF) and first appearance Position (Pos), which can represent the termhood of candidates, are used in this paper just like [21].

1. TF-IDF Candidates' TF-IDF features calculate the termhood with combination of word frequency within a document and word occurrence within a set of documents in a specific domain. High frequency candidates demonstrate more importance than lower frequency ones. On the other hand, candidates' document frequencies show their speciality. The TF-IDF value of term t within document d, $TF - IDF(t, d)$, is calculated as follows:

$$
\begin{aligned}
TF - IDF(t, d) &= TF(t, d) * IDF(t) \\
&= \frac{freq(t, d)}{|d|} * \log_2 \frac{|Corpus|}{numOfDoc(t)}
\end{aligned}
\tag{1}
$$

2. Pos Candidates' first appearance position contributes another kind of information to their termhood calculation besides TF-IDF values. The first appearance position of term t in document d, $Pos(t, d)$ is computed as:

$$
Pos(t, d) = \frac{wordsNumBefore(t, d)}{|d|}
\tag{2}
$$

where, $wordsNumBefore(t, d)$ is the number of words appeared before t's first appearance in document d.

---

[2] Stanford log-linear part-of-speech tagger :
   http://nlp.stanford.edu/software/tagger.shtml

Obviously, formula 1 and 2 show us that candidate terms' TF-IDF and Pos values are real numbers. Previous work show that supervised machine learning algorithm using discrete feature values outperforms that using continuous features , even the Equal Interval Width method [6]. We apply Equal Interval Width to discretize candidate terms' TF-IDF and Pos feature values into nominal values $tfidf, pos \in \{1, 2, 3, ..., DiscNum\}$ equally. Besides, $DiscNum$ is tuned experimentally as one parameter of our algorithm.

Naive Bayes Classification. With candidate terms' discrete feature values $tfidf$, pos prepared, a binary naive bayes classifier could be built. In the following, yes category denotes terminology category and no category denotes non-terminology category. Given a candidate term t in document d , according to Bayes' theorem the probability for being a terminology $P(yes|t)$ is calculated as:

$$P(yes|t) = \frac{P(yes)P(t|yes)}{P(t)} \propto P(yes)P(t|yes) \tag{3}$$

where, $P(yes)$ is the prior probability of terminology category, calculated in formula 4, $P(t|yes)$ means the probability of term t in terminology category, computed in formula 5.

$$P(yes) = \frac{|YesSet|}{|YesSet| + |NoSet|} \tag{4}$$

$$P(t|yes) = P_{tfidf}(tfidf|yes)P_{pos}(pos|yes) \tag{5}$$

$YesSet$ and $NoSet$ are the candidate term set with their item belongs to Yes category and No category respectively. $P_{tfidf}(tfidf|yes)$ and $P_{pos}(pos|yes)$ mean the probability of discretized TF-IDF and Pos value in terminology category, which are calculated as follows:

$$P_{tfidf}(tfidf|yes) = \frac{Num_{tfidf}(tfidf|yes) + 1}{Num_{tfidf}(yes) + |YesSet|} \tag{6}$$

$$P_{pos}(pos|yes) = \frac{Num_{pos}(pos|yes) + 1}{Num_{pos}(yes) + |YesSet|} \tag{7}$$

Similarly, we could compute $P_{tfidf}(tfidf|no)$, $P_{pos}(pos|no)$ and $P(no|t)$ respectively. Then we would get an overall probability of candidate term t as $P(t)$ with formula 8, which would be a representative statistics for termhood.

$$P(t) = \frac{P(yes|t)}{P(yes|t) + P(no|t)} \tag{8}$$

3.3   Post-processing

Though candidate terms are ranked by $P(t)$ value, post-processing is still essential. Post-processing part consists of "Re-ranking by TF-IDF" and "Remove Sub Terms with Low Termhood" making the terminology list more accurate.

Algorithm 1. Naive bayes Chinese terminology extraction method

---

Input: Training corpus D, a test document Doc, terminology set $TermSet$ and parameters $discNum$, N

Output: Terminology list of test document

1: **procedure** NB_CLASSIFIER_TRAINING(D)
2:     $TrainCTermSet \leftarrow$ candidate terminologies in D
3:     $YesSet \leftarrow \{t|t \in TermSet, t \in TrainCTermSet\}$
4:     $NoSet \leftarrow \{t|t \notin TermSet, t \in TrainCTermSet\}$
5:     **for all** d in D, t in d **do**
6:         Compute $TF-IDF(t, d)$ and $Pos(t)$ with formulas 1 and 2, discrete into $tfidf, pos$
7:         **if** t in $YesSet$ **then**
8:             $Num_{tfidf}(tfidf, yes)$++, $Num_{tfidf}(yes)$++, $Num_{pos}(pos, yes)$++, $Num_{pos}(yes)$++
9:         **else**
10:             $Num_{tfidf}(tfidf, no)$++, $Num_{tfidf}(no)$++, $Num_{pos}(pos, no)$++, $Num_{pos}(no)$++
11:     **for all** $tfidf, pos$ **do**
12:         Compute $P_{tfidf}(tfidf|yes)$ and $P_{pos}(pos|yes)$ with formula 6 and 7
13:         Compute $P_{tfidf}(tfidf, no)$ and $P_{pos}(pos, no)$ similarly
14: **procedure** NB_TERMINOLOGY_VERIFICATION
15:     $TestCTermSet \leftarrow$ candidate terminologies in Doc
16:     **for all** t in $TestCTermSet$ **do**
17:         Compute $TF - IDF(t, Doc)$ and $Pos(t)$, discrete into $tfidf, pos$
18:         Compute $P(t)$ with formula 8
19:     $TestTermSet \leftarrow \{(t, P(t))| t$ in $TestCTermSet\}$
20:     Rank $TestTermSet$ with $P(t)$ and do post-processing
21:     **Return** Top N terms in $TestTermSet$

- Re-ranking by TF-IDF

  As the feature discretization causes some of candidate terms' termhood to be the same value, original TF-IDF feature values would provide a basis to rank candidates with same value of $P(t)$.

- Remove sub terms with low termhood

  Sub terms with low termhood have smaller probability to be a terminology as they are easier to construct a new terminology together with other words than a longer term. After this, top N terms would be taken as final terminology list.

Overall, the process of naive bayes Chinese terminology extraction method is shown in algorithm 1. The algorithm takes a labelled training data set and a test document as input and outputs terminologies appeared in a test document. Procedures NB_classifier_training and NB_terminology_verification are the core parts of the algorithm, shown in section 3.2.

4   Chinese Terminology Extraction Using EM-Based Transfer Learning Method

Like many text classification transfer learning methods, which applied EM algorithm to transfer knowledge of source domain into target domain, our method also make full use of EM algorithm's high estimation ability to locally optimize hidden parameters. Differently, for the first time, our method focus on terminology extraction task on phrase level rather than text classification transfer task on document level like work in [2].

4.1   EM-Based Transfer Learning Method

Expectation Maximization (EM) algorithm was proposed to deal with parameters' log likelihood estimation problem [5]. Given a labelled data set D_l, an unlabelled data set D_u and an initial hypothesis h, EM algorithm makes a two step iteration to updating h to fit D_u. Step 1, a hypothesis related target function $Q(h)$ is calculated within current hypothesis h; step 2, hypothesis h is updated with a new one h' by maximizing $Q(h)$. EM algorithm ends up with a local optimized hypothesis h.

In transfer task, we have obtained a labelled data set D_l in one domain (source domain) and an unlabelled data set D_u in another domain (target domain). EM-based transfer learning method tries to transfer knowledge in source domain to assist learning task in target domain. Firstly, model parameters are initiated with data set D_l. Then, EM algorithm is used to iterate model parameters to adapt them in unlabelled data set D_u achieving best hidden parameters in D_u.

4.2   Chinese Terminology Extraction with EM-Based Transfer Learning Method

In our terminology extraction transfer task, it's uncertain whether candidate terminologies would be terminologies in target domain. Thus these uncertain probabilities of candidate terms can be regarded as hidden parameters in EM-based transfer learning method.

Initiate Model Parameters. Firstly, this method trains a naive bayes terminology extraction classifier, shown in section 3.2 in source domain to obtain the initial model parameters for target domain. Parameters such as $P(yes)$, $P(no)$, $P_{tfidf}(tfidf|yes)$, $P_{pos}(pos|yes)$, $P_{tfidf}(tfidf|no)$ and $P_{pos}(pos|no)$ are obtained.

Iteration of EM Algorithm. Secondly, EM algorithm is used to update model parameters by maximizing hidden parameters ending up with local optimization value.

- E step:
  Calculate $P(yes|t)$, $P(no|t)$ and $P(t)$ for candidate terms in unlabelled data set D_u with current parameters and formulas 3 and 8.

– M step:

Update parameters with $P(yes|t)$, $P(no|t)$ and $P(t)$ values of candidate terms in D_u and D_l with the following formulas.

$$P(yes) = \sum_{i \in l,u} P(D_i)P(yes|D_i) \tag{9}$$

where, $P(D_i)$ is a trade-off parameter for D_l and D_u. Under the unknown data distribution which D_u obeyed, we have assumption that $P(D_u) > P(D_l)$. Kullback-Leibler divergence was used to measure this trade-off parameter, calculated as follows [12].

$$KL(D_l|D_u) = \sum_{w \in W} P(w|D_l) \log_2 \frac{P(w|D_l)}{P(w|D_u)} \tag{10}$$

Then $P(D_l), P(D_u)$ can be obtained with formulas 11 and 12.

$$P(D_l) = \frac{KL(D_u|D_l)}{KL(D_l|D_u) + KL(D_u|D_l)} \tag{11}$$

$$P(D_u) = 1 - P(D_l) \tag{12}$$

Besides, $P(yes|D_i)$ is computed as follows:

$$P(yes|D_i) = \sum_{d \in D_i} \sum_{t \in d} P(yes|t)P(t|d)P(d|D_i) \tag{13}$$

Parameters $P_{tfidf}(tfidf|yes, D_i)$ and $P_{pos}(pos|yes, D_i)$ are computed in formulas 14 and 15.

$$P_{tfidf}(tfidf|yes, D_i) = \frac{num_{tfidf}(tfidf|yes, D_i) + 1}{num_{tfidf}(tfidf|yes, D_i) + |YesSet|} \tag{14}$$

$$P_{pos}(pos|yes, D_i) = \frac{num_{pos}(pos|yes, D_i) + 1}{num_{pos}(pos|yes, D_i) + |YesSet|} \tag{15}$$

Similarly, we can calculate respective parameters for non-terminology category. While calculating, we found a unexpected but reasonable phenomenon that for all candidate terms in D_u, term's $P(yes|t)$ value is less than $P(no|t)$ value, leading to zero terminologies in each iteration process. So we choose Top M largest candidate terms according to be potential terminologies after ranked by $P(yes|t)$ values. M is empirically taken according to number of terminologies out of all candidates number in training corpus.

Chinese terminology extraction using EM-based transfer learning method is shown in algorithm 2.

Algorithm 2. EM-based Chinese terminology extraction transfer learning method

---

Input: Labeled training corpus $D_l = \{d_1, \ldots, d_{|D_l|}\}$, unlabeled test corpus $D_u = \{d_1, \ldots, d_{|D_u|}\}$, terminology set $TermSet = \{t_1, \ldots, t_k\}$, maximum iteration number $MAXI$ and parameters $discNum, M, N$
Output: Terminology lists of test corpus

1: **procedure** PARAMETERS_INITIATION($D_l, discNum$)
2:     $TrainCTermSet \leftarrow$ candidate terminologies in D_l
3:     Call **procedure** NB_classifier_training in algorithm 1 with D_l, acquire parameters:
    $P(yes), P(no), P_{tfidf}(tfidf|yes), P_{pos}(pos|yes), P_{tfidf}(tfidf|no), P_{pos}(pos|no)$
4: **procedure** EM_ITERATION(Parameter set, $TestTermSet$)
5:     $TestCTermSet \leftarrow$ candidate terminologies in D_u
6:     **while** $i <= MAXI$ **do**
7:         **if** $i = 1$ **then**
8:             $TestTermSet \leftarrow TestCTermSet$
9:         **for all** term t in $TestTermSet$ **do**   ▷ E-Step
10:            Compute $P(yes|t), P(no|t), P(t)$ with formulas 3 and 8
11:        $TermSet_p \leftarrow \{$Top M t with largest $P(t)$ in $TestTermSet\}$ ▷ M-Step
12:        $YesSet \leftarrow \{t|t \in D_l, t \in TermSet\} \cup \{t|t \in D_u, t \in TermSet_p\}$
13:        $NoSet \leftarrow \{t|t \in D_l, t \notin TermSet\} \cup \{t|t \in D_u, t \notin TermSet_p\}$
14:        Update model parameters with $YesSet, NoSet$ with formulas 9, 13, 14 and 15
15:     Post-process $TestTermSet$
16:     **Return** Top N terms in $TestTermSet$

---

5 Experiments and Result Analysis

5.1 Corpus Construction

To verify the effectiveness of the EM-based Chinese terminology extraction transfer learning method, we conducted several experiments. Corpus used in this paper is extracted from corpus organized by Fu Dan University [14]. Duplicate documents are removed. Two categories of computer and environment are chosen as source domain and target domain. 150 documents in computer and 50 documents in environment category are used in our experiments.

5.2 Evaluation Indicators

Precision is the most often used evaluation indicator for terminology extraction. Precision measures the number of manual labelled terminologies out of extracted terminologies.

Generally, it's unlikely for system generated terminologies being exactly same like those labelled by human experts because of the unsatisfied effect of automatic candidate terminology extraction by computer. Given this limitations, exact and cover evaluation are utilized to evaluate the terminology extraction methods. Exact evaluation only takes the terminologies exactly same as manually annotated terminologies into account. While in cover evaluation all terminologies overlapped with manually labelled terms are considered to be right terms. In this paper, Exact Precision (EP), Cover Precision (CP) and the average Precision (avgP) are utilized as evaluation indicators calculated as follows:

$$EP = \frac{|TermSet_S \cap TermSet_M|}{|TermSet_M|} \tag{16}$$

$$CP = \frac{|TermSet_S \sqcap TermSet_M|}{|TermSet_M|} \tag{17}$$

$$avgP = \frac{EP + CP}{2} \tag{18}$$

where, $TermSet_S$ represents system generated terminology set and $TermSet_M$ is the manually labelled terminology set. $\cap$ means the set intersection which get exact common elements. $\sqcap$ means the fuzzy set intersection which get almost likely common elements.

5.3   Experimental Setting

We selected some baseline methods to compare with the EM-based Chinese terminology extraction transfer learning method (EMTE). Mutual information based terminology extraction method (MITE) is used as a statistical baseline method. We take naive bayes Chinese terminology extraction method (NBTE) as another baseline method.

In MITE method, we divide all candidate terms into two parts: single word terminologies and multiple word terminologies. We use TF value as single word terminologies' termhood. A candidate term's mutual information is applied to evaluate multiple word terminology's termhood computed in formula 19.

$$MI(s) = \log_2 \frac{prob(s)}{prob(a)prob(b)} \tag{19}$$

where $s = (s_1, ..., s_n)$ is a multiple word candidate, $a = s_1, ...s_{n-1}$ and $b = s_2, ..., s_n$ are two substrings of term s. $prob(s)$ means the probability of term s. The larger MI value, stronger inner strength of term s has.

These three methods extract candidate terminologies with the same module, which has been introduced in section 3.1. MITE method, a statistical method could extract terminologies in computer domain and environment domain separately without any model training process. NBTE is a machine learning based method demanding training corpus and testing corpus in same data distribution. While EMTE is a transfer learning method in which the strong limitations on training and testing corpus is removed.

Two groups of experiments with different training and testing data sets are conducted shown in table 1. Experiments in group 1 aim to explore whether the machine learning method performs better than the statistical method or not in traditional learning tasks in one domain (computer domain). While in group 2, experiments are conducted to test how well does EM-based terminology extraction method work in transfer learning tasks between computer domain and environment domain.

Table 1. Experimental settings

Experiment method		#Training documents		#Testing documents	
		Computer	Environment	Computer	Environment
	MITE	-	-	50	-
Group 1	NBTE	100	-	50	-
	EMTE	100	-	50	-
	MITE	-	-	-	50
Group 2	NBTE	100	-	-	50
	EMTE	100	-	-	50

5.4 Experimental Results Analysis

In this paper, there are several special parameters in NBTE and EMTE methods: 1)$discNum$ represents how many intervals we use in feature discretization, 2)N represents how many system generated terms would be evaluated. In following experiments, N is defined to be several numbers (10, 20, 30, 40 and 50). Maximum iteration number $MAXI$ is another parameter for EMTE method.

To select suitable $discNum$ value for feature discretization, we carried out experiments with fixed $MAXI$ (20) and all pre-defined numbers of N. Figure 1 shows how the average precision value changes when varying $discNum$ in NBTE method in group 1. From results shown in Figure 1, we can learn that $discNum$ influences NBTE method a lot. Extra experiments on EMTE method in group 1 show that $discNum$ making little difference to EMTE method's experimental results. We set $discNum$ to be 10 in following experiments. After value of $discNum$ is chosen, experimental results in

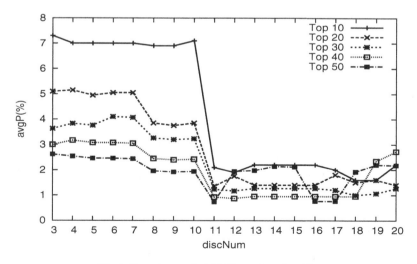

Fig. 1. Experiments in NBTE to select $discNum$

Table 2. Experimental results

Experimental results(%)		Group 1			Group 2		
		MITE	NBTE	EMTE	MITE	NBTE	EMTE
Top 10	EP	0.00	5.80	**31.80**	0.00	5.80	**12.40**
	CP	6.80	8.40	**49.20**	10.00	14.80	**42.80**
	avgP	3.40	7.10	**40.50**	5.00	10.30	**27.60**
Top 20	EP	0.00	3.00	**21.50**	2.90	4.40	**10.20**
	CP	6.70	4.70	**37.00**	11.70	12.90	**31.10**
	avgP	3.35	3.85	**29.25**	7.30	8.65	**20.65**
Top 30	EP	1.93	2.33	**18.27**	1.93	3.80	**8.67**
	CP	7.27	4.13	**32.33**	8.2	9.93	**25.87**
	avgP	4.60	3.23	**25.30**	5.07	6.87	**17. 27**
Top 40	EP	1.45	1.75	**16.50**	1.45	2.9	**7.90**
	CP	6.00	3.10	**29.05**	6.45	7.65	**22.45**
	avgP	3.73	2.43	**22.78**	3.95	5.28	**15.18**
Top 50	EP	2.20	1.4	**15.08**	1.16	2.32	**7.04**
	CP	5.92	2.48	**26.32**	5.96	6.36	**19.44**
	avgP	4.06	1.94	**20.70**	3.56	4.34	**13.24**

group 1 and 2 are presented in Table 2 with the maximum value in bold. Results show that: 1) NBTE method yields pretty good results, thus machine learning terminology extraction method performs not worse than statistical method; 2) EMTE method outperforms baseline methods significantly; 3) cross-domain terminology extraction task could be addressed using the transfer learning method.

From Table 2, we can see that EP values are very low even 0 in MITE method's results when Top 5 or Top 10 terms are evaluated. The worse results of Chinese candidate terminology extraction are mainly due to low EP value in which many terminologies are split into several words and not been recognized in repartition process. In MITE method, one word candidate terms' and multiple word terms' termhood values are calculated respectively. Only multiple word candidate terms are ranked by MI value while one word candidate terms are ranked by TF value. In test documents used in group 1, there only 19% of extracted terms are multi word terms on average.

For most documents, compared the generated top N term candidate lists by NBTE and by EMTE, we find that NBTE generated much fewer oracle terminologies than our EMTE. The main reason is due to the domain and knowledge mismatch. However in each iteration, EM algorithm in EMTE method updates probabilities of good terms to be terminology in target domain with those probabilities in source domain, and thus increases the probability of terms to be terminologies. The EM nature helps terminologies to achieve better position in EMTE method-generated terms list. That's why in Figure 5, EMTE method outperforms baseline methods greatly.

6   Conclusion and Future Work

In this paper we addressed terminology extraction transfer task with EM-based transfer learning method. Naive bayes terminology extraction method was used to deal with

Chinese terminology extraction task in a specific domain, in which TF-IDF and first appearance position features were applied. As for cross-domain terminology extraction task, EM algorithm was utilized to transfer knowledge leant from source domain task to target domain task. Related experiments showed that naive bayes terminology extraction method yields better results than statistical method and EM-based transfer method outperforms both baseline methods significantly.

In future work, Chinese candidate terminology extraction technology could be explored more deeply and contextual information like delimiters nearby or external resources like Wikipedia will be taken into consideration to improve extraction results.

Acknowledgement. This work is supported by the national natural science foundation of China (No. 61073130) and the project of National High Technology Research and Development Program of China (863 Program) (No. 2011AA01A207).

References

1. Blitzer, J., McDonald, R., Pereira, F.: Domain adaptation with structural correspondence learning. In: Proceedings of the 2006 Conference on Empirical Methods in Natural Language Processing, EMNLP 2006, pp. 120–128. Association for Computational Linguistics, Stroudsburg (2006)
2. Dai, W., Xue, G.R., Yang, Q., Yu, Y.: Transferring naive bayes classifiers for text classification. In: Proceedings of the 22nd National Conference on Artificial Intelligence, AAAI 2007, vol. 1, pp. 540–545. AAAI Press (2007)
3. Dai, W., Yang, Q., Xue, G.-R., Yu, Y.: Boosting for transfer learning. In: Proceedings of the 24th International Conference on Machine Learning, ICML 2007, pp. 193–200. ACM, New York (2007)
4. Ddjean, H., Gaussier, E., Sadat, F.: Bilingual terminology extraction: an approach based on a multilingual thesaurus applicable to comparable corpora. In: Proceedings of COLING, Taipei, Taiwan (2002)
5. Dempster, A.P., Laird, N.M., Rdin, D.B.: Maximum likelihood from incomplete data via the em algorithm. Journal of the Royal Statistical Society, Series B 39, 1–38 (1977)
6. Dougherty, J., Kohavi, R., Sahami, M.: Supervised and unsupervised discretization of continuous features. In: Machine Learning: Proceeding of the 12th International Conferences, pp. 194–202. Morgan Kaufmann (1995)
7. Erdmann, M., Nakayama, K., Hara, T., Nishio, S.: An Approach for Extracting Bilingual Terminology from Wikipedia. In: Haritsa, J.R., Kotagiri, R., Pudi, V. (eds.) DASFAA 2008. LNCS, vol. 4947, pp. 380–392. Springer, Heidelberg (2008)
8. Erdmann, M., Nakayama, K., Hara, T., Nishio, S.: Improving the extraction of bilingual terminology from wikipedia. ACM Trans. Multimedia Comput. Commun. Appl. 5(4), 31:1–31:17 (2009)
9. Frantzi, K., Ananiadou, S., Mima, H.: Automatic recognition of multi-word terms. the C-value/NC-value method. International Journal on Digital Libraries 3(2), 115–130 (2000)
10. Ji, L., Sum, M., Lu, Q., Li, W., Chen, Y.: Chinese Terminology Extraction Using Window-Based Contextual Information. In: Gelbukh, A. (ed.) CICLing 2007. LNCS, vol. 4394, pp. 62–74. Springer, Heidelberg (2007)
11. Justeson, J.S., Katz, S.M.: Technical terminology: some linguistic properties and an algorithm for identification in text. Natural Language Engineering 1, 9–27 (1995)

12. Kullback, S., Leibler, R.A.: On Information and Sufficiency. The Annals of Mathematical Statistics 22, 79–86 (1951)
13. Lee, L., Aw, A., Zhang, M., Li, H.: Em-based hybrid model for bilingual terminology extraction from comparable corpora. In: Proceedings of the 23rd International Conference on Computational Linguistics: Posters, COLING 2010, pp. 639–646. Association for Computational Linguistics, Stroudsburg (2010)
14. Li, R.L.: Corpus of fudan university (2011), `http://www.nlpir.org/download/tc-corpus-answer.rar`
15. Maynard, D., Ananiadou, S.: Identifying contextual information for multi-word term extraction. In: Terminology and Knowledge Engineering, pp. 212–221 (1999)
16. Mitchell, T.M.: Machine Learning. McGraw-Hill Science/Engineering/Math (1997)
17. Nigam, K., McCallum, A.K., Thrun, S., Mitchell, T.: Text classification from labeled and unlabeled documents using em. Machine Learning 39(2-3), 103–134 (2000)
18. Pan, S.J., Yang, Q.: A survey on transfer learning. IEEE Transactions on Knowledge and Data Engineering 22(10), 1345–1359 (2010)
19. Sharoff, S., Hartley, A.: Lexicography, terminology and ontologies. In: Mehler, A., Romary, L., Gibbon, D. (eds.) Handbook of Technical Communication (HAL 8), pp. 317–346. Mouton de Gruyter (2012)
20. Tsuruoka, Y., Tsujii, J.: Training a naive bayes classifier via the em algorithm with a class distribution constraint. In: Proceedings of the 7th Conference on Natural Language Learning at HLT-NAACL 2003, CONLL 2003, vol. 4, pp. 127–134. Association for Computational Linguistics, Stroudsburg (2003)
21. Witten, I.H., Paynter, G.W., Frank, E., Gutwin, C., Nevill-Manning, C.G.: Kea: practical automatic keyphrase extraction. In: Proceedings of the 4th ACM Conference on Digital Libraries, DL 1999, pp. 254–255. ACM, New York (1999)
22. Xue, G.R., Dai, W., Yang, Q., Yu, Y.: Topic-bridged plsa for cross-domain text classification. In: Proceedings of the 31st Annual International ACM SIGIR Conference on Research and Development in Information Retrieval, SIGIR 2008, pp. 627–634. ACM, New York (2008)
23. Yang, Y., Lu, Q., Zhao, T.: Chinese term extraction using minimal resources. In: Proceedings of the 22nd International Conference on Computational Linguistics, COLING 2008, vol. 1, pp. 1033–1040. Association for Computational Linguistics, Stroudsburg (2008)
24. Zhang, F., Xu, Y., Hou, Y., Fan, X.: Chinese Term Extraction System Based on Mutual Information. Application Research of Computers 22(5), 72–73 (2005)
25. Zhang, Z., Iria, J., Brewster, C., Ciravegna, F.: A comparative evaluation of term recognition algorithms. In: Proceedings of the Sixth Language Resources and Evaluation Conference, LREC 2008, Marrakech (2008)

Orthographic Transcription for Spoken Tunisian Arabic

Inès Zribi, Marwa Graja, Mariem Ellouze Khmekhem,
Maher Jaoua, and Lamia Hadrich Belguith

ANLP Research Group, MIRACL Lab., University of Sfax, Tunisia
ineszribi@gmail.com,
{marwa.graja,maher.jaoua,l.belguith}@fsegs.rnu.tn,
mariem.ellouze@planet.tn

Abstract. Transcribing spoken Arabic dialects is an important task for building speech corpora. Therefore, it is necessary to follow a definite orthography and a definite annotation to transcribe speech data. In this paper, we present OTTA, Orthographic Transcription for Tunisian Arabic. This convention proposes the use of some rules based on the standard Arabic transcription conventions and we define a set of conventions which preserve the particularities of Tunisian dialect.

Keywords: Tunisian dialect, orthographic transcription.

1 Introduction

Arabic is a Semitic language among the oldest in the world. It is recognized by its three main variants which are Classical Arabic (CA), Modern Standard Arabic (MSA) and colloquial or dialectal Arabic (DA) [1]. Arabic dialects are distinguished according to many levels. They can be classified according to geographical areas as they can be classified according to sociological and regional differences. In every Arab country, we usually find dialects used by urban residents, peasants / farmers and Bedouins [2]. Arabic dialects can also be classified into two groups the Eastern dialects (Levantine Arabic, Gulf Arabic and Egyptian Arabic) and Western dialects of the Arab world (the Arab Maghreb) [3]. The major difference between these dialects is located at the phonological level and mainly in the vowels and consonants interdentally. For example, the dialects of the west tend to neglect some short vowels and vowel lengthening to reduce the long vowels. Those in the east remain the same vowels of Classical Arabic [3].

Dialectal Arabic is essentially spoken and used in every day communication [1], [4]. It is used in Chat, public services, radio, telephone conversation, etc. So, it is so important to consider Dialectal Arabic in the new technologies like dialogue systems, telephone applications, etc. From this fact emerges the necessity to transcribe dialectal Arabic. But the transcription task is still difficult to achieve and the transcription output should use an orthographic convention to obtain a coherent data and consistent corpora. Few studies ([15], [16], [17]) have addressed the task of orthographic transcription for dialects. This is due to the lack of resources which are necessary to process dialects. Indeed, we still need resources and data to build main tools for

A. Gelbukh (Ed.): CICLing 2013, Part I, LNCS 7816, pp. 153–163, 2013.

natural language processing, extract features of dialects and test methods and approaches. In this context and to build a Tunisian Arabic corpus, we propose OTTA (Orthographic Transcription for Tunisian Arabic), a set of guidelines to orthography transcribe the spoken Tunisian dialect.

This paper is organized as follows. In section 2, we propose a comparison between Tunisian Arabic and MSA. In section 3, we present a background of existing orthography to other dialects and their limits. In section 4, we present our OTTA guidelines to orthography transcribe Tunisian Arabic.

2   Tunisian Arabic and Modern Standard Arabic: A Comparative Study

The Tunisian Arabic (TA) is a subset of Arabic dialects associated with the Arabic of the Maghreb (the west of Arab world). Like all Arabic dialects, it is characterized by morphology, syntax, phonology and lexicon which have differences and similarities compared to the MSA and even to other Arabic dialects. The TA, as the North African dialects, is strongly influenced by Berber, but also by other languages such as Turkish, Italian, Spanish and French. It has several main regional varieties but the Tunis variety (used in the capital city of Tunisia) is the most understood by all Tunisians [5].

To identify TA characteristics, we studied TuDiCoI corpus (Tunisian Dialect Corpus Interlocutor) [6]. It is a pilot corpus of spoken dialogue in Tunisian dialect. It is a collection of recorded conversations in railway station about request information such as the train schedule, ticket fare, booking, etc. It consists of 434 dialogues, which represent 3080 utterances.

2.1   Phonology

The phonological system of TA compared to the phonological system of MSA has several differences in vowel and consonant system [7]. Indeed, the vowel system of Tunisian dialect is distinguished, first, by a short vowel [8]. The MSA has only three short vowels [i, a, u] which can be doubled to their long corresponding. Tunisian dialect neglects short vowels especially when they are located at the end of a syllable [7].

Take the example of the verb ("شرب" [ʃariba], "he drank") in MSA which ends with the short vowel [a]. In TA, this verb is transformed into ("شرب" [ʃrib], "he drank"). We notice the deletion of the first and the last vowel [a]. Generally, deleting the first vowel changes the syllabic structure of lexical units which tend to monosyllables for certain words.

Tunisian Arabic is considered the closest one to MSA among other Arabic dialects especially during pronunciation of consonants. In most cases, the consonant "ق" of MSA is pronounced [q] in the dialects of urban and rural and, generally, is pronounced [q] (قال [qa:l], "he said") in the urban dialects and [g] (قال [ga:l], "he said") in the rural dialects. Sometimes, the word sense in the TA could be changed if we change the pronunciation of the letter "ق." For example, the word ("قرون" [QRU:n]) means "centuries" but the word ("قرون" [gru:n]) means "horns". The TA is characterized also by the presence of phonetic assimilation (the pronunciation of the letter "ج" [ʒ] in the word ("جزار" [ʒazza: r], "a butcher") often becomes "ز" [z] ("زازار", [zazza: r]) and metathesis (the word ("شمس" [ʃɛms], "sun"), is pronounced ("سمش" [semʃ])) [7]. Moreover, borrowing words from other languages have introduced new phonemes which are not used in the consonant system of MSA such as ('ڥ', [v]), ('ڨ' [g]) and (' پ' [p]) [9]. There are, for example ("ڥيسة" [vista], "jacket"), ("بڨرة" [Bagra], "cow") and ("پيست" "[pist]," track ") [10, 11].

2.2   Morphology

There are many differences between TA and MSA at morphological level. First, the suffixes relative to the dual form are generally absent. The noun suffixes "ين" and "ان" are usually replaced by the use of the numeral word "زوز" (two) located before or after the plural noun form. Also, we note the deletion of the feminine gender in the plural when we conjugate a verb.

MSA is characterized by the use of singular, dual and plural in both for masculine and feminine in the verb conjugation. However, in TA, we notice the total disappearance of the dual (male and feminine) and the feminine plural. Similarly, TA is characterized often by the disappearance of the feminine gender for the singular[1]. In addition, the fall of some markers and some conjugation suffixes are another difference from the MSA. Also, Tunisian verbs conjugation knows simplifications in affixation system [12].

The linguistic study of the corpus shows that the Tunisian Arabic knows the presence of new affixes and clitics compared to the MSA.

Negation in MSA can be marked with the use of one negation particle such as "ما/لا/لن/لم" [mɛ: / lɛ: / lɛm / lɛn] which is situated before the verb. In TA, the negation is generally marked with the presence of two particles. The first is located before the verb "ما" and the second is agglutinated in the end position of the verb [7]. The negation takes the following structure ما /[mɛ:] + verb +ش /[ʃ].

(a) Example: " مـــا كليتـــش" [mɛ:kli:tiʃ] "I have not eaten".

Among these news enclitics, we note, also, the introduction of ("و" [w] "his") as a new pronominal clitics. Likewise, the enclitic dual pronoun is replaced by the masculine plural pronoun. We note, also, the deletion of the verbal interrogation prefix "أ" which is transformed in a suffix "شي".

To transform trilateral MSA verbs in the passive form, we add some infixes to the verbal root. Indeed, we add the short vowels [ً, ,]/ [u, i, a] to generate the passive form. For example, the root "كتب" [ktb] (write) must follow the verbal model [fuʒila] to be in passive. The verb will be "كُتِب" [kutiba] (was written). While in TA, we add the prefix [t] to the verb. The passive form of the verb "كتب" [kteb] "write" in TA will be "تكتب" [tkteb] (was written) [13].

---

[1] We note that some Tunisian dialects distinguish between the masculine and the feminine in the singular form.

2.3   Orthography

Like other Arabic dialects, TA has known a variation in its orthographic transcription. This variation is caused by the absence of dialectal orthographic rules and also by the phonological differences between MSA and TA.

Indeed, the MSA common part of TA lexicon knows same orthographic variations. For example, the noun "تفاحة" can be transcribed with the suffix "ـة","ة" "tah marbuta" or with the "ـه","ه" "ha".

Also, the word "آش " appears in certain case as a proclitic "ش" and in certain cases is transcribed as a separate word.

Sometimes, speakers do not pronounce some letters and others pronounce them. For example, the word "قلت لك", (I said to you) is often pronounced "قتلك" by deleting the letter ل.

2.4   Lexicon and Code Switching

The current linguistic situation in Tunisia is both diglossic[2] and bilingual[3] [14]. The Tunisian people code switch easily and frequently between MSA, Tunisian dialect and French language in the same conversation. This is due to the lack of knowledge and the facility of use either Arabic or French language in a certain subject [14].

Code switching between Arabic and French language affects the lexical level of Tunisian dialect. In fact, it allows introducing new dialectal words which are derived from foreign languages.

Indeed, the verbal system of Tunisian dialect contains several French verbs [12]. These verbs are integrated into the system via the application of Arabic schemes derivation and the phoneme substitution for the phonetic integration of these verbs. Generally, the integration of these verbs conserves theirs original meaning. The application of Arabic schemes allows their conjugation [12]. Often, borrowed verbs are transformed via the scheme "فعْلِل" [faʕlil]. For example, the French verb "jongler" (to juggle) becomes "جنڤْل"[ʒang/i/l], and for the French verbal phrase "avoir sa maîtrise" (have his Master) is transformed into "مَتْرِز"[matr /i/z] [12]. In some cases, we just add affixes to the borrowed verb for its conjugation. For example, to conjugate the French verb "installer" (to install), the prefix "ي"[j] is added. The verb will be in Tunisian Arabic "ينستالي" [jansta:li] (he installs) [12].

The same rule is used for the generation of nouns and adjectives. From the borrowed verbs, nouns and adjectives are derived also. For example, the noun [4]"مريبوز" is derived from the French verb "reviser" (to revise). We add a simple prefix"م" [m].

When analyzing the TuDiCoI [6] corpus, we note that there are 2265 French words which represent 11.81% of the corpus and 328 words derived from French language which correspond to about 2% of the corpus.

---

[2]   Diglossic refers to the use of MSA and the dialectal form [14].
[3]   Bilingual involves the use of Arabic and the French [14].
[4]   A person who has revised some things.

These words are part of the vocabulary used in conversations and they must be added to any dictionary of Tunisian dialect lexicon. Table 1 gives some examples of loanwords which appear in Tunisian dialect lexicon.

Table 1. Examples of foreign words

Dialect	French origin	Translation
مرسي	merci	thank you
وي	oui	Yes
تران	train	Train
نريڤل	je règle	
نڤليدي	je valide	I validate

3 Background about Orthographic Conventions of Arabic Dialect

Few studies in the literature dealt with the transcription task of dialectal Arabic. Indeed, the transcription of Levantine dialects was the subject of Zawaydeh's work [15] and Maamouri et al. [16] who developed a set of rules to transcribe Levantine dialects in order to create a Levantine Arabic corpus. Their works are based on the MSA transcription conventions. The principle of their proposed method requires the transcription of dialectal Arabic using Arabic script without short vowels[5] (except the short vowel nunation[6]) by respecting the conventions of spelling and words segmentation of MSA.

Take the example of the Levantine word ("أنتلك" [ʔultilak], "I said to you"). This word is transcribed as two separate words. This segmentation is the result of applying the MSA rule which requires the separation between the verb and the prepositional object. As well, their convention converts the letter "أ"[7] to its origin "ق" in MSA. So the Levantine word "أنتلك" (I said to you) is transcribed into (قلت لك).

Maamouri et al. [16] justified their choice of using MSA-based orthography by reducing the cost of resource creation (the speed and ease of creation) for Levantine Arabic. Indeed, the annotators can use their knowledge of MSA instead of learning and using new phonetic symbols. Thus, transcription of Levantine Arabic using MSA-based orthography can use existing MSA tools.

Habash et al. [17] present also, a conventional orthographic Dialectal Arabic (CODA). It is designed mainly for developing computational models of Arabic dialects. CODA is an extension of the LDC guidelines. It proposes to develop a CODA for all Arabic dialects. Actually, Habash et al. [17] have developed a CODA map to cover only Egyptian Arabic.

---

[5] The work of [15] requires the use of the double consonant "shedda" when it raises the ambiguity of pronunciation.

[6] Nunation is a short vowel used in Arabic language (ٌ).

[7] The velar letter "ق" in Levantine dialect is pronounced "أ".

MSA-based orthographic transcription has several advantages. First, the transcription task is easy for annotators. Second, we can use the MSA existing tools to process Arabic dialects. Also, the corpus is readable for all Arabs. However, each dialect of the Arabic language has its own features which allow it to be distinguished from other dialects and even from MSA. Given the differences between Arabic dialects, existing works cannot be applied in totally to all Arabic dialects. That's why we propose, in this paper, a set of conventions to orthography transcribing the Tunisian Dialect.

4   Orthographic Transcription of Tunisian Arabic

To standardize the orthographic transcription of the Tunisian Arabic, we use some orthographic rules of MSA transcription and we define new rules for the specificities of the Tunisian Arabic.

We used this convention for the transcription of the corpus TuDiCoI. Indeed, we defined a set of annotations used to reflect the pronunciation of the Tunisian dialect in order to improve the transcription quality. Subsequently, the obtained corpus based on such transcription will be useful for the creation of processing tools for Tunisian Arabic such as the Tunisian dialect stemmer, morph-syntactic tagger, etc. Also, it is useful for the creation of automatic speech processing systems such as speech synthesis, automatic transcription systems, etc.

4.1   Transcription Rules

The Tunisian dialect lexicon consists of MSA words (with or without modification), dialectal Tunisian words and loanwords. The transcription of the words which are pronounced without modification compared to MSA must respect the orthography transcription standards of MSA. Therefore, we present, first, the main MSA-based rules to be respected in our transcripts. Next, we define specific rules based on the Tunisian Arabic specificities.

Tunisian Arabic transcription is based on the orthography of MSA in the case where the word is pronounced like in MSA or with the reduction of some short vowels [ˌ , ˎ ,]. Our transcription method keeps the main MSA orthographic rules in the following cases:

- The use of Arabic characters with short vowels.
- The conventions of word segmentation.

Thus the word in Tunisian Arabic ([kte:b], a book) is written "كتاب" even its reduced vocalism form compared to MSA [kitabon]. Similarly, we use word segmentation rules of MSA. Indeed, we transcribe some affixes as words. Take the example of the combination question mark ("آش" [ʔesh], "what") which is sometimes reduced to a single letter "ش" [sh] concatenated to the next word. This combination replaces the conjunction query ("ماذا" [ma: ða:], "what") in MSA. To make it closer to the sentence structure in MSA, we have chosen to transcribe this combination as a separate word with its extended form "شآ" and not its reduced form "ش". Also, in Tunisian Arabic, the preposition ("علىٰ" [ʕla], "on") is transformed into a single letter ("ع", [ʕ], "on"). In MSA, a letter must always be agglutinated to the word which follows it. So, this preposition is transcribed as an enclitic (see example (b) below). This is the same case for the conjunctive ("و" [w], "and"). In addition, we apply the segmentation of words in negation case and in the use of prepositional object case. For example, the word ("قلتلو"[qoltlou], "I said to him") must be transcribed in two words separated by spaces (see example (c)). This segmentation is justified by the fact that the prepositional object ("لو" [Lou], "to him") should not be agglutinated to the verb, similarly, for the conjunction of negation ("ما" [ma:]) (see example (d)).

(b) "عالطاولة" [ʕattawla], "on the table."

(c) "قلت لو" [qolt / lou], "I said," and not "قلتلو".

(d) "ما قلتش" [ma: qoltech], "I did not say" and not "ماقلتش" or "مقلتش".

The pronunciation of "Hamza[8]" presents a difference compared to the MSA. The "Hamza" is sometimes replaced by one of these vowels (ا, و or ي). For example, we replace the phoneme [ʔi] of the word ("فائدة" [faʔida], "profit") by the phoneme [j] ("فايدة" [fajda], "profit"). Moreover, speakers of Tunisian Arabic pronounce the "Hamza" only if it is located at the beginning of a word. For example, in the noun ("أستاذ", [osteth], professor) the hamza letter is pronounced.

Moreover, we can identify in Tunisian Arabic new phonemes such as [g]. The use of Arabic letters to transcribe these phonemes produces meaning ambiguities. Therefore, we propose to use the non-Arabic letters (('ڥ', [v]), ('ڨ' [g]) and ('پ' [p])) to transcribe these new phonemes.

Obviously, the lexicon of Tunisian Arabic knows the presence of words which have no roots in the Standard Arabic language. These are words specific to Tunisian dialect. The transcription of these words is different from one to another transcriber. In order to have a homogeneous corpus with words transcribed in a unique way, we define rules which combine rules of the MSA and phonemic compositions. Table 2 summarizes the proposed transcription rules.

Table 2. Transcription rules of Tunisian Arabic

N°	Transcription rule	Example
1	If the last phoneme of a word is a short vowel [a], then the word is spelled with the silent letter ة. Apply this rule only for names.	كرهبة [karhba] (car) بسكلة [pascla] (bicycle).
2	If the last phoneme of a word is characterized by a vowel lengthening ([a:], [u:] or [i:]), then the word is spelled with one of these vowels ي, ا or و.	راهو[ja:xi:],ياخي [ra:hu:]. We do not use the simple alif (ا) after (و). We do not write راهوا.

---

[8] Hamza is a letter in the Arabic alphabet representing the glottal stop [ʔ].

Table 2. *(Continued)*

3	Use the non-Arabic letters (ڥ, [v]), (ڨ [g]) and (پ [p]) to transcribe new phonemes used in Tunisian Arabic. We do not use the letters (ف, ب and ق).	Transcribe بڨرة, [bagra] and not بقرة [baqara] (cow).
4	The letter "Hamza" is transcribed only when it is pronounced.	Transcribe فايدة and do not فائدة.
5	Some words are transcribed as affixes: the preposition (على, on) is reduced to a single letter ع transcribed as a prefix.	Transcribe عالطاولة and do not ع الطاولة.
6	Some suffixes are transcribed as words: question mark آش is sometimes reduced to a single letter ش, but we must write it correctly.	Transcribe ش(آ) قريت and do not شقريت.

4.2   Annotation

We use and adapt transcription conventions of TOE [18] which are made by LPL laboratory. These conventions are developed for the transcription of conversational French corpus. We add some precisions and we modify some rules for the transcription of Tunisian Arabic.

Our transcription method consists of an orthographic transcription which reflects the phonetic pronunciation; therefore, we do not use acronyms in the transcripts. Thus, we should transcribe word as it is spelled. Our transcription doesn't note any abbreviation.

To annotate named entities, we propose adding an opening and closing tag before and after the proper name indicating its type (see example (e)). We use a ملك/شخ code (شخ for patronym and ملك for toponym). The form is: < شخ/ ملك ,Ortho[9]>.

(e) دكتور ‹المنصف المرزوقي, شخ›

Sometimes, dialect speakers do not pronounce some letters. For example, the word موش [mouʃ] ("not at all") is often pronounced ("مش" [meʃ], "not at all") by reducing vowel lengthening "و". Therefore, we propose to transcribe words into their correct forms and make non pronounced characters between parentheses. We should transcribe "م (و) ش" and not "مش". The same case for this example ("عمتلك" [ʕmetlek] "I've done for you") should be transcribed as "عم(ل)ت لك".

Given the existence of loanwords in Tunisian, we aim to annotate these words etymologically. This allows us to perform its automatic processing. We use this annotation [lan: language, orthography, pronunciation] for marking loanwords. We put in square brackets, and separated by comma, the language, the standard transcription in the foreign language, and the speaker's pronunciation (in SAMPA alphabet).

---

[9] "Ortho" is the orthographic transcription of named entity.

(f) يعيشك [lan:Fr, merci, d2] (Thank you very much)

(g) [lan:ASM ،ركِبَ] (he rides)

(h) « مشى محمد علي لتونس» (Mohamed Ali went to Tunis) is transcribed:
مشى <محمد علي, شخ> <لتونس, مك>

Besides, we aim to transcribe and annotate spontaneous spoken language. Thus, we are confronted with the problem of disfluencies which are defined as a phenomenon occurring frequently throughout spontaneous speech, and consist of the interruption of the normal course of speech [19]. As well, we propose to improve the annotations used to mark these facts. For marking incomplete words, we propose to use the sign "-". They are noted by a final dash just after the final sound of the truncated word, and followed by a blank (example: ع- علاش ما مشيتش).

Also, we identified a set of hesitations frequently used by Tunisian dialect speakers as "إمم!" which is transcribed by adding the symbol "÷" in the left of a word (إمم÷). Some of these hesitations are from other languages. We used a standard lexical list of hesitations.

The transcription is principally an orthographic transcription. We use punctuation marks for delimiting phrase boundaries in the transcription. We added precisions about particular pronunciations, and some other details. The most usual cases are described in the table 3.

Table 3. Annotation rules of Tunisian Arabic

	Annotation rules	Examples
Numbers	Numbers have to be written in letters.	واحد (one) عشرة ألاف (ten thousands)
Titles	Movies, books, newspaper titles are written between quotation marks.	تفرجت على برنامج " عندي ما نق(و)ل لك "
Undeterminable morphologic variants	Graphic variants are noted between braces, separated by commas.	{قتلوا, قتلو}{they killed, he killed him}
Atypical accords	We do not correct the transcription. We should write it as it is pronounced.	باهي الخلاعة (instead of " "باهية الخلاعة)
Liaisons	Links which are specific to Tunisian dialect are annotated. We also annotate the missing links between words.	أربعطاش ن=ألف (Tunisian dialect liaison) كراسة # الطفل (not used liaison)
Reported speech	Direct reported speech sequences are noted between the characters "\".It is preceded and followed by a blank.	أنا قلت لو \ أسكت وما عادش تحكي \ وهو ما هواش باش يقعد ساكت
Incomprehensible sequences	Long and short incomprehensible sequences are always noted by one star (*).	

Table 3. *(Continued)*

Laughers	Laughers are transcribed with the symbol "&". A speech sequence produced while laughing is coded between "&&".	توة جيناهاشي & ➜ means that the speaker said the words, and then laughed. && جيناهاشيتوة&& ➜ means that the speaker laughed when he is saying "توة جيناهاشي".
Pauses	Short pauses (less than 200 ms) are notated with "+".	عسلامة + مرحبا بيك خويا
Non-linguistic events	They are incomprehensible sequences. We suggest to precise what kind of event they are.	— Breathing (تنفس) — Puffing (نفخ) — Noise by the mouth (فم) — Cough (سعال) — Sneeze (عطاس) — Wistle (صفير)

5   Conclusion

Transcription of a spoken language which is in constantly evolution and known by the use of borrowing from other languages presents a challenge for its standardization. In this context, we tried to propose a set of rules which allow the transcription of Tunisian Arabic with respect to the MSA rules and also not to neglect the characteristics of such a dialect. We plan, also, to continue the enhancement of our conventions for the transcription of Tunisian dialect.

References

1. Al-Saidat, E., Al-Momani, I.: Future Markers in Modern Standard Arabic and Jordanian Arabic: A Contrastive Study. European Journal of Social Sciences 12(3) (2010)
2. Diab, M., Habash, N.: Arabic Dialect Processing Tutorial. In: Proceedings of the Human Language Technology Conference of the North American Chapter of the ACL, Rochester, pp. 5–6. Association for Computational Linguistics (April 2007)
3. Alorifi, F.S.: Automatic identification of Arabic dialects using Hidden Markov Models. In mémoire de thèse, Université de Pittsburgh (2008)
4. Almeman, K., Lee, M.: Towards developing a Multi-dialect Morphological analyzer for Arabic. In: 4th International Conference on Arabic Language Processing, Rabat, Morocco, May 2-3 (2012)
5. Khalfaoui, A.: A cognitive approach to analyzing demonstratives in Tunisian Arabic. In PhD thesis of university of Minnesota (November 2009)
6. Graja, M., Jaoua, M., HadrichBelguith, L.: Lexical Study of A Spoken Dialogue Corpus in Tunisian Dialect. In: ACIT 2010: The International Arab Conference on Information Technology, Benghazi - Libya, December 14-16 (2010)

7. Mejri, S., Said, M., Sfar, I.: Pluringuisme et diglossie en Tunisie. In: Synergies Tunisie, vol. (1), pp. 53–74 (2009)

8. Tilmatine, M.: Substrat Et Convergences: Le Berbère Et L'arabe Nord-Africain. Estudios de Dialectologia Norteafricana y Andalusl 4, 99–119 (1999)

9. Kirchhoff, K., Bilmes, J., Das, S., Duta, N., Egan, M., Ji, G., He, F., Henderson, J., Liu, D., Noamany, M., Schone, P., Schwartz, R., Vergyri, D.: Novel approaches to Arabic speech recognition: report from the 2002 Johns-Hopkins Summer Workshop. In: Proceedings of IEEE International Conference on Acoustics, Speech, and Signal Processing (ICASSP 2003), Missouri, USA, vol. 1, pp. 344–347 (April 2003)

10. Mejri, S., Baccouche, T.: L'atlas linguistique de Tunisie: repères méthodologiques pour la description du système dialectal. In: Lentin, J., Lonnet, A. (eds.) Mélanges David Cohen, pp. 47–54. Maisonneuve & Larose, Paris (2003)

11. Quitout, M.: Parlons l'arabe tunisien. In book edited by L'Harmattan (2006)

12. Ouerhani, B.: Interférence entre le dialectal et le littéral en Tunisie: Le cas de la morphologie verbale. In: Synergies Tunisie, vol. (1), pp. 75–84 (2009)

13. Maalej, Z.: Passives in modern standard and Tunisian Arabic. Matériaux Arabes et Suda-rabiques-Gellas 9, 51–76 (1999)

14. Bouzemni, A.: Linguistic situation in Tunisia: French and Arabic code switching. In: INTERLINGÜISTICA, vol. 16(1), pp. 217–223 (2005) ISSN 1134-8941

15. Zawaydeh, B., Stallard, D., Makhoul, J. (2003),
    `http://ldc.upenn.edu/Catalog/docs/LDC2005S08/BBN-Babylon-transcription-guidelines.pdf`

16. Maamouri, M., Buckwalter, T., Cieri, C.: Dialectal Arabic Telephone Speech Corpus: Principles, Tool Design, and Transcription Conventions. In: NEMLAR International Conference on Arabic Language Resources and Tools, Cairo, September 22-23 (2004)

17. Habash, N., Diab, M., Rambow, O.: Conventional Orthography for Dialectal Arabic. In: Proceedings of the Language Resources and Evaluation Conference (LREC), Istanbul (2012)

18. Bertrand, R., Blache, P., Espesser, R., Ferré, G., Meunier, C., Priego-Valverde, B., Rauzy, S.: Le CID - Corpus of Interactional Data - Annotation et Exploitation Multimodale de Parole Conversationnelle. Traitement Automatique des Langues 49(3), 105–134 (2008)

19. Heeman, P., Allen, J.: Detecting and correcting speech repairs. In: Proceedings of the 32nd Annual Meeting on Association for Computational Linguistics, Las Cruces, New Mexico, pp. 295–302 (1994)

An Improved Stemming Approach Using HMM for a Highly Inflectional Language

Navanath Saharia[1], Kishori M. Konwar[2], Utpal Sharma[1], and Jugal K. Kalita[3]

[1] Department of CSE, Tezpur University, India
{nava_tu,utpal}@tezu.ernet.in
[2] Department of MI, University of British Columbia Canada
kishori82@yahoo.com
[3] Department of CS, University of Colorado at Colorado Springs, USA
jkalita@uccs.edu

Abstract. Stemming is a common method for morphological normalization of natural language texts. Modern information retrieval systems rely on such normalization techniques for automatic document processing tasks. High quality stemming is difficult in highly inflectional Indic languages. Little research has been performed on designing algorithms for stemming of texts in Indic languages. In this study, we focus on the problem of stemming texts in Assamese, a low resource Indic language spoken in the North-Eastern part of India by approximately 30 million people. Stemming is hard in Assamese due to the common appearance of single letter suffixes as morphological inflections. More than 50% of the inflections in Assamese appear as single letter suffixes. Such single letter morphological inflections cause ambiguity when predicting underlying root word. Therefore, we propose a new method that combines a rule based algorithm for predicting multiple letter suffixes and an HMM based algorithm for predicting the single letter suffixes. The combined approach can predict morphologically inflected words with 92% accuracy.

1 Introduction

Most information retrieval systems represent documents as a set of words. The efficiency of such systems is adversely affected by the abundance of words appearing in various morphological forms either as a result of inflection or derivation. To reduce this detrimental effect of morphological variations, one common method is to represent the text in a normalized form. One such approach is the process of finding the root word from an inflected form. It is an initial step in analyzing the morphology of words. A number of approaches have been proposed by researchers for stemming, e.g., affix stripping, co-occurrence computation, dictionary look-up, longest suffix matching and probabilistic. Most approaches are first developed for English, and later adapted for other languages. So these approaches may not work properly for highly inflectional Indic languages.

Assamese, is a rarely studied low resource language, spoken in the north-eastern parts of India. Approximately 30 million people speak Assamese. In this

A. Gelbukh (Ed.): CICLing 2013, Part I, LNCS 7816, pp. 164–173, 2013.

study, we address the problem of stemming Assamese texts. Stemming in Assamese is difficult due to the common appearance of single letter suffixes as morphological inflections. Our experiments show that more than 50% of inflections in Assamese appear as single letter suffixes. Such single letter morphological inflections cause ambiguity when one predicts the underlying root word.

The rest of the paper is organized as follows. In Section 2, we describe previous work related to stemming followed by brief linguistic characterisation of Assamese, our experimental test-bed in Section 3. Section 4 describes our approach and Section 5 provides the results and analysis of the approach. Section 6 concludes our paper.

2   Previous Work

Porter stemmer [1], an iterative rule based approach has found great success and is used widely in various applications such as spell-checking, and morphological analysis. In the Indian language context, a few hand-crafted rule-based stemmers have been reported to strip off suffixes. Among these, [2] use a hand crafted suffix list and strip off longest suffixes for Hindi and report 88% accuracy using a dictionary of size 35,997. [3] learn suffix stripping rules from a corpus and use clustering to discover the nearest class of the root word for Bengali, English and French. They describe a centroid based approach that rewards the longest common prefix to form similar word clusters based on a threshold value. [4] focus on heuristic rules for Hindi and report 89% accuracy. [5] propose a hybrid form using approaches reported in [3] and [4] for Hindi and Gujarati with precisions of 78% and 83%, respectively. Their approach takes both prefixes as well as suffixes into account. They use dictionary and suffix replacement rules, and claim that the approach is portable and fast.

Kumar and Rana [6] use a dictionary of size 52,000 and obtain 81.27% accuracy in Punjabi using a brute-force approach. Majgaonker and Siddiqui [7] describe a hybrid method (rule based + suffix stripping + statistical) for Marathi and claim 82.50% precision for their system. Sharma et. al [8], [9], [10] describe an unsupervised approach, that learn morphology from unannotated Assamese corpus and report 85% precision value. The method discussed by Saharia et al. [11] and [12] for parts-of-speech tagging has three basic steps: brute-force determination of suffix sequences, suffix sequence pruning and suffix stripping. Table 1 enumerates the statistics reported by the different methods. In this paper, we extend this method for stemming inflected words in Assamese by using HMM for single character inflections.

3   Suffixes in Assamese

In the context of stemming, the most common property of Indic languages is that, they take a sequence of suffixes after the root words. We give an example from Assamese below.

নাতিনীয়েককেইজনীমানেহে→ নাতিনী + য়েক + কেইজনী + মান + ে + হে

Table 1. Reported performance of stemmers in some highly inflectional languages (except English)

Report	Language	Dictionary Size	Accuracy	Used technique
[1]	English		90.00%	Suffix Stripping
[13]	Arabic		96.00%	Rule base
[14]	Dutch	45000	79.23%	Porter Stemmer
[10]	Assamese		85%	Unsupervised approach
[3]	Bengali		90.00%	Suffix Stripping
[6]	Punjabi	52,000	81.27%	Brute Force Approach
[7]	Marathi		82.50%	Rule based + Statistical
[15]	Gujarati		90.00%	Unsupervised + Rule based
[16]	Malayalam	3,000	90.5%	Finite State Machine
[2]	Hindi	35,997	88.00%	Suffix Stripping
[4]	Hindi		90.00%	Unsupervised

nAtinIyekkeijanImAnehe → *nAtinI* +*yek* +*keijanI* +*mAn* +*e* +*he*
nAtinIyekkeijanImAnehe → noun root+ inflected form of kinship noun[1] + indefinite feminine marker + plural marker + nominative case marker + emphatic marker. (Approximate English meaning: only a few granddaughters)

These sequences of suffixes can easily be stripped off using algorithm proposed by [12]. A major drawback of the prior method is that it is not able to identify the single letter suffix well. For example, the method removes ৰ from the words অমৰ (*amar* : immortal) and মানুহৰ (*mAnuhr* : man+genitive marker), whereas the first word is a root word form, but the second word is inflected, with -ৰ (*ra*) as an genitive case marker. We have found that, in Assamese, a noun root word may potentially take more than 15,000 different inflections and up to 5 sequential suffixes after the noun root. Likewise, a verb may potentially also have more than 10,000 different inflectional forms. The frequency of appearance of single-letter inflections in Assamese is higher than multiple-letter inflections.

Among major Indic languages, Bengali is the closest to Assamese in terms of spoken and written forms. Table 2 tabulated an important observation around 2000 words collected from different news articles of English, Assamese, Bengali and Hindi. The forth column describes the inflected unique words in terms of number.

We observe that the compression rates for English, Assamese, Bengali and Hindi are 41.89%, 59.75%, 56.50% and 36.77% respectively. We also see that among the languages Assamese has the highest single letter inflectional suffixes. This behoove us to develop an algorithm to improve the accuracy of detecting

---

[1] All relational nouns in Assamese have the inflection যেক (*yek*) in 3rd person. For example in 3rd person relational noun ভাই (*bhAi : younger brother*) is inflected to ভায়েক (*bhAyek*), ককাই (*kakAi : elder brother*) is inflected to ককায়েক (*kakAyek*). Bora [17] reports that Assamese has the highest numbers of kinship nouns among Indo-Aryan languages.

Table 2. A random survey on single letter inflection. *MS**: Suffix sequence or multiple suffix end-with single letter suffix.

Language	Sent.	Words		Inflection type			Source of text
		Total	Unique	Single	MS*	Multiple	
English	82	2012	843	06.88%	-	18.50%	Times of India[2]
Assamese	132	2164	1293	28.21%	09.49%	13.06%	Dainik Janasadharan[3]
Bengali	202	2205	1246	17.97%	07.22%	18.37%	Anandabazar Patrika[4]
Hindi	116	2162	795	12.07%	03.14%	12.82%	Dainik Jagaran[5]

single-letter suffixes and use it in combination with the algorithm in [12]. The next section discusses the an Hidden Markov Model based approach we use to handle single-letter suffixes better.

4   HMM Based Approach

In this paper, we extend the algorithm in [12] to classify Assamese nouns and verbs. In this previously published work, Saharia et al. could automatically detect sequences of suffixes from inflected nouns and verbs and stem correctly with an accuracy 81%. Experimental result from [12] are given in Table 3. The algorithm accurately stems multiple character suffixes, but fails to handle well single character suffixes such as ৰ (*ra : genitive case marker*), and ক (*ka : accusative case marker*). These single letter morphological inflections, in Assamese are similar to post-positions in the English language.

We model Assamese text as a sequence of words produced by a generator with two possible states, *non-morphological* and *morphological*. When a morphological affix is present in a word, the state determines whether the affix is a part of the root word (in state *non-morphological*) or is a morphologically inflected word (in state *morphological*). In the current study, we present an HMM based algorithm to predict the hidden states of the generator. Our experiments show that our approach can stem inflected word with single character suffixes with an accuracy of 91%.

Our formulation of the problem in the form of a Hidden Markov Model parallels the well-known problem of "*Fair Bet Casino*", where a sequence of rolls of a dice find whether a dealer uses a fair dice or a loaded one. We model the commentator or writer as a generator of a sequence of words, $w_0, w_1, \cdots, w_{n-1}$, i.e., the words of a corpus in the order it is intended to be read. Each word w_i can be broken down as $p_i \circ s_i$, where p_i is a root word; s_i an inflectional suffix and $\circ$ the concatenation operation between two strings. We denote the set of inflectional suffixes by S, including the empty string ϵ. If w is a root word, $p \circ \epsilon$ is also the root word. For any word $w \equiv p \circ s$ if $s = \epsilon$, we say word w is a *root*

---

[2] http://timesofindia.indiatimes.com (*Access date : 22-Nov-2012*)
[3] http://janasadharan.in   (*Access date : 22-Nov-2012*)
[4] http://www.anandabazar.com   (*Access date : 22-Nov-2012*)
[5] http://www.jagran.com   (*Access date : 23-Nov-2012*)

Table 3. Calculated result for Assamese using [12] approach

		Accuracy in %
Correctly stemmed	81%	
No inflection (ϵ)		43%
One character inflection (S_1)		36%
Multiple character inflection (S_m)		21%
Wrongly stemmed	19%	
There is no inflection but stemmed as inflected	66%	
Mark as one character inflection (S_1)		62%
Mark as multiple character inflection (S_m)		38%
There is one character inflection, but stemmed wrongly	27%	
Mark as no inflection (ϵ)		83%
Mark as multiple character inflection (S_m)		17%
There is multiple character inflection, but stemmed wrongly	17%	
Mark as one character inflection (S_1)		32%
Mark as multiple character inflection (S_m)		12%
Mark as no inflection (ϵ)		56%

word. On the other hand, $s \in S$ and $s \neq \epsilon$, for any word w $(= p \circ s)$ implies that word w ends with an inflectional suffix but does not necessarily mean that $p \circ s$ is not a root word or the converse. We model this problem of predicting if a word in a sentence is morphologically inflected or not as being able to model the sense of the generator of the sentence when the word was written. Suppose we are given a set of inflections S in the language, not necessarily all inflections in the language. We can represent any given word w as $p \circ s$ such that $s \in S$. If $s = \epsilon$ is the only possible string of S that satisfies $w = p \circ s$, we say the generator G does not produce meaning leading to a morphological inflection for the word. On the other hand, if there is an inflection $s \in S$ and $s \neq \epsilon$ such that $w = p \circ s$, we say w is morphologically inflected whether the generation is meaningful. Therefore, we define two states of the generator at the time of generating the word, *viz.*, *morphologically inflected* (M) and *morphologically not inflected* (N). We associate with a corpus of some length ℓ, $w_0, w_1, \cdots w_{\ell-1}$ a series of states with labels N and Ms as $q_0, q_1, \cdots, q_{\ell-1}$ such that $q_i \in Q \equiv \{N, M\}$. For example in Table 4, we described the series of states of a sentence, "নবীনহঁতৰ ঘৰ আমাৰ ঘৰৰ পৰা এমাইলমান দূৰত"

TF: *nabinhatar ghar aAmAr gharar parA emAilmAn durat.*

WT: nabin's(plural) house our house from one-mile distance

Therefore, for a corpus generated by G the problem of deciding if a word is morphologically inflected, boils down to determining the state of G (N or M) at the exact moment of generating the word. We construct an HMM based algorithm to predict the states of G corresponding to the words of the corpus. Therefore, the problem has two steps: (a) training the HMM parameters with a training corpus and (b) applying the calibrated algorithm on a test corpus to detect morphologically inflected words.

Table 4. An example sentence as modelled using our generative model of the text for the morphological inflections

w	w_0	w_1	w_2	w_3	w_4	w_5	w_6
words	নবীনহঁতৰ	ঘৰ	আমাৰ	ঘৰৰ	পৰা	এমাইলমান	দুৰত
	(nabinhatar)	(ghar)	(aAmAr)	(gharar)	(parA)	(emAilmAn)	(durat)
p	নবীন	ঘৰ	আমাৰ	ঘৰ	পৰা	এমাইল	দুৰ
	(nabin)	(ghar)	(aAmAr)	(ghar)	(parA)	(emAil)	(dur)
s	-হঁতৰ	ϵ	ϵ	-ৰ	ϵ	-মান	-ত
q	M	N	N	M	N	M	M

We know that the inaccuracy of the method in [12] comes mostly from single letter inflections. For multiple letter inflections, the ambiguity of being a true inflection versus a coincidental match of the word with the set of inflections is significantly low. We denote by S_1 and S_m the set of single letter and multi-letter inflections, respectively. In order to simplify our analysis, we consider the following partition of the set of inflections S as $\{\epsilon\}$, S_1 and S_m. Therefore, the appearance of a multi-inflection suffix on a word almost definitely generates the presence of morphological inflection. Hence, we can safely assume that if $s_i \in S_m$ for a word w_i, $q_i = M$. We can state the same notion as for $q_i = N$, $e_{q_i}(s) = 0$ for $s \in S_m$. Since we are essentially trying to predict the correct state of G for only single letter inflections (i.e., S_1), we assume that all inflections in S_1 are equivalent and, similarly the inflections in S_m are also equivalent to one another. So, we assume that our alphabet S in the Hidden Markov Model as $S' = \{\epsilon, s_1, s_m\}$, where s_1 and s_m are single-letter and multi-letter morphological inflections, respectively.

Estimating $a_{k\ell}$ and $e_k(b)$. We estimate the two needed parameters $a_{k\ell}$ and $e_k(b)$ from the training corpus. First we mark the states of the generator, G for every word in the corpus. Next, we identify the inflections, for every word, as belonging to $\{\epsilon\}$, S_1 and S_m if it has no inflection, has a single letter inflection or has a multi-letter inflection, respectively. Then, we calculate the number of times each particular transition and emission occurs in the training corpus. Let us denote these counts by $A_{k\ell}$ and $E_k(b)$. Then estimate the the parameters $a_{k\ell}$ and $e_k(b)$ as

$$\hat{a}_{k\ell} = \frac{A_{k\ell}}{\sum_{\ell'} A_{k\ell'} + \delta} \ and \ \hat{e}_k(b) = \frac{E_k(b)}{\sum_{b'} E_k(b') + \delta}$$

where δ is a very small positive number to avoid division by 0.

5   Results and Discussion

Preparation of training data. For our experiment, we used text from the EMILLE[6] Assamese corpus. We labelled approximately 2,000 words (144 sen-

---

[6] http://www.emille.lancs.ac.uk/

tences) with 4 tags: words with multi-character inflection (M_{sm}), words with single character inflection (M_{s1}), words with no inflection (N_e) and words that have no inflection, but end with single character inflection marker (N_{s1}). Table 5 gives the details suffixes present in the training set. We found the suffix 'ৰ', *genitive case marker* and the suffix symbol ৹, *nominative case marker* are most frequently among single character suffixes.

Table 5. Training corpus details used for experiment

Words with single character inflection (S_1)	34%
Words with multiple character inflection (S_m)	21%
Words with no inflection (ϵ)	43%
Number of foreign words, numbers and symbols	2%

Result and Analysis. The results obtained using the prior approach [12] have already been given in Table 3. Out of 19% words that the published method stems incorrectly, 27% of the words have single character inflection. After applying HMM to detect single character inflection, the overall accuracy increases by approximately 11.43%. The results obtained by combining previous approach with HMM are given in Table 6. Our test data set contains 1542 words (108 sentences) taken from EMILLE corpus. We manually evaluate the correctness of the output. We have to keep in mind that the previous approach used a frequent word lists of around twenty thousand words and a rule-base. In Table 6, "Stemmed as no inflection" means, either a single letter or multiple letter suffix was attached with the word and marked incorrectly as no inflection. "Stemmed as single character inflection" means that there was no inflection or multiple inflection, but the program separated the last character from the word incorrectly. Similarly "Stemmed as multiple inflection" means that there was no inflection or

Table 6. Comparison of obtained result

	[12]	Current paper	Morfessor
Correctly stemmed	81%	92%	82%
Incorrectly stemmed	19%	8%	18 %
Stemmed as no inflection	23%	36%	29%
Stemmed as single character inflection	57%	33%	19%
Stemmed as multiple inflection	20%	31%	52%

a single character inflection and the program separated a sequence of characters from the word incorrectly. The same test data was used to run the experiment with Morfessor [18] as well.

The "transition probability" controls the way a state at time t is chosen over a given state at time $t - 1$. Table 7 gives transition probabilities for the training set, where S_0 is the initial state, M is the inflected form of the word and N is the root form of a word. The "emission probability" is the probability of observing the input sentence or sequence of words W given the state sequence T, that is $P(W|T)$. Table 8 describes the emission probabilities for the training set, where S_0 is the initial state, M is the inflected form of the word, N is the root form of a word, ϵ is the zero inflectional form, s_1 is the single character inflection and s_m is the multiple character inflectional form.

Table 7. Transition probabilities for the training set

	S_0	M	N
S_0	0.0000	0.5000	0.5000
M	0.0000	0.4269	0.5716
N	0.0000	0.4739	0.5261

Table 8. Emission probabilities for the training set

	ϵ	s_1	s_m
S_0	1.0000	0.0000	0.0000
M	0.0000	0.6705	0.3295
N	0.5557	0.4443	0.0000

Evaluation. Comparison of Table 1 and Table 6, demonstrates that the performance of the current approach is better, for Assamese. We evaluate stemmer strength using [19]. Table 9 shows the evaluation results for both stemming techniques. According to [19], a conflation class is, the number of unique words before stemming (N) divided by the number of unique stems after stemming (S), i.e., the average size of groups of words converted to a particular stem. The index compression factor, $(N - S)/N$ takes into account the collection of unique words compressed by the stemmer. Thus high index compression factor

Table 9. Evaluation of stemmer strength using [19]

	[12]	Current Paper	Morfessor
Words in the test file	1542	1542	1542
Unique words before stemming	1010	1010	1010
Unique words after stemming	859	810	721
Min./Max. words length after stemming	1/18	1/18	1/18
Number of words per conflation class	1.17	1.24	1.40
Mean stemmed word length	5.36	6.03	4.94
Index compression factor	0.15	0.20	0.29

represents a heavy stemmer. A heavy stemmer produces over-stemming,as it removes sequences of characters from words that are do not contain any suffix. For example, Morfessor separates words such as প্রয়োজন (*prayojan* : need), আয়োজন (*aAyojan* : arrangement) and মানুহজন (*mAnuhjan* : the man) into a single group removing the suffix -*jan* from each of the words, Whereas the first two words are not inflected and are root words, the last word *mAnuhjan* is inflected with the definitive marker -*jan*, although all the words are ends with -*jan* suffix.

6   Conclusion

In this paper, we have presented a stemmer for Assamese, a morphologically rich, agglutinating, and relatively free word order Indic language. In this language, the presence of single letter suffixes is the most common reason for ambiguity in morphological inflections. Therefore, we propose a new method that combines a rule based algorithm for predicting multiple letter suffixes and an HMM based algorithm for predicting single letter suffixes. The resulting algorithm uses the strengths of both algorithms leading to a higher overall accuracy of 92% compared to 81% for previously published methods for Assamese. The 92% result is better than the published results for all other Indian languages(see Table 1). Future work will include calibrating the parameters of the HMM model with a much larger training corpus. In addition, it would be interesting to explore the possibility of modelling all possible morphological variations using Conditional Random Fields, which has been very successful in similar situations. It will also be useful to apply the method to other highly inflectional languages.

References

1. Porter, M.F.: An algorithm for suffix stripping. Program 14, 130–137 (1980)
2. Ramanathan, A., Rao, D.: A lightweight stemmer for Hindi. In: Proceedings of the 10th Conference of the European Chapter of the Association for Computational Linguistics (EACL), on Computatinal Linguistics for South Asian Languages, Budapest, pp. 43–48 (2003)
3. Majumder, P., Mitra, M., Parui, S.K., Kole, G., Mitra, P., Datta, K.: Yass: Yet another suffix stripper. ACM Trans. Inf. Syst. 25(4) (October 2007)
4. Pandey, A.K., Siddiqui, T.J.: An unsupervised Hindi stemmer with heuristic improvements. In: Proceedings of the Second Workshop on Analytics for Noisy Unstructured Text Data, AND 2008, Singapore, pp. 99–105 (2008)
5. Aswani, N., Gaizauskas, R.: Developing morphological analysers for South Asian Languages: Experimenting with the Hindi and Gujarati languages. In: Proceedings of the Seventh Conference on International Language Resources and Evaluation (LREC), Malta, pp. 811–815 (2010)
6. Kumar, D., Rana, P.: Design and development of a stemmer for Punjabi. International Journal of Computer Applications 11(12), 18–23 (2010)
7. Majgaonker, M.M., Siddiqui, T.J.: Discovering suffixes: A case study for Marathi language. International Journal on Computer Science and Engineering 04, 2716–2720 (2010)
8. Sharma, U., Kalita, J., Das, R.: Unsupervised learning of morphology for building lexicon for a highly inflectional language. In: Proceedings of the ACL 2002 Workshop on Morphological and Phonological Learning, Philadelphia, pp. 1–6 (2002)
9. Sharma, U., Kalita, J., Das, R.: Root word stemming by multiple evidence from corpus. In: Proceedings of 6th International Conference on Computational Intelligence and Natural Computing (CINC 2003), North Carolina, pp. 1593–1596 (2003)
10. Sharma, U., Kalita, J.K., Das, R.K.: Acquisition of morphology of an indic language from text corpus. ACM Transactions of Asian Language Information Processing (TALIP) 7(3), 9:1–9:33 (2008)

11. Saharia, N., Sharma, U., Kalita, J.: Analysis and evaluation of stemming algorithms: a case study with Assamese. In: Proceedings of the International Conference on Advances in Computing, Communications and Informatics, ICACCI 2012, Chennai, India, pp. 842–846. ACM (2012)
12. Saharia, N., Sharma, U., Kalita, J.: A suffix-based noun and verb classifier for an inflectional language. In: Proceedings of the 2010 International Conference on Asian Language Processing, IALP 2010, Harbin, China, pp. 19–22. IEEE Computer Society (2010)
13. Al-Shammari, E.T., Lin, J.: Towards an error-free Arabic stemming. In: Proceedings of the 2nd ACM Workshop on Improving Non English Web Searching, iNEWS 2008, pp. 9–16. ACM, New York (2008)
14. Gaustad, T., Bouma, G.: Accurate stemming of Dutch for text classification. Language and Computers 14, 104–117 (2002)
15. Suba, K., Jiandani, D., Bhattacharyya, P.: Hybrid inflectional stemmer and rule-based derivational stemmer for Gujrati. In: 2nd Workshop on South and Southeast Asian Natural Languages Processing, Chiang Mai, Thailand (2011)
16. Ram, V.S., Devi, S.L.: Malayalam stemmer. In: Parakh, M. (ed.) Morphological Analysers and Generators, LDC-IL, Mysore, pp. 105–113 (2010)
17. Bora, L.S.: Asamiya Bhasar Ruptattva. M/s Banalata, Guwahati, Assam, India (2006)
18. Creutz, M., Lagus, K.: Induction of a simple morphology for highly-inflecting languages. In: Proceedings of the 7th Meeting of the ACL Special Interest Group in Computational Phonology: Current Themes in Computational Phonology and Morphology, SIGMorPhon 2004, Barcelona, Spain, pp. 43–51. ACL (2004)
19. Frakes, W.B., Fox, C.J.: Strength and similarity of affix removal stemming algorithms. SIGIR Forum 37(1), 26–30 (2003)

Semi-automatic Acquisition of Two-Level Morphological Rules for Iban Language

Suhaila Saee[1,2], Lay-Ki Soon[2], Tek Yong Lim[2],
Bali Ranaivo-Malançon[1], and Enya Kong Tang[3]

[1] Faculty of Computer Science and IT, Universiti Malaysia Sarawak,
Jalan Datuk Mohd Musa, 94300 Kota Samarahan, Sarawak, Malaysia
[2] Faculty of Computing and Informatics, Multimedia University,
Persiaran Multimedia, 63100 Cyberjaya, Selangor, Malaysia
[3] School of Computer Science and Information Technology,
Linton University College, Mantin, Negeri Sembilan, Malaysia

Abstract. We describe in this paper a semi-automatic acquisition of morphological rules for morphological analyser in the case of under-resourced language, which is Iban language. We modify ideas from previous automatic morphological rules acquisition approaches, where the input requirements has become constraints to develop the analyser for under-resourced language. This work introduces three main steps in acquiring the rules from the under-resourced language, which are morphological data acquisition, morphological information validation and morphological rules extraction. The experiment shows that this approach gives successful results with 0.76 of precision and 0.99 of recall. Our findings also suggest that the availability of linguistic references and the selection of assorted techniques for morphology analysis could lead to the design of the workflow. We believe this workflow will assist other researchers to build morphological analyser with the validated morphological rules for the under-resourced languages.

Keywords: Morphological rules, Rules extraction, Under-resourced language, Morphological analyser.

1 Introduction

Morphological analyser is a first processing task requires in Natural Language Processing (NLP). Morphological rules are crucial components in the analyser in order to analyse and generate the input word. The conventional method in acquiring the rules for morphological analyser was done using handcrafted, which has led to an ambiguity [1]. Therefore, the acquisition of the rules has received much attention from the researchers to automate the acquisition of the rules [2]. To automate the acquisition of the rules, there are two main components required as input: a) sufficient of linguistic references i.e. dictionary with stems and inflected words, the classification of words and affixes as well as a training data set and b) the selected techniques to acquire the rules that are depending

A. Gelbukh (Ed.): CICLing 2013, Part I, LNCS 7816, pp. 174–188, 2013.

on the availability of input types. The aim to acquire the rules automatically has been achieved with the sufficient linguistic references and precise techniques. However, we encountered the needed requirements became constraints to develop the analyser for under-resourced language (U-RL) by taking Iban language as a study case. This is because, at the morphological level, Iban language has insufficient linguistic references in terms of morphological rules, no morphological analyser for Iban yet and lack of linguists. While, selection of various techniques available for morphological analyser and generator that accommodate with Iban language has led to the other constraint at the linguistic level. Iban language is spoken by the Iban, a largest ethnic group in Sarawak, Malaysia formerly known as the Sea Dayaks [3]. Since 2008, Iban language has been adapted as one of the Malaysian Certificate of Education (fifth form) examination subject due to the significant of the language [4].

An objective of our work is to determine the morphological rules for building Iban morphological analyser. Previous works have only focused on the method of acquiring the rules automatically from the sufficient resources. Since this is a first research work conducted for under-resourced languages in Sarawak, we are hoping to fill the gap in term of acquiring morphological rules by inducing morphological information from raw text in a semi-automatic way, in the case of Iban language. The result from the induction process will be validated to ensure the correctness of the acquired morphological information in the later stage. The validated information will be used to extract the candidate rules and later to be applied in the two-level morphology.

Section 2 surveys the works related to acquisition of corpus and morphological rules and formalism to build the morphological analyser. Our proposed semi-automatic workflow and its components from the acquisition of corpus to Iban morphological analyser are presented in section 3. Section 4 describes standard metrics to be used in evaluation, and discusses the results from the analyser.

2   Related Works

Corpus and morphological rules are main requirement input for the morphological analyser to analyse and generate a structure of word. As one could derive the rules from the corpus; therefore, corpus acquisition plays an important role in building the automatic morphological analyser. For resourced languages like English, the corpus can be obtained freely from the Internet or any trusted sources such as linguists or web crawler. However, a main problem encountered for under-resourced languages is no resources at all in term of unavailability of the internal structure when word is analysed. For the case of under-resourced languages, the corpus can be obtained from the three types of sources that are dictionary, written text, and reference grammar book. The sources can be in either hardcopy or softcopy version. According to [5], if the sources are available in the hardcopy version, transformation into digitisation needs to be done urgently in order to get the softcopy version. Then, once can proceed to the next process, which is acquiring the morphological rules including morphographemic rules and morphotactics (also known as morphosyntax) information.

There are two possible ways to acquire the morphographemic rules, for under-resourced languages and resourced languages, either manual or automatic. For a manual acquisition, the linguists are required to hand-craft the rules rather applying machine learning technique in an automatic way. The two-level morphology invented by [1] had successfully implemented the rule-based approach that the acquisition of the rules is hand-crafted by the linguists in two-level format. Since the process requires a lot of linguistic works and expertise in preparing hundred or more input, [2] proposed an automatic acquisition of two-level morphological rules. The proposed approach involving segmentation of a set of pairs stem and inflected word, and from that determines the desirable two-level rule set. The word pairs were extracted from machine-readable dictionaries of Xhosa, agglutinative language, and African. In this approach, the dictionary that provides stems and inflected words was the main requirement. More recently, [6] adapted and upgraded [2] model into different languages in their works. They had implemented the model for agentive nouns in two languages i.e. English and Macedonian. They also used the word pairs of base and derived words and segmented it using Brew edit distance, the best result gained after compared with the other edit distance algorithms, rather sequence edit distance as implemented in the model. Besides, [7] also segmented the word pairs of base and inflected word of Tagalog, a morphologically complex language, to acquire the rules. The word list used to supervise the analyser was derived from Tagalogs Handbook.

For the morphotactics information, the acquisition could be in either hand-crafted rules/ lexicon building or re-write for learned. There have been a number of studies on lexicon building of the information. In [8] work, the requirement of large lexicon building for the prototype of morphological analyser for Aymara, a highly agglutinating language, was the crucial part. Indeed, the lexicon included two types of dictionaries that written in XML format which were root and suffix. In contrast, [9] approach has applied morphological paradigms in their morphological analysis for Hindi, Telugu, Tamil and Russian languages. As our work is closest to Bharati's approach, we thus adapted the paradigm method into our work. However, in his work, the paradigms were including the morphographemic and morphotactis (suppletion) processes. Meanwhile, we only adapted the morphotactics information in the paradigm class without part of speech as we only have stem and segmented word at the moment.

3   Our Approach

Four main steps involved in this work including wordlist acquisition, stems and prefixes acquisition, converting stems and prefixes into the two-level format and Iban morphological analyser. Fig. 1 depicts the morphological rules acquisition methods applied in the existing approaches that lead to the proposed approach.

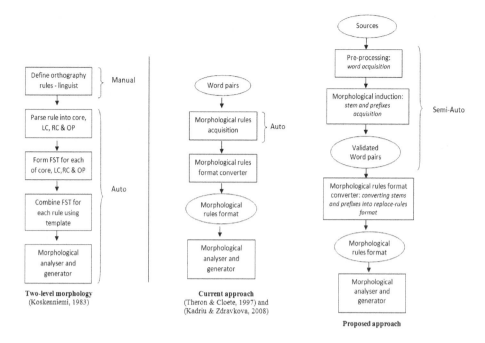

Fig. 1. Comparison between the existing and proposed approaches

In this section, we shall discuss further on each of the steps required from the proposed approach.

3.1 Step 1: Wordlist Acquisition

The first step in this work requires three activities in preparing the wordlist, which are data collection, data transformation, and data compilation. The purpose is to create a corpus and wordlist from the raw text, which will be used as input in the later processes. The activities shall be discussed as follows.

Data Collection. The required linguistic references to construct the corpus are a dictionary with derivation words, written texts, and a reference grammar book with examples of sentences. Similar to this work, [10] and [11] also used reference grammar books to avoid labour-intensive resources creation. Moreover, Feldman discovered that the reference grammar books are a perfect starting point for automatic morphology analysis.

Transformation. Digitisation including Optical Recognition Characters (OCR) and conversion is the process taken to deal with either a hardcopy or softcopy format of materials.

Compilation. Since our research is on building a morphological analyser, our linguistic units are words. Thus, we need to extract the words from all previously acquired linguistic references. We extracted the specific information from each of the linguistic references, then, we created new text files for each of the extracted information. Lastly, we compiled the information all together into one text file to get a wordlist form. To ensure the quality of our wordlist, we corrected all typos or unknown characters. The created wordlist should be error free, in other words, cleaned data. This wordlist will be the input of the morphology induction. Table 1 indicates a summary of the required references.

Table 1. Linguistic references for wordlist acquisition

References	Electronic version	Hardcopy version
Dictionary	Extraction of entry and sub-entry	Scanning, OCR & post-editing
Grammar book	Words extraction from example of sentences	Manually type the examples
Written text	Words extraction	Scanning, OCR & post-editing

3.2   Step 2: Acquiring Candidate Stems and Prefixes

At this second step, we shall present only the work done on the automatic recognition of prefixes and stems. Therefore, the morphology induction process has been settled to acquire stems and prefixes from the wordlist. This process enabled us to induce morphological information without prior to linguistic knowledge. Indirectly, the process could minimize human expertise control. This sub-section described our morphology induction workflow.

Feeding the Wordlist to *Linguistica*. The software application used to induce the morphological information from the wordlist was *Linguistica*, an open source tool [12]. Similar with other softwares i.e. Morfessor [13], Paramorph [14], *Linguistica* also applies an unsupervised machine learning technique that based on the frequencies of patterns of words that can be calculated at different levels of granularity. The requirement size of data from *Linguistica* is minimum 5000 words up to 500,000 words in order to get accuracy result. In this work, we have chosen *Linguistica* as we would like to show that our workflow is able to acquire the morphological rules without considering a good or poor result produced by one tool.

In general, there were two types of results produced by *Linguistica* which were unsegmented and segmented words. For the unsegmented words, it was produced due to its low frequency of words in the segmentation and detected to appear only one time. Thus, this file would be a potential test set for the morphological analyser later. While, the segmented words file used and analysed according to users need. For example, the previous works used *Linguistica* to find the allomorphy [15] and to see the morphological patterns from the generated signature of the desired languages [16]. More details on the *Linguistica* can be found in [12]. Fig. 2 shows the interface of *Linguistica* which its screen is divided into two parts. The first part, on the left, is known as the tree that showing general information of the selected corpus. While, the second part, on the right, is known as the collections and text. From the collections, we could get detailed information of corpus like list of stems, affixes and signatures. In contrast, the text is used for feedback to user in a few cases.

Fig. 2. An interface of *Linguistica*

In this work, we demanded the information of prefixes and stems therefore; *Linguistica* generated four main files which were list of *signatures*, *words*, *stems* and *prefixes*. The information from the *signature* file returned a maximal set of stems and suffixes with the property that all combinations of stems and suffixes were found in the wordlist. While, the *list of words* file held frequency of one word appeared in the wordlist. Since our work focused on the acquisition of prefixes and stems, we could extract the required information from the *list of stems* file by referring to the information in the *corpus count* and *affix count* columns. The *list of prefixes* file gave us the number of prefixes extracted from the wordlist are those that we have considered as candidate prefixes. We used the term candidate because, at this stage, we are still unsure about their real status yet. Fig. 3 shows a result generated from the *list of stems* file.

Extracting Result from *Linguistica*. To get a list of possible candidate prefixes and stems, we automatically extracted the morphological information, namely, list of stems and prefixes from the list of *stems* file in the *Linguistica* result [17]. The extracted information will be input to the validation process for a later stage. Fig. 4 shows an example of the list of candidate.

3.3   Step 3: Validating Candidate Stems and Prefixes

We proceeded with an automatic validation with a purpose to ensure the generated rules later were already validated. The availability of the list of candidate

```
# Stem Count
# ------------
   477

# Index | Stem        | Confidence      | Corpus Count | Affix Count | Affixes
# -------------------------------------------------------------------------------
   1       gagai         PF_1              447            3            NULL be n
   2       ngajar        PF_1              215            2            NULL pe
   3       undan         PF_1              118            2            NULL ng
   4       nataika       PF_1              98             3            NULL nge ngea
   5       ngasuh        PF_1              73             2            NULL pe
   6       diaka         PF_1              38             2            nye nyen
   7       gumpul        PF_1              38             3            NULL be n
   8       ngarang       PF_1              31             2            NULL pe
   9       nyamai        PF_1              23             2            NULL pe
  10       engkebang     PF_1              16             2            n t
```

Fig. 3. Example of a result from the *list of stems* file

INDEX	STEM	PREFIXES
1	gagai	NULL be n
2	ngajar	NULL pe
3	undan	NULL ng
4	nataika	NULL nge ngea

Fig. 4. Example of the list of candidate stems and prefixes

prefixes and stems as listed in Fig. 4 above would be checked with existing morphology resources. The morphology resources are referring to the Iban dictionary
and list of prefixes [1].

Table 2. Cases for Validation

File type	Cases
ValidatedWordList	Case A: stem = Yes & prefix = Yes Case B: stem = Yes & prefix = NULL Case C: stem = No & prefix = Yes
ValidatedWord	Case D: stem = No & (prefix + stem) =Yes
InvalidatedWordList	Case E: stem = Yes & prefix = No Case F: stem = No & (prefix + stem) = No Case G: stem = No & prefix= NULL

Table 2 above highlights seven cases used in validating the list of candidate
stems and prefixes. The seven cases resulted three types of validated files that are

---

[1] The list of prefixes is obtained from the *list of prefixes* file as discussed in the
section 3.2

ValidatedWordList, ValidatedWord and *InvalidatedWordList*. From the *Validated WordList* file, we obtained morphological information for simple concatenation morphology. The information could be used as input in generating the morphological rules. In the *ValidatedWord* file, a rule of thumb is a combination of stem and prefix are accepted as the word existed in the dictionary. Meanwhile, the *InvalidatedWordList* file considers the existence stem should be rejected because the prefix either was not existed or NULL from the list of prefixes. We attempted to minimize the number of invalid data in this file to get maximum data of *ValidatedWordList*. In fact, we have found the morphological information for non-concatenation morphology like alternation phenomena from the last two files. For instance, phonological changes of *n* to *t* from two words: *n*aban and *t*aban.

3.4   Step 4: Representing the Validated Rules in Two-Level Format

Similar to the previous works [2,6], we were applying the two-level approach and implementing the Iban morphological analyser in *XFST* tool, *Xeror* software. An objective of this step is to show that the validated morphological rules could be applied in the two-level morphology. Thus, our next step was to convert the list of candidate stems and prefixes from the *Linguistica* result in the two-level format in an automatic mean. Then, manual correction of the converted rules in the required format was taking place. Following were activities taken in the process:

Morphological Rules Extraction. After the validation process, we now consider the validated list of stems and prefixes as the validated morphological rules. The information that we can derive from the rules is morphographemic and morphosyntactic information. However, the acquisition process will take place to differentiate the information. At this step, we are required to decide which formalism suits best with the information. Lastly, the information will be fed into *XFST* tool. The activities involve shall be discussed as follows.

Morphological Acquisition. The morphographemic information was acquired from the construction of stems and prefixes, which obtained from the *Validated-WordList*. Besides, we could derive the morphotactics information from both two validated files that were *ValidatedWord* and *InvalidatedWordList*. We applied morphological paradigm in acquiring the morphotactics information since we only have the stems and construction of the stem and prefix. In this work, we applied paradigm prefixes that are considered similar to paradigm classes with the purpose to put in one group of prefix the stem and derivation words. This is consequent to the insufficient morphological information. See Fig. 5. We categorised the list of morphotactics information based on the type of prefixes. An objective is to identify the similarities of the words in term of its pattern.In fact, the prefixes are obtained from the list of prefixes with the highest frequent occurrence. The list of morphotactics information consists of the prefixes that attached to its stems and derivation words.

Stem	Construction
1. andau	ŋgandau
2. ajar	ŋgajar
3. udah	ŋgudah
4. antam	ŋgantam

Fig. 5. Example of morphological paradigm

Morphological Formalism. Two-level morphology formalism is used in this experiment and implemented in the *XFST* tool as mentioned earlier. Thus, the morphographemic information would be represented in a replace-rules format. Fig. 6 depicts the morphographemic rules in the replace-rules format. Meanwhile, the morphotactics information would be written in lexc, a lexicon to describe the spelling structure, which will be represented in a pattern-root format. See Fig. 7.

```
clear stack
define lbraces "^[" | "[" | "{" ;
define  IbanAssRules [f] ->l || .#. _
.o.
0 -> n g || %^N lbraces* _ ,, %^N -> 0 || _ lbraces* [ a | e | i
| u ] ! pengurus -> urus+peN- / ngarah -> arah+N-
.o.
[ g | k ] -> n g || %^N lbraces* _ ,, %^N -> 0 || _ lbraces* [ g
| k] ! pengurus -> kurus+peN- / ngurus -> kurus+N-
.o.
[ b | p ]-> n g e m || %^N lbraces* _ ,, %^N -> 0 || _ lbraces* [
b | p ] ! pengemesai -> besai+peN-/ ngemesai -> besai+N
```

Fig. 6. Morphographemic rules in replace-rules format

3.5   Step 5: Iban Morphological Analyser

The rules then would be fed into the *XFST* tool, to implement the Iban morphological analyser. The morphographemic rules were compiled into a rule transducer while, the morphotactics information compiled into a lexicon transducer. The lexicon and rule transducers were combined to create a single lexical transducer (a single FST) that could be applied to perform either analysis or generation. The analyser requires lexicographer to review the results in deciding which roots are valid since it analysed all possible stems. Fig. 8 shows a sample of output from the Iban morphological analyser.

```
! definitions of the Cons and Vows
Definitions
C = [b|c|d|f|g|h|j|k|l|m|n|p|q|r|s|t|v|w|x|y|z] ;
V = [a|e|i|o|u] ;

LEXICON Root
        FDPre3 ;

LEXICON FDPre3
< [ 0 .x. b e r ]^"@P.PREF.ber@" > Stems3 ;
< [ 0 .x. b e s ]^"@P.PREF.bes@" > Stems3 ;
< [ 0 .x. p e r ]^"@P.PREF.per@" > Stems3 ;

LEXICON                    Stems3
<[V C (V)] ^ {2,4}>        Pref3;       ! amal
<[(V) (C) C] ^ {2,4}>      Pref3;       ! antam
<[V (C) (C)] ^ {2,4}>      Pref3;       ! andau
<[C V (C)] ^ {2,4}>        Pref3;       ! lari
<[(V) (C) (C)] ^ {2,4}>    Pref3;       ! anchau

LEXICON    Pref3
        ber- ;
        bes- ;
        per- ;

LEXICON ber-
< "+ber-":0 "@R.PREF.ber@" >    #    ;

LEXICON bes-
< "+bes-":0 "@R.PREF.bes@" >    #    ;

LEXICON per-
< "+per-":0 "@R.PREF.per@" >    #    ;
```

Fig. 7. Morphosyntactic information in lexc format

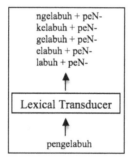

Fig. 8. Result from the Iban morphological analyser

4 Evaluation

In this section, we evaluated the Iban morphological analyser in terms of morphological rules generation and the performance of analyser.

4.1 Experimental Setup

Corpus Creation. We used list of entries under letter N in preparing data sets to induce the morphological information. The consequence was that letter N has the highest number of possible prefixes among other letters in alphabet.

Test Set Preparation. As mentioned earlier in the section 3.2, we are using the unsegmented data produced by *Linguistica*. Thus, we have two types of test sets that were a) 2400 unsegmented data and b) Top 100 words and 100 random words from the 2400 unsegmented data. The objectives of using a different test sets were because we attempted:

- To see:
    - the coverage in term of the number of morphological phenomena and word-formation that have been covered and yet to be covered with the generated rules.
    - the accuracy of the generated rules in analysing and generating the input word (test set (a)).
- To understand in details type of errors in the morphological analyser which are a) 100 top words and b) 100 random words both from the 2400 unsegmented data (test set (b)).

Evaluation Metric. We used two standard metrics that were precision and recall for evaluating the coverage and accuracy of the analyser.

4.2   Result Analysis

To discover types of morphological phenomena and word-formation analysed by the Iban morphological analyser, we have conducted a quantitative analysis. For the first analysis, we were using test set (a), 2400 unsegmented data. The following sub-sections shall discuss in details the overall analyses. See Fig. 9.

A pie chart is showing the number of correct and incorrect analyses with the 25 generated rules evaluated using 2400 unsegmented data. The 25 rules are *n, ny, ng, ngem, nge, ne, be, bes, ber, bete, beke, ke, per, dipe, che, te, se, me, en, eng, pe, pen, penge, sepe* and *sepen*. Out of 25 rules, six rules were never applied in the data, e.g. *be, ber, bes, beke, bete* and *dipe*. Although the validated rules were totally correct from the *Linguistica*, these rules have not been applied in either analysing or generating the unsegmented data. While, the other 19 validated rules have analysed 937 correct data and 1463 incorrect data. As we can see from Fig. 9, the data contains 47% of root words. This indicates that the root words were the majority in the test sets. On the other hand, 53% of words have been analysed as 41% of correct and 12% of incorrect analysis. From the analysis, it shows the morphological analyser was able to analyse simple concatenation and non-concatenation morphology. However, there were a number of morphological phenomena and word-formations that the morphological analyser still unable to cover when we see from the percentage of the incorrect analysis which will be further explained in error-analysis section.

Morphological analyser can be evaluated by using standard measures, which are precision and recall. In this work, we evaluated the accuracy of the extracted morphological rules from the precision. From the precision result, 0.7656, it shows the accuracy of the morphological analyser is nearly to 1.000. This was due to the nonexistent of the root word in the dictionary. For instance, the word

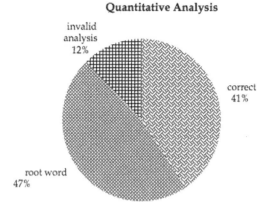

Fig. 9. Validated rules applied on 2400 unsegmented data

temegah should not be analysed as *te+megah* as it is a root word. Instead, the analyser should analyse the word as *temegah*. Since the morphological analyser found prefix *te-* existed in the list of prefix, thus, the analyser has analysed as *te+megah*. Therefore, there is still a room for improving the rules so that the morphological analyser able to differentiate between the one should be analysed and a root word. In order to achieve 1.000, we should enhance the lexical entry from the dictionary with the derivation words.

To determine the coverage of the Iban morphological analyser, we measured the coverage of the analyser from a recall. From the recall result, 0.9928, the analyser still has not covered yet for overall full reduplication and circumfixation, although it covers only two types of partial reduplication i.e. *che* and *ne*. After examined on the real data, the first test set (a), the morphological analyser could handle the non-concatenative morphology besides simple concatenation such as alternation, which are *peN* and *sepeN*.

We tested on the test set (b) to investigate the error analysis for the rules extraction in a quantitative analysis which were a) 100 top words and b) 100 random words both from the 2400 unsegmented data. From here, we have manually identified the reasons of the failures.

Results of the error analysis are shown in Fig. 10. There are two broad categories of error types which: *wrong segmentation* and *the rules were not applied* in the segmentation. Specifically, the 100 top data reached the highest peak of 23 data of the wrong segmentation at the circumfixation error type as the inexistent of the circumfix rules in the morphological analyser. Similar to the 100 random data, which has the same level of the highest peak of 23 data of the wrong segmentation at the root word error type. This was due to the insufficient of lexical entry in the Iban dictionary. From the graph, the top data felt down when there were no particular rules to be applied on the full and partial reduplications. For the random data, none of the rules had been applied on 7 root words as there were no prefixes matches to the inputs.

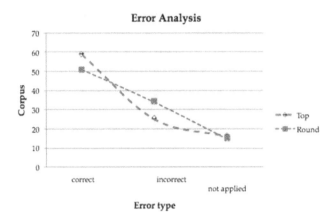

Fig. 10. Error Analysis for 100 top and 100 random unsegmented data

The error analysis we examined from the second test sets (b) is discussed as follows.

1. Wrong segmentation:
    (a) Unknown prefixes - The morphological analyser returned unknown prefixes for the simple concatenation. For example, *tebe, s, diper, sel* and *ter*. The analyser was able to analyse it because *te* existed in the list of prefixes. However, a word of *teberumpang* has been analysed as *berumpang + te-* which supposed to be segmented as *rumpang + tebe-*. When checked in the Iban grammar book, *tebe* is one of the existing prefix in the Iban language.
    (b) Unseen root word - *temegah* is a root word that does not list in the Iban dictionary. However, the analyser has analysed *temegah* as *megah+te-* because *te* exists in the list of prefix and *megah* exists in the Iban dictionary. This case happened due to the dictionary has a lack of Iban entries.
    (c) Unable to analyse - The analyser was unable to analyse a morphological phenomena with more than one combination of affixes. For instance, *ngenchuri* has been analysed as *ng + en + churi* instead *ng + enchuri*. This was because the phenomena do not exist in the list of segmented data.

2. Rules never applied in the analysis:
    (a) Full reduplication - The Iban morphological analyser was unable to analyse the full reduplication due to inexistent of the rule to recognise hyphen (-) from the input, e.g. *chamang-chamang*.
    (b) Partial reduplication - The analyser only analysed *che* and *ne* as the rules are listed in the list of rules.

5   Conclusions and Future Work

In this study, we presented a semi-automatic method for acquiring morphological rules of under-resourced language, in the case of Iban language. The workflow of acquiring the morphological rules plays an important role in developing the morphological analyser for under-resourced languages. We observed that the availability of linguistic references and the selection of assorted techniques for morphology analysis could lead to the design of the workflow. Furthermore, we discovered *Linguistica* tool could generate the non-concatenation morphology, besides simple concatenation. The results of this work indicate that the morphological induction and rules validation were the crucial processes. Although we have achieved 0.76 of precision but, there are limitations e.g. the unseen root words and the analyser was unable to analyse more than one combination of affixes. These errors occurred due to insufficient references of one language. To further our research we intend to improve our workflow in a number of ways. First, noisy data including typo errors and no standardization in the Iban spelling can be overcame by avoid erroneous data as this could lead to the inaccurate results. Second, improving the Iban morphological analyser to wider the coverage by forming new rules from the incorrect results, so that it able to handle reduplications, circumfixation and combination of more than one affixes. Third, amend the lexical entry and its derivation words in the dictionary to provide better results. Finally, enhance the computer involvement in the analyser at the evaluation phase owing to time consuming. Nevertheless, we hope our work will serve as a starting point for future studies mainly for the under-resourced languages.

Acknowledgments. We gratefully acknowledge Dr Kadriu A. for her valuable suggestions and discussions on the theoretical part of automatic rules acquisition. We also thank Dr Beesley K.R. who gave us much valuable advice in the early stages of this work.

References

1. Koskenniemi, K.: Two-Level Morphology: A General Computational Model for Word-form Recognition and Production. PhD thesis, University of Helsinki (1983)
2. Theron, P., Cloete, I.: Automatic acquisition of two-level morphological rules. In: Proceedings of the Fifth Conference on Applied Natural Language Processing, pp. 103–110. Association for Computational Linguistics (1997)
3. Sarawak Board Tourism:
   http://www.sarawaktourism.com/en/component/jumi/about-people
4. Ling, S.: Iban language for spm in 2008 (2008)
5. Karagol-Ayan, B.: Resource generation from structured documents for low-density languages. PhD thesis, University of Maryland, College Park (2007)
6. Kadriu, A., Zdravkova, K.: Semi-automatic learning of two-level phonological rules for agentive nouns. In: 10th International Conference on Computer Modelling Simulation (2008)

7. Yturralde, B.: Morphological rule acquisition for tagalog words using moving contracting window pattern algorithm. In: Proceedings of the 10th Philippine Computing Science Congress, Ateneo de Davao University (2002) ISSN 1908-1146

8. Beesley, K.R.: Computational Morphology and Finite-State Methods. IOS Press (2003)

9. Akshar, B., Rajeev, S., Dipti, M.S., Radhika, M.: Generic morphological analysis shell. In: SALTMIL Workshop on Minority Languages, (2004)

10. Feldman, A., Hana, J., Brew, C.: A cross-language approach to rapid creation of new morpho-syntactically annotated resources. In: LREC 2006, pp. 549–554 (2006)

11. Cucerzan, S., Yarowsky, D.: Bootstrapping a multilingual part-of-speech tagger in one person-day. In: Proceeding of the 6th Conference on Natural Language Learning - COLING 2002, pp. 1–7 (2002)

12. Goldsmith, J.: Unsupervised learning of the morphology of a natural language. Computational Linguistics 27(2), 153–198 (2001)

13. Creutz, M., Lagus, K.: Unsupervised morpheme segmentation and morphology induction from text corpora using morfessor 1.0, Helsinki University of Technology (2005)

14. Monson, C., Carbonell, J., Lavie, A., Levin, L.: Paramor: Finding paradigms across morphology (2009)

15. Karasimos, A., Petropoulou, E.: A crash test with linguistica in modern greek: The case of derivational affixes and bound stems. In: International Conference on Language Resources and Evaluation, LREC 2010 (2010)

16. Blancafort, H.: Learning morphology of romance, germanic and slavic languages with the tool linguistica. In: International Conference on Language Resources and Evaluation, LREC 2010 (2010)

17. Dasgupta, S., Vincent, N.: Unsupervised morphological parsing of bengali. Language Resources and Evaluation 40(3-4), 311–330 (2007)

Finite State Morphology for Amazigh Language

Fatima Zahra Nejme[1], Siham Boulaknadel[2], and Driss Aboutajdine[1]

[1] LRIT, Unité Associée au CNRST (URAC 29), Université Mohamed V-Agdal,
Rabat, Maroc
fatimazahra.nejme@gmail.com, aboutaj@fsr.ac.ma
[2] IRCAM, Avenue Allal El Fassi, Madinat Al Irfane, Rabat-Instituts, Maroc
boulaknadel@ircam.ma

Abstract. In the aim of safeguarding the Amazigh heritage from being threatened of disappearance, it seems opportune to equip this language of necessary means to confront the stakes of access to the domain of New Information and Communication Technologies (ICT). In this context, and in the perspective to build tools and linguistic resources for the automatic processing of Amazigh language, we have undertaken to develop a module for automatic lexical-analysis of the Amazigh which can recognize lexical units from texts. To achieve this goal, we have made in the first instance, a formalization of the Amazigh vocabulary namely: noun, verb and particles. This work began with the formalization of the two categories noun and particles by building a dictionary named "EDicAm" (Electronic Dictionary for Amazigh), in which entry is associated with linguistic information such as lexical categories and classes of semantics distribution.

Keywords: Amazigh language, Natural Language Processing, NooJ, lexical analysis, inflectional morphology.

1 Introduction

The Amazigh language in Morocco is considered as a prominent constituent of the Moroccan culture and this by its richness and originality. However it has been long discarded otherwise neglected as a source of enrichment cultural. Nevertheless, due to the creation of the Royal Institute of Amazigh Culture (IRCAM)[1], this language has been introduced in the public domain including administration, media also in the educational system in collaboration with ministries. It has enjoyed its proper coding in the Unicode Standard [9][28], an official spelling [4], appropriate standards for keyboard realization and linguistic structures that are being developed with a phased approach [5][13]. This process was initiated by the standardization, vocabularies construction [21][6][4][7], Alphabetical Arrangement [25], spelling standardization [4] and development of grammar rules [12].

However, this not sufficient for a less-resourced language [12] as the Amazigh to join the well-resourced language in information and Communication Technologies, mainly due to the lack of automatic language processing resources and tools. There-

---

[1] Institution responsible for the preservation of heritage and the promotion of the Moroccan Amazigh culture and its development (see http://www.ircam.ma/)

A. Gelbukh (Ed.): CICLing 2013, Part I, LNCS 7816, pp. 189–200, 2013.
© Springer-Verlag Berlin Heidelberg 2013

fore, a set of scientific and linguistic research are undertaken to remedy to the current situation. These researches are divided, on the one hand, on researches that are concentrated on optical character recognition (OCR) [8][18], and in the other hand, on those that are focused on natural language processing [20][10][15][11][16], which constitute the priority components of researches.

In this context, the present work deals with ongoing research efforts to build tools and linguistic resources for the Amazigh language. This paper focuses on the treatment of nouns and particles of Amazigh vocabulary with Finite-State programming environment. The description of nouns and particles was chosen, because nouns either particles constitute a major word category in Amazigh and play a major role in syntactic analysis. Secondly, the noun classification system in Amazigh is computationally interesting.

This paper is structured around five main sections: the first present a description of the Amazigh language particularities. The second expose the automatic Amazigh language processing, which includes an overview of programming environment, and the formalization of a set of rules. While the last section is dedicated to the conclusion and perspectives.

2   Amazigh Language Particularities

2.1   Historical Background

The Amazigh language also known as Berber or Tamazight (ⵜⵎⴰⵣⵉⵖⵜ [tamaziɣt]), is belonged to the African branch of the Afro-Asiatic language family, also referred to Hamito-Semitic in the literature [19][23]. It is currently presented in a dozen countries ranging from Morocco, with 50% of the overall population[2] [14], to Egypt, passing through Algeria with 25%, the Tunisia, Mauritania, Libya, Niger and the Mali [17].

In Morocco, we distinguish between three major Amazigh dialects. Tarifit is spoken in northern Morocco, Tamazight in the Middle Atlas and south-eastern Morocco, and Tashelhit in south-western Morocco and the High Atlas.

Today, the current situation of the Amazigh language is at a pivotal point. It holds official status in Morocco. Its morphology as lexical standardization process is still underway. At present, it represents the model taught in must schools and used on media and official papers published in Morocco.

2.2   Amazigh Morphology

Amazigh is a morphologically rich language. It is highly inflected and also shows derivation to high degree. The morphological word classes in Amazigh are [12][3]: noun, verb and particles. In this paper, we are interested in noun and particles morphology.

---

[2]   It presents the Amazigh population largest in number.

Noun Characteristics

In Amazigh language, noun is a lexical unit, composed from a root and a pattern. It could appear in several forms: simple form (ⴰⵔⴰⵣ [argaz] "the man"), compound form (ⴱⵓⵀⵢⵢⵓⴼ [buhyyuf] "the famine"), or derived one (ⴰⵎⵙⴰⵡⴰⴹ [amsawaḍ] "the communication"). Whether simple, compound or derived, the noun varies in gender (masculine or feminine), number (singular or plural), and state (free or construct) [12].

1. Gender: the Amazigh noun is characterized by one of grammatical gender: masculine or feminine.

- The masculine noun: begins with one of the initial vowels: ⴰ [a], ⵉ [i] or ⵓ [u]. However, there are some exceptions as: ⵉⵎⵎⴰ [imma] "(my) mother".
- The feminine noun: is marked with the circumfix ⵜ...ⵜ [t....t]. However, there are some exceptions such as nouns which have only the initial or the final ⵜ [t] of feminine morpheme: ⵜⴰⴷⵍⴰ [tadla] "the sheaf", ⵕⵕⵎⵓⵢⵜ [ṛṛmuyt] "the fatigue".

2. Number: the noun, masculine or feminine, has a singular and plural. This latter has four forms:

- The external plural: is formed by an alternation of the first vowel ⴰ/ⵉ [a/i] accompanied by a suffixation of ⵏ [n] or one of its variants (ⵉⵏ [in], ⴰⵏ [an], ⴰⵢⵏ [ayn], ⵡⵏ [wn], ⴰⵡⵏ [awn], ⵡⴰⵏ [wan], ⵡⵉⵏ [win], ⵜⵏ [tn], ⵢⵉⵏ [yin]): ⴰⵅⵅⴰⵎ [axxam] "house"-> ⵉⵅⵅⴰⵎⵏ [ixxamn] "houses", ⵜⴰⵔⴱⴰⵜ [tarbat] "the girl"-> ⵜⵉⵔⴱⴰⵜⵉⵏ [tirbatin] "girls").
- The broken plural: involves a change of internal vowels: ⴰⴱⴰⵖⵓⵙ [abaγus] "monkey"-> ⵉⴱⵓⵖⴰⵙ [ibuγas] "monkeys", ⵜⵉⵖⵎⵙⵜ [tiγmst] "tooth"-> ⵜⵉⵖⵎⴰⵙ [tiγmas] "teeth".
- The mixed plural: is formed by vowels' change accompanied, sometimes by the use of the suffixation by ⵏ [n]: ⵉⵣⵉⴽⵔ [izikr] "the rope" -> ⵉⵣⴰⴽⴰⵔⵏ [izakarn] "the ropes".
- The plural in ⵉⴷ [id]: this kind of plural is obtained by prefixing the noun with ⵉⴷ [id]. It is applied to a set of nouns including: nouns with an initial consonant, proper nouns, parent nouns, compound nouns, numerals, as well as borrowed nouns: ⴱⵓⵜⵅⵔⴰ [butgra] "the turtle" -> ⵉⴷ ⴱⵓⵜⵅⵔⴰ [id butgra] "the turtles".

3. State: we distinguish between two states: the free state and the construct one.

- The free state: is unmarked. The noun is in free state if it is a single word isolated from any syntactic context, a direct object, or a complement of the predictive particle ⴷ [d].
- The construct state: involves a variation of the initial vowel. In case of masculine nouns, it takes one of the following forms: initial vowel alternation ⴰ [a]/ⵓ [u]; adding of ⵡ [w] or adding of ⵢ [y]. For the feminine nouns, it consists to drop or maintaining of the initial vowel.

Particles

The particles are a set of Amazigh words that is not assignable to noun neither to verb. It consists of several elements, namely:

1. The pronouns

The pronoun refers to any element that could replace a noun or nominal group. It may represent a nominal group already employed or designate a person participates in the communication. The paradigm of pronouns includes: the personal pronouns; the possessive; the demonstratives; the interrogative and the indefinite ones: ⵏⴽⴽ [nkk] "me".

2. The prepositions

The preposition is a closed paradigm that combines simple and complex forms that express various semantic values. For the first form, it consists of several formats: ⵏ [n], ⵉ [i], ⵙ [s], ⴳ [g], ⴷⵉ [di], ⵣⴳ [zg], ⵅⵂ [xh], ⴳⵔ [gr], ⴰⵍ [al]/ⴰⵔ [ar], ⴱⵍⴰ [bla], ⵖⵔ [vr], ⴷⴰⵔ [dar], ⴰⴳⴷ [agd], ⴷ [d]. For the second one, it composed of two or three prepositions that one of which can be used adverbially: ⵉⵣⴷⴰⵔ ⵏ [izdar n] "below".

3. The adverb

The adverbs are the elements that change the meaning of a verb. Generally, they are classified according to their semantics. Thus we distinguish adverbs of place, time, quality and manner: ⴷⴰ [da] "here" (adverb of place).

Further these elements, there are also: aspectual, orientation and negative particles; subordinates and conjunctions.

Generally, the particles are known as uninflected words. However in Amazigh, some of these particles are inflected, such as the possessive and demonstrative pronouns: ⵡⴰⴷ [wad] "this one" → ⵡⵉⴷ [wid] "these ones".

3   Automatic Amazigh Language Processing: Development and Evaluation

To develop a module for automatic lexical-analysis of the Amazigh, we followed the following steps: (1) construction of electronic dictionary, (2) formalization of the two categories noun and particles and (3) evaluation.

To achieve this goal, we use finite state machines integrated in the linguistic development platform NooJ.

3.1   NooJ Platform

NooJ[3], released in 2002 by Max Silberztein [22], is a linguistic development platform that provides a set of tools and methodologies for formalizing and developing a set of Natural Language Processing (NLP) applications. It presents a package of finite state tools that integrates a broad spectrum of computational technology from finite state automata to augmented/recursive transition networks. Thus, it presents a complete platform for formalizing various types of textual phenomena (orthography, lexical and productive morphology, local, structural and transformational syntax). For each of these formalization levels, NooJ propose a methodology, one or more formalisms,

---

[3]   See http://www.nooj4nlp.net/ for information of NooJ.

tools, software development and a corresponding parser that can be used to test each piece of the linguistic formalization on a large corpora.

Currently, users of NooJ form a large community including linguists, teachers and computer specialists in NLP, who share the same scientific and technical goals, to develop a large-coverage of language resources in fifteen languages (Arabic, Armenian, Bulgarian, Catalan, Chinese, English, French, Hebrew, Hungarian, Italian, Polish, Portuguese, Spanish, Vietnamese and Belarusian).

Given these advantages, we have undertaken to adopt NooJ for formalization, description and analysis of Amazigh language. We begin our work by the formalization of the Amazigh language vocabulary. This formalization is described and stored into inflectional grammars, and can recognize all the corresponding inflected forms. To test these grammars, we built an electronic dictionary which the lexical entries are attached to a set of linguistic information automatically generated using inflectional grammars which will be used for lexical analysis of texts.

3.2   Building of Dictionary

As part of developing the analysis module for the Amazigh language, we elaborate our dictionary « EDicAM » (Electronic Dictionary for Amazigh) for simple nouns (ex. ₒXXₒⵞ [axxam]) and inflected particles (ex. ⵓⵉⵏⵓ [winu] "mine" (masc.) -> +ⵉⵏⵓ [tinu] "mine" (fem.)).

Each entry into Amazigh dictionary generally presents following details: the lemmas, Lexical category, Semantic feature, French translation and the inflexional paradigm.

Our dictionary contains, currently, 4480 entries of simple nouns as singular masculine and 44 particles. These particles include personal pronouns, demonstrative and possessive ones. Each entry in our dictionary is associated with an inflectional class allowing to generate all the corresponding inflected forms (feminine, plural and constructed state).

3.3   Formalization

Given that any linguistic analysis must go through a first step of lexical analysis, which consists in testing membership of each word of the text to the lexicon of the language, we began our work by formalization of the Amazigh vocabulary. This study is focus on the formalization of the tow sets: nouns and particles, using finite state machines integrated to linguistic development platform NooJ.

Formalization of Morphological Rules

This study presents the formalization of the noun and particles categories in the NooJ platform. For this, a set of rules has been defined allowing to generate from each entry, its inflectional information: gender, number and state for the nouns.

The formalization is based on use of some generic predefined commands such as:

— <LW> move at the beginning of lemma,
— <RW> move at the end of lemma,

— <S> delete current character,
— <B> delete last character,
— <L> go left,
— <R> go right.

The nouns

To formalize the inflexional rules of the noun, we have created, through graphs integrated in the linguistic development platform NooJ, an inflectional analysis system describing flexing models in standard Amazigh of Morocco.

1. Gender

To formalize the gender we built this rule that generate from a masculine entry its feminine correspondent.

Table 1. Example of gender rule

The rules in Nooj	Explanation	Examples
<LW>+<RW>+	This rule adds the morpheme "+" [t] at the beginning and at the end of the noun.	ⵉⵙⵍⵉ [isli] "married"(masc.) -> +ⵉⵙⵍⵉ+ [tislit] "married" (fem.).

2. Number

For the Amazigh plural, we have many plural forms which are generally unpredictable due to Amazigh complex morphology. We searched formal rules to unify the calculation of these plural forms. To achieve this goal, we have relied on the new grammar of Amazigh [12], on the works of Boukhris [13] and those of Oulhaj [24]. According to these works and to an heuristic study of the nouns in the Taifi dictionary [27] and those of Amazigh language vocabulary [5], we have raised, actually, 303 classes which 97 classes is for the external plural, 99 for the broken plural, 104 for the mixed plural and 3 classes for the plural in ⵉⴷ [id]. Each class corresponds to a scheme and contains a set of nouns, for example for the external plural, if the noun begins and ends with a 'o...o' [a...a], the plural forms will be built by a vowel alternation accompanied by suffixation of +ⵏ [tn]. Thereafter, we provided some examples of rules for each of these plural types.

• The external plural

Table 2. Plural form for the masculine nouns beginning and ends with ⴰ [a]

The rules in Nooj	Explanation	Examples
<LW>ⵉ<S><RW>+ⵏ	The initial vowel is transformed into ⵉ and the suffix +ⵏ [tn] is add at the end of the noun.	ⴰⵙⵉⵔⴰ [asira] "desk" -> ⵉⵙⵉⵔⴰ+ⵏ [isiratn] "desks".

• The broken plural

Table 3. Plural for the nouns of VCn form

The rules in Nooj	Explanation	Examples
<LW>ⵥ<S><RW><L>ⵄ	The rule changes the initial vowel into ⵥ [i] and include ⵄ [a] before the final consonant.	ⵄⵥⵉⵥⵀ [azgzl] "abbreviation" -> ⵥⵥⵉⵥⵄⵀ [izgzal] "abbreviations"

- The mixed plural

Table 4. Example of plural forms for the masculine nouns

The rules in Nooj	Explanation	Examples
<LW>ⵥ<S><RW>ⵖ<L2>ⵥ	The rule change the initial vowel into ⵥ [i], include the vowel ⵥ [i] before the last consonant and add a suffix ⵉ [n] at the end of the noun.	ⵄⵖⵚⵏⵙ [aḥudr] "the fact of lean" -> ⵥⵖⵚⵏⵥⵚⵓⵉ [iḥudirn].

- The plural in ⵥⵏ [id]

Table 5. Example of plural in ⵥⵏ [id]

The rules in Nooj	Explanation	Examples
<LW>ⵥⵏ" "	The rule adds ⵥⵏ [id] before the noun.	ⵣⵚ+ⵅⵔⵄ [butgra] "tortoise" -> ⵥⵏ ⵣⵚ+ⵅⵔⵄ [id butgra] "tortoises".

3. State

Table 6. Example of constructed state

The rules in Nooj	Explanation	Examples
<LW><R><B>ⵚ	The rule deletes the initial vowel and adds ⵚ [u] at the beginning of the noun.	ⵄⵀⵥⵔⵄⵔ [afiras] "pear" -> ⵚⵀⵥⵔⵄⵔ [ufiras].

The particles

Similarly to the noun, we formalized the inflectional particles, such as: demonstrative pronouns, possessive pronouns etc., with a collection of graphs and subgraphs presenting the inflectional grammars.

In the following, we will cite an example of possessive pronouns. The possessive pronouns are formed by the combination of determination supports with affixes pronouns of noun complement.

1. Gender

To formalize the gender of particles, we built five morphological rules that generates from a masculine entry its feminine correspondent.

Table 7. Example of gender graph of particles

The rules in Nooj	Explanation	Examples
<LW><S>+	This rule removes the first consonant and adds the discontinuous morpheme + [t] at the beginning of the particle.	ⵍⵉⵏⵓ [winu] "mine" (masc.) -> +ⵉⵏⵓ [tinu] "mine" (fem.).

2. Number

The plural of particles can be formed either by a suffixation ⵏ [n] or by a vowel alternation / final consonant sometimes accompanied by a suffixation ⵏ [n]. The following table shows the graph corresponding to the second case.

Table 8. Example of a graph for the particles plural

The rules in Nooj	Explanation	Examples
<RW><B>ⵓ	The final consonant is transformed into ⵓ [u] and the suffix ⵏ [n] is added at the end of the particle.	ⵍⵉⵏⵏ [winnk] "yours" -> ⵍⵉⵏⵏⵓ [winnun] "yours".

For the rest of non-inflectional particles, we construct a set of morphological grammars that allow these particles to be known on the texts. In the following, we will cite the example of adverb.

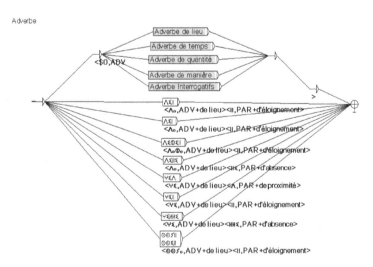

Fig. 1. Example for a formalization of Adverb

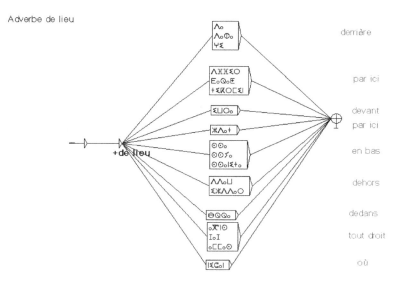

Fig. 2. Formalization of adverb of place

For the complex particles, we construct a set of local grammars which we cite an example:

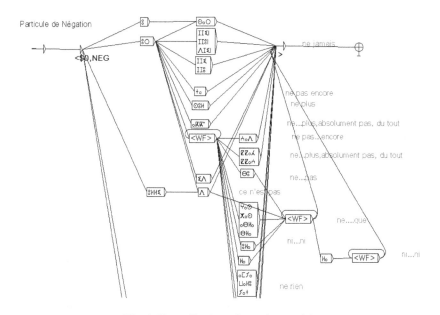

Fig. 3. Formalization of complex particles

3.4   Evaluation

To evaluate our resources, we extract a list of 1000 nouns based on lexicon school [1] and 119 particles from the new grammar of Amazigh [12].

We present in the table below, an example of the result of experiments obtained:

Table 9. Example of result of our experiments

Results	Number of recognized forms		Number of unrecognized forms	
	Number	%	Number	%
Noun	982	98,2%	18	1,8%
Particles	109	91,59%	10	8,4%

The results allow us to review, correct and complete all our resources.

As indicated in the table of evaluation, the applications of our resources have resulted a lexical coverage about 97%. The unrecognized word forms (from nouns and particles) include:

- 14 entries: corresponds to the foreign nouns that are not part of our electronic dictionary;
- 10 entries: corresponds to the foreign particles that are not part of our morphological grammar;
- 4 entries: corresponds to the nouns inflected with a rule that does not correspond to its exact form, for example if we take the feminine noun as follows: "ⴰⵎⵇⵓⵍⵉⵜ" [Timelwit - soft], normally the general form of inflected feminine nouns is as follows:
  - If the noun is in the form 'ⵜⴰ...vⵜ'   (v: vowel), the vowel 'ⴰ' is transformed into 'ⵉ' [i] and a suffix of 'ⵉⵏ' [in] is applied.
  - If the noun is in the form 'ⵜⴰ...cⵜ' (c: consonant), the vowel 'ⴰ' is transformed into ' ⵉ' [i], the last ' ⵜ' is deleted, and a suffix 'ⵉⵏ' [in] is applied.

Therefore, the noun inflexion in the vocabulary is ⵜⵉⵎⵇⵓⵍⵉⵏ. These types of nouns form the exceptions in Amazigh.

4   Conclusion and Future Works

The main objective of this work was to develop a module for automatic lexical-analysis of the Amazigh which can recognize lexical units from texts. To achieve this goal, we have made in the first instance, a formalization of the morpho-syntactic categories noun and particles. This work demonstrates the application of finite state approach in the analysis of Amazigh vocabulary.

Based on the results obtained, the preferred method seems to be appropriate and encouraging. However, other rules must be added to improve the result.

For future work we plane to formalize the rest of Amazigh word categories. The aim is to develop a morphological analyser of Amazigh that will be an input for many other applications such as learning systems and machine translation.

References

1. Agnaou, F., Bouzandag, A., El Baghdadi, M., El Gholb, H., Khalafi, A., Ouqua, K., Sghir, M.: Lexique scolaire. IRCAM, Rabat (2011)
2. Ameur M., Boumalk A.(DIR).: Standardisation de l'amazighe, Actes du séminaire organisé par le Centre de l'Aménagement Linguistique à Rabat, Décembre 8-9 (2003); Publication de l'Institut Royal de la Culture Amazighe, Série: Colloques et seminaries (2004a)
3. Ameur, M., Bouhjar, A., Boukhris, F., Boukouss, A., Boumalk, A., Elmedlaoui, M., Iazzi, E., Souifi, H.: Initiation à la langue amazighe. IRCAM, Rabat (2004b)
4. Ameur, M., Bouhjar, A., Boukhris, F., Boukouss, A., Boumalk, A., Elmedlaoui, M., Iazzi, E.: Graphie et orthographe de l'amazighe. IRCAM, Rabat (2006a)
5. Ameur, M., Bouhjar, A., Boukhris, F., Elmedlaoui, M., Iazzi, E.: Vocabulaire de la langue amazighe (Français-Amazighe). série: Lexiques (1). IRCAM, Rabat (2006b)
6. Ameur, M., Bouhjar, A., Boumalk, A., El Azrak, N., Laabdelaoui, R.: Vocabulaire des médias (Français-Amazighe-Anglais-Arabe). série: Lexiques (3). IRCAM, Rabat (2009a)
7. Ameur, M., Bouhjar, A., Boumalk, A., El Azrak, N., Laabdelaoui, R.: Vocabulaire grammatical. série: Lexiques (5). IRCAM, Rabat (2009b)
8. Amrouch, M., Rachidi, A., El Yassa, M., Mammass, D.: Handwritten Amazigh Character Recognition Based On Hidden Markov Models. International Journal on Graphics, Vision and Image Processing 10(5), 11–18 (2010)
9. Andries, P.: Unicode 5.0 en pratique, Codage des caractères et internationalisation des logiciels et des documents. Dunod, France, Collection InfoPro (2008)
10. Ataa Allah, F., Jaa, H.: Etiquetage morphosyntaxique: Outil d'assistance dédié à la langue amazighe. In: Proceedings of the 1er Symposium International sur le Traitement Automatique de la Culture Amazighe, Agadir, Morocco, pp. 110–119 (2009)
11. Ataa Allah, F., Boulaknadel, S.: Online Amazigh Concordancer. In: Proceedings of International Symposium on Image Video Communications and Mobile Networks, Rabat, Maroc (2010)
12. Berment, V.: Méthodes pour informatiser des langues et des groupes de langues peu dotées. Thèse de doctorat de l'Université J. Fourier - Grenoble I, France (2004)
13. Boukhris, F., Boumalk, A., Elmoujahid, E., Souifi, H.: La nouvelle grammaire de l'amazighe. IRCAM, Rabat (2008)
14. Boukous, A.: Société, langues et cultures au Maroc: Enjeux symboliques, Casablanca, Najah El Jadida (1995)
15. Boulaknadel, S.: Amazigh ConCorde: an appropriate concordance for Amazigh. In: Proceedings of the 1er Symposium International sur le Traitement Automatique de la Culture Amazighe, Agadir, Morocco, pp. 176–182 (2009)
16. Boulaknadel, S., Ataa Allah, F.: Building a standard Amazigh corpus. In: Proceedings of the International Conference on Intelligent Human Computer Interaction, Prague, Tchec (2011)
17. Chaker, S.: Le berbère, Actes des langues de France, 215-227 (2003)
18. Es Saady, Y., Rachidi, A., El Yassa, M., Mammas, D.: Printed Amazigh Character Recognition by a Syntactic Approach using Finite Automata. International Journal on Graphics, Vision and Image Processing 10(2), 1–8 (2010)

19. Greenberg, J.: The Languages of Africa. The Hague (1966)
20. Iazzi, E., Outahajala, M.: Amazigh Data Base. In: Proceedings of HLT & NLP Workshop within the Arabic World: Arabic Language and Local Languages Processing Status Updates and Prospects, Marrakech, Morocco, pp. 36–39 (2008)
21. Kamel, S.: Lexique Amazighe de géologie. IRCAM, Rabat (2006)
22. Silberztein, M.: An Alternative Approach to Tagging. In: Kedad, Z., Lammari, N., Métais, E., Meziane, F., Rezgui, Y. (eds.) NLDB 2007. LNCS, vol. 4592, pp. 1–11. Springer, Heidelberg (2007)
23. Ouakrim, O.: Fonética y fonología del Bereber, Survey at the University of Autònoma de Barcelona (1995)
24. Oulhaj, L.: Grammaire du Tamazight. Imprimerie Najah El Jadida (2000)
25. Outahajala, M.: Les normes de tri, Du clavier et Unicode. La typographie entre les domaines de l'art et de l'informatique, Rabat, Morocco, pp. 223–237 (2007)
26. Outahajala, M., Zekouar, L., Rosso, P., Martí, M.A.: Tagging Amazigh with AnCoraPipe. In: Proceeding of the Workshop on Language Resources and Human Language Technology for Semitic Languages, Valletta, Malta, pp. 52–56 (2010)
27. Taifi, M.: Le lexique berbère (parlers du Maroc central) (1988)
28. Zenkouar, L.: Normes des technologies de l'information pour l'ancrage de l'écriture amazighe. Etudes et Documents Berbères 27, 159–172 (2008)

New Perspectives
in Sinographic Language Processing
through the Use of Character Structure

Yannis Haralambous

Institut Télécom - Télécom Bretagne Lab-STICC UMR CNRS 6285
yannis.haralambous@telecom-bretagne.eu

Abstract. Chinese characters have a complex and hierarchical graphical structure carrying both semantic and phonetic information. We use
this structure to enhance the text model and obtain better results in
standard NLP operations. First of all, to tackle the problem of graphical
variation we define allographic classes of characters. Next, the relation of
inclusion of a subcharacter in a characters, provides us with a directed
graph of allographic classes. We provide this graph with two weights:
semanticity (semantic relation between subcharacter and character) and
phoneticity (phonetic relation) and calculate "most semantic subcharacter paths" for each character. Finally, adding the information contained
in these paths to unigrams we claim to increase the efficiency of text
mining methods. We evaluate our method on a text classification task
on two corpora (Chinese and Japanese) of a total of 18 million characters
and get an improvement of 3% on an already high baseline of 89.6% precision, obtained by a linear SVM classifier. Other possible applications
and perspectives of the system are discussed.

1 Introduction

The Chinese script is used mainly in the Chinese and Japanese languages. Chinese characters (or "sinographs") are notorious for their large number (over 84
thousand have been encoded in Unicode [1]) and their complexity (they can have
from 1 stroke, like 一, to as many as 64 strokes, like 𪚥). Despite its complexity, the Chinese script is quite efficient, since semantic and phonetic information
is stored in stroke patterns, easily recognizable by native readers. In this paper we will deal with a specific kind of stroke pattern, namely those that exist
also as stand-alone characters—we call them *subcharacters*. We will study the
phonetic similarity (called *phoneticity*) and the semantic relatedness (called *semanticity*) between a character and its subcharacters. After having built a graph
of subcharacter inclusions, and attaching various kinds of information to it, we
introduce an enhanced NLP task feature model: together with individual characters we use data contained in their subcharacter graphs. Indeed, in sinographic
language processing it is customary to combine character-level and word-level
processing. The advantage to our approach is that an additional level is added

A. Gelbukh (Ed.): CICLing 2013, Part I, LNCS 7816, pp. 201–217, 2013.

to these two: the level of subcharacters. By exploring this additional level, we go deeper into the inherent structure of sinographic characters—this inherent structure is completely lost in conventional NLP approaches.

We believe that this new feature model will prove useful in various branches of statistical NLP. As a first step in that direction, we evaluate our tools by applying them to a text classification task.

1.1 Related Work

There have been various attempts at describing sinographs in a systematic way: [2] and [3] use generative grammars, [4] uses a Prolog-like language, [5] uses the biological metaphor of gene theory, [6] describes sinographs as objects with features (the Chaon model), [7] defines a formal language type called planar regular language, [8] and [9] describe the topology of sinographs in XML, [10] uses projections of stroke bounding boxes, and [11] use combinatorics of IDS operators.[1]

The OCR community has also shown a strong interest in the structure of sinographs [12].

As for considering Chinese script as a network, this has been done in several papers, such as [13, 14] (where edges represent components combined in the same sinograph) or [15], where components and sinographs form a bipartite graph; [16], where edges represent sinographs combined into words; [17], dealing with purely phonemic networks. [14] give an example of a small phono-semantic graph (a bipartite graph where edges connect semantic and phonetic components), but do not enter into the calculations of phoneticity and semanticity. *Hanzi Grid* [18, 19] maps component inclusions to relations in ontologies.

Finally the cognitive psychology community is also heavily interested in the (cognitive) processing of phonetic and semantic components [20, 21], and even in the effects of stroke order and radicals on linguistic knowledge [22].

2 Definitions

2.1 Strokes

A sinograph consists of a number of *strokes*, arranged inside an (imaginary) square according to specific patterns. Strokes can be classified as belonging to 36 *calligraphic stroke classes*. The latter have been encoded in Unicode (table CJK STROKES). Furthermore, strokes are always drawn in a very specific order.

2.2 Components, Subcharacters, Radicals

In a manner similar to etymological roots in Western languages, readers of sinographs recognize patterns of strokes, so that the meaning of an unknown sinograph can be identified, more or less effectively, by the stroke patterns it contains.

---

[1] *Ideographic Description Sequences* do not describe sinographs per se, but provide operators ⿰ ⿱ ⿲ ⿳ ⿴ ⿵ ⿶ ⿷ ⿸ ⿹ ⿺ ⿻ for graphically combining existing sinographs in groups of two or three. IDS operators can be arbitrarily nested and have been encoded in Unicode (table IDEOGRAPHIC DESCRIPTION CHARACTERS).

Frequently appearing patterns of strokes are called *components*. In this paper we will deal only with components that also exist as isolated sinographs. Using the term "character" in the sense of "Unicode character," we will call such components, *subcharacters*. In other words, subcharacters are *components having a Unicode identity*.

Classification of sinographs in dictionaries traditionally uses a set of several hundred subcharacters called *radicals*. We do not discuss radicals in this paper.

2.3 Allographic Classes

An important property of components is that they can change shape when combined with other components (for example, 火 becomes 灬 when combined, like in 点). Some of these variant shapes have been encoded in Unicode. Also, some sinographs can have variant forms. In particular, during the Chinese writing reform [23], about 1,753 sinographs obtained simplified shapes (for example, 蔔 became 卜), which are encoded separately in Unicode.

As these variations in shape do not affect semantic or phonetic properties (at least not at the level of statistical language processing), we merge characters and their variants into sets called *allographic classes*. For example, 糸 belongs to class [糸,糹,糸,糸,糹] where the two first characters belong to the table of Unicode radicals and the others are graphical variants.

We obtain 18,686 allographic classes, out of which 87.356% are singletons, the highest number of characters per class is 15 and the average is 1.1382.

In the remainder of this paper we will use italics for characters ($c, s, \ldots$) and bold letters for allographic classes ($\mathbf{c}, \mathbf{s}, \ldots$). The term "subcharacter" will mean a character or an allographic class, depending on the context.

2.4 Semanticity and Phoneticity

Subcharacters can play a *semantic role* (when one of their meanings is close to one of the meanings of the sinograph) and/or a *phonetic role* (when one of their readings is identical or close, in a given language, to one of the readings of the sinograph). In this paper we will deal with Mandarin and Japanese.

These two properties of subcharacters in relation to characters can be quantified and are then called *semanticity* and *phoneticity* [24].

3 Resources

3.1 Frequency Lists

First some notation: let T be a sinographic text (or a corpus considered as a single text). A *frequency list* A, generated out of T, is an M-tuple of pairs $(c_i, f_A(c_i))$, where the frequency $f_A(c_i)$ of character c_i is defined as $\frac{\#c_i}{\#T}$, that is the number of occurrences of c_i in T divided by the length of T. The c_i must be pairwise different. A is sorted in decreasing order of frequencies: $f_A(c_i) \geq f_A(c_j)$ when $i < j$. If $N \in \mathbb{N}$, let $A_{1\ldots N}$ be the subtuple of the first N pairs.

Let char(A) be the underlying set of characters of A. Let $A_{1...N}$ be the sublist of the N first characters of A. Let

$$\mathrm{comchar}_N(A, A') :=$$
$$(\mathrm{char}(A_{1...N}) \cap \mathrm{char}(A'))$$
$$\cup (\mathrm{char}(A'_{1...N}) \cap \mathrm{char}(A))$$

be the *set of N-common characters* between two lists A and A'. In other words, among the first N characters of A we take those that also belong to A' and vice versa. We define the *N-common coverage factor* between A and A' as:

$$\mathrm{comcov}_N(A, A') := \frac{\#\mathrm{comchar}(A, A')}{N}.$$

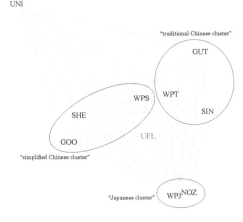

Fig. 1. Sinograph frequency lists.

Using $\mathrm{comchar}_N(A, A')$ as the underlying character set, we obtain sub-lists A_c of A and A'_c of A' (although not noted, A_c depends not only on A but also on A' and on N, and A'_c also on A and N).

Finally, we define a *distance of character frequency lists d_N* as follows:

$$d_N(A, A') := 1 - \mathrm{comcov}_N(A, A') \cdot \frac{\rho(A_c, A'_c) + 1}{2}$$

where ρ is the Spearman ranking correlation coefficient (ρ takes values in $[-1, 1]$).

We have used three publicly available frequency lists: UNI, the language-independent *Unihan* [25]; SIN, the traditional Chinese *Sinica* compiled by Academia Sinica Taipei; and NOZ, a Japanese list, taken from [26].

We have also compiled our own frequency lists out of five corpora: Chinese Wikipedia (WPS for simplified Chinese and WPT for traditional Chinese), Japanese Wikipedia WPJ, Chinese Project Gutenberg GUT, Chinese GoogleBooks GOO, Leeds Chinese Internet Corpus CIC [27].

As we needed a frequency list suitable for all sinographic languages, we calculated distance d_N between them (for $N = 3,000$). In Fig. 1 the reader can see a graph of frequency lists with edges proportional to values of d_N. One can identify three linguistic clusters, while the Unihan list can be considered as an outlier. After removing the Unihan list we aggregated the remaining lists to form a "Universal Frequency List," UFL, as the normalized average of NOZ, WPJ, WPS, WPT, SHE, GOO, SIN and GUT, defined as follows: if $f_X(c)$ is the frequency of character c in corpus X, and $\#X$ is the size of the corpus in characters, then:

$$\mathrm{char}(\mathrm{UFL}) = \bigcup_{X \in \mathcal{X}} \mathrm{char}(X), \qquad f_{\mathrm{UFL}}(c) = \sum_{X \in \mathcal{X}} \frac{f_X(c) \cdot \#\mathrm{char}(X)}{\#\mathrm{char}(\mathrm{UFL})},$$

where $\mathcal{X} = \{\mathrm{NOZ}, \mathrm{WPJ}, \mathrm{WPS}, \mathrm{WPT}, \mathrm{SHE}, \mathrm{GOO}, \mathrm{SIN}, \mathrm{GUT}\}$.

3.2   Character Descriptions and Subcharacter Inclusions

Wenlin Institute kindly provided us with the CDL database of sinographs. This XML file provides an ordered stroke list for each sinograph, and for each stroke, its calligraphic type and coordinates of endpoints. We modified stroke order for the few exceptional cases where components are overlapping.[2]

Because of the many affine transformations a component is subject to, this topological sinograph description is not suitable for effectively detecting subcharacters. On the other hand, use of topological properties is unavoidable, since some sinographs have the same strokes in the same order and combinatorial arrangement but differ by the (relative) size of strokes, the typical example being 士 (scholar) and 土 (earth), where the bottom stroke of the latter is longer than that of the former. For this reason we used a different representation of sinographs, based on relative size of strokes and extrapolated intersection locations. For example, here is the relation between the two first strokes of the sinograph 言, which are of type d (= "dot") and h (= "horizontal"):

```
d
(1.4,9.1,0.0,0.6)
h
```

(see Fig. 2 for the description of the numeric values).

Using this affine transformation invariant representation of sinographs, we extracted 868,856 subcharacter inclusions from the Wenlin CDL database, where inclusion $s \to c$ of subcharacter s into character c means that the code lines of our representation of s are contained, in their identical form, in the representation of c. After merging with data from the CHISE project [28] and with data kindly provided by Cornelia Schindelin [29], and after having removed identities, we got a list of 824,120 strict inclusions. The number is quite high because inclusion chains like: 丿 → 勹 → 匋 → 蜀 provide automatically all triangulations 丿 → 匋, 勹 → 蜀, etc., and there are exponentially many of them. We have detriangulated by taking systematically the longest path, and hence reduced the number of inclusions to 185,801.

3.3   The Inclusion Graph

We construct a graph, using all sinographs as vertices and representing the inclusions as edges, giving us 74,601 vertices and 185,801 edges. Both the in-degree and out-degree properties of this graph (see Fig. 3) follow a power law distribution of parameters $\alpha^- = 1.138$ (in-degree) and $\alpha^+ = 1.166$ (out-degree). Remarkably, these are a bit low compared to typical scale-free networks like the Web or proteins, which are in the 2–3 range [30].

---

[2] An important fact about components is that in the sequence of strokes drawn in traditional stroke order, *components form subsequences without overlapping*: the first stroke of a component is drawn after the last stroke of the preceding component. There are a few exceptions to this rule: sinographs like 団 (subcharacter □ containing 才) where the drawing of the lower horizontal stroke of □ is postponed until the drawing of the internal subcharacter 才 has been completed.

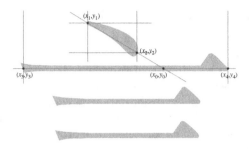

Fig. 2. Representation of sinographs based on relative sizes of strokes and extrapolated intersection locations. The blue lines are skeletons of strokes given by CDL. The values $(1.4, 9.1, 0.0, 0.6)$ correspond to: $\frac{d((x_0, y_0) - (x_1, y_1))}{d((x_2, y_2) - (x_1, y_1))}$ (that is: distance of the intersection point from (x_1, y_1) with stroke length as unit), $\frac{|x_4 - x_3|}{|x_2 - x_1|}$ (stroke box width ratio), $\frac{|y_4 - y_3|}{|y_2 - y_1|}$ (stroke box height ratio) and $\frac{d((x_0, y_0) - (x_3, y_3))}{d((x_4, y_4) - (x_3, y_3))}$ (again, distance of the intersection point from (x_3, y_3) with second stroke length as unit), where d is Euclidean distance. When strokes are parallel or orthogonal, some of the parameters are infinite and we write E instead.

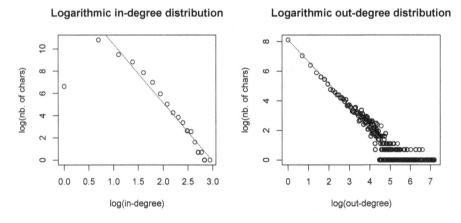

Fig. 3. Power-law fitting of the in- and out-degree distribution of the graph of allographic classes

As higher Unicode planes contain rare sinographs which are of little use to common NLP tasks, and for reasons of computational efficiency, we restricted ourselves to the subgraph of sinographs contained in Unicode's BMP (Basic Multilingual Plane). This subset covers only 91.65% of the sinographs contained in the frequency list, but its frequency-weighted coverage[3] is as high as 99.9995%, showing that our choice of BMP is justified.

By lifting[4] sinograph inclusions to allographic classes we obtain a graph $\mathcal{C}$ of 18,686 allographic classes and 39,719 class inclusions.

In all, 99.8% of the allographic classes have incoming inclusions (the highest in-degree is 12 for class [華]), 14 classes are "sources" (zero in-degree nodes): [一], [丨], [丶], [丿], [丨], [力], [巛], [丶、乁、乛、乚、乙], [乃], [勹], [又], [及], [巜] and [乀]. 87.05% of the classes are leaves (zero out-degree nodes) and the highest out-degree is 996 for class [氵,水]).

4   Phoneticity

Unihan provides phonetic data for sinographs in several languages. For this study, we used data for Mandarin Chinese, Japanese On (readings originating from China, and imported to Japan together with the writing system) and Japanese Kun (native Japanese readings). We define *phoneticity* as the degree of phonetic similarity between subcharacter and character. To calculate it we need a phonetic distance.

For Mandarin Chinese we implemented the phonetic distance described in [31].

For Japanese we have defined our own phonetic distance.We obtained this distance by applying a methodology given by [32] and [33]: indeed, we defined a distance for syllables, using seven features: consonant place of articulation (weight 4), consonant voicing (1), consonant manner of articulation (4), consonant palatalization (1), vowel frontness (5), vowel height (1), and vowel rounding (1). Distance between syllables is Euclidean in feature space. When sinograph readings have an unequal number of syllables[5] we use a sliding window approach to find the shortest phonetic distance between the shorter word and a subword of equal length to the longer one.

If d is a phonetic distance between sinographs, let $d_{\min}$ be the distance between allographic classes defined as the minimum distance between class members. We define the *phoneticity coefficient* $\varphi(\mathbf{c}_1, \mathbf{c}_2) := \frac{1 - d_{\min}(\mathbf{c}_1, \mathbf{c}_2)}{N}$ where N

---

[3] If standard coverage is $\frac{\#\{c_i \in \text{char}(\text{UFL})\}}{\#\{c_i \in \text{BMP}\}}$ where UFL is our Universal Frequency List, then *frequency-weighted coverage* is defined as $\frac{\sum_{c_i \in \text{char}(\text{UFL})} f_{\text{UFL}}(c_i)}{\sum_{c_i \in \text{BMP}} f_{\text{UFL}}(c_i)}$.

[4] If $\mathbf{c}_1, \mathbf{c}_2 \in \mathcal{C}$ we have $\mathbf{c}_1 \to \mathbf{c}_2$ iff there is at least one character pair $c_1 \in \mathbf{c}_1, c_2 \in \mathbf{c}_2$ such that $c_1 \to c_2$.

[5] Contrary to Chinese language where sinograph readings are monosyllabic, in Japanese they can have up to 12 syllables (the longest readings are *nuhitorioshi-tanameshigaha* for 轌 and *hitohenotsutsusodeudenuki* for 構, both being Kun readings).

is a normalization constant such that $Im(\varphi) = [0, 1]$. We will write $\varphi(\mathbf{s} \to \mathbf{c})$ for $\varphi(\mathbf{s}, \mathbf{c})$ when $\mathbf{s} \to \mathbf{c}$ is a class inclusion. In Fig. 4, we show the distribution density of φ for Mandarin, Japanese On and Kun. One can see that On mimics the distribution of Mandarin (with a lesser φ value for the right peak) while Kun has a completely different distribution with a single peak around $\varphi = 0.4$. Indeed, the historical relation of On and Mandarin is reflected in the similarity of distributions, while in Kun phonetic distance between subcharacter and character is random.

Here is an example:

	Mandarin	Jap. On	Jap. Kun
任	rèn	nin	makaseru, ninau, taeru
↑	↑ ≈	↑ =	↑ ≠
人	rén	nin	hito

where the difference between the phonetics of the two characters in Mandarin is at the tone level only, and since tones are not phonemic in Japanese, the sinographs are homophones in On. Incidentally, this inclusion has very low semanticity: 人 means "man" and 任, "to trust."

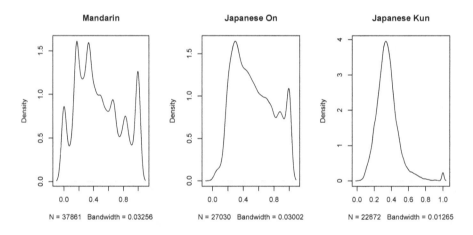

Fig. 4. Phoneticity coefficient distribution for Mandarin, Japanese On and Kun

4.1   The Least Phonetic Chain

Under the hypothesis that *subcharacters with higher phoneticity have statistically lower semanticity and vice versa*, we consider the subcharacters with the lowest phoneticity, as these have an increased potential of having higher semanticity. We define the *least phonetic chain* $(\mathbf{p}_i)_{i \geq 0}$ of a class $\mathbf{c}$ as follows: let $\mathbf{p}_0 = \mathbf{c}$ and given $\mathbf{p}_i$ let $\mathbf{p}_{i+1} := \text{argmin}_{\mathbf{z}} \varphi(\mathbf{z} \to \mathbf{p}_i)$, that is, the subcharacter of $\mathbf{p}_i$ with least phoneticity (see Fig. 5 for an example). We will use this construct in our text classification strategies.

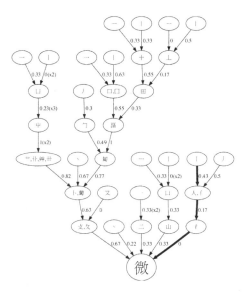

Fig. 5. The inclusion tree of allographic class [微], in bold: the least phonetic chain. Edge labels are φ values; between parentheses, the subcharacter's multiplicity.

5   Semanticity

As semantic resources for sinographic languages we used three WordNets: the Academia Sinica BOW [34] for traditional Chinese, the Chinese WordNet [35] for simplified Chinese and the Japanese WordNet [36]. All three provide English WordNet synset IDs. The single-sinograph entries they contain are rather limited: 3,075 for traditional Chinese, 2,440 for simplified Chinese and 4,941 for Japanese. From these we obtain a first mapping of allographic classes into synset IDs. In all, 2,852 allographic classes are covered, out of which 1,063 have a single synset ID, 581 have two IDs (IDs from different WordNets are not merged), etc. The highest number of IDs is 45, for [刺] (= "thorn"). Only 154 classes share the same synset ID in all three WordNets.

The Unihan database also provides meanings for a large number of sinographs, but these meanings are not mapped to WordNet. We used the following method to attach synset IDs to them: let w_i be a Chinese or Japanese word attached to synset ID σ_i in one of the three WordNets; let $e_{i,k}$ be one of the terms of the English WordNet synset with ID σ_i; if $e_{i,k}$ can be found in Unihan as meaning of a character c_j *and* if $c_j \in w_i$ then we attach synset ID σ_i to c_j. We lift this information to allographic classes.

Here is an example:

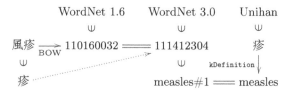

Since 疹 has the meaning "measles" in Unihan and is contained in the word 風疹 which belongs to the measles synset, we attach the measles synset ID to 疹.

Thanks to this method, 1,392 additional allographic classes were mapped to at least one WordNet synset ID, raising the number of semantically annotated classes to 4,244.

5.1 Extracting Semantic Relations

We attempted three methods of estimating semanticity of subcharacters:

1. By measuring distance between WordNet nodes using semantic similarity measures, such as those of [37–39] or simply the inverse of the shortest path length in the WordNet graph. This method was unfruitful.

2. By using the following algorithm: whenever there was a semantic relation (hyponymy, meronymy, antonymy, etc.) between two WordNet synsets σ_1 and σ_2, and we have a sinograph belonging to a word from σ_2 and one of its subcharacters in a word from σ_1, we added a unit of "semantic weight" to the inclusion. We counted these over the three WordNets and obtained 6,816 allographic class inclusions of various weights.

Here is an example:

"Greenland spar" is an hyponym of "mineral," 冰晶石 belongs to the synset of the former and 礦物 to that of the latter. Characters 石 and 礦 appear in these two words, and the former being a subcharacter of the latter, we attach a unit of "semantic weight" to it. The (log of the) total amount of units is our tentative semanticity measure.

3. To a lesser degree, we used the Kāng Xī radical (see next section) as an additional semanticity indicator. We did the following: for every inclusion, we compared the Kāng Xī radicals on both sides: when equal, then this may be a hint that the given subcharacter has higher semanticity than others.

Example: in the inclusion 每 → 毓, both sinographs have the same Kāng Xī radical 毋, so we assume that 每 → 毓 has higher semanticity than 㐬 → 毓 (which, incidentally, also has higher phoneticity since it is pronounced *liú* which is closer to *yù* (毓) than *měi* (每)).

6   Evaluation

To evaluate an application of our graph to a common NLP task we attempted text classification on two corpora:

1. The Sogou Corpus of Chinese news [40], a collection of 2.36 million online articles (611 million sinographs) taken from various sources. We removed all sinographs not in the CJK UNIFIED IDEOGRAPHS Unicode table and extracted 5,000 texts per class:

Category	Avg length	Min length	Max length
Sports	728.23	500	1,000
Finance	714.51	500	1,000
News	699.63	500	1,000
Entertainment	696.36	500	1,000
Global	709.68	500	1,000

2. A corpus we built from the online archives of Japanese Reuters [41], covering the period from 2007 to today. We removed all sinographs not in the CJK CHARACTERS Unicode table (including the kana) and extracted 5,000 texts per class. Because of the nature of Reuters news, these texts are shorter than the Chinese ones:

Category	Avg length	Min length	Max length
Sports	105.54	74	521
Finance	293.79	188	1,262
News	315.69	166	876
Entertainment	94.73	44	588
Global	202.44	44	1,262

Our baseline is obtained as follows: we take unigrams for all sinographs appearing at least 10 times in the corpus. This results in 4,275 unigrams for the Chinese and 2,010 for the Japanese corpus. Then we apply a linear SVM with 10-fold cross validation. Here are the results obtained:

	Accuracy	# of SVs
Baseline Chinese	89.605%	4,933
Baseline Japanese	86.925%	4,237

6.1   Strategy 1: The Most Semantic Chain

Let $\iota : \mathbf{s} \rightarrow \mathbf{c}$ be a class inclusion. We calculate three quantities:

1. $f_1(\iota)$: the frequency of character pairs $(s \in \mathbf{s}, c \in \mathbf{c})$ contained in words $s \in w_1, c \in w_2$ belonging to synsets $w_1 \in \Sigma_1, w_2 \in \Sigma_2$ such that there is a semantic relation $\Sigma_1 \rightarrow \Sigma_2$ in WordNet (see Section 5);
2. $f_2(\iota)$: the frequency of character pairs $(s \in \mathbf{s}, c \in \mathbf{c})$ contained in words $s \in w_1, c \in w_2$ belonging to synsets $w_1 \in \Sigma_1, w_2 \in \Sigma_2$ such that there is a two-step semantic relation $\Sigma_1 \rightarrow \Sigma' \rightarrow \Sigma_2$ in WordNet;
3. $r(\iota)$: let $r(s, c)$ be 1 when s and c share the same Kāng Xī radical, and 0 otherwise. $r(\iota)$ will be $\frac{1}{nm} \sum_{s \in \mathbf{s}} \sum_{c \in \mathbf{c}} r(s, c)$.

The semanticity $S(\iota)$ of inclusion $\iota : \mathbf{s} \to \mathbf{c}$ will be:

$$S(\iota) = \tfrac{1}{2}\log(1 + f_1(\iota)) + \tfrac{1}{4}\log(1 + f_2(\iota)) + \tfrac{1}{4}r(\iota),$$

normalized so that its values stay in the interval $[0,1]$. Here, coefficients $\frac{1}{2}, \frac{1}{4}$ and $\frac{1}{4}$ have been obtained heuristically by a grid method applied to a series of tests.

The *most semantic chain* $(\mathbf{s}_i)_{i \geq 0}$ of class $\mathbf{c}$ is calculated as follows: $\mathbf{s}_0 = \mathbf{c}$, and, given $\mathbf{s}_i$, $\mathbf{s}_{i+1} := \text{argmax}_\mathbf{z}\, S(\mathbf{z} \to \mathbf{s}_i)$, that is, the subcharacter of $\mathbf{s}_i$ with maximal semanticity.

Let $w(c)$ be the weight of unigram c, obtained by frequency in the baseline case. Let $w(\mathbf{c}) := \max_{c \in \mathbf{c}} w(c)$ be the weight of class $\mathbf{c}$. For every member $\mathbf{s}_i$ of the most semantic chain of $\mathbf{c}$, we added a weight $w(\mathbf{s}_i) = \frac{1}{i}S(\mathbf{s}_{i-1} \to \mathbf{s}_i)$ (with $\mathbf{s}_0 = \mathbf{c}$) to unigram c.

We obtained the following results:

	Accuracy	# of SVs	Un.w/ mod.	Un.added
Chinese	**92.62%**	3,287	594	20
Japanese	89.99%	3,728	524	152

where the two last columns contain the number of updated unigrams and the number of new unigrams. Notice that the increase in performance is similar for Chinese and Japanese, while the number of support vectors needed is higher in Japanese, probably due to the shorter length of texts making the classification task harder.

6.2 Strategy 2: Combining Most Semantic Chain and Least Phonetic Chain

With the notation of previous sections, let $\varphi(\mathbf{s} \to \mathbf{c})$ be the phoneticity of inclusion $\mathbf{s} \to \mathbf{c}$ (calculated as explained in Section 4). Recall that the least phonetic chain $(\mathbf{p}_i)_{i \geq 0}$ is obtained by taking $\mathbf{p}_0 = \mathbf{c}$, and, given $\mathbf{p}_i$, $\mathbf{p}_{i+1} := \text{argmin}_\mathbf{z}\, \varphi(\mathbf{z}, \mathbf{p}_i)$. For each $\mathbf{p}_i$ we define a new class weight $w'(\mathbf{p}_i)$ as follows:

$$w'(\mathbf{p}_i) = w(\mathbf{p}_i) + \tfrac{1}{i}\varphi(\mathbf{p}_{i-1} \to \mathbf{p}_i).$$

We obtained the following results:

	Accuracy	# of SVs	Un.w/ mod.	Un.added
Chinese	92.435%	3,299	851	245
Japanese	**90.125%**	3,737	745	453

The value for Chinese is a bit lower than in Strategy 1, but we get a better result for Japanese. We have an increase in the number of new unigrams, due to the fact that we have significantly more phonetically annotated inclusions than semantically annotated ones.

7 Unknown Character Semantic Approximation

Besides the usual NLP tasks where a semantic relatedness distance is needed, our system can also be applied to the processing of unknown sinographs. Let u be an unknown sinograph (*i.e.*, we know neither reading nor meaning), and **u** its class. In our graph, there is subgraph $G_\mathbf{u}$ generated by paths leading to **u**. The way the graph is built, our subgraph has no cycles. When we leave **u** and head towards the leaves, for every path $\mathbf{s}_i$ there will be a first node $\mathbf{n}_i$ for which we have semantic annotation (for example, a WordNet synset ID). Furthermore, we can attach weights to the $\mathbf{n}_i$ by using the product of semanticity expectations (calculated in a way similar to phoneticity expectations) of the edges of path $\mathbf{un}_i$ and the distance from **u** on the path $\mathbf{un}_i$.

We obtain a vector in the space of WordNet synset IDs. This is not a precise semantic, but it can be used in various statistical NLP tasks.[6]

8 Perspectives

Further work will involve three main areas of focus:

1. further analyzing the graph and extracting knowledge about sinographic languages;
2. similarly processing various higher-order graphs (*n*-grams, words, concepts, . . .) and studying their interactions;
3. last, but not least, in the frame of the HanziGraph Project, manually validating the semanticity of the 39,719 class inclusions of our graph by a team of native Chinese/Japanese speakers, using the dedicated Web site `http://www.hanzigraph.net`.

References

1. Allen, J.D., et al. (eds.): The Unicode Standard, Version 6.0. Unicode Consortium (2011)
2. Fujimura, O., Kagaya, R.: Structural patterns of Chinese characters. In: Proceedings of the International Conference on Computational Linguistics, Sånga-Säby, Sweden, pp. 131–148 (1969)
3. Wang, J.C.S.: Toward a generative grammar of Chinese character structure and stroke order. PhD thesis, University of Wisconsin-Madison (1983)
4. Dürst, M.J.: Coordinate-independent font description using Kanji as an example. Electronic Publishing 6(3), 133–143 (1993)
5. Chu, B.F.: 漢字基因 朱邦復漢字基因工程 (Genetic engineering of Chinese characters) (2003), `http://cbflabs.com/down/show.php?id=26`
6. Moro, S.: Surface or essence: Beyond the coded character set model. In: Proceedings of the Glyph and Typesetting Workshop, Kyoto, Japan, pp. 26–35 (2003)
7. Sproat, R.: A Computational Theory of Writing Systems. Studies in Natural Language Processing. Cambridge University Press (2000)
8. Peebles, D.G.: SCML: A Structural Representation for Chinese Characters. PhD thesis, Dartmouth College, TR2007–592 (2007)
9. Bishop, T., Cook, R.: Wenlin CDL: Character Description Language. Multilingual 18, 62–68 (2007)

---

[6] See also [42] where radicals are used to guess not semantics but rather a sinograph's POS tag.

10. Haralambous, Y.: Seeking meaning in a space made out of strokes, radicals, characters and compounds. In: Proceedings of ISSM 2010-2011, Aizu-Wakamatsu, Japan (2011)

11. Qin, L., Tong, C.S., Yin, L., Ling, L.N.: Decomposition for ISO/IEC 10646 ideographic characters. In: COLING 2002: Proceedings of the 3rd Workshop on Asian Language Resources and International Standardization. Association for Computational Linguistics (2002)

12. Dai, R., Liu, C., Xiao, B.: Chinese character recognition: history, status and prospects. Frontiers of Computer Science in China 1, 126–136 (2007)

13. Fujiwara, Y., Suzuki, Y., Morioka, T.: Network of words. Artificial Life and Robotics 7, 160–163 (2004)

14. Li, J., Zhou, J.: Chinese character structure analysis based on complex networks. Physica A: Statistical Mechanics and its Applications 380, 629–638 (2007)

15. Rocha, J., Fujisawa, H.: Substructure Shape Analysis for Kanji Character Recognition. In: Perner, P., Rosenfeld, A., Wang, P. (eds.) SSPR 1996. LNCS, vol. 1121, pp. 361–370. Springer, Heidelberg (1996)

16. Zhou, L., Liu, Q.: A character-net based Chinese text segmentation method. In: SEMANET 2002 Proceedings of the 2002 Workshop on Building and Using Semantic Networks, pp. 1–6. Association for Computational Linguistics (2002)

17. Yu, S., Liu, H., Xu, C.: Statistical properties of Chinese phonemic networks. Physica A: Statistical Mechanics and its Applications 390, 1370–1380 (2011)

18. Hsieh, S.K.: Hanzi, Concept and Computation: A Preliminary Survey of Chinese Characters as a Knowledge Resource in NLP. PhD thesis, Universität Tübingen (2006)

19. Chou, Y.-M., Hsieh, S.-K., Huang, C.-R.: Hanzi Grid: Toward a Knowledge Infrastructure for Chinese Character-Based Cultures. In: Ishida, T., R. Fussell, S., T. J. M. Vossen, P. (eds.) IWIC 2007. LNCS, vol. 4568, pp. 133–145. Springer, Heidelberg (2007)

20. Taft, M., Zhu, X.: Submorphemic processing in reading Chinese. Journal of Experimental Psychology: Learning, Memory and Cognition 23, 761–775 (1997)

21. Williams, C., Bever, T.: Chinese character decoding: a semantic bias? Read Writ. 23, 589–605 (2010)

22. Tamaoka, K., Yamada, H.: The effects of stroke order and radicals on the knowledge of Japanese Kanji orthography, phonology and semantics. Psychologia 43, 199–210 (2000)

23. Zhao, S., Baldauf Jr., R.B.: Planning Chinese Characters. Reaction, Evolution or Revolution? Language Policy, vol. 9. Springer (2008)

24. Guder-Manitius, A.: Sinographemdidaktik. Aspekte einer systematischen Vermittlung der chinesischen Schrift im Unterricht Chinesisch als Fremdsprache. Sino-Linguistica, vol. 7. Julius Groos Verlag, Tübingen (1999)

25. Jenkins, J.H., Cook, R.: Unicode Standard Annex #38. Unicode Han Database. Technical report, The Unicode Consortium, property kHanyuPinlu (2010)

26. Chikamatsu, N., Yokoyama, S., Nozaki, H., Long, E., Fukuda, S.: A Japanese logographic character frequency list for cognitive science research. Behavior Research Methods, Instruments, & Computers 32(3), 482–500 (2000)

27. Sharoff, S.: Creating general-purpose corpora using automated search engine queries. In: Baroni, M., Bernardini, S. (eds.) WaCky! Working papers on the Web as Corpus, Bologna, GEDIT (2006), http://wackybook.sslmit.unibo.it/pdfs/wackybook.zip

28. Morioka, T.: CHISE: Character Processing Based on Character Ontology. In: Tokunaga, T., Ortega, A. (eds.) LKR 2008. LNCS (LNAI), vol. 4938, pp. 148–162. Springer, Heidelberg (2008)

29. Schindelin, C.: Zur Phonetizität chinesischer Schriftzeichen in der Didaktik des Chinesischen als Fremdsprache. SinoLinguistica, vol. 13. Iudicium, München (2007)

30. Newman, M.J.: Networks. An introduction. Oxford University Press (2010)

31. Chang, C.H., Li, S.Y., Lin, S., Huang, C.Y., Chen, J.M.: 以最佳化及機率分佈判斷漢字聲符之研究 (Automatic identification of phonetic complements for Chinese characters based on optimization and probability distribution). In: Proceedings of the 22nd Conference on Computational Linguistics and Speech Processing (RO-CLING 2010), Puli, Nantou, Taiwan, pp. 199–209 (2010)

32. Sriram, S., Talukdar, P.P., Badaskar, S., Bali, K., Ramakrishnan, A.G.: Phonetic distance based cross-lingual search. In: Proc. of the 5th International Conf. on Natural Language Processing (KBCS 2004), Hyderabad, India (2004)

33. Kondrak, G.: A new algorithm for the alignment of phonetic sequences. In: NAACL 2000: Proceedings of the 1st North American Chapter of the Association for Computational Linguistics Conference (2000)

34. Huang, C.R.: Sinica BOW: Integrating bilingual WordNet and SUMO ontology. In: International Conference on Natural Language Processing and Knowledge Engineering, pp. 825–826 (2003)

35. Gao, Z., et al.: Chinese WordNet (2008), http://www.aturstudio.com/wordnet/windex.php

36. Isahara, H., Bond, F., Uchimoto, K., Utiyama, M., Kanzaki, K.: Development of the Japanese WordNet. In: Proceedings of the Sixth International Conference on Language Resources and Evaluation, LREC 2008 (2008)

37. Jiang, J.J., Conrath, D.W.: Semantic similarity based on corpus statistics and lexical taxonomy. In: Proceedings of International Conference on Research in Computational Linguistics, Taiwan (1997)

38. Lin, D.: An information-theoretic definition of similarity. In: Proceedings of 15th International Conference on Machine Learning, Madison WI (1998)

39. Resnik, P.: Using information content to evaluate semantic similarity in a taxonomy. In: Proceedings of the 14th International Joint Conference on Artificial Intelligence, Montréal, pp. 448–453 (1995)

40. Sogou: 互联网语料库 (SogouT) (2008), http://www.sogou.com/labs/dl/t.html

41. Reuters: 過去ニュース (2007-2012), http://www.reuters.com/resources/archive/jp/index.html

42. Zhang, H.J., Shi, S.M., Feng, C., Huang, H.Y.: A method of part-of-speech guessing of Chinese unknown words based on combined features. In: International Conference on Machine Learning and Cybernetics, pp. 328–332 (2009)

	あ	い	う	え	お	や	ゆ	よ	か	き	く	け	こ	きゃ	きゅ	きょ	さ	し	す	せ	そ	しゃ	しゅ	しょ	た	ち	つ	て	と	ちゃ	ちゅ	ちょ	な	に	ぬ	ね	の	にゃ	にゅ	にょ	は	ひ	ふ	へ	ほ	ひゃ	ひゅ	ひょ	ま	み	む	め	も	みゃ
あ	0	.55	.59	.35	.46	.58	.82	.74	.43	.70	.73	.56	.63	.48	.76	.67	.35	.65	.68	.49	.58	.94	.94	.94	.25	.60	.64	.43	.53	.93	.93	.93	.50	.74	.77	.61	.68	.55	.80	.71	.63	.83	.86	.72	.78	.99	.99	.99	.25	.60	.64	.43	.53	.33
い		0	.53	.20	.56	.80	.79	.80	.70	.43	.69	.48	.71	.73	.72	.74	.65	.35	.64	.41	.66	.73	.73	.73	.60	.25	.59	.32	.61	.69	.69	.69	.74	.50	.73	.54	.75	.77	.76	.78	.83	.63	.82	.66	.84	.85	.85	.85	.60	.25	.59	.32	.61	.64
う			0	.46	.16	.82	.58	.60	.73	.69	.43	.63	.46	.76	.48	.51	.68	.64	.35	.58	.39	.93	.93	.93	.64	.59	.25	.52	.30	.91	.91	.91	.77	.73	.50	.68	.53	.80	.55	.57	.86	.82	.63	.78	.65	.99	.99	.99	.64	.59	.25	.52	.30	.68
え				0	.43	.68	.74	.72	.56	.48	.63	.43	.61	.60	.67	.65	.49	.41	.58	.35	.55	.80	.80	.80	.43	.32	.52	.25	.49	.78	.78	.78	.61	.54	.68	.50	.66	.65	.71	.69	.72	.66	.78	.63	.76	.90	.90	.90	.43	.32	.52	.25	.49	.48
お					0	.74	.60	.58	.63	.71	.46	.61	.43	.67	.51	.48	.58	.66	.39	.55	.35	.95	.95	.95	.53	.61	.30	.49	.25	.93	.93	.93	.68	.75	.53	.66	.50	.71	.57	.55	.78	.84	.65	.76	.63	.99	.99	.99	.53	.61	.30	.49	.25	.57
や						0	.59	.46	.31	.63	.67	.47	.56	.38	.70	.60	.31	.63	.66	.46	.55	.93	.93	.93	.36	.65	.69	.50	.59	.94	.94	.94	.35	.65	.68	.49	.58	.41	.72	.62	.29	.62	.66	.45	.55	.93	.93	.93	.41	.68	.72	.54	.62	.47
ゆ							0	.16	.67	.62	.31	.55	.35	.70	.38	.42	.66	.62	.31	.55	.35	.92	.92	.92	.69	.64	.36	.58	.40	.93	.93	.93	.68	.64	.35	.58	.39	.72	.41	.44	.66	.61	.29	.54	.33	.92	.92	.92	.72	.67	.41	.61	.44	.75
よ								0	.56	.64	.35	.53	.31	.60	.42	.38	.55	.64	.35	.53	.31	.94	.94	.94	.59	.67	.40	.56	.36	.95	.95	.95	.58	.66	.39	.55	.35	.62	.44	.41	.55	.63	.33	.52	.29	.94	.94	.94	.62	.69	.44	.59	.41	.66
か									0	.55	.59	.35	.46	.22	.63	.51	.27	.61	.65	.44	.54	.93	.93	.93	.25	.60	.64	.43	.52	.93	.93	.93	.25	.60	.64	.43	.53	.33	.68	.57	.21	.58	.62	.40	.51	.92	.92	.92	.40	.67	.71	.53	.61	.45
き										0	.53	.20	.56	.59	.58	.60	.61	.27	.60	.34	.62	.70	.70	.70	.60	.25	.59	.32	.61	.69	.69	.69	.60	.25	.59	.32	.61	.64	.63	.65	.58	.21	.57	.29	.60	.68	.68	.68	.67	.40	.67	.45	.68	.71
く											0	.46	.16	.63	.22	.27	.65	.60	.27	.53	.32	.91	.91	.91	.64	.59	.25	.52	.30	.91	.91	.91	.64	.59	.25	.52	.30	.68	.33	.37	.62	.57	.21	.50	.26	.90	.90	.90	.71	.67	.40	.60	.43	.74
け												0	.43	.41	.51	.48	.44	.34	.53	.27	.51	.78	.78	.78	.43	.32	.52	.25	.49	.77	.77	.77	.43	.32	.52	.25	.49	.48	.56	.54	.40	.29	.50	.21	.47	.77	.77	.77	.53	.45	.60	.40	.58	.57
こ													0	.51	.27	.22	.54	.62	.32	.51	.27	.94	.94	.94	.52	.61	.30	.49	.25	.93	.93	.93	.53	.61	.30	.49	.25	.57	.37	.33	.51	.60	.26	.47	.21	.93	.93	.93	.61	.68	.43	.58	.40	.65
きゃ														0	.59	.46	.35	.65	.68	.49	.58	.94	.94	.94	.33	.64	.67	.48	.57	.94	.94	.94	.33	.64	.68	.48	.57	.25	.64	.53	.30	.62	.66	.46	.55	.93	.93	.93	.45	.71	.74	.57	.65	.40
きゅ															0	.16	.68	.64	.35	.57	.38	.93	.93	.93	.67	.63	.33	.56	.37	.92	.92	.92	.68	.63	.33	.56	.37	.64	.25	.30	.66	.61	.30	.55	.34	.92	.92	.92	.74	.70	.45	.64	.48	.71
きょ																0	.58	.66	.38	.55	.35	.95	.95	.95	.57	.65	.37	.54	.33	.95	.95	.95	.57	.65	.37	.54	.33	.53	.30	.25	.55	.63	.34	.52	.30	.94	.94	.94	.65	.72	.48	.62	.45	.61
さ																	0	.55	.59	.35	.46	.91	.91	.91	.11	.56	.60	.36	.47	.91	.91	.91	.44	.70	.73	.56	.64	.49	.77	.67	.38	.66	.70	.51	.60	.95	.95	.95	.31	.63	.67	.47	.56	.38
し																		0	.53	.20	.56	.65	.65	.65	.56	.11	.55	.23	.57	.66	.66	.66	.70	.44	.69	.48	.71	.73	.73	.74	.66	.38	.65	.43	.67	.74	.74	.74	.63	.31	.62	.37	.64	.67
す																			0	.46	.16	.89	.89	.89	.60	.55	.11	.47	.20	.89	.89	.89	.73	.69	.44	.63	.47	.77	.49	.52	.70	.65	.38	.59	.41	.93	.93	.93	.67	.62	.31	.55	.35	.70
せ																				0	.43	.74	.74	.74	.36	.23	.47	.11	.44	.75	.75	.75	.56	.48	.63	.44	.61	.60	.67	.65	.51	.43	.59	.38	.57	.81	.81	.81	.47	.37	.55	.31	.53	.52
そ																					0	.92	.92	.92	.47	.57	.20	.44	.11	.92	.92	.92	.64	.71	.47	.61	.44	.67	.52	.49	.60	.67	.41	.57	.38	.96	.96	.96	.56	.64	.35	.53	.31	.60
しゃ																						0	.29	.23	.91	.66	.89	.75	.92	.05	.35	.29	.96	.76	.95	.83	.97	.97	.96	.98	.95	.74	.93	.81	.96	.19	.48	.42	.94	.72	.92	.79	.94	.95
しゅ																							0	.08	.91	.66	.89	.75	.92	.35	.05	.14	.96	.76	.95	.83	.97	.97	.96	.98	.95	.74	.93	.81	.96	.48	.19	.27	.94	.72	.92	.79	.94	.95
しょ																								0	.91	.66	.89	.75	.92	.29	.14	.05	.96	.76	.95	.83	.97	.97	.96	.98	.95	.74	.93	.81	.96	.42	.27	.19	.94	.72	.92	.79	.94	.95
た																									0	.55	.59	.35	.46	.91	.91	.91	.40	.67	.71	.53	.61	.45	.74	.65	.41	.68	.72	.54	.62	.96	.96	.96	.25	.60	.64	.43	.53	.33
ち																										0	.53	.20	.56	.65	.65	.65	.67	.40	.67	.45	.68	.71	.70	.72	.68	.41	.67	.46	.69	.75	.75	.75	.60	.25	.59	.32	.61	.64
つ																											0	.46	.16	.89	.89	.89	.71	.67	.40	.60	.43	.74	.45	.48	.72	.67	.41	.61	.44	.94	.94	.94	.64	.59	.25	.52	.30	.68
て																												0	.43	.74	.74	.74	.53	.45	.60	.40	.58	.57	.64	.62	.54	.46	.61	.41	.59	.82	.82	.82	.43	.32	.52	.25	.49	.48
と																													0	.92	.92	.92	.61	.68	.43	.58	.40	.65	.48	.45	.62	.69	.44	.59	.41	.96	.96	.96	.53	.61	.30	.49	.25	.57
ちゃ																														0	.29	.23	.95	.75	.94	.81	.96	.97	.95	.97	.96	.75	.94	.82	.96	.20	.50	.43	.93	.69	.91	.78	.93	.94
ちゅ																															0	.08	.95	.75	.94	.81	.96	.97	.95	.97	.96	.75	.94	.82	.96	.50	.20	.29	.93	.69	.91	.78	.93	.94
ちょ																																0	.95	.75	.94	.81	.96	.97	.95	.97	.96	.75	.94	.82	.96	.43	.29	.20	.93	.69	.91	.78	.93	.94
な																																	0	.55	.59	.35	.46	.22	.63	.51	.35	.65	.69	.50	.58	.94	.94	.94	.37	.66	.70	.51	.59	.43
に																																		0	.53	.20	.56	.59	.58	.60	.65	.35	.64	.41	.66	.73	.73	.73	.66	.37	.65	.43	.67	.70
ぬ																																			0	.46	.16	.63	.22	.27	.69	.64	.35	.58	.39	.93	.93	.93	.70	.65	.37	.59	.41	.73
ね																																				0	.43	.41	.51	.48	.50	.41	.58	.35	.55	.80	.80	.80	.51	.43	.59	.37	.57	.56
の																																					0	.51	.27	.22	.58	.66	.39	.55	.35	.95	.95	.95	.59	.67	.41	.57	.37	.63
にゃ																																						0	.59	.46	.42	.69	.72	.54	.62	.96	.96	.96	.43	.70	.73	.56	.63	.37
にゅ																																							0	.16	.72	.68	.42	.62	.45	.94	.94	.94	.73	.69	.43	.63	.46	.70
にょ																																								0	.62	.70	.45	.60	.42	.96	.96	.96	.63	.71	.46	.61	.43	.59
は																																									0	.55	.59	.35	.46	.91	.91	.91	.56	.78	.81	.66	.72	.60
ひ																																										0	.53	.20	.56	.65	.65	.65	.78	.56	.77	.59	.79	.81
ふ																																											0	.46	.16	.89	.89	.89	.81	.77	.56	.72	.58	.84
へ																																												0	.43	.74	.74	.74	.66	.59	.72	.56	.70	.69
ほ																																													0	.92	.92	.92	.72	.79	.58	.70	.56	.76
ひゃ																																														0	.29	.23	.99	.82	.98	.87	.99	.99
ひゅ																																															0	.08	.99	.82	.98	.87	.99	.99
ひょ																																																0	.99	.82	.98	.87	.99	.99
ま																																																	0	.55	.59	.35	.46	.22
み																																																		0	.53	.20	.56	.59
む																																																			0	.46	.16	.63
め																																																				0	.43	.41
も																																																					0	.51
みゃ																																																						0
みゅ																																																						
みょ																																																						
ら																																																						
り																																																						
る																																																						
れ																																																						
ろ																																																						
りゃ																																																						
りゅ																																																						
りょ																																																						
わ																																																						
ゐ																																																						
ゑ																																																						
を																																																						
ん																																																						
が																																																						
ぎ																																																						
ぐ																																																						
げ																																																						
ご																																																						
ぎゃ																																																						
ぎゅ																																																						
ぎょ																																																						
ざ																																																						
じ																																																						
ず																																																						
ぜ																																																						
ぞ																																																						
じゃ																																																						
じゅ																																																						
じょ																																																						
だ																																																						
ぢ																																																						
づ																																																						
で																																																						
ど																																																						
ぢゃ																																																						
ぢゅ																																																						
ぢょ																																																						
ば																																																						
び																																																						
ぶ																																																						
べ																																																						
ぼ																																																						
びゃ																																																						
びゅ																																																						
びょ																																																						
ぱ																																																						
ぴ																																																						
ぷ																																																						
ぺ																																																						
ぽ																																																						
ぴゃ																																																						
ぴゅ																																																						
ぴょ																																																						

Fig. 6. (On two pages) Values of Japanese phonetic distance for all syllable combinations

The Application of Kalman Filter Based Human-Computer Learning Model to Chinese Word Segmentation

Weimeng Zhu[1], Ni Sun[2], Xiaojun Zou[2], and Junfeng Hu[1,2,*]

[1] School of Electronics Engineering & Computer Science, Peking University
[2] Key Laboratory of Computational Linguistics, Ministry of Education, Peking University, Beijing, 100871, P.R. China
{zhuweimeng,sn96,xiaojunzou,hujf}@pku.edu.cn

Abstract. This paper presents a human-computer interaction learning model for segmenting Chinese texts depending upon neither lexicon nor any annotated corpus. It enables users to add language knowledge to the system by directly intervening the segmentation process. Within limited times of user intervention, a segmentation result that fully matches the use (or with an accurate rate of 100% by manual judgement) is returned. A Kalman filter based model is adopted to learn and estimate the intention of users quickly and precisely from their interventions to reduce system prediction error hereafter. Experiments show that it achieves an encouraging performance in saving human effort and the segmenter with knowledge learned from users outperforms the baseline model by about 10% in segmenting homogenous texts.

Keywords: Chinese Word Segmentation, Unsupervised Learning, Human-Computer Interaction, Kalman Filter.

1 Introduction

Chinese text is written without natural delimiters, so word segmentation is an essential first step in Chinese information processing[1]. Approaches to Chinese segmentation fall roughly into two categories: dictionary-based methods and statistical methods. In dictionary-based methods, a predefined dictionary is introduced along with hand-generated rules for segmenting input sequence[2][3]. To solve the ambiguity problem and out-of-vocabulary (OOV) problem of dictionary-based methods, many techniques have been developed, including various statistical methods[4][5][6]. Statistical learning approaches are more desirable and have been successful in both unsupervised learning[7][8] and supervised learning[9][10][11].

Supervised statistical learning approaches, which relay on statistical models or features learned automatically from labeled corpus, are more adaptive and robust in processing unrestricted texts than the traditional dictionary-based

---

* To whom all correspondence should be addressed.

A. Gelbukh (Ed.): CICLing 2013, Part I, LNCS 7816, pp. 218–230, 2013.

methods. However, in some domain specific applications, say, rare ancient Chinese processing, there is neither enough homogeneous corpus to train an adequate statistical model, nor redefined dictionary readily available for such text. In these tasks, unsupervised word segmentation is preferred to utilize the linguistic knowledge derived from the raw corpora itself. Many researches also enable the users to take part in the segmentation process and add expert knowledge to the system[12][13]. This is reasonable since the standard of word segmentation is dependent on a user and the destination of use in many applications[14]. From this point of view, learning and adapting from users is also an important ability for pragmatic word segmentation systems.

In this paper, we presents a human-computer interaction learning model for segmenting Chinese texts depending upon neither lexicon nor any annotated corpus. It enables users to add language knowledge to the segmenter and the segmenter learns and adapts to these knowledge recursively, and finally a segmentation result that fully matches the destination of a specific use (or an accurate rate of 100% by human judgement) is returned. The learning model is equipped with a well-designed Kalman filter to make it learn language knowledge quickly and precisely, and eventually, to reduce expert interventions to the most extent and save the efforts of experts.

The rest of this paper is organized as follows. The next section reviews the related work, and the details of the Kalman filter based human-computer learning model will be elaborated in the third section. In Section 4, the evaluation will be presented and in the final section, we conclude the proposed model and discuss the future work.

2   Related Work

This section reviews the related work in three aspects: unsupervised segmentation approach, human-computer interaction segmentation approach, and the using of Kalman filter in promoting user experience.

2.1   Unsupervised Segmentation Approach

As the preparation of linguistic resources is very hard and time consuming due to the particularity of Chinese. Furthermore, even the lexicon is large enough, and the corpus annotated is balanced and huge in size, the word segmenter will still face the problem of data incompleteness, sparseness and bias as it may be utilized in different domains. To make best of the raw corpus and reduce the efforts of human is the main starting point of unsupervised strategy.

An unsupervised segmentation strategy has to follow some predefined criteria, e.g., *mutual information* (*mi*), to recognize a subsequence as a word. [15] was an early comprehensive study in this direction using mutual information. Many successive research applied mutual information with different ensemble methods[16][17][18]. Sun et al. proposed *difference of t-score* (*dts*)[7] as a useful complement to mutual information and designed an algorithm based on the linear combination of *mi* and *dts*, which is called *md*[19].

Kit et al. proposed a compression-based unsupervised segmentation algorithm, named after description-length-gain (DLG) based segmentation[20]. Feng et al. proposed a statistical criterion called *assessor variety* (AV) to measure how likely a sub-sequence is a word, and then to find the best segmentation pattern that maximizes a target function of accessor variety and the length of the sub-sequence as variants[21]. Jin et al. proposed *branch entropy* as another criterion for unsupervised segmentation[22]. Both criteria share a similar assumption as in the fundamental work by Harris[23]: if the uncertainty of successive tokens increases, then the location is at a border.

2.2   Human-Computer Interaction Segmentation Approach

All segmentation models have limitations in some way and far from fully match the particular need of users. Thus human-computer interaction segmentation approaches are explored to learn and adapt the users, whose linguistic knowledge can be passed to the segmenter by directly intervening the segmentation.

Wang et al. developed a sentence based human-computer interaction inductive learning method[12]. Their method suppose that a common character string appearing repeatedly in text has high probability as a word, based on which unknown words are predicted by extracting such strings from input sentences recursively. The segmentation is conducted by these words, and then the result is proofreaded by the user and the confirmed new words are added to the dictionary for the next iteration.

Feng et al. proposed a certainty-based active learning segmentation algorithm, which use EM (Expectation Maximization) algorithm to train an n-multigram language model in an unsupervised learning framework[24]. In this algorithm, the sentences with low confidence are submitted to user for manual tagging. Li et al. further explored a candidates words based human-computer interaction segmentation strategy[13]. In their algorithm, candidates words are extracted from text recursively, then judged and edited by the user. Thus, a lexicon of the text is gained and applied to segmenting the text.

2.3   Kalman Filter

The Kalman filter, proposed by R. E. Kalman in 1960[25], is an efficient recursive filter that estimates the internal state of a linear dynamic system from a series of noisy measurements. It is used in a wide range of engineering and econometric applications from radar and computer vision to estimation of structural macroeconomic models, and is an important topic in control theory and control systems engineering.

Many recent researches have introduced Kalman filter model to promote the user experience (UE) of internet applications, by estimating click-through rate (CTR) of available articles (or other objects on web pages) in near real-time for news display systems[26] or recommender systems[27], for instance. These Kalman filter based CTR tracking algorithm yields good indicator quality and popularity temporally. In this paper, we adopt Kalman filter model to learn and estimate user intentions quickly and precisely from their interventions (which may contain noise).

3   Kalman Filter Based Word Segmentation Model

3.1   The Basic Model

The model from [19] is a reference for our basic model before interaction. It is an unsupervised model without dictionary by using two statistical measures of a bigram, namely, *mutual information* and *difference of t-test*. In terms of a linear combination of them, denoted by md, and a threshold Θ, whether a bigram should be combined or seperated can be determined. When the value of md is greater than Θ, the bigram has more chance to be in a word. The formulas for calculating md are as follows (where $mi(x, y)$ and $dts(x, y)$ denote the *mutual information* and *difference of t-test* between x and y respectively, and they are normalized as $mi^*(x, y)$ and $dts^*(x, y)$):

$$mi^*(x, y) = \frac{mi(x, y) - \mu_{mi}}{\sigma_{mi}}, \tag{1}$$

$$dts^*(x, y) = \frac{dts(x, y) - \mu_{dts}}{\sigma_{dts}}, \tag{2}$$

$$md(x, y) = mi^*(x, y) + \lambda \times dts^*(x, y), \tag{3}$$

where λ is set as an empirical value 0.6 in [19], while is calculated by simulated annealing algorithm in [28].

Meanwhile, there is a optimization when a local maximum or local minimum of md appears[19]. Considering a character string $vxyw$, when $md(x, y) > md(v, x)$ and $md(x, y) > md(y, w)$ at the same time, it is called a *local maximum*; and if $md(x, y) < md(v, x)$ and $md(x, y) < md(y, w)$, it is a *local minimum*. Obviously, even if the $md(x, y)$ of a local maximum does not reach the threshold Θ, the xy may be combined, while even if the $md(x, y)$ of a local minimum is greater than Θ, the xy is likely to be separated.

3.2   The Dynamic Process

The Kalman filters are based on linear dynamic systems discretized in the time domain. Given parameters, the unobserved state-process can be estimated via the Kalman filter or Kalman smoother.

Kalman filter is adopted in our human-computer interaction learning model for Chinese word segmentation. In this model, we define the manual judgement of segmentation results of a bigram as a *human interaction process* and map it to a time series process, which means the state of the time step k in our model will move to the state of step $k + 1$ each time this bigram appears. Hence, the interaction process is a dynamic process. We take the user interactions as a form of measurements (which may contain noise) to estimate the genuine intentions of the user. The linguistic knowledge is gradually learned and accumulated from this process of user interactions, and eventually, we acquire a segmentation result that fully matches a specific use (or with a accurate rate of 100% by manual judgement) within limited times of intervention as the state in the model has been updated.

3.3   The Process State

Based on the the model mentioned in Section 3.1, we can estimate the initial state of the process. Our basic hypothesis is that each bigram (of different characters) is independent, i.e., if the state of one bigram changes, the other different bigrams will not be affected. Given a bigram xy, we suppose the $md(x, y)$ in the corpus follows a stable distribution and denotes it as $md'(x, y)$. Assume that the corpus is huge enough, and the distribution of $md'(x, y)$ in it will be distributed normally as

$$md'(x, y) \sim N(\mu_{md(x,y)}, \sigma^2_{md(x,y)}) , \tag{4}$$

where $\mu_{md(x,y)}$ and $\sigma^2_{md(x,y)}$ are the expectation and variance of $md(x, y)$ respectively.

We assume $md'(x, y)_t$ is the *system state* of the bigram xy at time t, which indicates the combination strength between x and y. The state of a discrete-time controlled process in Kalman filter is governed by the linear stochastic difference equation,

$$md'(x, y)_{t+1|t} = md'(x, y)_t + w(x, y)_t , \tag{5}$$

where the random variable $w(x, y)_t$ represents the process noise which obeys a normal distribution.

$$w(x, y)_t \sim N(0, Q(x, y)_t) , \tag{6}$$

where $Q(x, y)_t$ is the autocovariance of bigram xy at time t. In our model, the autocovariance is introduced as the covariance of the variable against a time-shifted version of itself and has plenty of applications [29][30][31]. Thus, the $Q(x, y)_t$ in Eq. 6 can be calculated as

$$Q(x, y)_t \\ = E[(md'(x, y)_{t-2} - \mu(x, y)_{t-2})(md'(x, y)_{t-1} - \mu(x, y)_{t-1})] , \tag{7}$$

where

$$\mu(x, y)_t = E[md'(x, y)_t] \;\; for \; each \; t . \tag{8}$$

Also, the predicted covariance of state change is estimated to be

$$P(x, y)_{t+1|t} = P(x, y)_t + Q(x, y)_t , \tag{9}$$

where $P(x, y)_t$ represents the state covariance of time t.

At the time 0, which means the interaction has not yet started, the initial state should be the expectation of $md'(x, y)$,

$$md'(x, y)_0 = \mu_{md(x,y)} . \tag{10}$$

Besides, because of the existence of local maximum and local minimum, the parameters will change as time goes on. It contradicts our independent assumption. Suppose the bigram xy is in the context of v and w (that is, appears as a character string $vxyw$ in text), we use the $md(v, x)$ and $md(y, w)$ instead of $md'(v, x)$ and $md'(y, w)$ to judge if $md'(x, y)_t$ is a local maximum or local minimum.

3.4 The Measurement and the Measurement Space

In practice, the manual judgements are binary-valued, namely, 0 or 1. To simplify the model and make calculation easier, we introduce the *measurement system*. The manual judgements will be the inputs of this system and the outputs will be the measurement of the dynamic model. To avoid confusion, we call the measurement to the model *system measurement*, and call the manual judgements *manual measurement*.

The measurement system is aimed at mapping the manual measurements to the system measurements, that is, mapping a binary space to a continuous space of md' ranging from $-\infty$ to $+\infty$. Apparently, some uncertainty (or observation noise) is inevitable in this mapping, and it obeys a normal distribution. To guarantee that the mapping value corresponds to the manual measurements, we can take a certain interval as a high confidence interval (99%, for instance).

There are four cases of the manual judgements, i.e., correct separation, wrong combination, correct combination and wrong separation. The corresponding interval for the former two cases is $[min(separate(x, y)_t), max(separate(x, y)_t)]$, and that for the latter two is $[min(combine(x, y)_t), max(combine(x, y)_t)]$. The $separate(x, y)_t$ denotes a set of all the possible md to separate xy at time t and $combine(x, y)_t$ denotes a set of all the possible md to connect. Because of the existence of local maximum and local minimum, the range of intervals should be changed.

For those unbounded intervals, to make sure the computability of the expectation and variance, we also adopt the following method—the upper or lower bound are the maximum or minimal value of md in the nearby context according to statistical results:

$$max'(combine(x, y))$$
$$= \forall_{v, w \in context(x, y)}(max(md(v, x), md(y, w), \Theta_{loc\_min})) , \quad (11)$$
$$min'(separate(x, y))$$
$$= \forall_{v, w \in context(x, y)}(min(md(v, x), md(y, w), \Theta_{loc\_max})) , \quad (12)$$

where Θ_{loc_min} and Θ_{loc_max} are thresholds for a local minimum and a local maximum.

At time t, the system measurement (or observation) $z(x, y)_t$ of the true state $md_{t|z}(x, y)$ is made according to

$$z(x, y)_t = md'_{t|z}(x, y) + v(x, y)_t , \quad (13)$$

where $v(x,y)_t$ is the observation noise which is assumed to be zero mean Gaussian white noise with standard deviation $\sigma(x,y)_t$.

$$v(x,y)_t \sim N(0, \sigma(x,y)_t^2) \tag{14}$$

If it is a correct separation, $\sigma(x,y)_t$ is estimated as

$$\hat{\sigma}(x,y)_t = -\frac{1}{2z_{\frac{1-\alpha}{2}}}(max(separate(x,y)) - min(separate(x,y))), \tag{15}$$

where $z_{\frac{1-\alpha}{2}}$ is the quantile of the normal distribution and α is the confidence level (we take 0.99, for instance); if a correct connection, estimated as

$$\hat{\sigma}(x,y)_t = -\frac{1}{2z_{\frac{1-\alpha}{2}}}(max(combine(x,y)) - min(combine(x,y))). \tag{16}$$

With the method above, the manual judgements are transformed to the system observation, and the covariance of the observation noise is

$$R(x,y)_t = var(v(x,y)_t). \tag{17}$$

3.5   States Update

As is mentioned in Section 3.2, we can update the state using the predicted state and the measurement result. We update the state using an optimal Kalman gain

$$K(x,y)_t = \frac{P(x,y)_{t+1|t}}{P(x,y)_{t+1|t} + R(x,y)_{t+1}}. \tag{18}$$

The updated state of time $t+1$ is estimated as

$$md'(x,y)_{t+1} = md'(x,y)_{t+1|t} + K(x,y)_t(z(x,y)_{t+1} - md'(x,y)_{t+1|t}), \tag{19}$$

and the updated state covariance is estimated as

$$P(x,y)_{t+1} = (1 - K(x,y)_t)P(x,y)_{t+1|t}. \tag{20}$$

Then, we will use the updated state $md'(x,y)_{t+1}$ to do segmentation next time this bigram appears.

4   Experimental Results

In this section, we conducted several experiments to evaluate the performance of our segmentation model in three aspects. Firstly, we verified the effectiveness of human interaction model and the improvement after introducing the Kalman filter based learning model. Secondly, the reusability of knowledge learned from human interaction is tested. Finally, we took a bigram as an example to track the states in the human interaction process. All experiments are based on the *People's Daily* corpus from Jan.1998 to Mar.1998 provided by Institution of Computational Linguistics of Peking University.

4.1   Simulating Human-Computer Interaction Word Segmentation

In this part, we simulated the human-computer interaction by using the correct segmentation.We define a *binary prediction rate* (BPR) to describe the conformity of the prediction in the model with user intentions as

$$BPR[\%] = \frac{\#\ of\ predictons\ being\ correct}{\#\ of\ predictons\ in\ interaction\ process} \times 100\% . \qquad (21)$$

To compare our model with other interaction word segmentation model, we designed a dynamic memory based human interaction approach (shorted as *Memory Approach*). It is a bigram based human interaction model similar to our Kalman filter based model (shorted as *Kalman Approach*). The initial segmentation result of a bigram in this model is exactly the same as that of this bigram in Kalman Approach. The prediction of segmentation for a bigram is taken from human interaction (that is, the correct segmentation judged by human) for this bigram last time it appears.

Apart from the Memory Approach, we used another two models without human interaction support as baseline. One is a model based on the approach proposed by Sun in [19], but we just use the segmentation result of this approach as the prediction for each bigram in human interaction process. (This model is shorted as *Sun's Approach*.) The other is a model based on a simplified version of Kalman Approach (that is, without dynamic process), which we use the initial state of a bigram to predict its segmentation each time it appears. (This model is shorted as *Non-Dynamic Approach*.)

In this experiment, we simulated the human-computer interaction process of word segmentation on the raw corpus of *People's Daily* from Jan. 1998 to Mar. 1998 respectively. The result of experiment is shown in Table 1. It is shown that the application of interaction help to increase the BPR by approximately 10% from the models without human interaction support. Meanwhile, our model (Kalman Approach) gains a slightly higher BPR than Memory Approach, which means our model can help decrease the times of human intervention in errors of word segmentation (that is, gaining a better conformity to user intentions to save human efforts).

Table 1. The BPR[%] of different models for each month

Corpus	Sun's Approach	Non-Dynamic Approach	Memory Approach	Kalman Approach
Jan.1998	84.13%	84.23%	94.53%	**94.68%**
Feb.1998	84.25%	84.34%	94.73%	**94.88%**
Mar.1998	84.43%	84.52%	95.10%	**95.26%**

4.2   Knowledge Usefulness Test

After the experiment in Section 4.1, we obtained the value of final state of each bigram, and we assume these state values as a kind of knowledge of the bigrams that can be used to help word segmentation on homogeneous corpus.

We took the knowledge learnt from the *People's Daily* Jan.1998 and applied it to the word segmentation on the *People's Daily* Feb.1998 and Mar.1998. We used the original approach from [19] as baseline (shorted as *Sun's Orig. Approach*). Then, we overwrote the value of *md* of the bigram with the learnt knowledge (if this bigram exists in the knowledge), which we call *Sun's Orig. Approach with Knowledge*. Table 2 presents the accuracy of segmentation of the two approaches, defined in [7] as

$$\frac{\# \ of \ locations \ being \ correctly \ marked}{\# \ of \ locations \ in \ texts} . \tag{22}$$

The correct word segmentation is provided by Institution of Computational Linguistics of Peking University.

Table 2. The accuracy of different models

Corpus	Sun's Orig. Approach (Baseline)	Sun's Orig. Approach with Knowledge
Feb.1998	84.25%	**94.26%**
Mar.1998	84.43%	**94.31%**

It is shown that after applying the learnt knowledge, the accuracy of segmentation increased 10.01% and 9.88% respectively processing the corpus of Feb. 1998 and Mar. 1998. This significant increase in accuracy indicates that the knowledge learnt from human interaction process is reusable on homogeneous corpus.

4.3   Filter Process Analyzing

We tracked the state sequences in the human interaction process of some bigrams. In data analysis, we found some complicated cases may be hard for a segmenter to deal with, but some of them can be well handled by our model. We take one of the cases (the bigram "及其") as an example in the following.

Fig. 1 depicts the state values of bigram "及其" changing through time, where X_{post} is the updated state value for each time, Θ is the threshold of word segmentation, $\Theta + 0.5$ is threshold for local minimum, and $\Theta - 0.5$ for local maximum. It is shown that our model can control the state values in the interval of Θ and $\Theta + 0.5$ to adapt to various context situations. The following two examples illustrate two main possible situations of bigram "及其".

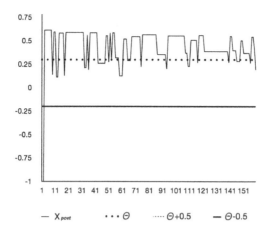

Fig. 1. Results for updated state values of bigram "及其" for each time

Example 1: 但维护它的自身经济利益及其在亚洲的战略利益是根本出发点。
(*But to maintain its own economic interests and its strategic interests in Asia is the fundamental starting point.*)

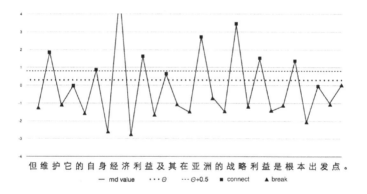

Fig. 2. The distribution of *md* for the sentence in Example 1

As is shown in Fig. 2, the *md* of bigram "及其" is greater than Θ, so it is judged combined and this is a correct combination as a word "及其"(*and its*). Most cases in this situation have similar context to this example and they should be judged combined by threshold Θ (the state value is greater than Θ).

Example 2: 有关单位应当如实反映情况，提供必需的账簿、文件以及其他资料。
(*The department concerned should report the situation truthfully, providing necessary account books, documents and other materials.*)

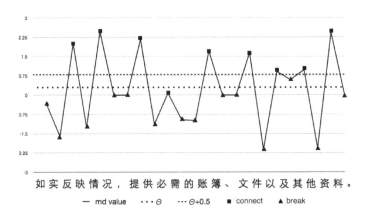

Fig. 3. The distribution of md for the sentence in Example 2

As is shown in Fig. 3, the md of bigram "及其" is less than $\Theta+0.5$ (it is a local minimum here), so it is judged separated and this is a correct separation between "以及"(and) and "其他"($other$). Most cases in this situation have similar context to this example and they should be judged separated by the local minimum threshold $\Theta + 0.5$ (the state value is less than $\Theta + 0.5$).

To sum up, our model can find a balance point in the interval of Θ and $\Theta + 0.5$ in the process of interaction to deal with these two significantly different situations of bigram "及其". This kind of balance point can be found in many bigrams with complicated situations, which can hardly be well handled by Memory Approach mentioned in Section 4.1.

5   Conclusions and Future Work

In this paper, we present a Kalman filter based human-computer interaction learning model for segmenting Chinese texts depending upon neither lexicon nor any annotated corpus. It enables users to add language knowledge to the system by directly intervening the segmentation results. Within limited times of user intervention, a segmentation result that fully matches the use (or with an accurate rate of 100% by manual judgement) is returned. Experiments show that it achieves an encouraging performance in saving human effort and the segmenter with knowledge learned from users significantly outperforms the baseline model in segmenting homogenous corpus.

Our future work will concentrate on the following two aspects. One is to extend the bigram based model with single parameter to multigram based model with multiparameter, which may cover more complicated cases. The other is to apply our human interaction learning model to other natural language processing fields, e.g., POS tagging.

Acknowledgments. We would like to thank several colleagues who have been involved in various aspects of this project. Yi Tong has contributed to the basic model coding. Tongfei Chen has implemented an online version[1] of our model. We thank Zhongwei Tan for his brief introduction of Kalman filter to the authors which made the beginning of this project. This work is partially supported by the Chiang Ching-kuo Foundation for International Scholarly Exchange under the project "Building a Diachronic Language Knowledge-base" (RG013-D-09) and the Open Project Program of the National Laboratory of Pattern Recognition (NLPR).

References

1. Liang, N.Y.: CDWS: An Automatic Word Segmentation System for Written Chinese Texts. Journal of Chinese Information Processing 1 (1987) (in Chinese)
2. Nie, J.Y., Jin, W., Hannan, M.L.: A Hybrid Approach to Unknown Word Detection and Segmentation of Chinese. In: Proceedings of the International Conference on Chinese Computing, pp. 326–335 (1994)
3. Wu, Z.: LDC Chinese Segmenter,
   `http://www.ldc.upenn.edu/Projects/Chinese/segmenter/mansegment.perl`
4. Luo, X., Sun, M., Tsou, B.K.: Covering Ambiguity Resolution in Chinese Word Segmentation Based on Contextual Information. In: COLING 2002, pp. 1–7 (2002)
5. Li, M., Gao, J., Huang, C.N., Li, J.: Unsupervised Training for Overlapping Ambiguity Resolution in Chinese Word Segmentation. In: Proceedings of the 2nd SIGHAN Workshop on Chinese Language Processing, pp. 1–7 (2003)
6. Sun, C., Huang, C.N., Guan, Y.: Combinative Ambiguity String Detection and Resolution Based on Annotated Corpus. In: Proceedings of the 3rd Student Workshop on Computational Linguistics (2006)
7. Sun, M.S., Shen, D.Y., Tsou, B.K.: Chinese Word Segmentation Without Using Lexicon and Hand-Crafted Training Data. In: COLING/ACL 1998, pp. 1265–1271 (1998)
8. Goldwater, S., Griffiths, T.L., Johnson, M.: Contextual Dependencies in Unsupervised Word Segmentation. In: COLING/ACL 2006, pp. 673–680 (2006)
9. Xue, N.: Chinese Word Segmentation as Character Tagging. Computational Linguistics and Chinese Language Processing 8, 29–48 (2003)
10. Zhang, H., Liu, Q., Cheng, X., Zhang, H., Yu, H.: Chinese Lexical Analysis Using Hierarchical Hidden Markov Model. In: Proceedings of the Second SIGHAN Workshop, pp. 63–70 (2003)
11. Peng, F., Feng, F., Mcallum, A.: Chinese Segmentation and New Word Detection Using Conditional Random Fields. In: COLING 2004, pp. 23–27 (2004)
12. Wang, Z., Araki, K., Tochinai, K.: A Word Segmentation Method with Dynamic Adapting to Text Using Inductive Learning. In: Proceedings of the First SIGHAN Workshop on Chinese Language, vol. 18, pp. 1–5 (2002)
13. Li, B., Chen, X.H.: A Human-Computuer Interaction Word Segmentation Method Adapting to Chinese Unknown Texts. Journal of Chinese Information Processing 21 (2007) (in Chinese)

---

[1] `http://klcl.pku.edu.cn/clr/ccsegweb/kalman_segmenter.aspx`

14. Sproat, R., Shih, C., Gale, W., Chang, N.: A Stochastic Finite-State Word-Segmentation Algorithm for Chinese. Association for Computational Linguistics 22, 377–404 (1996)

15. Sproat, R., Shih, C.: A Statistical Method for Finding Word Boundaries in Chinese Text. Computer Processing of Chinese and Oriental Languages 4, 336–351 (1990)

16. Chien, L.F.: Pat-Tree-Based Keyword Extraction for Chinese Information Retrieval. In: Proceedings of the 20th Annual International ACM SIGIR Conference on Research and Development in Information Retrieval, pp. 50–58 (1997)

17. Zhang, J., Gao, J., Zhou, M.: Extraction of Chinese Compound Words–an Experimental Study on a Very Large Corpus. In: Proceedings of the Second Chinese Language Processing Workshop, pp. 132–139 (2000)

18. Yamamoto, M., Church, K.W.: Using Suffix Arrays to Compute Term Frequency and Document Frequency for All Substrings in a Corpus. Computational Linguistics 27, 1–30 (2001)

19. Sun, M., Xiao, M., Tsou, B.K.: Chinese Word Segmentation without Using Dictionary Based on Unsupervised Learning Strategy. Chinese Journal of Computers 6, 736–742 (2004)

20. Kit, C., Wilks, Y.: Unsupervised Learning of Word Boundary with Description Length Gain. In: Proceedings of the CoNLL 1999 ACL Workshop, pp. 1–6 (1999)

21. Feng, H., Chen, K., Deng, X., Zheng, W.: Accessor Variety Criteria for Chinese Word Extraction. Computational Linguistics 30, 75–93 (2004)

22. Jin, Z., Tanaka-Ishii, K.: Unsupervised Segmentation of Chinese Text by Use of Branching Entropy. In: COLING/ACL 2006, pp. 428–435 (2006)

23. Harris, Z.S.: Morpheme Boundaries within Words. In: Papers in Structural and Transformational Linguistics, pp. 68–77 (1970)

24. Feng, C., Chen, Z.X., Huang, H.Y., Guan, Z.Z.: Active Learning in Chinese Word Segmentation Based on Multigram Language Model. Journal of Chinese Information Processing 1 (2004) (in Chinese)

25. Kalman, R.E.: A New Approach to Linear Filtering and Prediction Problems. Transaction of the ASME-Journal of Basic Engineering, 35–45 (1960)

26. Agarwal, D., Chen, B., Elango, P., Motgi, N., Park, S., Ramakrishnan, R., Roy, S., Zachariah, J.: Online Models for Content Optimization. Advances in Neural Information Processing Systems 21, 17–24 (2009)

27. Chu, W., Park, S.T.: Personalized Recommendation on Dynamic Content Using Predictive Bilinear Models. In: Proc. of the 18th International World Wide Web Conference, pp. 691–700 (2009)

28. Tong, Y.: Chinese Word Segmentation Based on Statistical Method with General Dictionary and Component Information. Bachelor Degree Thesis. Peking University (2012)

29. Odelson, B.J., Rajamani, M.R., Rawlings, J.B.: A New Autocovariance Least-Squares Method for Estimating Noise Covariances. Automatica 42, 303–308 (2006)

30. Åkesson, B.M., Jørgensen, J.B., Poulsen, N.K., Jørgensen, S.B.: A Generalized Autocovariance Least-Squares Method for Kalman Filter Tuning. Journal of Process Control 18, 769–779 (2008)

31. Rajamani, M.R., Rawlings, J.B.: Estimation of the Disturbance Structure from Data Using Semidefinite Programming and Optimal Weighting. Automatica 45, 142–148 (2009)

Machine Learning for High-Quality Tokenization Replicating Variable Tokenization Schemes

Murhaf Fares[1], Stephan Oepen[1], and Yi Zhang[2]

[1] Institutt for Informatikk, Universitetet i Oslo
{murhaff,oe}@ifi.uio.no
[2] LT-Lab, German Research Center for Artificial Intelligence
yizhang@dfki.de

Abstract. In this work, we investigate the use of sequence labeling techniques for tokenization, arguably the most foundational task in NLP, which has been traditionally approached through heuristic finite-state rules. Observing variation in tokenization conventions across corpora and processing tasks, we train and test multiple CRF binary sequence labelers and obtain substantial reductions in tokenization error rate over off-the-shelf standard tools. From a domain adaptation perspective, we experimentally determine the effects of training on mixed gold-standard data sets and make a tentative recommendation for practical usage. Furthermore, we present a perspective on this work as a feedback mechanism to resource creation, i.e. error detection in annotated corpora. To investigate the limits of our approach, we study an interpretation of the tokenization problem that shows stark contrasts to 'classic' schemes, presenting many more token-level ambiguities to the sequence labeler (reflecting use of punctuation and multi-word lexical units). In this setup, we also look at partial disambiguation by presenting a token lattice to downstream processing.

Keywords: Tokenization, Domain Variation, Sequence Labeling.

1 Background, Motivation, and Related Work

Tokenization is the process of splitting a stream of characters into smaller, word-like units for downstream processing—or, in the definition of Kaplan [9], breaking up *"natural language text ... into distinct meaningful units (or tokens)"*. As such, tokenization constitutes a foundational pre-processing step for almost any subsequent natural language processing (NLP), and token errors will inevitably propagate into any downstream analysis. With a few notable exceptions, however, tokenization quality and adaptability to (more or less subtly) different conventions have received comparatively little research attention over the past decades.

A recent survey by Dridan and Oepen [3], for example, observes that state-of-the-art statistical constituency parsers perform relatively poorly at replicating even the most common tokenization scheme, viz. that of the venerable Penn

A. Gelbukh (Ed.): CICLing 2013, Part I, LNCS 7816, pp. 231–244, 2013.

Treebank (PTB; [12]). In a similar spirit, Foster [6] reports an improvement in statistical parsing accuracy of 2.8 points F_1 over user-generated Web content, when making available to the parser gold-standard token boundaries.

While the original PTB scheme is reasonably defined and broadly understood, today there exist many variants. These often address peculiarities of specific types of text, e.g. bio-medical research literature or user-generated content, or they refine individual aspects of the original PTB scheme, for example the treatment of hyphens and slashes.[1] Variations on the PTB scheme are at times only defined vaguely or extensionally, i.e. through annotated data, as evidenced for example in resources like the Google 1T n-gram corpus (LDC #2006T13) and the OntoNotes Initiative [8].

Furthermore, it is quite possible that different types of downstream processing may call for variation in tokenization, e.g. information retrieval vs. machine translation could prefer different notions of *"meaningful units"*. In § 2 below, we observe that in the context of a broad-coverage grammar couched in a detailed theory of syntax, very different views on tokenization can be motivated, including the recognition of some types of multi-word expressions. These lead to higher token boundary ambiguities and, thus, make tokenization a task that is intricately entangled with downstream analysis.

One of the 'classic' tokenization difficulties is punctuation ambiguity; for example, the period is a highly ambiguous punctuation mark because it can serve as a full-stop, a part of an abbreviation, or even both (as in 'U.S.' when it occurs at the end of a sentence in the PTB). Parentheses and commas typically form individual tokens, but they can be part of names and numeric expressions, e.g. in 'Ca(2+)' or '390,926'. Handling contracted verb forms and the Saxon genitive of nouns is also problematic because there is no generally agreed-upon standard on how to tokenize them. But corner cases in tokenization are by no means restricted to English; for other languages and with different tokenization conventions, abstractly rather similar problems emerge.

Notwithstanding its essential role, it is our impression that tokenization for languages with (somewhat) explicit word boundary marking[2] is widely considered a relatively uninteresting, 'trivial' sub-problem of NLP. However, we believe that tokenization comprises a range of challenges and opportunities that should be equally amenable to careful engineering and experimentation as other tasks—especially so as NLP increasingly turns from formal, edited texts towards more informal, 'non-canonical' language, showing greater variation and dynamics—e.g. Web 2.0 data with non-standard punctuation, emoticons, hashtags, and the like. From [3] it appears that tokenization is predominantly approached through

---

[1] The 2008 Shared Task of the Conference on Natural Language Learning, for example, produced variants of PTB-derived annotations with most hyphens as separate tokens, to match NomBank annotation conventions of propositional semantics [18].

[2] For CJK languages where the writing systems have no explicit marker for word boundaries, however, tokenization is a well acknowledged challenge commonly known as the word segmentation task. Despite the relatedness, we restrict our discussion on English tokenization in this paper.

rule-based techniques, typically finite-state machines or (cascades of) regular expressions. As they observe, these methods offer great flexibility about how to specifically interpret the tokenization task, where in practice it seems common to mesh (a) decisions on token boundaries with (b) some amount of string-level normalization.[3] Quite possibly owing to this somewhat diffuse interpretation of the task, there has been preciously little work on machine learning based study for high-quality tokenization.

Tomanek et al. [19] apply Conditional Random Field (CRF) learners to sentence and token boundary detection in bio-medical research literature, essentially modeling sub-task (a) above as a sequence labeling problem. Our work is similar in spirit but broader in several respects: First, [19] defined their own uniform tokenization scheme for bio-medical texts, whereas we experiment with a range of tokenization conventions and text types—in all cases using generally available, 'standard' resources. Second, we include in our experiments a 'heretically' different and far more challenging view on tokenization—motivated by a large-scale computational grammar—which to a higher degree blends token boundary detection and aspects of syntactic analysis. Third, we provide empirical results on the utility of various types of features and sensitivity to genre and domain variation, as well as in-depth error analyses for the most pertinent experiments. Forth, we investigate possibilities and trade-offs in *ambiguous* tokenization, i.e. lattices representing n-best lists of token sequences; For downstream integration with syntactic parsing, for example, this setup affords greater flexibility in balancing accuracy and efficiency, abstractly parallel to work in interweaving sequence labeling for lexical category assignments and structured prediction for syntactic analysis (e.g. [2], [21]). Finally, we complement our search for premium tokenization accuracy in practical NLP with a 'quality control' perspective on existing or emerging resources, suggesting to utilize mismatches between learned generalizations and individual annotated instances as indicators of inconsistencies or problematic 'grey' areas in annotation conventions.

In our view, separating the two sub-tasks of tokenization explicitly (boundary detection vs. text normalization) is a methodological advantage, making it possible to experiment more freely with different techniques.[4] Our results in all cases improve over previously reported (or reconstructed) 'state-of-the-art' tokenization and suggest an advantage of supervised machine learning over finite-state techniques, both in terms of available accuracy, adaptability to variable schemes, and resilience to domain and genre variation. For these reasons, we hope to

---

[3] One compelling instance of the second sub-tasks discussed in [3] is the disambiguation of quote marks (into opening and closing, or left and right delimiters). Assuming a tokenizer input not making this distinction already, i.e. text using non-directional, straight ASCII quotes only, such disambiguation can be performed with relative ease at the string level—based on adjacency to whitespace—but not later on.

[4] Maršík and Bojar [13] present a related but again interestingly different view, by combining tokenization with sentence boundary detection to form one joint classification task, which they approach through point-wise Maximum Entropy classification; their experiments, however, do not address tokenization accuracy for English.

encourage fresh research into very high-quality tokenization and thus improved performance of downstream analysis.

The rest of this paper is organized as follows: §2 and §3 review extant tokenization schemes and the use of sequence labelers for tokenization, respectively. In §4, we provide experimental results for PTB-style tokenization and derivatives; followed by parallel experiments on the tokenization scheme of a large computational grammar in §5.

2   Review: Different Tokenization Conventions

As observed earlier, there is substantive variation in existing schemes for tokenization, even for English. This section contrasts two very different tokenization schemes, viz. that of the Penn Treebank (and derivatives) and that of the English Resource Grammar (ERG; [4]), highlighting main design principles and key differences. We choose to replicate PTB-style tokenization because of its central position for much NLP reserach, and ERG-style tokenization because of its roots in formal syntactic theory and, of course, its relative intricacy.

2.1   PTB-Style Tokenization and Derivatives

According to the original PTB tokenization documentation, the PTB-style tokenization is "fairly simple".[5] The key principles in PTB-compliant tokenization are:

- Whitespaces are explicit token boundaries.
- Most punctuation marks are split from adjacent tokens.
- Contracted negations are split into two different tokens.
- Hyphenated words and ones containing slashes are not split.

Even though PTB tokenization seems to be uncomplicated, not so many standard NLP tools can *accurately* reproduce PTB tokenization (from raw strings) [3]. However, [3] develop a cascade of regular, string-level rewrite rules that provide state-of-the-art results in PTB tokenization, with almost 99% sentence accuracy. We will use their results as our main point of reference.[6]

Among other problems (to which we will return in §4), one recurring issue in the PTB tokenization are sentence-final abbreviations, especially the U.S. Observe how the preceding sentence, in standard typography, ends in only one period. In the PTB, however, there seems to be a special treatment for the 'U.S.' abbreviation, so whenever it happens at the end of a sentence, an extra period is added to that sentence; however, if, for example, 'etc.' ends the sentence no extra period would be added.

Further, the PTB-style tokenization constitutes the basis for other broadly used resources such as the GENIA Treebank [10]. Somewhat parallel to the assessment

---

[5] For more detailed specifications see
http://www.cis.upenn.edu/~treebank/tokenization.html
[6] We are indebted to Rebecca Dridan of the University of Oslo for sharing her version of the 'raw' WSJ text, aligned to sentences from the PTB for gold-standard tokenization.

of common tools in [3], Øvrelid et al. [15] observe that tokenizing `GENIA` text using the `GENIA` tagger [20] leads to token mismatches in almost 20% of the sentences in the treebank.

Finally, the NLP community is moving toward producing—and processing—new data sets other than the `PTB`. At the same time, much current work uses `PTB` tokenization conventions with "some exceptions".[7] A consequence of these developments, in our view, is the decreasing community agreement about 'correct' tokenization and, thus, greater variety in best practices and types of input expected by downstream processing. For these reasons, we believe that ease of adaptability to subtle variation (e.g. retraining) is of growing importance, even in the realm of `PTB`-style tokenization.

2.2 `HPSG`-Style Tokenization in the `ERG`

In the context of detailed syntacto-semantic analyses couched in the framework of Head-Driven Phrase Structure Grammar (`HPSG`; [17]), the broad-coverage English Resource Grammar (`ERG`) defines a tokenization regime that diverges from `PTB` legacy in three key aspects. First, the `ERG` treats most punctuation marks as pseudo-affixes (i.e. akin to inflectional morphology, rather than as separate tokens that attach syntactically), i.e. commas, periods, quote marks, parentheses, et al. are *not* tokenized off. While Adolphs et al. [1] offer a linguistic argument for this analysis, we just note that it eliminates the dilemma presented by periods in sentence-final abbreviations and in general appears to predict well the interactions of whitespace and punctuation in standard orthography. As a consequence, however, some punctuation marks are ambiguous between being units of 'morphology' or syntax, for example colons and hyphens (see below). Second, like in some `PTB` derivatives, hyphens (or dashes) and slashes in the `ERG` introduce token boundaries, e.g. ⟨open-, source⟩ or ⟨3, -, 4⟩. Observe, however, that the functional distinction between intra-word hyphens and inter-token dashes projects into different tokenizations, with the hyphen as a pseudo-affix in the first example, but the n-dash a separate token in the second. As both functions can be encoded typographically as either a single hyphen or an n-dash ('-' or '--', respectively), the functional asymmetry introduces token-level ambiguity. Third, the `ERG` includes a lexicon of some classes of multi-word expressions, most notably so-called 'words with spaces' (e.g. ⟨ad, hoc⟩ or ⟨cul, de, sac⟩), proper names with syntactic idiosyncrasies (⟨New, Year's, Eve⟩, for example, can be a temporal modifier by itself), multi-word prepositions and adjectives (e.g. ⟨as, such⟩ or ⟨laid, back⟩), and of course 'singleton' words split at hyphens (e.g. ⟨e-, mail⟩ or ⟨low-, key⟩).[8] Finally, the `ERG` rejects the implied parallelism and defective paradigms for contracted negations suggested in the `PTB` splits

---

[7] The 2012 Shared Task on Parsing the Web, for example, used portions from Ontonotes 4.0 which is tokenized with "slightly different standards" from the original `PTB` conventions [16].

[8] The 2012 release of the `ERG` includes a hand-built lexicon of some 38,000 lemmas (i.e. uninflected stems), of which almost 10% are multi-word lexical entries.

of ⟨do, n't⟩, ⟨ca, n't⟩, et al. Observing the sharp grammaticality contrast between, say, *Don't you see?* vs. **Do not you see?*, [1] argue linguistically against
tokenizing off the contracted negation in these cases.

Taken together, these conventions make ERG-style tokenization a more intri-
cate task, calling for a certain degree of context-aware disambiguation and lexical
knowledge. To date, parsers working with the ERG typically operate off an ambiguous token lattice [1], and token boundaries are determined as a by-product
of syntactic analysis. In § 5 below, we seek to determine to what degree CRF
sequence labeling in isolation scales to this conception of the tokenization task,
considering both one-best and n-best decoding.

3   Tokenization as a Sequence Labeling Problem

The specifications of tokenization as a sequence labeling problem might vary
on five different dimensions, some are specific to the tokenization task itself,
others are common to various machine learning problems. In general, to recast
tokenization as a sequence labeling problem we need to define:

1. **Target tokenization scheme.**   The tokenization conventions which the
   model is expected to learn. This can be any coherently defined or practised
   tokenization scheme such as PTB or ERG.
2. **Basic processing unit.**   The smallest unit that can make up a single
   token. Said differently, the instances the ML model is supposed to label. We
   will elaborate more on this below.
3. **Tokenization labels.**   The set of classification labels. We perceive the
   tokenization of an input as a sequence of binary classifiaction, hence we
   define two classification labels, namely 'SPLIT' and 'NONSPLIT'.
4. **Machine learning models and features.**   In the tokenization task as
   such, there isn't any restriction on the type of sequence labeling algorithm
   one can employ. We experiment with conditional random fields (CRFs) using
   Wapiti's toolkit[9] [11].
5. **Data split.**   The train-development-test data split.

The core of the tokenization task is to split raw untokenized text and into smaller
units. To identify the potential splitting points, we need to first identify the
atomic units that cannot be further split.

One can look at 'tokens' from two perspectives: First, what might be a "token
separator" or what is not a token. Second, what constitutes a token. In the
what-is-not-a-token view, there is one main approach that is to define a set of
characters as token separators such as whitespace, period, and comma. While in
the *what-is-a-token* perspective, we can distinguish at least two approaches.

First, the character-based method where each character in the raw string is
a candidate token (thus followed by a candidate tokenization point). While this
is the common practice for word segmentation of CJK languages which have (to

---

[9] See http://wapiti.limsi.fr/

varying degrees) grapheme-based writing systems, this is probably not suitable for languages with phoneme-based writing system where single characters are typically representing basic significant sounds rather than a word, morpheme or semantic unit. An alternative approach would be to group the characters into different character classes and, accordingly, define the concept of **sub-tokens** where each one consists of a sequence of homogeneous characters belonging to the same character class.

In this approach we define the candidate tokenization points as the points between each pair of sub-tokens, and a token consists of one or more sub-tokens. The rationale behind this notion of character classes is that a consecutive sequence of homogeneous characters is most likely to constitute one single token and should not be further split. Table 1 lists some of the character classes we define for English tokenization.

More concretely, a word like 'well-educated' consists of three sub-tokens: 'well', '-' and 'educated', where these three sub-tokens belong to three character classes, respectively: 'alpha', 'hyphen' and 'alpha'. Hence, the candidate tokenization points are: 1) Between alpha and hyphen, 2) Between hyphen and alpha.

Table 1. Examples of the character classes

Character Class	Description
alphaC	A sequence of alphabetical characters with the initial capitalized
alpha	A sequence of a alphabetical characters
num	A sequence of numerical characters
Hyphen	A hyphen
SQ	Single quote
OQ	Open quote

4   Experiment 1: PTB-Style Tokenization

First, we carried out in-domain and out-of-domain tests of our machine learning based tokenizers with the PTB-style tokenized data.

For the training of the base PTB model, we followed the standard data split used in the part-of-speech tagging experiments. We train our model on PTB WSJ sections 0 to 18, improve it on sections 19 to 21 and test it on sections 22 to 24. Moreover, we extract the 'gold-tokens' from the treebank and align it with raw string provided with the 1995 release of the PTB (LDC #1995T07).

Assuming the gold sentence boundaries, we split each sentence into a sequence of sub-tokens as defined in § 3, hence the classifier's task is to decide whether or not to split each sub-token. However, our notion of character classes as such doesn't cover all possible token boundaries in the PTB conventions. More concretely, according to the PTB tokenization conventions the word 'cannot' must be split into two tokens, 'can' and 'not', but our *character classes* approach would recognise this as a single sub-token, eliminating the possibility of a split within a homogenous sequence of characters. We see a similar problem with contractions in the PTB, but we believe this is a restricted phenomena peculiar to PTB tokenization conventions. Extra regular rules were added to recognise the sub-tokens in these special cases.

As list in Table 2, we construct a set of 30 features for each sub-token exploiting their lexical and orthographical information as well as their local context in a bidirectional-window of six sub-tokens (denoted by W_i with $i \in [-3..3]$). The following is an example sentence with *sub-token* indices below and the corresponding *character classes* above each unit, respectively.

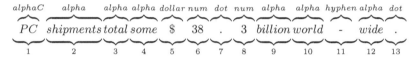

Table 2. Features used for tokenization classifiers

Feature	Feature	Feature
W_0 surface form (FORM)	W_{-1} FORM	W_{-1} FORM & CC
W_0 followed by space (S)	W_{-2} FORM	W_{-3} FORM
W_0 character class (CC)	W_0 FORM & S	W_0 FORM & CC & LEN
W_0 surface form length (LEN)	W_0 S & W_{+1} S	W_0 S & W_{-1} S
W_0 first character (FC)	W_0 FC & LEN	W_0 & W_{-1} & W_{+1} FORMs
W_0 last character (LC)	W_0 CC & W_{+1} CC	W_0 & W_{-1} CCs
W_0 FC & W_{-1} FC	W_{+1} FORM	W_{+1} FORM & CC
W_0 FC & W_{+1} FC	W_{+2} FORM	W_{+3} FORM
W_0 LC & W_{+1} FC	W_0 & W_{-1} & W_{-2} FORMs	W_0 & W_{-1} & W_{-2} & W_{-3} FORMs
W_{-1} LC & W_0 FC	W_0 & W_{+1} & W_{+2} FORMs	W_0 & W_{+1} & W_{+2} & W_{+3} FORMs

4.1 Results and Error Analysis

To allow fair comparison with other tokenization tools, we measure tokenization performance on the sentence level, i.e. will consider a sentence erroneous when it contains at least one token mismatch against the gold standard.[10]

Testing our PTB model on the last three sections of PTB led to an error rate of **0.93%**. As a reference point we compared our system with the best-performing set of rules from Oepen & Dridan [3] (henceforth the REPP tokenizer). Evaluating their rules on the last three sections of PTB resulted in **1.40%** error rate.[11]

Examining the errors of our PTB model revealed that about 45% of the sentence mismatches are due to tokenization inconsistencies within the PTB (gold-standard errors). More precisely, almost 30% of the PTB model's *"errors"* are blamed on the 'U.S.' idiosyncrasy in the PTB (discussed in § 2.1). Inconsistencies in splitting hyphenated words are the source of another 4% of the total error rate, for example 'trade-ethnic' is split into three tokens in the gold-standard in contrast to the PTB tokenization scheme. Yet another problem with hyphens

---

[10] Thus, unless explicitly said otherwise, all experimental results we report indicate sentence-level metrics i.e. full-sentence accuracy or full-sentence error rate.

[11] We had to replicate the experiments of Oepen & Dridan because results reported in [3] are against the full PTB.

in the PTB are cases like *"... on energy-, environmental- and fair-trade-related ..."* where the hyphens are separated from 'environmental-' and 'energy-', while 'fair-trade-related' is *correctly* kept all together. Other types of inconsistency, such as not splitting off some punctuation and splitting periods from acronyms (e.g. 'Cie.', 'Inc.', and 'Ltd.') sum up to almost 11% of the errors.

Interestingly enough, our PTB model shares 77% of the errors with the REPP tokenizer (apart from 'U.S.'-related mismatches).

Finally, to build a better understanding of the non-trivialty of the PTB tokenization task, we present a unigram baseline model which assigns to each sub-token its most frequent label and for unseen sub-tokens the most occurring label in the training data. This method results in a merely 40.04% sentence-level accuracy.

In (the right of) Figure 1 we observe the CRF learner's in-domain behavior with the increasing amount of training data. We see that the high accuracy can be established with relatively small amount of training data, and no steep improvement in accuracy thereafter, which can be partially due to the noisy (inconsistent) annotations in the PTB.

4.2 Domain Variation

To further validate our system, and also to investigate the robustness of ML-based and rule-based approaches for tokenization, we test the two systems on a different genre of text, the Brown Corpus.[12] Although the Brown annotations strictly follow the PTB tokenization scheme, there are many sentences where the mapping from the raw text to the tokenization in the treebank cannot be established, mainly because of notorious duplication of punctuation marks like semicolons and question and exclamation marks in the PTB annotations. As both tokenizers would be equally affected by these artifacts, we excluded 3,129 Brown sentences showing spurious duplicates from our evaluation.

In this experiment, the ML-based approach (i.e. our PTB model trained with WSJ sections) outperforms the REPP tokenizer by a good margin: it delivers a sentence accuracy of **99.52%**, contrasting with **97.13%** for the REPP rules of [3]. While for the ML-based tokenizer, performance on Brown is comparable to that on the WSJ text, the REPP tokenizer appears far less resilient to the genre variation. Although speculative, we conjecture that REPP's premium performance in our WSJ tests may in part be owed to tuning against this very data set, as discussed in [3] and evidenced by the explicit coding for sentence-final 'U.S.', for example.

Furthermore, we experiment with tokenizing out-of-domain texts from the GENIA Corpus. Using 90% of its sentences for training and the rest for testing, we carry out a new set of experiments: First, testing 'pure' PTB- or GENIA-only models on GENIA text; second, training and testing adapted models combining the PTB

---

[12] For this experiment, we again could rely on data made available to us by Rebecca Dridan, produced in a manner exactly parallel to her WSJ test data: a raw-text version of the Brown Corpus was aligned with the gold-standard sentence and token boundaries available in the 1999 release of the PTB (LDC #1999T42).

with either a small portion of GENIA (992 sentences; dubbed PTB+GENIA(1k)), or the entire GENIA (dubbed PTB+GENIA). The results of these experiments are presented in Table 3 together with the result of testing the REPP tokenizer on GENIA.

Table 3. The results of the GENIA experiments

Model	Error rate
REPP Tokenizer	8.48%
PTB	25.64%
GENIA	2.41%
PTB+GENIA(1k)	3.36%
PTB+GENIA	2.36%

From the results above, we see that domain adaptation works quite well even with limited in-domain annotation. We further contrast the learning curve of the adapted model with that of the PTB in-domain experiment (see Figure 1). The extended training with more in-domain annotation continues to improve the tokenization accuracy, contributing to the best performing tokenizer using all available annotations.

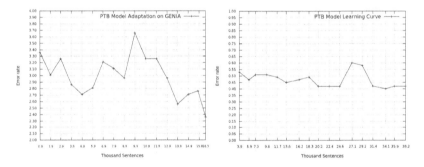

Fig. 1. Learning curves of the tokenizers. Left: effect of retraining the PTB model on GENIA corpus (tested on GENIA); Right: the PTB model (tested on PTB).

5   Experiment 2: ERG-Style Tokenization

For experimentation with ERG-style tokenization, we rely on the Redwoods Tree-bank [14], a collection of some 70,000 sentences annotated with gold-standard, ERG-conformant HPSG analyses. Redwoods samples a varity of domains, including transcribed dialogue, e-commerce email, tourism information, and a sample of Wikipedia. Recently, about 26,000 sentences from the WSJ portion of the PTB (Sections 0 – 15) were added to the Redwoods Treebank, also with gold-standard

ERG analyses (dubbed DeepBank; [5]). We deployed the standard splits of the treebank (as published with the ERG), for 55,867 sentences of training data, and 5,012 and 5,967 sentences for development and testing, respectively. Some of the Redwoods texts include residual markup, which the ERG interprets, for example using italics as a cue in recognizing foreign-language phrases in Wikipedia.

To focus on the 'pure' linguistic content of the text, our experiments on ERG-style tokenization actually start from a pre-tokenized, markup-free variant recorded in the treebank, essentially the result of a very early stage of parsing with the ERG [1]. In ERG terminology, this tokenization is dubbed *initial* tokenization, and it actually follows quite closely the PTB scheme.[13] During lexical analysis, the ERG parser combines the 'token mapping' rules of [1] with lexical look-up in the grammar-internal lexicon to arrive at the ERG-internal tokenization that we characterized in § 2.2 above (dubbed *lexical* tokenization). In standard ERG parsing, this mapping from initial to lexical tokenization is a by-product of full syntactic analysis only, but in the following experiments we seek to disambiguate lexical tokenization independent of the parser.[14]

As such, we train a binary classifier (ERG model henceforward) to reproduce ERG-like tokenization (lexical tokens) starting from PTB-compliant initial tokens. As noted in § 2.2 above, however, mapping from initial to lexical tokens can require additionally candidate splitting points around hyphens and slashes. Somewhat parallel to our notion of sub-tokens in the PTB universe (see § 4 above), we employ three heuristic rules to hypothesize additional candidate token boundaries, for example breaking a hyphenated initial token 'well-educated' into 'well-' and 'educated'.[15]

Green et al. [7] study multi-word expression (MWEs) identification using syntactic information and argue that MWEs cannot in general be identified reliably by "surface statistics" only. We largely agree with this point of view, but note that the task of ERG tokenization (as sequence labeling) only comprises a very limited range of MWEs, viz. multi-token lexical entries (see § 2.2 above). In analogy to the (practically) circular interdependencies observed between sequence labeling vs. full syntactic analysis for the task of lexical category disambiguation (i.e.

---

[13] The standard parsing setup for the ERG relies on a PoS tagging pre-processing stage, to provide the grammar with candidate lexical categories for unknown words. For compatibility with external, off-the-shelf tools, the ERG adopts PTB tokenization at this stage.

[14] We see at least two candidate applications of stand-alone ERG tokenization, viz. (a) to speed up parsing with the ERG, reducing or eliminating tokenization ambiguity; and (b) to enable 'pure' data-driven parsing of ERG structures, where recent work by Zhang & Krieger [23] and Ytrestøl [22], for example, assumes either idealized, gold-standard ERG tokenization, or undisambiguated full token lattice as inputs.

[15] Our three splitting heuristics are actually borrowed literally from the 'token mapping' rules that come with the ERG [1]. We also experimented with the introduction of sub-token splits points based on character classes, as in our PTB-style experiments, but found that—given PTB-style initial tokens as our starting point—the more specific heuristics targeting only hyphens and slashes led to fewer candidate boundaries and mildly superior overall results.

Table 4. N-best list for ERG-style tokenization

N	Accuracy
2	97.97%
3	98.77%
4	99.02%
5	99.16%

part-of-speech tagging), we hypothesize that the arguments of [7] do not apply (in full) to our ERG tokenization experiments.

5.1   Results and Error Analysis

Our ERG tokenization model deploys the same set of features presented in Table 2, complemented by a couple of additional features recording the length, in characters, of adjacent words. The accuracy of the ERG model, trained, tuned, and tested on the data sets and splits described above, is 93.88%, i.e. noticeably below PTB-style tokenization performance, but still rather decent (at the level of full sentences). Looking at tokenization mismatches of the ERG model unveiled that 41% of the errors involve MWEs unseen in training, such as 'a priori'. The rest of the errors contain ambiguous multi-word lexical units e.g. 'as well as', which should sometimes split and sometimes not. Another source of errors are hyphenated multi-word lexical units, such as 'south-west', which our ERG model regards as a hyphenated word and hence wrongly splits into two tokens.

To try and compensate for the higher error rate in ERG tokenization, we investigate limited degrees of remaining tokenization ambiguity, i.e. apply n-best (or list Viterbi) CRF decoding. The results in Table 4 suggest that quite small values of n lead to substantial accuracy gains, with 5-best decoding reaching sentence accuracies approaching our PTB results. As discussed above, at least for the purpose of pruning the search space of the full ERG parser, n-best decoding offers attractive flexibility in trading off accuracy and efficiency.

Finally, to gauge cross-domain sensitivity in these results (over the full, relatively diverse Redwoods Corpus), we train and test our ERG model on the WSJ portion of the data only. Using Sections 0 – 13 to train, and sections 14 and 15 to test, this WSJ-only ERG model delivers a moderately improved 94.06% per-sentence accuracy; its performance in 2-best mode is 98.79%, however, substantially better than the more general, cross-domain ERG model.

6   Reflections and Outlook

The subtleties of the tokenization task, variability in common tokenization schemes, and their potential impact on the downstream processing have inspired our investigation of a data-driven approach towards tokenization. We have shown that domain-adaptable tokenization models in a sequence labeling approach can achieve very high accuracies, and generally outperform state-of-the-art rule-based systems.

While a certain degree of feature engineering has been invested during our experiments, we believe that further (empirical) improvements can be made in the design of the sequence labeler, e.g. to differentiate different quotation conventions by introducing additional character classes, or even to copy the period following certain abbreviations (if strict compliance to the WSJ section of the PTB were the goal). Also, extra labels could be introduced in the sequence labeler to achieve certain rewriting operations (in a sense similar to that of a finite-state transducer). These remain to be investigated in future work.

In addition, examining the generalizations of our models and mismatches against gold-standard annotations, already has proven a useful technique in the identification of inconsistencies and errors within existing resources. For replicability and general uptake, the tokenization toolkit and models developed for our experiments will be made available as open source software.

Acknowledgements. We are grateful to Rebecca Dridan and Emanuele Lapponi at the University of Oslo for practical help and invaluable discussions and suggestions. The third author thanks BMBF for their support of the work through the funded project Deependance (01IW11003).

References

1. Adolphs, P., Oepen, S., Callmeier, U., Crysmann, B., Flickinger, D., Kiefer, B.: Some fine points of hybrid natural language parsing. In: Proceedings of the 6th International Conference on Language Resources and Evaluation, Marrakech, Morocco (2008)
2. Curran, J.R., Clark, S., Vadas, D.: Multi-tagging for lexicalized-grammar parsing. In: Proceedings of the 21st International Conference on Computational Linguistics and the 44th Meeting of the Association for Computational Linguistics, pp. 697–704. Association for Computational Linguistics, Sydney (2006)
3. Dridan, R., Oepen, S.: Tokenization. Returning to a long solved problem. A survey, contrastive experiment, recommendations, and toolkit. In: Proceedings of the 50th Meeting of the Association for Computational Linguistics, Jeju, Republic of Korea, pp. 378–382 (July 2012)
4. Flickinger, D.: On building a more efficient grammar by exploiting types. Natural Language Engineering 6(1), 15–28 (2000)
5. Flickinger, D., Zhang, Y., Kordoni, V.: DeepBank: A dynamically annotated treebank of the Wall Street Journal. In: Proceedings of the 11th International Workshop on Treebanks and Linguistic Theories (TLT 2011), Lisbon, Portugal (2012)
6. Foster, J.: "cba to check the spelling": Investigating parser performance on discussion forum posts. In: Human Language Technology Conference: The 2010 Annual Conference of the North American Chapter of the Association for Computational Linguistics, pp. 381–384. Association for Computational Linguistics, Los Angeles (2010)
7. Green, S., de Marneffe, M.C., Bauer, J., Manning, C.D.: Multiword expression identification with tree substitution grammars: A parsing tour de force with french. In: Proceedings of the 2011 Conference on Empirical Methods in Natural Language Processing, pp. 725–735. Association for Computational Linguistics, Edinburgh (2011)

8. Hovy, E., Marcus, M., Palmer, M., Ramshaw, L., Weischedel, R.: Ontonotes. The 90% solution. In: Proceedings of the Human Language Technology Conference of the North American Chapter of the Association for Computational Linguistics, New York City, USA, pp. 57–60 (June 2006)
9. Kaplan, R.M.: A method for tokenizing text. Festschrift for Kimmo Koskenniemi on his 60th birthday. In: Arppe, A., Carlson, L., Lindén, K., Piitulainen, J., Suominen, M., Vainio, M., Westerlund, H., Yli-Jyrä, A. (eds.) Inquiries into Words, Constraints and Contexts, pp. 55–64. CSLI Publications, Stanford (2005)
10. Kim, J.D., Ohta, T., Teteisi, Y., Tsujii, J.: GENIA corpus — a semantically annotated corpus for bio-textmining. Bioinformatics 19, i180–i182 (2003)
11. Lavergne, T., Cappé, O., Yvon, F.: Practical very large scale CRFs. In: Proceedings of the 48th Meeting of the Association for Computational Linguistics, Uppsala, Sweden, pp. 504–513 (July 2010)
12. Marcus, M., Santorini, B., Marcinkiewicz, M.A.: Building a large annotated corpora of English: The Penn Treebank. Computational Linguistics 19, 313–330 (1993)
13. Maršík, J., Bojar, O.: TrTok: A Fast and Trainable Tokenizer for Natural Languages. Prague Bulletin of Mathematical Linguistics 98, 75–85 (2012)
14. Oepen, S., Flickinger, D., Toutanova, K., Manning, C.D.: LinGO Redwoods. A rich and dynamic treebank for HPSG. Research on Language and Computation 2(4), 575–596 (2004)
15. Øvrelid, L., Velldal, E., Oepen, S.: Syntactic scope resolution in uncertainty analysis. In: Proceedings of the 23rd International Conference on Computational Linguistics, pp. 1379–1387. Association for Computational Linguistics, Stroudsburg (2010)
16. Petrov, S., McDonald, R.: Overview of the 2012 shared task on parsing the web. In: Notes of the First Workshop on Syntactic Analysis of Non-Canonical Language, SANCL (2012)
17. Pollard, C., Sag, I.A.: Head-Driven Phrase Structure Grammar. Studies in Contemporary Linguistics. Contemporary Linguistics. The University of Chicago Press, Chicago (1994)
18. Surdeanu, M., Johansson, R., Meyers, A., Màrquez, L., Nivre, J.: The CoNLL 2008 shared task on joint parsing of syntactic and semantic dependencies. In: Proceedings of the 12th Conference on Natural Language Learning, Manchester, England, pp. 159–177 (2008)
19. Tomanek, K., Wermter, J., Hahn, U.: Sentence and token splitting based on conditional random fields. In: Proceedings of the 10th Conference of the Pacific Association for Computational Linguistics, Melbourne, Australia, pp. 49–57 (2007)
20. Tsuruoka, Y., Tateishi, Y., Kim, J.-D., Ohta, T., McNaught, J., Ananiadou, S., Tsujii, J.: Developing a Robust Part-of-Speech Tagger for Biomedical Text. In: Bozanis, P., Houstis, E.N. (eds.) PCI 2005. LNCS, vol. 3746, pp. 382–392. Springer, Heidelberg (2005)
21. Yoshida, K., Tsuruoka, Y., Miyao, Y., Tsujii, J.: Ambiguous part-of-speech tagging for improving accuracy and domain portability of syntactic parsers. In: Proceedings of the 20th International Joint Conference on Artifical Intelligence, pp. 1783–1788. Morgan Kaufmann Publishers Inc., Hyderabad (2007)
22. Ytrestøl, G.: Cuteforce. Deep deterministic HPSG parsing. In: Proceedings of the 12th International Conference on Parsing Technologies, Dublin, Ireland, pp. 186–197 (2011)
23. Zhang, Y., Krieger, H.U.: Large-scale corpus-driven PCFG approximation of an HPSG. In: Proceedings of the 12th International Conference on Parsing Technologies, Dublin, Ireland, pp. 198–208 (2011)

Structural Prediction
in Incremental Dependency Parsing

Niels Beuck and Wolfgang Menzel

Fachbereich Informatik
Universität Hamburg
{beuck,menzel}@informatik.uni-hamburg.de

Abstract. For dependency structures of incomplete sentences to be fully connected, nodes in addition to those corresponding to the words in the sentence prefix are necessary. We analyze a German dependency corpus to estimate the extent of such predictive structures. We also present an incremental parser based on the WCDG framework and describe how the results from the corpus study can be used to adapt an existing weighted constraint dependency grammar for complete sentences to the case of incremental parsing.

Keywords: Parsing, Incrementality, Weighted Constraints, Dependency Grammar.

1 Motivation

A dependency structure of a sentence is a fully connected, single headed, directed acyclic graph, where every node corresponds to a word in the given sentence. For incomplete sentences, as encountered during incremental parsing, achieving connectedness is often not possible when the dependency structure is limited to the nodes corresponding to words in the sentence prefix. This problem occurs for every word that is attached to the right, as there will be at least one sentence prefix where the word but not its head is part of the prefix.

There are two possible approaches to deal with this problem: Either to drop the connectedness requirement for the dependency structures of incomplete sentences (from now on called partial dependency analyses, PDA) or allow additional nodes in the structure as needed to establish connectedness. In the first variant, words are left unattached until a suitable head appears. While this approach to incremental dependency parsing has been successfully used in parsers like MaltParser [1], we argue that it is unsatisfactory when partial dependency structures are of interest, for several reasons:

From a psycholinguistic point of view: Humans not only have interpretations of partial sentences, but also have strong expectations about the integration of words even if their syntactic head is still missing [2].

From an application point of view: Information already contained in the perceived part of the sentence is missing in the dependency structure if words are left unconnected. In the incomplete sentence "Pick up the red",

A. Gelbukh (Ed.): CICLing 2013, Part I, LNCS 7816, pp. 245–257, 2013.

the head of 'red' is still missing, but we can already argue that the object of 'pick' is something red. This is not limited to words close to the end of the prefix, like in a noun phrase under construction. It also applies to long distance dependencies like in German subclauses, where the verb comes last and all its arguments would need to stay unconnected. [3] proposed explicit dependencies between the arguments, but these are unnecessary if we provide an extra node for the expected verb, allowing early attachment of the arguments.

From a parsing perspective: Parsers choose between alternative structures based on some measure of grammaticality or probability. The WCDG parser used in this work uses a set of weighted constraints to measure the grammaticality of dependency structures. To be able to reuse measures optimized for judging analyses of complete sentences, a structure similar to the final structure is required to provide a rating similar to the rating for the final structure. Otherwise, unintuitive structures with no possible counterpart in a complete sentence analysis might be rated unreasonably high and lead the parser astray.

On the one hand, additional nodes in partial dependency analyses are desirable because they allow more informative structures. This leaves open the question of how to construct such structures in a parser. On the other hand, a constraint based parser would penalize partial dependency structures even if the constraint violation could and probably will be remedied by further input, making it hard to fairly rate and select a suitable dependency structure for a given sentence prefix.

We propose that the second problem holds the answer to the first one. The very constraint violations that unnecessarily penalize partial dependency analyses lacking structural prediction can be used as indicators for how to extend these structures with the needed additional nodes.

The remainder of this paper consists of two parts: First, we will present a constraint based incremental parsing algorithm able to make structural predictions about the upcoming input. It uses an existing weighted constraint dependency grammar and an existing transformation based parsing algorithm as well as a set of potential placeholders, so called virtual nodes (VNs), for structural prediction.

Secondly, we will analyze a given corpus of dependency annotations to assess the additional structure needed to keep PDAs connected and to fulfill all valency requirements. From this result we can infer the set of VNs we have to provide to our algorithm to achieve a certain coverage. In addition, we analyze to what extend the given set of weighted constraints can be used to judge such structurally predictive partial analyses and where it has to be adapted, e.g. by adding suitable exceptions.

2   Related Work

While to our knowledge this is the first strictly incremental dependency parser (strict in the sense of providing a fully connected structure for every input increment), strict incrementality has been explored already for the Tree Adjoining

Grammar (TAG) formalism. [4] investigated the amount of structure needed for strictly incremental parsing by building gold standard annotations incrementally with a simulated parser. They introduced the term *connection path* to denote the set of nodes of the (connected) tree for increment n that are not part of the tree for increment $n-1$ and not part of the subtree provided by the new word. While they worked with phrase structures for English and only regarded predictions needed to achieve connectedness, our work is on dependency structures for German and regards valency requirements in addition to connectedness.

PLTAG (PsychoLinguistically motivated Tree Adjoining Grammar)[5] is a TAG variant, that allows strictly incremental parsing. There, connectedness is facilitated by so called prediction trees, non-lexicalized elementary trees whose nodes later need to be instantiated by nodes from lexicalized elementary trees. Prediction trees are not anchored in the input directly, but are integrated when necessary to connect the elementary tree of the new material to the structure built so far. As the number of possible connection paths grows polynomially with the number of prediction trees contained, the parser presented in [5] restricts connection paths to one prediction tree.

The PLTAG parser deals with temporary ambiguity by keeping a beam of possible derivation trees, ranked by their probability learned from a training corpus. Thus, the best structure needs not be a monotonic continuation of the previously best ranked structure. The output of the incremental WCDG parser is non-monotonic, too, but instead of working with a beam of monotonically extended structures, it works by successively transforming a single structure. In contrast to the PLTAG parser, incremental WCDG achieves state of the art final accuracy (see [5], p.230 and [6]).

3   The WCDG Framework

Weighted Constraint Dependency Grammar (WCDG) [7] is a framework for dependency parsing by means of constraint optimization. The grammar consists of a set of constraints regarding single or pairs of dependency edges. So called context sensitive predicates allow constraints to access properties of the complete tree, e.g. to express the existence requirement for the subject of a verb. The constraints are weighted with a penalty score between 0 and 1, where 0 denotes a hard constraint. By applying all constraints of the grammar to a dependency structure, a grammaticality score for the structure can be calculated by multiplying the penalties of all constraint violations. Thus, parsing in WCDG is the search for the highest scored dependency structure. Every word can, in principle, be attached to any of the other words in the sentence, including the generic root node via any dependency type. The only restriction is that a word can only be attached to one head.[1] All further restrictions, like projectivity, are expressed via constraints. The number of possible edges grows quadratic with

---

[1] WCDG supports multiple levels, i.e. building multiple dependency structures at once, including constraints mediating between them. Yet, for a given level, a word can have only one head.

the number words in the sentence, resulting in an exponential number of possible dependency structures. Thus, for non-trivial sentences, a complete search is usually not feasible.

[7] presented a repair based heuristic algorithm called frobbing. The algorithm starts with an arbitrary initial structure and identifies candidates for replacement by looking at the constraint violation with the highest penalty. Replacement variants for the edges violating the selected constraint are then evaluated for whether they would prevent this conflict and one of them is applied. Several of these repair steps might have to be chained to escape from a local score maximum.

This algorithm is inherently non-incremental, as it starts with a structure for the whole sentence. An incremental processing mode can be introduced by applying the procedure on prefixes of the input and using the resulting dependency structure as the initial structure for analyzing the extended prefix. In this incremental mode, edges selected in a previous step can be replaced later on, if necessary. This can happen when, given the additional input, they violate additional constraints or when an old constraint violation can be resolved given the new input. The algorithm guarantees no monotonicity at all, but in practice an attachment stability of around 70% has been observed [6].

4   Structural Prediction with Virtual Nodes

When parsing a sentence prefix instead of a full sentence with WCDG, several constraints might prevent the parser from choosing a prefix of the structure which would be selected for the full sentence, i.e. a structure only containing dependency edges also present in the complete structure. This is foremost due to 'fragmentation' constraints, penalizing unattached words. They force the parser to select an unsuitable attachment point that might be penalized by other constraints, but not as hard as the fragmented reading. Also, valency constraints are commonly violated in such prefix structures.

There are two possible approaches to deal with these kinds of constraints when applied to partial dependency analyses: Either we prevent them from being applied to PDAs or we add as many additional nodes and edges to the dependency structure as needed to prevent the constraint violation. The first approach has the severe drawback of over-generation. Disabling constraints (or adding exceptions to them) will remove their ability to penalize ungrammatical input.

In this paper we follow the second approach by providing a set of so called virtual nodes that can be integrated into the dependency structure if a constraint violation can be avoided by doing so, but there is no penalty if they stay unconnected. If being unconnected, they are not considered to be part of the structure.

To prevent the parser from choosing a structure that includes virtual nodes even if another structure of equal or slightly worse score would be available, we add a slight penalty to every analysis where a virtual node is used, i.e. connected. This way, the structurally predictive reading is chosen only if necessary,

Fig. 1. Example for bottom-up prediction: The determiner is unattachable to any word directly, but attachable if predicting a noun

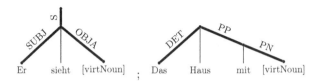

Fig. 2. Examples for top-down predictions of an object and the kernel noun of a preposition ("he sees" and "the house with")

but not otherwise. No further modification to the algorithm is needed. During constraint optimization the virtual nodes are treated like other input nodes. The special handling of virtual nodes (i.e., the slight penalty for connected and no penalty for unconnected virtual nodes, in contrast to the harsh penalty for other unconnected nodes) is defined solely in the constraints of the WCD-Grammar.

We can distinguish two different scenarios leading to structurally predictive readings. The scenario seen in Figure 1 is a bottom-up prediction, where the necessity to connect a new word leads to the prediction of a suitable head. As seen in Figure 2, dependents can be predicted, too, via top down prediction. Whenever a word has a valency that can not be satisfied by any of the other available word, e.g. missing subject or a transitive verb missing an object, filling the valency with a VN makes the corresponding constraint violation disappear.

If we would allow some words to stay unattached, it cannot be guaranteed that there is any possible continuation of the sentence, where these unattached words could be attached without changing the already established part of the structure. For fully connected structurally predictive structures, in contrast, the absence of severe constraint violations licenses not only the integration of the otherwise unattachable words but also the acceptability of the whole structure. It anticipates a potential continuation of the sentence prefix, which does not introduce any additional constraint violations. The structural prediction serves as a kind of "certification" for the grammaticality of all the attachments in the PDA.

It has to be assured that for every sentence prefix the grammar licenses a partial dependency analysis free of additional constraint violations. Besides providing a suitable set of virtual nodes, grammar modifications might be necessary to achieve this goal, as will be discussed in Section 6.

To be able to predict structural relationships for yet unseen words, the incremental frobbing algorithm needs to be provided with a finite set of virtual nodes. The larger this set, the higher the coverage will be. Additional VNs, however, introduce a quadratic increase of potential dependency edges into the search space and, as a consequence, an exponential increase in the number of possible dependency structures.

For the sake of computational feasibility, the number of VNs should be kept as small as possible. This can be achieved by using general VNs (which stand for whole classes of words such as nouns or verbs). Aside from the combinatoric problems of providing too many VNs, it also avoids the danger of many variants receiving the same grammaticality score and thus being indistinguishable by the parser. Even the order among VNs will remain unspecified. While the existing grammar is prepared to deal with underspecified values for many lexical features, attributes like the word form or the linear precedence relations among words are always considered being given explicitly. To allow constraints to ignore the arbitrary order/adjacency relations between VNs as well as their particular lexical instantiations, respective exceptions need to be added.

Thus, the questions we have to answer are:

- Which set of virtual nodes is needed to achieve a certain coverage
- What changes to a constraint dependency grammar for full sentences are necessary?

In the second half of the paper we will present an empirical investigation to answer these questions.

5   Data Preparation

Not only we have no annotations for sentence prefixes available, but also it would be difficult to specify them uniquely, since they always depend on a hypothesis for a possible sentence continuation. Therefore, we generate them from complete sentence annotations. We want to find a monotonic sequence of prefixes resulting in the final annotation. Those prefixes are therefore not necessarily the most plausible for any given prefix.

For every word in a given sentence, we generate an annotation up to that word by retaining certain nodes and dependency edges while dismissing others.[2]

5.1   Prefix Baseline Annotation

The basic variant of the procedure that generates connected PDAs is quite straightforward:

- all words inside the prefix and dependencies between them are kept
- all words that are heads of retained words are retained themselves, including the dependency edge between the head and the dependent. This rule applies recursively.

Words not belonging to the prefix but nonetheless retained need to be 'virtualized' to produce generic representatives of their part-of-speech category, the form of the virtual nodes used by the incremental parsing algorithm.

---

[2] The software to reproduce this data preparation as well as the evaluation can be found at http://www.CICLing.org/2013/data/274

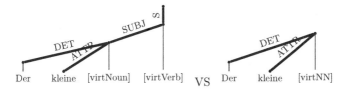

Fig. 3. Example for assuming a complete sentence vs a fragment given the prefix "der kleine" ("the little")

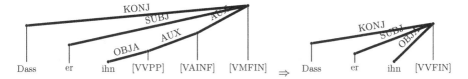

Fig. 4. Example for folding: When generating a PDA for the first three words of the clause "dass er ihn gesehen haben soll" (literally: "that he him saw has should" meaning:"that he is supposed to have seen him") . All three verbs would be kept for connectedness, but are then folded into a single virtual verb.

The procedure of virtualization removes word form and lemma, keeps part-of-speech information, removes all other lexical features, and blurs the original linear order among the nodes.

5.2   Prefix Annotation Variants

Three parameters for the prefix generation algorithm have been considered:
1. Assuming complete sentences or not, i.e. whether we retain all nodes needed to connect up to the root or only retain those needed to connect all words in the prefix to each other (see Figure 3).
2. Top-down prediction. The connectedness criterion only provides for bottom-up prediction. Additional nodes could be retained via top-down prediction, i.e. to fill obligatory valencies (see Figure 2).
3. Folding of connection paths. The connection paths extracted directly from the full sentence annotations might not represent the shortest possible connection path. This is especially true for chains of equally labeled dependencies which correspond to nested phrase structures. Folding these structures by only keeping one of the dependencies will lead to more plausible prefix annotations (see Figure 4). They correspond to the minimally recursive structures [4] investigated in their second experiment where they anticipated parser miss-analyses in their simulated parsing procedure.

These three parameters have different consequences for the parser. Unless there are clear hints for a nested structure, the minimal tree, i.e. the one with the least number of virtual nodes, should always be preferred. WCDG achieves this due to the prediction penalty, selecting the PDA with a minimal amount of VNs among otherwise equally scored alternatives. The parser will not be able to utilize additional VNs set aside for additional nesting depth. Folding can be expected to provide a better estimation of the amount of VNs needed.

While an incremental dependency parsing algorithm that provides bottom-up but no top-down prediction is possible, the parsing algorithm used here has no distinct mechanisms for the two prediction modes. If respective virtual nodes are provided, they will be used to satisfy valency constraints, unless those constraints are disabled for incomplete input. Therefore, by adding top-down prediction to the prefix generation algorithm, we get a better estimation for the VNs needed by the parser. Note that folding reduces the number of VNs, while top-down prediction increases it.

In contrast to folding, where there is no choice for parser behavior, the parser can easily be changed between a full sentence and a fragment mode. Which mode is activated depends solely on whether there is a constraint for penalizing words other than finite verbs to be the root of the tree. Such a constraint would incite the prediction of a virtual finite verb (and possibly its subject) even if none is needed to establish connectedness. Thus, the choice between these modes affects the number of required virtual nodes. In this paper we only explored the fragment variant, as did [4] and [5]. The predictions in this mode are expected to be smaller and less speculative.

6   Experiments

To determine the set of virtual nodes needed, we generate a partial but connected structure for every prefix of every sentence in an annotated corpus. A subset of the dependency edges of the complete tree is selected by using the procedure presented above. Given these structures, we can count how many additional nodes are needed to keep them connected. Also, by evaluating these PDAs with the WCD-grammar, we can determine whether they violate any constraints in addition to those violated in the complete sentence, as such additional constraint violations would prevent the prefix structure to be selected by the parser. Based on these results we can then modify the grammar accordingly.

Using the 500 sentences from the Negra corpus [8] transformed to the dependency annotation scheme described in [9] and a broad coverage WCD-grammar for German [7] we perform the following procedure:

1. generate a set of prefix annotations from the gold standard annotation.
2. Calculate constraint violations that only appear on prefix analyses, but not in the corresponding complete sentence
3. check:
   (a) whether changes to the constraints themselves are needed
   (b) whether additional VNs corresponding to words in the complete sentence could avoid any of the additional constraint violations
   (c) whether VNs can be removed from the connection path systematically without triggering additional constraint violations
4. repeat from step 1 with different variants of the PDA generation algorithm

Table 1. Results of experiment 1: The number of VNs needed for prefix structures, only regarding connectedness but no valencies; total VNs: number of VNs per PDA; CP length: only those VNs needed to connect the most recent word; note that the numbers in the columns don't add up: the 4 VNs of a PDA might consist of 3 virtual verbs and 1 virtual nominal

	0	1	2	3	4	5+
total VNs	56.5%	30.1%	10.8%	2.4%	0.2%	0%
verbs	68.8%	26.0%	4.6%	0.7%	<0.1%	0%
nominals	81.2%	18.7%	0.1%	0%	0%	0%
adjectives	97.5%	2.5%	<0.1%	0%	0%	0%
other	99.1%	0.8%	<0.1%	0%	0%	0%
CP length	80.8%	17.0%	2.0%	0.1%	0%	0%
WSJ (CP len)	79.4%	17.8%	2.7%	0.2%	0%	0%

6.1   Prediction with Baseline Annotations

We start with the most simple variant of the prefix generation algorithm, i.e. without folding or top-down prediction. The constraint violations encountered in the prefix annotations but not in the respective complete sentence annotation can be roughly categorized as follows:

Prediction Penalties. As described in Section 4, there is a small penalty for integrating VNs into a structure. The respective constraint violations are expected in the difference lists, and can safely be ignored here.

Order Constraints, Especially Projectivity. These constraint violations occur due to VNs being unordered. These constraints should simply not be applied to VNs. Hence, respective exceptions need to be added to them.

Unsatisfied Valencies. As we did not account for valencies when deciding on what to keep in the prefix structures of this iteration, all kinds of valency constraints are violated, e.g. missing subject, missing object, missing kernel noun for prepositions, missing determiners for nouns, missing conjunctions and missing conjuncts. These are prime examples for top-down prediction, i.e. these conflicts will disappear when attaching a VN with a respective PoS and dependency type. For VNs missing determiners or infinitive markers 'zu', PoS variants without these valencies (NE and VVIZU respectively) will be selected instead of top-down-prediction.

Punctuation Several constraints expect a comma or quotation marks between two words. If at least one of the words is virtual, the missing punctuation might still show up later on. Thus we will add respective exceptions to these constraints.

Word Form. There are constraints accessing the word forms or lemmas. Many of these occurrences are exceptions for certain words. We will ignore them for now, as we neither want too detailed predictions (exploding the search space of the parser) nor do we want to allow idiosyncratic constructions before the licensing word appears in the input. For constraints deemed general enough, exceptions which prevent them to be applied to VNs are added.

Table 1 shows the percentage of prefix structures containing a certain number of VNs, in total and coarsely grouped by part of speech. We also added a line for the number of VNs in the connection path, i.e. only the VNs needed in addition to those already present in the previous prefix, to connect the most recent word to the rest of the structure. This line is roughly comparable to the numbers presented in [4]. Their headless projections in the connection path correspond to our VNs. In both cases connection paths up to length 1 cover 98% of the observed prefixes and 2 cover over 99%. No occurrences of path longer than 4 are observed. Instead of 82.3% of prefixes not requiring any headless projections there, 80.8% of the words can be connected without VNs here.

To estimate whether the minor difference results from differences in language (English VS German) or grammar formalism (phrase structure VS dependency structure), we ran our evaluation on 500 sentences from the WSJ corpus, i.e. the same corpus used in [4], converted to dependency structures. The results for connection path length are shown in the last line of Table 1 and are closer to our results, suggesting that the differences result from the grammar formalism.

6.2   Valency Respecting Prediction

In the previous experiment, a large class of constraint violations were related to unsatisfied valencies. We therefore modified the prefix generation algorithm to keep words in the structure for the following cases:
– Subjects of all finite verbs, virtual or not
– Kernel nouns of prepositions
– Dependents of a conjunction
– Objects of non-virtual words
– The second part of a truncation like 'an- und verkaufen' if the first is observed
– The last part of a conjunction chain, if the first two are observed, e.g. in the sequence 'A, B, C und D', 'und D' would be kept and C skipped, if 'A,B' was observed
– Certain adverbs that fill the role of conjunction in subclauses are kept, if the verbal head of the subclause is already kept for other reasons.
Since the valencies of virtual words are not available to the parser, they all have to be considered optional. Therefore, no objects for virtual verbs are kept.

6.3   Prediction with Folded Structures

The results for PDAs with folding are shown in Table 3. Compared to Table 1, the most noticeable difference is that the percentage of prefixes with two virtual verbs went down by around 3.5%, while the number for 1 virtual verb increased. As expected, there is no change in the zero column number, as folding will not remove the last virtual verb remaining.

Table 2. Results of experiment 2: The number of VNs needed for prefix structures, regarding connectedness and valencies

	0	1	2	3	4	5+
total VNs	37.4%	34.6%	20.7%	5.8%	1.3%	0.2%
verbs	62.6%	30.8%	5.8%	0.8%	<0.1%	0%
nominals	58.5%	37.5%	3.9%	0.1%	0%	0%
adjectives	96.1%	3.8%	0.1%	0%	0%	0%
conjunction	97.9%	2.1%	0%	0%	0%	0%
adverb	99.4%	0.4%	0.2%	0%	0%	0%
preposition	98.3%	1.7%	0%	0%	0%	0%
other	99.2%	0.9%	0%	0%	0%	0%

Table 3. Results of experiment 3: The number of VNs needed for prefix structures, regarding connectedness while folding recursive verb chains

	0	1	2	3	4	5+
total VNs	56.5%	33.3%	9.3%	0.9%	<0.1%	0%
verbs	68.8%	30.2%	1.0%	0.1%	0%	0%
nominals	81.2%	18.8%	0.1%	0%	0%	0%
adjectives	97.5%	2.5%	<0.1%	0%	0%	0%
other	99.1%	0.8%	<0.1%	0%	0%	0%
CP length	81.15%	17.2%	1.6%	<0.1%	0%	0%

6.4 Prediction with Valencies and Folding

The results for the combined algorithm, folding plus valencies, can be seen in Table 4. As expected, the numbers are generally above those of Table 3 and below those of Table 2. Three VNs are sufficient to achieve a coverage of 99%.

One line in the table that might surprise is that prepositions as VNs were observed (1.7%), since a preposition is the leftmost word in a PP. There are three different ways prepositions are predicted: to connect an adverb (arguably a very ambiguous and hard to predict case), to fill a verb's valency for a prepositional object, and to complete a coordination of prepositional phrases, when the first PP and a conjunction like 'und' was already observed.

The question remains, what part-of-speech category combinations among the VNs achieve which coverage. Some possible combinations are shown in Table 5. A set consisting of two nominals and one verb, as used in [6], covers nearly 90%. Adding a VN for adjectives, conjunctions, prepositions and adverbs improve the coverage towards 99%. Whether the parser can actually benefit from this improved coverage, especially of the latter two categories, will have to be evaluated, but is beyond the scope of this paper.

Table 6 shows how VNs are distributed over the different prediction constellations. If a VN serves to connect two different words, e.g. it is the head of a determiner the object of a verb, the leftmost word is selected. The most common reason for prediction is predicting the kernel noun of a PP (18%), followed by

Table 4. Results of experiment 4, the complete algorithm: The number of VNs needed for PDAs, regarding connectedness and valencies while folding nested structures

	0	1	2	3	4	5+
total VNs	37.4%	36.7%	20.6%	4.3%	0.9%	0.1%
verbs	62.6%	35.0%	2.3%	0.1%	0%	0%
nominals	58.5%	37.5%	3.9%	0.1%	0%	0%
adjectives	96.1%	3.8%	0.1%	0%	0%	0%
conjunction	97.9%	2.1%	0%	0%	0%	0%
adverb	99.4%	0.4%	0.2%	0%	0%	0%
preposition	98.3%	1.7%	0%	0%	0%	0%
other	99.1%	0.9%	0%	0%	0%	0%

Table 5. Percentage of sentence prefixes that can be represented with a given set of VNs. N: nominal, V: verb, A: adjective, C: conjunction, P: preposition, Av: Adverb

VN-Set	coverage (%)
no VNs	37.4
$2*N+V$	89.4
$2*N+V+A$	93.0
$2*N+2*V+A$	94.7
$2*N+2*V+A+C$	96.7
$2*N+2*V+A+C+P$	98.3
$2*N+2*V+A+C+P+Av$	98.7

predicting the subject of a finite verb (11%), predicting a clause initial conjunction (10%) and connecting a determiner (6%).

6.5 Discussion

This study only covers part of the problem of grammar development: We made sure for every prefix of every sentence there is a dependency structure that is scored at least as good as the dependency structure for the final sentence. So far, however, we did not consider the complementary case, namely to make sure that for the best scored partial dependency structure of every sentence prefix there actually is a continuation that would indeed be scored as good. This cannot be examined by a corpus study but only by inspecting actual parser behavior.

7 Conclusions

In this paper we presented a parsing system based on the WCDG framework capable of incrementally parsing into connected dependency structures. The system uses a grammar consisting of weighted constraints as well as a set of virtual nodes. While there is an existing state of the art broad coverage WCD-grammar for German, that grammar is optimized for complete sentences, not for sentence

Table 6. An overview over common reasons for why VNs where included in a PDA; bottom-up prediction to the left, top-down to the right; determined for the PDAs of experiment 4;

reason	% of VNs
connecting a	
- determiner	6.0
- subject	5.9
- clause initial conjunction	5.6
- adverb	4.5
- accusative obj.	3.8
- preposition	3.4

reason	% of VNs
filling the valency:	
- kernel noun of a prep.	18.0
- subject	11.4
- the conjunct under a conjunction	10.3
- accusative obj	5.3
- conjunction to finish an incomplete coord.	3.5
- auxiliary verb	2.7
- predicate	3.4
- subject clause	2.7
- object clause	1.1

prefixes. We conducted a corpus study to identify what changes to the existing grammar are needed, as well as to identify sets of virtual nodes suitable to cover the observed sentence prefixes. Our experiments complement those conducted by [4] by exploring another grammar formalism and valency driven top-down prediction in addition to connectedness-driven bottom-up prediction.

References

1. Nivre, J., Hall, J., Nilsson, J., Chanev, A., Eryigit, G., Kübler, S., Marinov, S., Marsi, E.: Maltparser: A language-independent system for data-driven dependency parsing. Natural Language Engineering 13, 95–135 (2007)
2. Sturt, P., Lombardo, V.: Processing coordinated structures: Incrementality and connectedness. Cognitive Science 29, 291–305 (2005)
3. Bornkessel, I.: The Argument Dependency Model: A neurocognitive approach to incremental interpretation. Max Planck Institute of Cognitive Neuroscience (2002)
4. Lombardo, V., Sturt, P.: Incrementality and Lexicalism: a Treebank Study. In: Lexical Representations in Sentence Processing, pp. 137–154. John Benjamins (2002)
5. Demberg, V.: A Broad-Coverage Model of Prediction in Human Sentence Processing. PhD thesis, The University of Edinburgh (2010)
6. Beuck, N., Köhn, A., Menzel, W.: Incremental parsing and the evaluation of partial dependency analyses. In: Proceedings of the 1st International Conference on Dependency Linguistics, DepLing 2011 (2011)
7. Foth, K.A.: Hybrid Methods of Natural Language Analysis. PhD thesis, Uni Hamburg (2006)
8. Brants, T., Hendriks, R., Kramp, S., Krenn, B., Preis, C., Skut, W., Uszkoreit, H.: Das negra-annotationsschema. Negra project report, Universität des Saarlandes, Computerlinguistik, Saarbrücken, Germany (1997)
9. Daum, M., Foth, K., Menzel, W.: Automatic transformation of phrase treebanks to dependency trees. In: 4th Int. Conf. on Language Resources and Evaluation, LREC 2004 (2004)

Semi-supervised Constituent Grammar Induction Based on Text Chunking Information*

Jesús Santamaría and Lourdes Araujo

U. Nacional de Educación a Distancia, NLP-IR Group, Madrid, Spain
{jsant,lurdes}@lsi.uned.es

Abstract. There is a growing interest in unsupervised grammar induction, which does not require syntactic annotations, but provides less accurate results than the supervised approach. Aiming at improving the accuracy of the unsupervised approach, we have resorted to additional information, which can be obtained more easily. Shallow parsing or chunking identifies the sentence constituents (noun phrases, verb phrases, etc.), but without specifying their internal structure. There exist highly accurate systems to perform this task, and thus this information is available even for languages for which large syntactically annotated corpora are lacking. In this work we have investigated how the results of a pattern-based unsupervised grammar induction system improve as data on new kind of phrases are added, leading to a significant improvement in performance. We have analyzed the results for three different languages. We have also shown that the system is able to significantly improve the results of the unsupervised system using the chunks provided by automatic chunkers.

1 Introduction

The aim of Grammar Induction (GI) is to extract a grammar from a collection of texts. There are three main approaches to GI: (1) supervised (SGI), which requires syntactically annotated examples — not available for many languages— (2) unsupervised (UGI), which only requires raw texts[1], and usually achieves a lower performance than the supervised approach, and (3) semi supervised (SSGI), which is an intermediate approach that requires some degree of supervision but not fully syntactically annotated texts. Much research has been developed within the three types of GI. Recent works [9,4,16,8] on SGI report results ranging between 90% and 93% when applied to the WSJ section of the Penn Treebank[2].

Some of the most recent works devoted to the UGI approach are due to Klein and Manning [5], Bod [2], Seginer [12] and Santamaria and Araujo [11]. The UGI system by Santamaria and Araujo [11] is based on the statistical detection of certain POS tag patterns that appear in the sentences. The goal is to find a POS tag classification according to the role that the POS tag plays in the parse trees. This system, which considers

---

* Supported by projects Holopedia (TIN2010-21128-C02) and MA2VICMR (S2009/TIC-1542).

[1] It is also considered an unsupervised approach if you use the Part-of-Speech (POS) tags associated with the words of the sentence.

[2] The Wall Street Journal section of Penn Treebank corpus [6].

A. Gelbukh (Ed.): CICLing 2013, Part I, LNCS 7816, pp. 258–269, 2013.

any kind of tree — not only binary ones— has achieved an F-measure of 75.82%, without including constituents spanning the whole sentence, and 80.50% including them, as the other works do.

Finally, there have also been an increasing interest on the semi supervised approach (SSGI), though most works in this line are devoted to dependency parsing. Klein and Manning have also tried to include some supervision in the work mentioned previously [5]. Haghighi and Klein [3] have also proposed a model to incorporate knowledge to the CCM model by Klein and Manning [5]. In this case, prior knowledge is specified declaratively, by providing a few canonical examples of each target phrase type.

The supervised approach requires a syntactically annotated corpus which is not available for many languages. Moreover, current treebanks are usually composed of a particular kind of text (usually newspaper articles) and may not be appropriate for others. This makes the unsupervised approach worth considering. In order to bridge the performance gap between supervised and unsupervised systems we propose to introduce a very low degree of supervision which could be achieved without requiring fully syntactic annotation of texts. Accordingly, we have resorted to the so-called *shallow parsers*, which aim at identifying the constituents of a sentence, but do not specify their internal structure. In the experiments carried out in this work we have first analyzed the results obtained using the chunks annotated in a treebank for English. We have later evaluated the system using automatic chunkers and other languages. Our proposal is considered a semi-supervised approach, since it relies either on the chunk annotations or on an accurately trained chunker. Several chunkers reports results above 95% [10,13,14] and even higher for some particular type of phrase, such as NPs [1], for which a precision of 97.8% is achieved. Accordingly, we expect that the results would only be slightly affected by using a chunker instead of chunking annotations. Certainly the high performance mentioned chunkers are supervised systems, but the amount of data needed, for instance, to train a finite state automaton for NP detection, VP detection, etc. is much smaller than the one required for supervised parsing.

We have tested the effect of adding different degrees of supervision to the underlying UGI system. Specifically we have checked how the performance is affected as the *chunks* corresponding to different type of phrases —noun, verb and prepositional phrases— are identified prior to the parsing process. We have chosen the Santamaria and Araujo model [11] as the underlying UGI system because it extracts patterns which can be easily combined with others representing the introduced supervision — it is not restricted to binary trees— and at the same time provides competitive results. The addition of this knowledge has improved the performance of the underlying UGI system from 75.82% to 81.51% with the lower level of supervision, thus proving the usefulness of the proposal. However, the approach described in this work could also be applied to other UGI systems.

The paper is organized as follows: section 2 briefly describes the underlying UGI system, Section 3 explains two strategies to introduce the chunking information to the model, that thus becomes a SSGI system, Section 4 discusses the results, and finally the conclusions are outlined in Section 5.

2   Pattern Based Unsupervised Grammar Induction

The underlying system [11] for UGI relies on the idea that the different POS tags play particular roles in the parse trees. Thus, the POS tags can be classified in a small number of POS tag classes, whose elements have a tendency to appear at certain positions of the parse tree. Furthermore, this feature is shared by many languages. This proposal applies some heuristics based on the frequencies of the POS tag sequences to classify the POS tag set of a corpus into POS tag classes. This classification is then used to parse any sentence by means of a deterministic and fast procedure which selects the constituents and the branching of the parse tree depending on the POS tag classes of the tags in the sentence.

Let us consider a parse tree example to illustrate the different types of POS tag classes considered in the model. Examining the example appearing in Figure 1 we can observe that the POS tags are distributed in different levels which divide the POS tag sequence. Some POS tag sequences tend to appear at the boundaries of the constituents. This is the case of (DT, NNP) and (DT, NN), where DT is a determiner, NNP a proper singular noun, and NN a singular noun. Other POS tags such VBZ (verb, 3rd ps. sing. present) and TO (To) give rise to a new level in the parse tree since a new subtree appears after them. The proposed model tries to capture these ideas, identifying a number of POS tag classes which correspond to particular structural patterns.

The procedure to identify the different classes of POS tags relies on the detection of the minimum constituent (MC):

The MC is the most frequent sequence of two POS tags in the corpus.

It is expected that at least for the most frequent constituent the number of occurrences overwhelms the number of sequences appearing by chance. In the WSJ10 corpus the MC is DT NN*, where NN* represents any type of noun (NN, NNS, NNP, NNPS).

The model considers the following set of POS tag classes, whose detection is based on the POS tag behavior with regard to the MC:

- *Expanders*: POS tags that have a tendency to appear inside the MC. This is often the case of the tag JJ (adjective) which appears between a determiner (DT) and a noun (NN).
- *Separators*: POS tags with a statistical tendency to appear outside the MC. This kind of POS tag usually gives rise to a new level in the parse tree. This is the case of VBZ (verb, 3rd person singular present) in the parse tree of Figure 1.
- *Subseparators*: POS tags which do not have a tendency to be inside or outside the MC. They usually appear delimiting constituents, but without giving rise to new levels in the parse tree as the separators do.

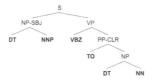

Fig. 1. Parse tree for the sentence *The U.S.S.R. belongs to neither organization*

A POS tag is identified as an expander, a separators or subseparators by computing frequencies of occurrence at each side of the boundaries of the CM. A detailed description of the process can be found in [11].

The set of separators obtained applying the described procedure to the WSJ10 corpus is (MD, PRP, IN, RB, RBR, CC, TO, VB, VBD, VBN, VBZ, VBP,VBG, EX, LS, RP, UH, WP, WRB, WDT), and the set of subseparators is (DT, PDT, POS, SYM, NN, NNS, NNP, NNPS). Sub-separators can be grouped to their right or to their left. The bias of each of them for one direction or another is determined by comparing the number of occurrences of the most frequent sequence composed of the sub-separator and a POS tag on the right and on the left. Then, the preference direction for a sub-separator is the one corresponding to the most frequent sequence. According to this criteria DT, PDT, SYM tend to appear on the left of the constituent, whereas POS, NN, NNS, NNP and NNPS on the right.

The classification of the POS tag set in separators, sub-separators and expanders allow defining a deterministic parsing procedure based on the behavior expected for each POS tag in the sentence. The steps of this procedure are the following:

– Identify the separators. In the sentence example *The (DT) computers(NNS) were(VBD) crude(JJ) by(IN) today(NN) 's(POS) standards(NNS)* separators appear in boldface:

$$[DT\ NNS\ \textbf{VBD}\ JJ\ \textbf{IN}\ NN\ POS\ NNS]$$

– Split the sentence according to the separators. The first separator which is a verb, if any, is used to split the sentence into two parts.

$$[[DT\ NNS]\ [\textbf{VBD}\ [JJ\ [\textbf{IN}\ [NN\ POS\ NNS]]]]]$$

Each separator can give rise to two groups: one composed of the tag sequence between the separator and the next separator, and another one which includes the separator and the POS tags up to the end of the part of the sentence in which it appears.

– Identify the subseparators, underlined in the sentence:

$$[[\underline{DT}\ \underline{NNS}]\ [\textbf{VBD}\ [JJ\ [\textbf{IN}\ [\underline{NN}\ \underline{POS}\ \underline{NNS}]]]]]$$

– Split the sentence according to the sub-separators forming groups according to their preference direction.

$$[[\underline{DT}\ \underline{NNS}]\ [\textbf{VBD}\ [JJ\ [\textbf{IN}\ [[\underline{NN}\ \underline{POS}]\ \underline{NNS}]]]]]$$

At this point the parse tree for the sentence has been obtained.

3   Introducing Text Chunking Information

Text chunking is a partial parsing which amounts to dividing sentences into nonoverlapping segments. Chunkers assign a partial syntactic structure to a sentence, yielding flatter structures than full parsing. They usually provide chunks for a fixed tree depth, instead of arbitrarily deep trees.

The SSGI approach we have adopted assumes that the chunks corresponding to some particular types of phrases and tree depths have been prior identified. We have designed two different ways of applying the identified chunks. The first one, called *constraining strategy*, applies chunks in advance to the parsing process, as a kind of constraint. The second one, called *error-correcting strategy*, generates the parse tree applying the UGI system and then uses the chunks to correct some possible errors. In the next subsections we described each of them.

3.1 Constraining Strategy

The idea underlying this strategy is to apply the UGI system to infer the different sub-trees of the parse tree resulting from the chunks identification. One of these subtrees is the top subtree obtained by substituting the identified chunks by special tags representing some of the POS tag classes. The other subtrees are the ones corresponding to each identified chunk. Specifically, we perform the following process:

1. Extract the constituents of the kind (noun phrases, verb phrases, etc.) and level under consideration from the parse tree.
2. Replace in the sentence the constituents extracted in the previous step by special POS tags which belong to some of the POS tag classes considered in the UGI system. Then, apply the UGI system to generate the tree for the new sentence.
3. Apply the UGI system to build the sub-tree for the constituents extracted in step 1.
4. Replace the special POS tags of the tree generated in step 2, for the sub-trees generated in step 3.

The result of this process depends on the types of constituents being identified in advance. In order to clarify the described procedure let us see an example. Assume that we consider only noun phrases (NP) as the type of constituent being extracted in advance. Then NPs have to be substituted by a special POS tag which is being assigned to one of the POS tag classes of the UGI system: separators, subseparators and expanders. In the case of NPs we have chosen the subseparator class since the head of this kind of phrase, the nouns (NN, NNS, NNP and NNPS), have been identified as subseparators by the UGI system. Accordingly we have defined the special POS tag SUBS to substitute the NPs. Now, we apply the described procedure to the sentence appearing in Figure 2(a), considering only top level NPs:

1. The top level NPs are identified: NP = [DT NN IN PRP$ NN NN NN].
2. The NP identified in the previous step is replaced by SUBS, giving rise to the tree appearing in Figure 2(b).
3. The UGI system is also applied to generate the parse tree for the replaced NP. The result appears in Figure 2(c). Notice that we are not assuming to have information on the deep parsing of these constituents, but only on the chunking process, which is a much more weak requirement.
4. Finally, the special POS tag SUBS is replaced by the parse tree found for the constituent in step 3, obtaining the tree shown in Figure 2(d).

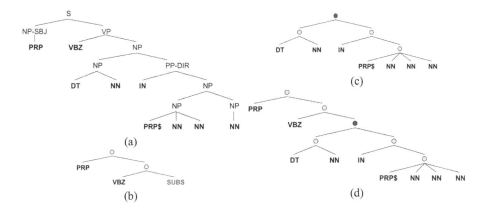

Fig. 2. (a)Parse tree for the sentence *It has no bearing on our work force today.* from the Wall Street journal part from the Penn treebank. (b) Tree provided by the UGI system for the sequence of POS tags obtained after replacing the top level NP in the tree in Figure (a) by the special POS tag SUBS. (c) Tree generates by the UGI system for the sequence of POS tags corresponding to the NP replaced in Figure (a). (d) Final tree obtained by the semisupervised GI system for the sentence in Figure (a).

For other types of phrases the process is analogous. In the case of verb and prepositional phrases the replacing special POS tag is called SEP, which is considered part of the separator class. The reason is that the head of the verb phrases, the verb (VB, VBD, VBN, VBZ, etc.), belongs to the separator class. We can also observe that the preposition (IN), the head of the prepositional phrases, also belongs to this class.

3.2   Error-Correcting Strategy

This strategy applies the UGI system to generate the initial parse tree for a sentence. Then, the identified chunks are applied to correct possible errors in the generated parse tree. The gold chunks are compared to each constituent in the UGI tree sharing some word (POS tag corresponding to the word). The UGI constituent is considered wrong if there is a partial overlap between it and the gold chunk. In this case the constituent containing the overlapping sequence of POS tags is removed and the overlapping sequence is considered a constituent. Once all the conflicts have been solved, the gold chunks are included as constituents, provided they had not been identified by the UGI system. Let us consider the following sentence *Pierre Vinken, 61 years old, will join the board as a nonexecutive director Nov. 29*, from the Penn Treebank. According to the gold standard the higher level VP for this sentence is:

$$\begin{array}{cccccccccc} MD & VB & DT & NN & IN & DT & JJ & NN & NNP & CD \\ 6 & 7 & 8 & 9 & 10 & 11 & 12 & 13 & 14 & 15 \end{array}$$

where the number below the POS tags represents the position of the corresponding word in the sentence. Let us assume that the parse tree generated by the UGI system is the one appearing in Figure 3(a). This tree presents the following constituent at the beginning of the sentence:

NNP NNP CD NNS JJ MD
1   2   3   4   5 6

We can observe that there is a conflict with the POS tag MD at position 6, since it belongs to two different constituents. Then MD is included as a new constituent, and [NNP NNP CD NNS JJ MD] is not considered a constituent any more. Notice that there is not conflict between the gold VP chunk and the UGI constituent

VB DT NN IN DT JJ NN NNP CD
7   8   9   10 11   12 13   14   15

because the latter is completely included in the chunk. Figure 3(b) presents the final tree obtained by applying the gold chunks.

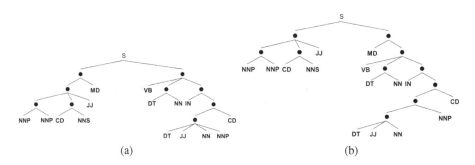

(a)   (b)

Fig. 3. (a) Tree obtained by the UGI system for the sentence *Pierre Vinken, 61 years old, will join the board as a nonexecutive director Nov. 29..* (b) *Final tree obtained after correcting the errors in the UGI tree of Figure (a) by applying the gold chunks.*

This approach can take better advantage of the result of the UGI system, because it is applied to the whole sentence, instead of to short segments of it, as the constraining approach does.

4   Experimental Results

We have used the WSJ10 corpus, with 7422 sentences, in our first experiments. It is composed of the sentences of up to ten words from the Wall Street Journal section from the Penn Treebank. Though this corpus is syntactically annotated, we have used this information only to obtain the chunks that are expected to be provided by a shallow parser, thus not relying on the performance of a particular system, and for evaluation purposes. In order to evaluate the quality of the obtained grammar we have used the most common measures for parsing and grammar induction evaluation: recall, precision, and their harmonic mean (F-measure).

The underlying UGI system achieves an UF of 75.82%, which is used as our baseline. Next, we discuss the results obtained when a perfect shallow parser[3] provides us

---

[3] At this point we use the treebank annotations for the chunks corresponding to the different kind of phrase.

the boundaries of the noun phrases, verb phrases and prepositional phrases. We have studied the effect of using information on the chunking corresponding to each kind of constituent separately, before evaluating the results of applying all of them simultaneously. For each of them we have also investigated two different levels of supervision. In subsection 4.3 we present results obtained for different languages. Finally, subsection 4.4 shows results obtained using automatic chunkers.

4.1   Isolated Effect of Some Kind of Phrases

Table 1(a) compares, for different kind of phrases and supervision levels, the UF achieved by the two strategies, constraining (SSGI_Cons) and error-correcting (SSGI_EC) of the SSGI system, and the baseline given by the UGI system. In considering the different types of phrase separately, supervision levels refer only to tree depths in which the type of phrase under consideration appears. The two first rows show the results using the chunks corresponding to the NPs appearing in the sentence at level 1 and 2, respectively. In the case of the constraining strategy, NP chunks are replaced by the special SUBS POS tag, which behaves as a subseparator. The supervision introduced by the NP chunking significantly improves the results. Even for the lower supervision level, i.e., considering only the chunks corresponding to the minimum depth NPs appearing in the sentence (at a supervision degree of 17,55%) we obtain an improvement of 3,12% over the UGI baseline. The two considered strategies significantly improve the results for all levels, though the error-correcting one is slightly better. We think that the reason may be that with the error correcting strategy the UGI system is applied to the whole sentence, taking advantage of more information that the constraining strategy.

The next two rows in Table 1(a) show the results for VPs annotated at level 1 and 2, respectively. In the case of the constraining strategy, VP chunks are replaced by the special SEP POS tag, which behaves as a separator. Again the error-correcting strategy performs better. The improvement achieved in this case is higher than for NPs. The reason may be that the verb phrases are more difficult to be captured in an unsupervised manner, and therefore the VP information is more helpful. This information also helps to know the point that separates the subject and the predicate. For the lower supervision level, i.e., considering only the chunks corresponding to the minimum depth VPs appearing in the sentence (at a supervision degree of 21,81%) we obtain an improvement of 6,11% over the UGI baseline.

The last two rows in Table 1(a) show the results for PPs annotated at level 1 and 2, respectively. In the case of the constraining strategy, PP chunks are replaced by the special SEP POS tag, which behaves as a separator. Once again the error-correcting strategy performs better. The rate of improvement in this case is lower, as well as the supervision degree, indicating that PPs appear less often. Considering only the chunks corresponding to the minimum depth PPs appearing in the sentence (at a very low supervision degree of 5,51%) we obtain an improvement of 1.59% over the UGI baseline.

4.2   Combined Effect

We can consider different kinds of phrases annotated simultaneously. Table 1(b) shows the results obtained the first two supervision levels. The error-correcting strategy is

Table 1. Applying chunking for some kind of phrases in WSJ10: F-measure corresponding to the baseline (UGI) and the SSGI systems, the constraining strategy (SSGI_Cons) and the error-correcting strategy (SSGI_EC) per supervision level, (a) considering each kind of phrase and (b) all together. The number in bracket next to the SSGI results indicates the improvement rate with respect to the UGI F-measure.

(a)

P. kind	Superv. Level	UGI	SSGI_ Cons	SSGI _EC
NP	1	75,82	77,67(2,43)	78,19(3,12)
NP	2	75,82	78,85(3,99)	79,07(4,28)
VP	1	75,82	79,87(5,34)	80,46(6,11)
VP	2	75,82	80,19(5,76)	80,77(6,52)
PP	1	75,82	76,44(0,81)	77,03(1,59)
PP	2	75,82	76,47(0,85)	77,05(1,62)

(b)

Superv. Level	UGI	SSGI_ Cons	SSGI _EC
1(26,40%)	75,82	81,18(7,06)	81,51(7,50)
2(47,93%)	75,82	83,31(9,87)	83,66(10,34)

Table 2. Results for WSJ10 (a) and WSJ30(b), considering only minimum depth chunks, of unlabeled recall (UR), precision (UP), and F-measure (UF) for the baseline provided by the underlying UGI system, the semi-supervised system (error-correcting strategy) which uses the NPs chunks (SSGI-NP), the one which uses the VP chunks (SSGI-VP), the one which uses the PP chunks (SSGI-PP) and the whole semisupervised system (SSGI).

(a)

System	UR	UP	UF(% Improv.)
UGI	77.99	73.77	75.82
SSGI-NP	81.44	75.18	78.19(3.12%)
SSGI-VP	82.89	78.17	80.46(6.11%)
SSGI-PP	79.46	74.74	77.03(1.59%)
SSGI	84.28	78.92	81.51(7.50%)

(b)

System	UR	UP	UF(% Improv.)
UGI	61.35	64.07	62.68
SSGI-NP	70.59	70.09	70.34(12.22%)
SSGI-VP	68.68	71.26	69.95(11.59%)
SSGI-PP	66.21	67.57	66.88(6.70%)
SSGI	71.23	71.43	71.33(13.80%)

again the better. The combined effect of identifying NPs, VPs, and PPs in advance improves the effect obtained for each of them separately. The results at lowest level of supervision, that can also be achieved by automatic chunker, are particularly interesting. Table 2(a) shows the results (recall, precision and F-measure, as long as the rate of improvement achieved for the last one) for the error-correcting strategy of the different systems considered, and for the minimum tree depth in which the phrases under consideration appear. We can observe that all proposed systems improve the three measures, being the F-measure improvement achieved by the final system more than 7%. Of course, the rate of improvement is lower than the degree of supervision introduced because many constituents had also been identified by the UGI system.

We have also investigated how our SSGI system behaves for longer sentences. We have considered the WSJ30 corpus, composed of 41227 sentences of up to 30 words from the WSJ part of the Penn Treebank. Table 2(b) shows the results for the error-correcting approach and for the minimum depth chunks. We can see that for longer sentences the improvement is larger, leading to high performance results even for the smaller level of supervision (22.21% for the SSGI). Another interesting observation is that for longer sentences the NP supervision becomes more relevant. As sentences get longer, NPs are more complex and difficult to detect in an unsupervised way.

4.3   Other Languages

We have also evaluated the semisupervised system for other languages, specifically, German and Spanish. We have used the UAM Spanish Treebank [7], with 1501 syntactically annotated sentences, and the German corpus NEGRA [15], with 20,602 sentences from German newspaper texts. For comparison, we have used a restricted part of the corpora with sentences of up to 10 words, which we call UAM10, and NEGRA10. Table 3(a) shows the results obtained using NPs, VPs and PPs at two different supervision levels. We can observe that the two proposed approaches are able to obtained a significant performance improvement even for the minimum supervision level and in both languages. These results prove the viability of the method on different languages.

Table 3. (a) Results for Spanish (UAM10) and German (NEGRA10) using NP, VP and PP annotations for different levels of supervision. (b) Results for automatic chunkers. The first column indicates the languages considered: English (Engl.), Spanish (Span.) and German (Germ.). The second column shows the unsupervised system results for comparison purposes. The third and fourth columns show the results of the semi-supervised system using the chunk annotations provided by automatic chunkers (Treetagger for English and German, and Freeling for Spanish). The last two columns correspond to the results obtained with the two proposed approaches using the treebank annotations, also for comparison purposes.

Lang.	Superv. Level	UGI	SSGI_ Cons	SSGI _EC
Spanish	1(32.7)	70.68	75.78	76.03
Spanish	2(62.9)	70.68	80.35	81.07
German	1(60.2)	49.67	56.98	55.87
German	2(75.9)	49.67	59.01	58.85

(a)

Lang.	UGI	Tagger SSGI_ Cons	Tagger SSGI _EC	TB annot. SSGI_ Cons	TB annot. SSGI _EC
Engl.	75.82	72.17	72.10	81.18	81.51
Span.	70.68	75.05	74.21	75.78	76.03
Germ.	49.67	52.77	53.52	56.98	55.87

(b)

4.4   Experiments with Automatic Chunking

Finally, we provide evaluation results for automatically identified chunk boundaries. We have used the Treetagger chunker [4] for English and German, and the Freeling chunker [5] for Spanish. Table 3(b) shows the results obtained for the minimum supervision level, i.e., without using nested chunks. We can observe a significant improvement over the unsupervised results (first column) for Spanish and German. This improvement is slightly lower than the one obtained using the treebank annotations (two last columns), but is close to it for both languages. The English case is different. The chunker extracts "bare NPs" without postmodifiers and verb groups (VGs not VPs) without complements. These constituents do not match those of the Penn Treebank, and thus the

---

[4] http://www.ims.uni-stuttgart.de/projekte/corplex/TreeTagger/.
[5] http://www.lsi.upc.edu/~nlp/freeling.

semisupervised system worsens the unsupervised results for English. However, this is simply a mismatch in the annotation schemes of the chunker and the treebank used for evaluation, and not an inherent problem of the proposed approaches, as the results for other languages show. The results obtained for Spanish and German show the viability of the proposed approaches to take advantage of current automatic chunkers.

5   Conclusions

We have presented a semi supervised method for grammar inference. The proposed model extends an unsupervised GI model based on the detection of POS tag classes classified according to the tag role in the structure of the parse trees. The supervision added, which amounts to knowing the *chunks* corresponding to the some types of phrase (noun, verb and prepositional phrases) at a certain tree depth, is introduced in the unsupervised system in a very natural way following two different strategies. The error-correcting strategy, which uses the chunks to correct possible errors in the tree generated by the UGI system, has proven to be slightly better. We have investigated the effect of knowing in advance the NPs, VPs, and PPs. Results obtained by this semi-supervised system using gold chunks improve those obtained by the underlying UGI system in all cases. The constituents that have exhibited the greater effect on the results are VPs, showing the difficulty of detecting this kind of constituent in an unsupervised manner. However, all of them have improved the results, even for the lowest supervision degree. Furthermore, results improve with the length of the sentences. We have tested the system for English, Spanish and German, significantly improving the unsupervised system performance for all of them. We have been able to improve the performance of the UGI system with the supervision provided by some of the automatic chunkers that currently exist. In fact, we have observed significant improvement using automatic chunkers for Spanish and German, for which the chunker annotation scheme is similar to the one of the treebank used in the evaluation.

References

1. Araujo, L., Serrano, J.I.: Highly accurate error-driven method for noun phrase detection. Pattern Recognition Letters 29(4), 547–557 (2008)
2. Bod, R.: Unsupervised parsing with u-dop. In: CoNLL-X 2006: Proceedings of the Tenth Conference on Computational Natural Language Learning, pp. 85–92. Association for Computational Linguistics, Morristown (2006)
3. Haghighi, A., Klein, D.: Prototype-driven grammar induction. In: Proceedings of the 21st International Conference on Computational Linguistics and 44th Annual Meeting of the Association for Computational Linguistics, pp. 881–888. Association for Computational Linguistics (2006)
4. Huang, L.: Forest reranking: Discriminative parsing with non-local features. In: Proc. of ACL (2008)
5. Klein, D., Manning, C.D.: Natural language grammar induction with a generative constituent-context model. Pattern Recognition 38(9), 1407–1419 (2005)
6. Marcus, M.P., Santorini, B., Marcinkiewicz, M.A.: Building a large annotated corpus of english: The penn treebank. Computational Linguistics 19(2), 313–330 (1994), citeseer.ist.psu.edu/marcus93building.html

7. Moreno, A., Grishman, R., Lopez, S., Sanchez, F., Sekine, S.: A treebank of spanish and its application to parsing. In: LREC 2000: Proceedings of the International Conference on Language Resources & Evaluation, pp. 107–111. European Language Resources Association, ELRA (2000)

8. Petrov, S.: Products of random latent variable grammars. In: Human Language Technologies: The 2010 Annual Conference of the North American Chapter of the Association for Computational Linguistics, HLT 2010, pp. 19–27 (2010)

9. Petrov, S., Klein, D.: Improved inference for unlexicalized parsing. In: HLT-NAACL, pp. 404–411 (2007)

10. Sang, E.F.T.K.: Memory-Based Shallow Parsing. ArXiv Computer Science e-prints (April 2002)

11. Santamaría, J., Araujo, L.: Identifying patterns for unsupervised grammar induction. In: Proceedings of the Fourteenth Conference on Computational Natural Language Learning, pp. 38–45. Association for Computational Linguistics, Uppsala (2010), http://www.aclweb.org/anthology/W10-2905

12. Seginer, Y.: Fast unsupervised incremental parsing. In: Proceedings of the 45th Annual Meeting of the Association of Computational Linguistics, pp. 384–391 (2007)

13. Sha, F., Pereira, F.: Shallow parsing with conditional random fields. In: Proceedings of HLT-NAACL 2003, pp. 213–220 (2003)

14. Shen, H., Sarkar, A.: Voting Between Multiple Data Representations for Text Chunking. In: Kégl, B., Lee, H.-H. (eds.) AI 2005. LNCS (LNAI), vol. 3501, pp. 389–400. Springer, Heidelberg (2005)

15. Skut, W., Krenn, B., Brants, T., Uszkoreit, H.: An annotation scheme for free word order languages. In: Proceedings of the Fifth Conference on Applied Natural Language Processing, ANLC 1997, pp. 88–95. Association for Computational Linguistics (1997)

16. Zhang, H., Zhang, M., Tan, C.L., Li, H.: K-best combination of syntactic parsers. In: Proceedings of the 2009 Conference on Empirical Methods in Natural Language Processing, EMNLP 2009, vol. 3, pp. 1552–1560 (2009)

Turkish Constituent Chunking
with Morphological and Contextual Features

İlknur Durgar El-Kahlout and Ahmet Afşın Akın

TÜBİTAK-BİLGEM
Gebze, KOCAELİ, Turkey
{ilknur.durgar,akin.ahmet}@tubitak.gov.tr

Abstract. State-of-the-art phrase chunking focuses on English and shows high accuracy with very basic word features such as the word itself and the POS tag. In case of morphologically rich languages like Turkish, basic features are not sufficient. Moreover, phrase chunking may not be appropriate and the "chunk" term should be redefined for these languages. In this paper, the first study on Turkish constituent chunking using two different methods is presented. In the first method, we directly extracted chunks from the results of the Turkish dependency parser. In the second method, we experimented with a CRF-based chunker enhanced with morphological and contextual features using the annotated sentences from the Turkish dependency treebank. The experiments showed that the CRF-based chunking augmented with extra features outperforms the baseline chunker with basic features and dependency parser-based chunker. Overall, we developed a CRF-based Turkish chunker with an F-measure of 91.95 for verb chunks and 87.50 for general chunks.

1 Introduction

Chunking, the first step of shallow parsing, is defined as dividing a sentence into non-overlapping and syntactically meaningful word groups [1]. These meaningful word groups (or chunks) provide useful information to many natural language processing (NLP) applications such as question answering, information extraction and information retrieval. In addition, chunk information can be used as an additional feature in some NLP applications such as machine translation and dependency parsing. The main motivation behind preferring shallow parsing to full parsing in some NLP applications is two-fold: 1) Shallow parsing gives very limited information about the sentence and does not deal with the sentence ambiguity. Thus, it is more accurate with high coverage for large domains. 2) As the problem is relatively simpler than full parsing, shallow parsing algorithms deal with less complexity.

During the past two decades, several researchers showed interest in text chunking problem where most efforts were focused on English. Ramshaw and Marcus [2] was the first work that applied machine learning methods for the base-NPs (base noun phrases). Following their work, different machine learning methods

A. Gelbukh (Ed.): CICLing 2013, Part I, LNCS 7816, pp. 270–281, 2013.

were applied to chunking and/or shallow parsing such as memory-based systems [3–5] and Hidden Markov Models (HMMs) [6, 7]. Handling several correlated and overlapped features were addressed by Maximum Entropy Markov Models (MEMMs) [8], Support Vector Machines (SVMs) [9, 10], modified Winnow as a classifier [11], and Conditional Random Fields (CRFs) [12, 13]. Even though researchers were mainly interested in NP chunks, some research explored different types of chunks such as VP, PP, ADJP and ADVP [14]. As a result of the intensive interest on the problem, the CoNLL-2000 shared task [15] was dedicated to text chunking for all chunk types of English.

In English, chunks are defined as phrases such as Noun Phrase (NP), Verb Phrase (VP), Prepositional Phrase (PP), etc. The non-recursiveness constraint of the chunk definition [1] may not suit in morphologically rich languages thus, should be adapted depending on the language properties. For example, Goldberg et al. [16] replaced base-NPs with simple-NPs which allows nested NPs. Despite the intensive interest in English chunking in the last decades, a very limited work was reported on less resourced and/or morphologically rich languages such as Korean [17, 18], Hindi [19], and Hebrew [16]. Moreover, the features (e.g., word and POS) used for English phrase chunking are not sufficient for an accurate chunking for morphologically rich languages. Morphological richness means very valuable information within the words and exploiting that information can help the chunking task.

Turkish is a morphologically rich language where words are formed by attaching inflectional and derivational suffixes to the root word or other suffixes. The derivational suffixes can change the part-of-speech (POS) of the word. In the sub-word level, NP chunks can be nested by derivational processes. As opposed to English, several function words including prepositions and coordinations are bound morphemes attached to the root word or other bound morphemes, which makes the phrase chunking an ambiguous task for Turkish. Nouns with derivational suffixes can turn into verb chunks, in this case, some NPs cannot be determined at the word level. Figure 1 illustares the sub-word level chunking of the Turkish sentence "altı odalı bir evdeydik" and shows the ambiguity of "phrase" chunk labeling in this sentence. The morpheme-level decomposition of the sentence and the corresponding English counterparts are as follows:

altı$_1$ oda$_2$ +lı$_3$ bir$_4$ ev$_5$ +de$_6$ +ydi$_7$ +k$_8$
we$_8$ were$_7$ in$_6$ a$_4$ six$_1$-room$_2$ed$_3$ house$_5$

To our best knowledge, Kutlu [20] implemented the first Turkish NP chunker which uses dependency parser with handcrafted rules. NPs were defined as two sub-classes as main NPs (base NPs) and all NPs (including sub-NPs). Noun phrases with relative clauses were excluded in his work. However, that type of NPs has a wide usage in Turkish such as "kapıdan gelen rüzgar"[1]. For Turkish, we introduced a new chunk definition by replacing phrase chunks with constituent chunks. We used the Turkish dependency treebank [21] for annotation. We only labeled sentence (verb) chunks (for detailed Turkish dependency labels see [22])

---

[1] English translation: The wind coming from the door.

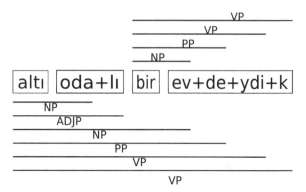

Fig. 1. Phrase Chunking of a sample Turkish sentence

explicitly. We left the other chunks as "general" chunks in order to make the annotation easier as naming the "general" chunk labels can be done automatically by the dependency parser.

In this work, we explore the first Turkish constituent chunker with both dependency parser-based chunking and CRF-based chunking. We augment CRF-based chunking with morphological and contextual features. We report the results of both chunking methods and show that the best chunking method is the CRF-based chunking with additional features.

This paper is organized as follows: Section 2 describes the corpus annotation process, Section 3 introduces Turkish dependency parser-based chunking, Section 4 explains the CRF-based chunker, Section 5 reports the experimental results of our approaches and Section 6 concludes with future work.

2   Turkish Chunking Corpus

Turkish, a member of the Ural-Altay language family, is an agglutinative language with very productive inflectional and derivational morphology. Through, derivational processes, (theoretically) infinite number of possible word forms can be generated where the POS might be changed after each derivation. From the dependency point of view, Turkish is generally a head-final language and the dependency relations do not cross each other except in some rare cases [23].

2.1   Annotation Process

For Turkish, there is no syntactically annotated corpus. In the CoNLL-2000 shared task, the training data is automatically produced from the Penn Treebank which is fully annotated with all syntactic information. For Turkish, we may use Turkish dependency parser to extract chunks as explained in Section 3.2. However, it is clear that dependency parser is erroneous and results in head and dependency relation mistakes especially on unseen data.

In the lack of an explicitly chunked Turkish corpus, to obtain a gold data with chunk information we used 4758 sentences[2] of the METU-Sabancı Turkish Treebank [21]. We selected this corpus in order to encrich the Turkish dependency treebank with explicit chunk information. As the test set, we used 300 sentences of the ITU validation set [24].

13 annotators worked in the annotation process. The annotation process is divided into two parts: First, annotators are asked to determine the sentence constituent (generally same as the verb phrase) and then segment the other parts of the sentence by asking the questions to the sentence constituent such as "who/what", "when", "where", etc. to extract the chunks for "Subject", "Object", "Adjunct", etc. There are some sentences where no explicit verb can be found, such as *Sorunun yanıtı "zorunluluktan"*.[3]. In these cases, annotators only separated the chunk boundaries, if possible. In case of multiple sub-sentences or nested sentences such as *"Ben de baloya gitmek istiyordum" demiş hıçkırarak Külkedisi*.[4], the annotators were asked to separately chunk each sentence. For the punctuation marks, the annotators left them out of the chunk boundaries unless the punctuation marks are surrounded by other words in the same chunk.

Annotators only labeled the verb chunk explicitly and just put the chunk boundaries for the others. The reason behind leaving the label annotation as a future work is two-fold: first, even though the annotators are Turkish-natives, they are not linguists. When we take Turkish agglutinative structure into consideration, the label attachments can turn the chunking problem to a hard task. Second, we argue that after determining the chunk boundaries it is easier to attach labels by using the dependency relations taken from the Turkish dependency treebank automatically[5]. After the annotation process, the test set is double-checked and corrected by two other annotators via cross validation. Table 1 shows some statistics of the annotated corpus and the test set. An annotated Turkish sentence with chunks[6] is shown as follows:

Buradaki üst düzey görüşmelerimizde $|_1$, $|_2$ turizm için ellerinden geleni yapacaklarını $|_3$ *söylediler* $|_4$.

In our high level talks here $|_1$, $|_2$ *they said* $|_4$ they would do their best for tourism $|_3$.

3   Dependency Parser-Based Chunking

Turkish dependency treebank suffers from data sparseness as the corpus is relatively small for machine learning techniques where accurate estimation of the model parameters is important. On the other hand, this corpus contains

---

[2] After excluding the non-projective sentences.

[3] English translation: The answer of the question is "necessity".

[4] English translation: "I wanted to go to the ball too" said Cinderella sobbingly.

[5] The chunk takes its label from the head word's label.

[6] Sentence chunks are labeled with *.

Table 1. Corpus Statistics

Statistics	Train	Test
Sentences	4758	300
Words	43126	3610
Avg. Sentence Length	9.06	12.03
Total Chunks	16674	1255
Sentence Chunks	5424	368
Avg. Chunk Length	2.16	2.50

valuable information which enables automatic extraction of chunks from dependency relations and labels.

3.1 Turkish Dependency Parser

Turkish dependency parser [25] is developed with the MaltParser [26] by using Turkish specific parameters. Turkish dependency parser uses the Turkish dependency treebank [21], a corpus of 5635-sentences from eight different domains. It contains several features including word, lemma, POS, coarse POS[7], inflectional group (IG)[8], and 23 different labels [22]. In our work, we explored the same version (i.e, CoNLL-X [27] format) of the dependency treebank that is used in [25]. We performed some modifications on the dependency treebank, as for constituent chunking, the IG-based representation is not needed. We turned the dependency treebank automatically into a word-based representation by removing IGs and their relations while keeping the relations and attachments consistent. Figure 2[9] shows the constituent chunks of a Turkish sentence with word-based representation that is used in chunking experiments in addition to the IG-based (dotted) representation which is the original dependency treebank format.

3.2 Chunk Extraction

Abney [28] stated that chunks can be defined as parse tree fragments. We applied this definition to the Turkish dependency parser output. Turkish is a (predominantly) head-final language that allows us to extract chunks by using the word's head and its relation in one-pass. When we look at the dependency graph of a Turkish sentence, it is clear that each dependent of the verb defines a chunk boundary. We apply the chunk extraction procedure described in Figure 3.

---

[7] Turkish morphological analyzer gives a two-layered POS information such as *Noun+Proper* and *Num+Ordinal*. Coarse POS is the first part of this POS information. In the absence of the second layer, POS is assigned same as the Coarse POS.

[8] To represent the morphology, words are separated at the derivational boundaries. The inflectional morphemes with derivation (or the root word for the first IG) is called as inflectional group.

[9] English translation: My mother was happy with the sudden change of me.

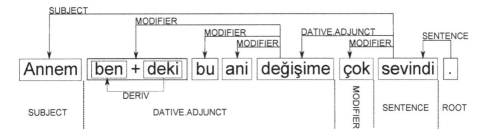

Fig. 2. Constituent chunks and dependency relations of a sample Turkish sentence

---

Algorithm: Chunk extraction by using the Turkish dependency parser

---

```
Extract the dependency tree of the sentence using the Turkish dependency
parser
Convert the IG representation to the word representation
for each word of the sentence
    if (word is not tagged as "Sentence" or the word's head is not
        "Sentence")
      Add the word to the chunk list;
    else
      if (the word's head is "Sentence")
           Mark it as a regular chunk boundary;
           Define the chunk as all words in the accumulated chunk list;
           Clear the chunk list;
      else
           Mark it as Sentence chunk;
```

Fig. 3. Algorithm for chunk extraction by using the Turkish dependency parser output

It is important to note that this procedure only works for projective sentences. Because non-projective sentences have a very small percentage in the Turkish corpus, we did not apply a special procedure is applied for non-projective sentences.

4   CRF-Based Chunking

Text chunking can be seen as a sequential labeling (or prediction) task, where every token in a sequence is associated with the chunk-tags [2]. In early approaches, Markov models, specifically HMMs, were widely used in the labeling and segmentation tasks. The biggest problem in labeling process is the intractability of the inference step when all dependencies between the input (observation) variables are defined. Markov models solves the problem by assuming independence relations between the observations but it causes the problem to be an approximation of the task. We can conclude that the labeling problem cannot be modeled properly with HMMs. Later, MEMMs were introduced that solved the tractability issue but this model has a disadvantage which is known as label biasing.

In the MEMM structure, transitions are compared within the state instead of the whole network which causes choosing the states with few successors.

4.1   Conditional Random Fields

Recently, Conditional Random Fields (CRFs) [12] became the state-of-the-art framework by outperforming the previous models for several labeling tasks such as text segmentation, named entity recognition, part-of-speech tagging and shallow parsing. Both HMM and MEMM, which model the joint probability $P(x, y)$, are called as generative models. CRFs are introduced as discriminative (undirected graphical) models and solves the intractability and label biasing problems. As CRFs directly estimate the conditional probability $P(y|x)$ without modeling $P(x)$ explicitly, they allow relaxation in the input variables. This allows the model to benefit from many overlapping and correlated feature functions. CRFs also have several advantages over the previous probabilistic models, as they can be arbitrarily structured.

The input (sentence) is denoted by $x = \{0...n\}$ and the output (labels) is denoted by $y = \{0...n\}$. Since x_i is known, the goal of the probability model is to estimate the conditional probability $P(y_i|x_i)$ that maximizes the log probability:

$$p(y|x) = \frac{1}{Z(x)} \prod_{c \in C} \omega_c(x_c, y_c)$$

where C is the set of maximal cliques, $\omega_c(x, y) = \exp(\lambda_c f_c(x_c, y_c))$ is the potential function with weights λ_c and feature functions $f(x, y)_c$, and $Z(x) = \sum_y \prod_{c \in C} \omega_c(x_c, y)$ is the normalization factor.

4.2   Chunk Identification

The state-of-the-art method of labeling chunk is the so-called IOB tagging which was introduced by Ramshaw and Marcus [2]. With this tagging, each token in the sentence is tagged with either B (beginning), I (inside) or O (outside) tags. As we only annotated the sentence chunks explicitly and preferred a general representation for the rest of the chunks, we only used two subclasses for the tagging the sentences where C denotes the general chunks and V denotes the sentence chunks. A sample for Turkish sentence chunk tagging is shown below:

Buradaki	üst	düzey	görüşmelerimizde	,	turizm	için	ellerinden	geleni
B-C	I-C	I-C	I-C	O	B-C	I-C	I-C	I-C

yapacaklarını	söylediler	.
I-C	B-V	O

Morphological and Contextual Feature Functions for CRF: The general features used for the chunking process (especially for English) are the word itself, POS and possibly the lemma. Baseline experiments shown in Section 5 showed that using these features are not sufficient to obtain a high accuracy in Turkish chunking due to rich morphology. In our work, we enhanced the feature set with morphological analyzer and dependency treebank features to exploit more information for a more accurate chunk estimation. The set of all features that we used in the experiments are listed as follows:

- **F1 - Word:** The word itself.
- **F2 - Lemma:** How the word is shown in a dictionary entry.
- **F3 - CPOS:** The *main* POS of the word.
- **F4 - POS:** The second layer of the POS information of the word as explained above.
- **F5 - IG:** The combined inflectional morpheme information of the word's last IG.
- **F6 - Num:** The number of the word for specific POS such as *Noun, Verb, Pron* and *Ques*. The feature can take singular values *A1sg ,A2sg or A3sg* if the word is singular. A "_" is assigned if the word is plural.
- **F7 - Case:** The case of the word when its POS is *Noun* or *Pronoun*. The feature can take the values *Acc, Dat, Nom, Loc* or *Abl*.
- **F8 - Distance:** The word's distance to its head. If the word is attached to a head within a distance one or two words then the feature is assigned as **short**, otherwise it is assigned as **long**.
- **F9 - L4Char:** The last four characters of the word.
- **F10 - D-Label:** The dependency relation of the word to its head word.

Table 2 shows the complete set of features that were used for the Turkish sentence shown in Figure 2.

Table 2. Turkish Chunking Features

Word	Lemma	CPOS	POS	IG	Num	Case	Distance	L4Char	D-Label	Tag
Annem	anne	Noun	Noun	A3sg|P1sg|Nom	A3sg	Nom	Long	nnem	SUBJECT	B-C
bendeki	ben	Adj	Adj	Rel	_	_	Long	deki	MODIFIER	B-C
bu	bu	Pron	DemonsP	A3sg|Pnon|Nom	A3sg	Nom	Short	bu	MODIFIER	I-C
ani	ani	Adj	Adj	_	_	_	Short	ani	MODIFIER	I-C
değişime	değişim	Noun	Noun	A3sg|Pnon|Dat	A3sg	Dat	Short	şime	DATIVE.ADJUNCT	I-C
çok	çok	Adv	Adv	_	_	_	Short	çok	MODIFIER	B-C
sevindi	sevin	Verb	Verb	Pos|Past|A3sg	A3sg	_	Short	indi	SENTENCE	B-V
.	.	Punc	Punc	_	_	_	Long	.	ROOT	O

5   Experiments

We used the CRF++[10] tool, a linear-chain CRF implementation, to train and test the chunker. The window size for the feature functions is 5 covering the preceding two words, the word itself and the following two words. For training the

---

[10] CRF++: Yet Another CRF toolkit.

CRF-based constituent chunker, we used all training corpus that is segmented by the annotators. As the test data, we used the ITU validation set. We used the precision, recall, and F-measure metrics that are widely used in the chunking task [15].

5.1 Results

We initially ran the CRF-based chunking experiments with the standard features (e.g., word and POS that were used in [13]). We then incrementally included the other features introduced in Section 4.2. On the other hand, we also extracted the chunks by the Turkish dependency parser as mentioned in Section 3.2. Tables 3 and 4 compares the performances of these chunkers. For both sentence and general chunks, extracting chunks from the treebank performs worse than the CRF models. The features help general chunking significantly but the performance improvement for the sentence chunks is limited. The reason is that the sentence chunks are determined with a high accuracy even with the standard features.

Specifically, we observed that morphological features such as **IG, coarse POS, Number** and **Case**, and the dependency parser feature **Distance** harm the sentence chunk performance, while they improve the general chunk performance. On the other hand, the contextual feature **L4Char** has a positive effect on both chunk types. As expected, adding the **D-label** feature results in a high improvement on both chunk types. For sentence chunks, the dependency parser-based chunker performs worse than the baseline chunker.

Table 3. Sentence Chunk Results

Feature Template	Precision	Recall	F-measure
(1):Word+POS	87.53	90.59	89.03
(2):Word+Lemma+POS	87.94	92.20	90.02
(3):(2)+IG	88.05	91.12	89.56
(4):(3)+CPOS	88.54	91.39	89.94
(5):(4)+Sing+Case	89.15	90.59	89.86
(6):(5)Distance	89.38	90.59	89.98
(7):(6)+L4Char	88.97	91.12	90.03
(8):(7)+D-Label	**91.71**	**92.20**	**91.95**
(9):DepParser:	88.40	88.17	88.29

The CRF models for Turkish constituent chunking gave promising results especially for general chunking. Experiments supported our claim that exploiting features that give more information about the word are crucial for morphologically rich languages.

Table 4. General Chunk Results

Feature Template	Precision	Recall	F-measure
(1):Word+POS	87.09	73.97	80.00
(2):Word+Lemma+POS	87.92	74.75	80.80
(3):(2)+IG	88.01	77.68	82.52
(4):(3)+CPOS	88.26	78.99	83.37
(5):(4)+Sing+Case	87.93	82.70	85.23
(6):(5)+Distance	88.36	83.24	85.78
(7):(6)+L4Char	88.68	83.55	86.04
(8):(7)+D-Label	**90.23**	**84.94**	**87.50**
(9):DepParser:	89.14	85.01	87.03

6   Conclusion and Future Work

Despite the efforts for developing many NLP tools and resources for well-studied languages, Turkish as a less-resourced language lacks many NLP resources and tools such as a chunker. In this work, we presented the initial explorations on Turkish constituent chunking by using CRFs and the Turkish dependency parser. We stated that the non-recursiveness constraint of the state-of-the-art chunk definition is not suitable for morphologically rich languages. As a solution, we adapted the chunk definition for Turkish by labeling constituents instead of phrases. We annotated the Turkish dependency treebank for chunking. Moreover, we claimed that standard (e.g., word and POS) features are not sufficient to obtain an accurate chunker and introduced several morphological and contextual feature functions including coarse POS, number, last four characters, and several features from the dependency treebank. As a result of the experiments, a CRF-based constituent chunker enhanced with morphological and contextual features outperformed both the baseline (e.g., word and POS features) CRF-based chunker and the dependency parser-based chunker. We reported a performance of F-measure 91.95 (an improvement of 3.3% over baseline) for verb chunks and 87.50 (an improvement of 9.4% over baseline) for general chunks. We believe that our work on Turkish chunking will be the first step for Turkish NLP tools and motivation for similar languages.

As a future work, we will experiment with constituent chunk labeling by assigning dependency relation labels as the chunk label to the general chunks both manually (as gold-data) and automatically. We plan to annotate the Turkish dependency treebank in IG-level for phrase chunks and compare the performances with constituent chunking. Finally, we are going to use this chunk information as an additional feature in several Turkish NLP tasks to improve their performances such as the Turkish dependency parser and in Turkish-English machine translation systems.

Acknowledgments. Authors would like to thank Şeniz Demir and Coşkun Mermer from TÜBİTAK-BİLGEM Speech and Natural Language Processing Group for their contributions and valuable comments.

References

1. Abney, S.P.: Parsing by chunks. In: Principle-Based Parsing: Computation and Psycholinguistics, pp. 257–278 (1991)
2. Ramshaw, L.A., Marcus, M.P.: Text chunking using transformation-based learning. In: Proceedings of the Third ACL Workshop on Very Large Corpora, pp. 88–94 (1995)
3. Argamon, S., Dagan, I., Krymolowsky, Y.: A memory-based approach to learning shallow natural language patterns. In: Proceedings of 36th Annual Meeting of the Association for Computational Linguistics (ACL), Montreal, Canada, pp. 67–73 (1998)
4. Tjong Kim Sang, E.F., Veenstra, J.: Representing text chunks. In: Proceedings of the EACL, pp. 173–179 (1999)
5. Cardie, C., Pierce, D.: Error-driven pruning of treebank grammars for base noun phrase identification. In: Proceedings of the COLING/ACL, Montreal, Canada, pp. 218–224 (1998)
6. Church, K.: A stochastic parts program and noun phrase parser for unrestricted texts. In: Proceedings of the Second Conference on Applied Natural Language Processing, pp. 136–143 (1988)
7. Freitag, D., McCallum, A.: Information extraction with hmm structures learned by stochastic optimization. In: Proceedings of AAAI (2000)
8. McCallum, A., Freitag, D., Pereira, F.: Maximum entropy markov models for information extraction and segmentation. In: Proceedings of International Conference on Machine Learning, California, CA, pp. 591–598 (2000)
9. Kudo, T., Matsumoto, Y.: Chunking with support vector machines. In: Proceedings of the NACCL 2001, pp. 192–199 (2001)
10. Ratnaparkhi, A.: A maximum entropy model for part-of-speech tagging. In: Proceedings of the Conference on Empirical Methods in Natural Language Processing (EMNLP), Philadelphia, PA (1996)
11. Zhang, T., Damerau, F., Johnson, D.: Text chunking based on a generalization of winnow. Journal of Machine Learning Research 2, 615–637 (2002)
12. Lafferty, J., McCallum, A., Pereira, F.: Conditional random fields: Probabilistic models for segmenting and labeling sequence data. In: Proceedings of the 18th International Conference on Machine Learning, pp. 282–289 (2001)
13. Sha, F., Pereira, F.: Shallow parsing with conditional random fields. In: Proceedings of Human Language Technology-NAACL, Edmonton, Canada, pp. 134–141 (2003)
14. Veenstra, J.: Memory-based text chunking. In: Workshop on Machine Learning in Human Language Technology, Crete, Greece (1999)
15. Sang, E.F.T.K., Buchholz, S.: Introduction to the conll-2000 shared tasks: Chunking. In: Proceedings of the CONLL-2000 and LLL-2000, pp. 127–132 (2000)
16. Goldberg, Y., Adler, M., Elhadad, M.: Noun phrase chunking in hebrew: influence of lexical and morphological features. In: Proceedings of the 21st International Conference on Computational Linguistics and the 44th Annual Meeting of the Association for Computational Linguistics, Sydney, Australia, pp. 689–696 (2006)

17. Park, S.B., Zhang, B.T.: Text chunking by combining hand-crafted rules and memory-based learning. In: Proceedings of the 41st Annual Meeting of the Association for Computational Linguistics (ACL), pp. 497–504 (2003)
18. Lee, Y.-H., Kim, M.-Y., Lee, J.-H.: Chunking Using Conditional Random Fields in Korean Texts. In: Dale, R., Wong, K.-F., Su, J., Kwong, O.Y. (eds.) IJCNLP 2005. LNCS (LNAI), vol. 3651, pp. 155–164. Springer, Heidelberg (2005)
19. Gune, H., Bapat, M., Khapra, M.M., Bhattacharyya, P.: Verbs are where all the action lies: experiences of shallow parsing of a morphologically rich language. In: Proceedings of the 23rd International Conference on Computational Linguistics: Posters, Beijing, China, pp. 347–355 (2010)
20. Kutlu, M.: Noun phrase chunker for turkish using dependency parser, Ms. Thesis (2010)
21. Oflazer, K., Say, B., Hakkani-Tür, D.Z., Tür, G.: Building a turkish treebank. In: Treebanks: Building and Using Parsed Corpora, pp. 261–277 (2003)
22. Say, B.: Metu-sabancı turkish treebank user guide (2007)
23. Oflazer, K.: Dependency parsing with an extended finite-state approach. Computational Linguistics 29, 515–544 (2003)
24. Eryiğit, G.: Itu validation set for metu-sabancı turkish treebank (2007)
25. Eryiğit, G., Nivre, J., Oflazer, K.: Dependency parsing of turkish. Computational Linguistics 34, 357–389 (2008)
26. Nivre, J., Hall, J., Nilsson, J., Chanev, A., Eryiğit, G., Kübler, S., Marinov, S., Marsi, E.: Maltparser: A language-independent system for data-driven dependency parsing. Natural Langauge Engineering Journal 2, 99–135 (2007)
27. Buchholz, S., Marsi, F.: Conll-x shared task on multilingual dependency parsing. In: Proceedings of the 10th Conference on Computational Natural Language Learning (CONLL-X), New York, NY, pp. 149–164 (2006)
28. Abney, S.P.: Chunks and dependencies (1991)

Enhancing Czech Parsing
with Verb Valency Frames

Miloš Jakubíček and Vojtěch Kovář

Natural Language Processing Centre, Faculty of Informatics
Masaryk University, Czech Republic
{jak,kovar}@fi.muni.cz
http://nlp.fi.muni.cz

Abstract. In this paper an exploitation of the verb valency lexicons for the Czech parsing system `Synt` is presented and an effective implementation is described that uses the syntactic information in the complex valency frames to resolve some of the standard parsing ambiguities, thereby improving the analysis results. We discuss the implementation in detail and provide evaluation showing improvements in parsing accuracy on the Brno Phrasal Treebank.

Keywords: parsing, syntactic analysis, verb valency, Czech.

1 Introduction

The key combination principles in natural language syntax are *valency, agreement* and *word order* [1, s. 303]. While morphological agreement and word order can be seen a dual principles (languages with weak agreement, e.g. English, possess fixed word order and vice versa, e.g. Czech), verb valency[1] represents a common principle and basically stands for a bridge between syntax and semantics. In further text we use the notion of the *valency frame* (also called *subcategorization frame*) which consists of a given verb and its valency positions, i.e. gaps which need to be filled with arguments (also called *fillers*).

Hence a general valency frame schema of arity n looks as follows:

$$\mathbf{verb}(\text{argument}_1, \text{argument}_2, \dots, \text{argument}_n)$$

A verb frame instantiated with corresponding arguments is also called a *functor-argument* or *predicate-argument* structure. A verb cannot be combined with its arguments arbitrarily, but imposes strict requirements on what combinations of word forms represent valid argument structure for the verb.

Not unlike many other NLP problems, while valency frame can be formalized trivially, it is not easy to provide a verb valency lexicon for a particular language.

---

[1] In further text we focus only on verbs as primary valency carriers, though other part of speech (like e.g. prepositions) might be accompanied with a valency structure as well.

A. Gelbukh (Ed.): CICLing 2013, Part I, LNCS 7816, pp. 282–293, 2013.
© Springer-Verlag Berlin Heidelberg 2013

Opinions about what is and what is not a valency of a given verb differ heavily, since in many cases the boundary between an obligatory argument (valency, complement) and an optional modifier (adjunct) is rather fuzzy.

The exploitation of verb valencies in parsing has been theoretically studied in [2] and their positive impact on the accuracy of a Czech statistical parser has been evaluated in [3]. In this work we focus on a practical implementation in the Czech rule-based parser Synt using the verb valencies obtained from the valency lexicons VerbaLex [4] and Vallex [5] and evaluate the impact of using valencies on parsing quality.

2   Synt Parser

Syntactic parser Synt [6,7] is based on a context-free backbone and performs a stochastic agenda-based head-driven chart analysis using a hand-crafted meta-grammar for Czech, employs several contextual actions and disambiguating techniques and offers ranked phrase-structure trees as its primary output.

2.1   Parsing Workflow

After basic head-driven CFG analysis, all parsing results are collected in the so called *chart* structure which is built up during the analysis. The chart can encode up to exponential number of parse trees in a polynomial space.

Upon the basic CFG analysis, we build a new graph structure called *forest of values*.[2] This structure originates from applying contextual actions on the resulting chart, which are programmed functions that are called for a given rule and may modify the analysis result (adjust parse ranking or even prune an analysis) and are mostly used for covering contextual phenomena such as grammatical agreement. From the implementation point of view, the forest of values comes more handy for post-processing than the original chart (while it still keeps the polynomial size) – among other features, it enables efficient extraction of n best trees. A sample forest of values is given in Figure 1: nodes (values) in one row build a value node list which represents children within a single analysis. One value may have links to multiple value node lists, i. e. multiple alternative analyses.

Finally, Synt offers three output possibilities from the resulting forest of values: a phrase-structure tree, a dependency graph or a set of syntactic structures.

2.2   Grammar

To prevent maintenance problems which a hand-written grammar is susceptible to, the grammar used in Synt is edited and developed in form of a small meta-grammar (an approach first described [8]) which at this time contains 239 rules.

---

[2] The name is actually misleading and we keep using it just to be compatible with former work – a more fitting notion would be e. g. a *value graph*.

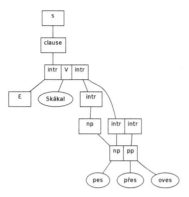

Fig. 1. A sample packed shared forest. Note that for visualization purposes this is a slightly simplified version.

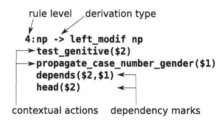

Fig. 2. A meta-grammar rule example

From this meta-grammar, a full grammar can be automatically derived (having 3,867 expanded rules). In addition, each rule can be associated with several so called **contextual actions** (as illustrated in Figure 2), i.e. programmed functions that are called when the given rule matches and may modify the analysis result (adjust parse ranking or even prune an analysis) and are mostly used for covering contextual phenomena such as grammatical agreement.

- a set of one or more so called **actions**, i.e. programmed functions that are called when the given rule matches and may modify the analysis result (adjust parse ranking or even prune an analysis) and are mostly used for covering contextual phenomena such as grammatical agreement,
- a **derivation template** which is used to derive the full grammar and which can define various operations on the rule which are used to obtain additional alternations, such as permutation of the right-hand side of the rule, inserting a particular non-terminal between each two non-terminals on the right-hand side or expanding each non-terminal on the right-hand side to its own right-hand side in the grammar (i.e. sort of "skipping" the non-terminal),
- a **rule level** which allows stratification of the grammar into separate levels of structural complexity which can be enabled or disabled at runtime and

make it possible to improve the parsing precision or add domain specific rules (where by domain we refer e. g. to uncommon text like tables, mathematical formulas and similar "chunks" of strings) into the grammar that are employed only if they are necessary to parse the sentence,
- a **head or dependency marker** which can be used to build a dependency graph from the chart analysis.

3   Complex Valency Frames

An illustrative example of a valency frame for the verb "skákat" (to jump) is provided in Figure 3. Its structure is as follows:

| 1 | přehoupnout se$_1$, přehupovat se$_1$, přešvihnout se$_1$, přeskakovat$_1$, přeskočit$_1$, skákat$_2$, skočit$_2$ ≈

-**frame:** $\mathbf{AG}$<person:1|animal:1>$_{kdo1}^{obl}$ $\mathbf{VERB}^{obl}$ $\mathbf{LOC}$<location:1>|$\mathbf{ENT}$<stream:1>$_{pres+co4}^{obl}$

-**example:** skákal přes příkop **(impf)**
-**example:** kůň přeskočil přes potok **(pf)**
-**synonym:** přehoupnout$_1$, přehupovat$_1$
-**use:** fig (přehoupnout se, přehupovat se); prim (přešvihnout se, přeskakovat, přeskočit, skákat, skočit)
-**reflexivity:** no (přeskakovat, přeskočit, skákat, skočit); refl (přehoupnout se, přehupovat se, přešvihnout se)

Fig. 3. Example valency frame for the Czech verb "skákat" (to jump)

- on the first line, the synset is provided – all verbs in the synset share the information given below,
- the second line contains the frame itself: a first-level semantic role of all arguments is denoted by upper-case letters (AG – agent, LOC – location, ENT – entity), each of which is accompanied by a list of second-level roles. Finally, all valency "gaps" end with the syntactic information – in the example with direct case ("kdo1" – animate nominative) and a prepositional case specification ("přes+co4" – over + inanimate accusative), each of which can be either obligatory (marked as OBL) or optional (OPT),

4   Exploiting Valencies

In [9] it has been shown how a parsing system can help in building a valency lexicon. The work described in this paper goes into opposite direction: how can the valency lexicon contribute to the analysis performed by Synt?

As briefly outlined in Section 1, valency frames carry important syntactic information in that they distinguish obligatory arguments of a verb from its optional modifiers. This kind of information has a (theoretically straightforward) exploitation in parsing for resolving various attachment ambiguities.

On Figure 4, a standard PP-attachment ambiguity issue is shown. Without additional information there is no way for the parser to know what analysis should be the preferred (or even right) one. The preposition "přes" (over) can

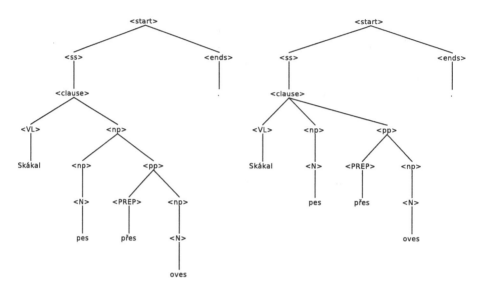

Fig. 4. A PP-attachment ambiguity example

introduce an argument (as presented on the figure), an adjunct or even a part of a noun-phrase argument.

However, if we could efficiently lookup the valency frame of the verb "skákat" (which has been demonstrated in Figure 3) and provide its content to the parsing process, we would be able to determine the right (or at least more probable) hypothesis and prefer it over other analyses found in the preceding parsing steps. This can be basically done in two ways:

– **pruning unsatisfied hypotheses**
   This is an aggressive but straightforward way: all parses that fulfill some condition (e. g. that they do not fully match any of the valency frame for the given verb) will be removed from the set of possible analyses, effectively reducing number of resulting parse trees. Naturally, the main disadvantage is that we might prune correct hypotheses that may not match any valency frame for several reasons:
   - because of an *error in the valency lexicon* (and, since manually built language resources often suffer from consistency problems, this must be taken into consideration as a standard situation),
   - because of an *elliptical construction* which can cause one or more arguments of the verb to be missing within the scope of the clause,

– **altering ranking of unsatisfied hypotheses**
   Altering the parse ranking on one hand doesn't reduce the ambiguity of the analysis, but on the other hand it is not susceptible to the possible overpruning effect described above and, moreover, if combined with e. g. selection of n-best trees, it might effectively lead to the same results.

However, it immediately introduces the question how the current ranking should be modified in order to account for the valency frames reasonably. A first necessary step that needs to be done is that all hypotheses will be ranked according to how much they satisfy valency frames of the particular verb – an algorithm to solve this problem is proposed in the next section.

5   Ranking Valency Hypotheses

A valency structure found in a parse hypothesis might differ from the valency frame as given in the lexicon in that it contains more or less arguments and/or that some of the arguments do not satisfy the restrictions imposed by the valency frame.

Moreover, several hypotheses may completely match different valency frames.[3] In such cases we would like to prefer the hypothesis that covers the largest (with respect to the number of arguments) valency frame.

Therefore, our ranking algorithm (which presumes that we have already found all valency structures – hypotheses – in the sentence) proceeds as follows:

1. Let V be the set of all found valency structures and F the set of all valency frames obtained from the lexicon for the particular verb. For each $v \in V, f \in F$, let $found$ be the number of all found arguments, $correct$ the number of them which are satisfied in the valency frame and $size$ be the number of all arguments in the frame, we define the scoring function s:

$$s : V \times F \to \mathbb{R}$$

given as:

$$s(v, f) = \begin{cases} log(size) \cdot size \cdot va & \text{if } found = correct = size \\ \frac{2 \cdot precision \cdot recall}{precision + recall} & \text{otherwise} \end{cases}$$

where

$$precision = correct/found$$
$$recall = correct/size$$
$$va = \text{valency aggressivity constant}$$

The overall score s_v of a found valency structure $v \in V$ is then defined as:

$$s_v = \max_{f \in F}(s(v, f))$$

Basically, we compute the maximum f_1 measure of the found valency structure and given valency frames and among those structures that achieve $f_1 = 1$, we prefer larger structures.

2. As next step, we may multiply the probability of the analysis by the score of the associated valency structure or set a valency score threshold and prune all analyses that achieve lower score.

---

[3] This typically occurs for verbs with monovalent nominative valency as well as e. g. a divalent nominative–accusative (or any other case) one.

5.1   Filtering by Negative Valency Frames

The content of the valency lexicons can be used not only to suggest possible (positive) valency candidates, but also to limit optional verb modifiers (adjuncts), i.e. to say that a phrase *cannot* be attached to verb as its adjunct. In Czech this is the case of 4 direct cases (nominative, genitive, dative and accusative), which cannot be attached to the verb unless they are verb valency arguments (i.e. not adjuncts).

This fact is exploited in the system in a straightforward way: if, after matching a found valency frame with those from lexicon, there are remaining valency slots consisting of noun phrases in one of the four cases and these could not be matched with any slot in the lexicon, this frame is considered invalid and is either downscored to minimum or (optionally) directly removed (i.e. all parses involving this frame are pruned).

6   Implementation

In this section, the preparation and format of the data from the valency lexicons is described, followed by an explanation of algorithm used for retrieving valency structures from (partial) analyses and completed by the description of the matching and scoring procedures.

6.1   Preparation of the Valency Data

Valency lexicon represents a large database that needs to be converted into a comprehensive, easily readable (for a program) format that will contain only the information that is indeed used during the analysis. The lexicon is primarily stored in an XML format with a plain text counterpart which we use to convert from.

As for now, only the syntactic information from the valency lexicons is used to contribute to the analysis in Synt, therefore the information we actually need to extract from each frame is rather limited. We converted the valency lexicon source files into what we call *inverted valency format* which can be efficiently read and stored in memory by Synt.

The format itself is very simple: it is line based, each line starting with a (unique) valency frame, followed by a comma-delimited list of verbs that share this valency frame. The valency frame itself is reduced on a hyphen-delimited list of valencies, each of which is either a number (denoting a direct case) or a number followed by a preposition name (denoting a prepositional case) or a list of alternatives of both prepositional and direct cases (delimited by a vertical line). The file created by the converter from current VerbaLex lexicon has the length of only 352,125 bytes. An example line in the inverted valency format is shown in Figure 5.

The inverted valency format is read by Synt on startup and effectively stored in the memory into a hash containing the verb name as key and a list of possible

```
1-7za|3ke-2z,přistěhovat se,imigrovat
```

Fig. 5. An example line in the inverted valency format for the verb "přistěhovat se" (to move in) and "imigrovat" (to immigrate)

valency frames as value. Using this approach all the strings (verb names as well as valency frames) are stored in the memory only once, all together requiring a space of only 1,768 kB.[4] Thanks to the hashing, the valency frame list can be looked up in constant time for a given verb. The valency hash is allocated only at the beginning in case of batched processing of multiple sentences (though reading the file and allocating the hash takes only very little time – 0.0351 s).[5]

6.2 Searching for Valencies

The search algorithm itself is run on the forest of values structured as described in Section 2. Although the forest of values represents an effective and compact packing of all analyses into a polynomial structure, its processing and subsequent gathering information is not trivial – a naïve exhaustive search would still lead to exponential explosion and is hence unfeasible.

The valency frames are looked up within the scope of a single clause and we need to retrieve all possible valency structures ("frames candidates") within the relevant clause. To achieve this, we employ a dynamic technique on a depth-first search of forest of values. The algorithm itself starts top-down (from the clause edge), but a recursive implementation of depth-first search allows us to collect the results bottom-up.

What are actually the valencies that we try to find? For our purposes, valencies are values (in the forest of values) that correspond to a noun-phrase or prepositional phrase edge in the original chart structure. Note that we do not necessarily need to search the whole forest of values to find all valencies: basically, we follow all descendants of the clause edge to find the first noun- or prepositional phrase on the way – after that there is no need to continue the search.

However, there are several caveats that need to be taken into consideration:

- Since we cannot perform an exhaustive search, we must not visit a value (in the graph – forest of values) twice and need to remember partial results – valencies that have been already found by a previous visit of this value,
- The results collected for separate analyses (i. e. separate value node lists) must be merged to the parent value,
- The results collected *within a single analysis* for separate children values (i. e. children of a value within a particular tree) must be combined to the

---

[4] For VerbaLex, which is the bigger of the two lexicons (10,564 verbs and 29,092 verb-frame pairs), on a Linux machine with page size of 4 kB.

[5] The time is an average over 10 runs when starting Synt on empty input and it has been measured with and without loading valencies (using the system utility time).

parent value while the dynamic algorithm must ensure that the valencies of a particular value remain independent for recurrent visits of the same node. Note that this step still suffers from possible combinatoric explosion but the number of valencies is usually small and hence it does not represent a problem in practice.

Triggering the Search Action. Searching for valencies, matching and subsequent reranking must occur when all the necessary information is available (i. e. the analysis has been completed up to the clause level) but we can still influent the ranking of all hypotheses (i. e. it is not a post-processing step). Therefore triggering the valency actions has been implemented as a contextual action (as described in Section 2) in the grammar, tied to particular `clause` rules.

Hereat the design and expressional power of the meta-grammar formalism used in `Synt` was proven to be an outstanding concept. Also, since `clause` rules constitute the core of the grammar, they are carefully designed and achieve high coverage what concerns possible verb(s) structure of a Czech sentence within a single clause.

Search Algorithm. A pseudo-code of the search algorithm is given as Algorithm 1. It shows the `dfsSearch` function that is run recursively on the forest of values and dynamically collects possible valency frames. Note the `setValue` function that stores a pointer for each found valency frame to the value node list (there might be several of them) where the analysis diverted from another one. This is necessary to be able to modify the ranking or prune the hypothesis after all valency frames are calculated.

Matching and Scoring. After the search algorithm, the matching and scoring procedures are run on resulting valency structures. For each structure its scoring functions s is computed and afterwards negative valencies are evaluated. It is possible to customize the aggressivity of the valencies at runtime, the agressivity is defined as a number which can take following values:

0 **dry-run** mode: no ranking modifications and no pruning is performed, just the valency search is performed and its output is displayed,

1 − 4 **normal** mode: analyses ranking is multiplied by the value of the scoring function s and the given value,

5 **aggressive** mode: the ranking of all analyses with score function less than 1 is set 0,

6 **pruning** mode: all analyses with score function less than 1 are pruned.

7   Evaluation

As the testing set, the Brno Phrasal Treebank (BPT) [10] was used, which was originally created in years 2006–2008 and is still under continuous development.

Algorithm 1. A pseudo-code of the depth-first search of forest of values

```
result = ∅
for all analysis ∈ analyses do
    for all child ∈ children of analysis do
        if !child→hasBeenVisited then
            child→hasBeenVisited = 1
            if isNounPhrase OR isPrepPhrase then
                child→valencies = createValency()
            else
                child→valencies = dfsSearch(child)
            end if
        end if
        analysis→valencies = combine(analysis→valencies, child→valencies)
    end for
    if multipleAnalysis then
        setValue(analysis→valencies)
    end if
    result = merge(result, analysis→valencies)
end for
return  result
```

Currently the corpus contains in overall 86,058 tokens and 6,162 syntactically annotated sentences. The main source of the sentences is the Prague Dependency Treebank [11] which may allow future comparisons on parallel sentences.

As a similarity metric between a "gold" tree in the treebank and a parse tree from the output of the parser, we use a metric called *leaf-ancestor assessment* (LAA) [12]. This metric was shown to be more reliable than the older PARSEVAL metric which is however (and unfortunately) still used more frequently [13].

The LAA metric is based on comparing so called *lineages* of two trees. A *lineage* is basically a sequence of non-terminals found on the path from a root of the derivation tree to a particular leaf. For each leaf in the tree, the lineage is extracted from the candidate parse as well as from the gold tree. Then, the edit distance of each pair of lineages is measured and a score between 0 and 1 is obtained. The mean similarity of all lineages in the sentence represents the score of the whole analysis.

In [13], it is argued that the LAA metric is much closer to human intuition about the parse correctness than other metrics, especially PARSEVAL. It is shown that the LAA metric lacks several significant limitations described also in [14], especially that it does not penalize wrong bracketing so much and it is not so tightly related to the degree of the structural detail of the parsing results.

In the test suite, an implementation of the LAA metric by Derrick Higgins is used.[6]

---

[6] Which is publicly available at http://www.grsampson.net/Resources.html

Using the test suite for Synt, several evaluations have been performed on the Brno Phrasal Treebank that are shown below. In Table 1, an overall performance of the valency algorithm is demonstrated. It can be seen that the average number of retrieved valencies is relatively low (8.2) and the search procedure does not significantly worsen time performance of the parser.

Table 1. A comparison of three test suite runs with regard to the impact on parsing time

run	#verbs	#verb frames	time
without valencies	–	–	0.147 s
VerbaLex valencies	10,564	29,092	0.162 s
Vallex valencies	4,743	10,925	0.158 s

In Table 2, an overview of contribution of the valency information to the parsing precision is provided (the LAA figure is shown for the first tree selected by the automatic ranking functions) – as we can see the performance was best for the valency aggressivity 6 setting. In general we can see that the more weight the valencies had, the better were the obtained results. Moreover, the tree space was pruned significantly in case of valency agressivity settings 5 and 6.

The results are very similar for both valency lexicons, which is surprising with regard to the fact that Vallex is about half the size of VerbaLex.

Table 2. A performance overview of the valency information – first column is the valency aggressivity, second and third contain the LAA metric score for the first tree on output for the respective valency dictionary

VA	LAA First - VerbaLex	LAA First - Vallex
0	86.41	86.41
1	87.03	87.04
2	87.05	87.06
3	87.05	87.07
4	87.05	87.08
5	87.06	87.08
6	87.67	87.70

8   Conclusions

We have presented an extension of the Czech parser Synt that exploits the information about verb valencies as given by two available valency lexicons for Czech. An effective implementation of the underlying algorithms has been described and the measurements we have performed are showing improvements in both parsing precision and ambiguity.

Acknowledgments. This work has been partly supported by the Ministry of Education of CR within the LINDAT-Clarin project LM2010013 and by the Czech Science Foundation under the project P401/10/0792.

References

1. Hausser, R.: Foundations of Computational Linguistics, 2nd edn. Springer, Heidelberg (2001)
2. Hlaváčková, D., Horák, A., Kadlec, V.: Exploitation of the VerbaLex Verb Valency Lexicon in the Syntactic Analysis of Czech. In: Sojka, P., Kopeček, I., Pala, K. (eds.) TSD 2006. LNCS (LNAI), vol. 4188, pp. 79–85. Springer, Heidelberg (2006)
3. Zeman, D.: Can subcategorization help a statistical dependency parser? In: Proceedings of the 19th international Conference on Computational Linguistics, COLING 2002, Stroudsburg, PA, USA, vol. 1, pp. 1–7. Association for Computational Linguistics (2002)
4. Hlaváčková, D., Horák, A.: VerbaLex – New Comprehensive Lexicon of Verb Valencies for Czech. In: Computer Treatment of Slavic and East European Languages, Bratislava, Slovakia, pp. 107–115 (2006)
5. Lopatková, M., Žabokrtský, Z., Skwarska, K.: Valency Lexicon of Czech Verbs: Alternation-Based Model. In: Proceedings of the Fifth International Conference on Language Resources and Evaluation (LREC 2006), vol. 3, pp. 1728–1733. ELRA (2006)
6. Kadlec, V.: Syntactic Analysis of Natural Languages Based on Context-Free Grammar Backbone. PhD thesis, Faculty of Informatics, Masaryk University, Brno, Czech Republic (2007)
7. Jakubíček, M., Horák, A., Kovář, V.: Mining Phrases from Syntactic Analysis. In: Matoušek, V., Mautner, P. (eds.) TSD 2009. LNCS, vol. 5729, pp. 124–130. Springer, Heidelberg (2009)
8. Horák, A., Kadlec, V.: New Meta-grammar Constructs in Czech Language Parser synt. In: Matoušek, V., Mautner, P., Pavelka, T. (eds.) TSD 2005. LNCS (LNAI), vol. 3658, pp. 85–92. Springer, Heidelberg (2005)
9. Jakubíček, M., Kovář, V., Horák, A.: Measuring coverage of a valency lexicon using full syntactic analysis. In: RASLAN 2009: Recent Advances in Slavonic Natural Language Processing, Brno, pp. 75–79 (2009)
10. Kovář, V., Jakubíček, M.: Test Suite for the Czech Parser Synt. In: Proceedings of Recent Advances in Slavonic Natural Language Processing 2008, Brno, pp. 63–70 (2008)
11. Hajič, J.: Building a syntactically annotated corpus: The Prague Dependency Treebank. In: Issues of Valency and Meaning, Prague, Karolinum, pp. 106–132 (1998)
12. Sampson, G.: A Proposal for Improving the Measurement of Parse Accuracy. International Journal of Corpus Linguistics 5(01), 53–68 (2000)
13. Sampson, G., Babarczy, A.: A Test of the Leaf-Ancestor Metric for Parse Accuracy. Natural Language Engineering 9(04), 365–380 (2003)
14. Bangalore, S., Sarkar, A., Doran, C., Hockey, B.A.: Grammar & parser evaluation in the XTAG project (1998),
http://www.cs.sfu.ca/~anoop/papers/pdf/eval-final.pdf

An Automatic Approach to Treebank Error Detection Using a Dependency Parser

Bhasha Agrawal[1], Rahul Agarwal[1], Samar Husain[2], and Dipti M. Sharma[1]

[1] IIIT-Hyderabad, India
[2] University of Potsdam, Germany

Abstract. Treebanks play an important role in the development of various natural language processing tools. Amongst other things, they provide crucial language-specific patterns that are exploited by various machine learning techniques. Quality control in any treebanking project is therefore extremely important. Manual validation of the treebank is one of the steps that is generally necessary to ensure good annotation quality. Needless to say, manual validation requires a lot of human time and effort. In this paper, we present an automatic approach which helps in detecting potential errors in a treebank. We use a dependency parser to detect such errors. By using this tool, validators can validate a treebank in less time and with reduced human effort.

1 Introduction

Treebanks are an essential resource for developing solutions to various NLP related problems. As a treebank provides important linguistic knowledge, its quality is of extreme importance. The treebank used for experiments in this work is a part of the new multi-layered and multi-representational Hindi Treebanking project [8,19] (the dependency scheme is based on Computational Paninian Grammar (CPG) model, first proposed by [7]) whose target is 450k words. As is generally the case, most of the annotation process is either completely manual or semi-automatic, and the validation is completely manual.[1]

The process of developing a treebank involves manual or semi-automatic annotations of linguistic information. A semi-automatic procedure involves annotating the grammatical information using relevant NLP tools (eg. POS taggers, chunker, parsers, etc.). The output of these tools is then manually checked and corrected. Both these procedures may leave errors in the treebank on the first attempt. Therefore, there is usually another step called *validation* in which these errors are manually identified and corrected. But the validation process is as time-consuming as annotation process. As the data has already been annotated carefully, we need tools that can supplement the validators' task with a view of making the overall task fast, without compromising on reliability. Using such a

---

[1] Dependency annotation scheme of this treebank conforms to [6].

A. Gelbukh (Ed.): CICLing 2013, Part I, LNCS 7816, pp. 294–303, 2013.

tool, a validator can directly go to error instances and correct them. So, the tool must have high recall. A human validator can reject unintuitive errors without much effort, so one can compromise a little bit on precision.

In this paper, we propose such a strategy. The identified errors are classified under two categories (*attachment error* and *labeling error*) for the benefit of the validators, who may choose to correct a specific type of error at one time. Our method to identify such errors involves using a dependency parser.

If we can demonstrate considerable benefit/relevance of such a strategy, it might then be a good idea to incorporate it in a treebank development pipeline.

The rest of the paper is organized as follows: Section 2 describes related work. Section 3 describes data used in the experiments. In section 4, we describe our approach to error detection using a dependency parser. Here, we also show the results obtained using this approach. In section 5, we compare our results with older techniques. Section 6 concludes the paper.

2   Related Work

Error detection has become an active topic of research over the last decade due to increase in demand for high quality annotated corpora. Some earlier methods employed for error detection in syntactic annotation (mainly POS and chunk), are by [9] and [10]. [18] employed a method similar to [10], to automatically correct errors at the dependency level. Their main idea was to reproduce the gold standard data using MSTParser, MALTParser and MDParser. They use the outputs of any two parsers to detect error nodes, and mark a node as an error if tags predicted by the two parsers differ from the one in the gold data. In case the predicted tags are different, they use the output of the third parser to check if the tag predicted by the third parser matches with any of the tags predicted by the other two parsers. If the tag matches with any of the other two tags, changes are made in the gold data. Using large corpora, [17] and [13] employed error mining techniques. Other efforts towards detection of annotation errors in treebanks include [12] and [14]. Most of the aforementioned techniques work well with large corpora where the word-type frequencies are high. Thus, none of them account for data sparse conditions except for [13]. Moreover, the techniques employed by [17], [13] and [18] rely on the output of a reliable state-of-the-art parser.

Work by [3], [5] and [2] detects errors in a Hindi dependency treebank. They used a combination of both rule-based and hybrid systems to detect annotation errors. A rule-based system contributes towards increasing the precision of the system; it uses robust rules formed using annotation guidelines and the CPG framework, whereas a hybrid system is a combination of a statistical module with rule-based post-processing. The statistical module detects potential errors from the treebank and the rule-based post-processing module prunes out false positives, with the help of robust and efficient rules to increase the precision of the overall system.

[2] improved over the PBSM (Probability Based Statistical Module) used by [5]. They extended PBSM used by [5] by adding more linguistically motivated features like dependency labels of the parent, grandparent and count of its children. They also improved the hybrid system by placing a new module 'Rule based correction' before the statistical module.

Previous works show that parser outputs may help in detecting errors in Treebank [17,13] and sometimes, automatic error correction as well [18]. Hence, in our approach, we use MaltParser[2] [16] to detect potential errors in Hindi Dependency Treebank. Baseline accuracy with MaltParser for Hindi are 77.58%, 88.97% and 80.48% for LAS[3], UAS[4] and LA[5] respectively [4]. Using the parser output alone does not help much because the state-of-the-art parsers for free word order languages like Hindi perform lower than those for fixed word order language like English. So we also use an algorithm which uses the output produced by MaltParser to detect potential errors in the treebank.

3   Our Approach

In this section we describe our approach to error detection using a dependency parser. We detect errors at dependency level annotation. We use inter-chunk dependency trees rather than expanded trees, i.e., dependency relations are marked only among chunk heads in a sentence and dependency relations within a chunk[6] are not considered.

3.1   Using Parser Output to Detect Potential Errors

In this strategy, we use the output of the parser to detect potential errors in the annotated data. We make an assumption that most of the decisions taken by the parser and the humans are similar, i.e., similar types of errors are made by both. Both, the parser and the annotators find it difficult to parse certain type of structures and both are good at other constructions.

The question may arise why we are using a parser to detect errors made by humans when human performance is better, as compared to a parser's in all respects? The answer is simple. Decisions taken by human beings may sometimes vary but a statistical parser (e.g. MaltParser) gives consistent decisions. Even if the parser gives a wrong output, it will always give the same wrong result for a specific case/instance.

Reasons of human error can be:

---

[2] MaltParser (version 1.4.1).

[3] LAS - Labelled Attachement score (Percentage of words that are assigned both correct head and correct label).

[4] UAS - Unlabelled Attachment Score (Percentage of words that are assigned correct head).

[5] LA - Labelled Accuracy (Percentage of words that are assigned correct label).

[6] A typical chunk consists of a single content word surrounded by a constellation of function words [1]. Chunks are normally taken to be a 'correlated group of words'.

1. *Random errors*: Decisions of annotators may change according to their state of mind. They may make random errors if they are not able to concentrate on their work. Though they know the correct decision, they may unknowingly make some errors becuase of some psychological conditions. Everything being right, an annotator can also by mistake make wrong decisions.

2. *Errors due to unawareness of a concept/Misinterpretation of guidelines decision*: Annotators may make the same mistake again and again if they do not know the concept properly or do not know the correct decision to be taken. Misinterpretation of certain guidelines decision may also cause consistent errors.

3. *Errors in the guidelines*: Some sections of the guidelines are always evolving. Some other sections might have errors due to oversight. Annotators will make consistent mistakes in such a case unknowingly.

The types of arcs for which the parser (MaltParser) consistently gives wrong decisions are[7]:

1. *Long distance arcs*: MaltParser is greedy in marking dependencies and searches for the head of a word in the vicinity of the word. It adds a node as the parent of another node as soon as it finds a suitable candidate for it. Hence, it generally marks long distance dependencies also as short distance dependencies.

2. *Non-projective arcs*: The algorithm used in MaltParser always produces projective graphs and cannot handle non-projective dependencies. [15]

The idea here is to use consistent decisions of the parser to mark the human errors. If the parser performs well in some constructions, it will, most of the time, give a correct decision and we can compare annotator's decision with that of the parser and if the decisions made by them arc diffcrent, we can infer that the decision taken by the annotator may be incorrect (as the parser's decision was right and annotator's decision does not match). Even if the parser takes wrong decisions in other types of constructions, it will consistently mark them wrong. If we know these cases, we can again compare such decisons with annotator's decisions and if they match, we can say that decision taken by the annotator is incorrect (as the parser was wrong and annotator's decision is similar to the parser's). Here, we are only marking the potential errors, so even if we mark a correct decision as an incorrect one, it will not create much problem as it can be checked further during the process of validation. But we must not leave any error unmarked because this would lead to an erroneous treebank.

Our classification of parser errors is taken from [11]. Here, parser errors were classified based on:

---

[7] These errors are becuase of the algorithm used (Arc-eager [15]) for parsing. We have used this algorithm because it outperformed other algorithms for hindi [4].

1. *Edge[8] Type and Non-Projective Edge*: The edges were categorized based on the distinct dependency labels on the edges.
2. *Edge Depth*: Depth of a tree is the total number of levels of links that it has. Consequently, the depth of an edge is basically the level at which it exists in the tree.

Examples of edge types from [11] are:

```
1. Main
2. Intra Clausal
     a) Verb Argument Structure
          i)  Complement
          ii) Adjunct
     b) Non-verbal
          i)   Noun-modifier
          ii)  Adjective-modifier
          iii) Apposition
          iv)  Genitive
     c) Others
          i)   Co-ordination
          ii)  Complex Predicate
          iii) Others
3. Inter Clausal
     a) Co-ordination
     b) Sub-ordination
          i)   Conjunction
          ii)  Relative Clause
          iii) Clausal Complement
          iv)  Apposition
          v)   Verb Modifier
```

First of all, Maltparser was trained using manually validated training data and development data was parsed using the trained model. Then using the above mentioned classification, the data was classfied according to different categories (edge type and edge depth). After that, the parsed output was analyzed by comparing it with gold data to find out the classes of arcs for which the parser works well and for which it doesn't. Using this analysis, a knowledge base was prepared, which contains the information about the arcs being correclty or incorrelty parsed for various edge type and edge depth.

With the help of this knowledge base, confidence level for different categories of arcs (different edge-types and different edge-depths) was set as 1, if the parser was confident on giving correct decisions for that type of arcs and 0 if it was less confident of giving correct decisions for that type of arcs. For example:

---

[8] An edge in a dependency tree is the arc that relates two nodes (which are basically words). These are binary asymmetric relations with labels that specify the relation type.

- The parser gives wrong attachments and labels for most of the inter-clausal arcs. They are parsed correctly only if they are at depth 1 or 2. e.g. 'Inter Clausal Relative Clause' is only parsed correctly if it occurs at depth 1, so confidence level for this class of arcs is set to 1 if it occurrs at depth 1 and is set to 0 if the arc occurrs at any other depth.
- Most of the Intra Clausal arcs are parsed correctly but only if the sentence is not very long. Intra Clausal arcs in general occur at the leaves of a dependency tree. For Intra Clausal arcs, confidence level is 1 for short depths and 0 when depth is high. e.g. Confidence level is 1 for 'Intra Clausal, Verb Argument Structure (Complement)' upto depth 10 and confidence level is 0 when this class of arcs occur at a depth greater than 10.

For boundary conditions, confidence level was set using the results obtained from development data. e.g. for arcs of category 'Intra Clausal, Verb Argument Structure (Complement)', confidence level was set 1 for depth 9, 1 for depth 10, 0 for depth 11 and 0 for depth 12 because confidence level 1 for depth 9 gave better results for development data than confidence level 0. Confidence level 0 for depth 11 was better than 1 and so on. In this manner, database for confidence level (knowledge base) for each category of arcs was prepared.

After preparing the confidence level database (knowledge base) using the development data, testing data was also parsed using the trained model and was flagged with potential errors with the help of confidence level mentioned in the knowledge base. Testing was performed as follows:

For each arc, its category was checked. For that category, confidence level was taken from the knowledge base. If confidence level was 1 and parser output was different from the annotated output, the arc was marked to be a potential error. Again, if confidence level is 0 and parser output matches annotated output, the arc was marked as a potential error. This can be explained in detail as follows:

1. We extract cases for which the parser is highly confident (confidence level 1) on giving correct labels and attachments.
   Here, the parser gives correct decisions and if parser output doesn't match the annotated one, there is a possibility that the anntator annotated it wrong. So, we mark that node as a potential error node.
   For example, let there be a category of arcs containing the labels 'k1', 'k2', 'k4', etc. and the parser is confident enough on giving correct attachment and label for this class of arcs. Then, if an arc x → y is labeled as 'k1' by parser and the annotators marked it 'k2', it is flagged as a potential error.
2. Next, we extract cases for which the parser is less confident (confidence level 0) on giving correct labels.
   In this, the decision taken by the parser might be incorrect and if the parser output matches the annotated output, there is a possiblity that the annotator also made the same mistake. Hence, we flag that node as a potential error node.
   For example, again let there be a category of arcs containing the labels 'k1', 'k2', 'k4', etc. and the parser is less confident here, of giving correct attachment and label for this class of arcs. Then, if an arc x → y is labeled

as 'k1' by the parser and the annotators also mark it 'k1', it is flagged as a potential error.

After this precision and recall are calculated which are shown in Table 1.

Table 1. Results of validation using parser output

Data Precision/Recall	Attachment	Labeling
Development	56.72/79.72	60.38/82.41
Test	44.56/78.44	63.45/89.71

4   Data Used in Our Approach

The data used for the experiments is part of a larger Hindi Dependency Treebank. The size of training, development and testing data is 47k, 5k and 7k respectively. The training data used was manually validated and hence we assume it to be free of any errors. Hence the patterns learnt by the dependency parser will generally be consistent, thereby improving parser accuracy.

5   Comparison with other Techniques

Here we compare the results of our strategy with one of the earlier tools [2]. They detected errors in Hindi Dependency Treebank using a hybrid approach, as described earlier in Section 2. As showin in Table 3, our approach significantly outperforms [2] both in terms of precision and recall. Also, a detailed comparison between their approach and our approach is shown in Table 2. There is an overalap of 73.46% of the total errors between our approach and [2].

As is visible from Table 2, the earlier approach [2] flagged a large number of nodes as error nodes (1724 nodes were flagged as error nodes while actual errors were 490) out of which 395 errors were correct. So, the precision of this approach was little low. Our approach flagged 672 errors, out of which 434 were correct. This leads to high precision and recall. Correct errors common to both the approaches were 360. Further,both these approaches when used together were able to detect a total of 483 out of 490 errors. It might be a good idea to consider errors flagged by both approaches to cover almost all the errors in the Treebank. Also, [2] tried to capture only labelling errors while we capture attachment errors too, which are difficult to capture as compared to labelling errors.

Our approach outperformed [2] though it was a hybrid approach, because they mostly concentrated on capturing as much errors as possible while we have tried to maintain balance between both precision and recall. However, it is very likely that the approach by [2] might not have performed well because of lack of a good size of training data and their rule based approach to increase precision could not cover all concepts of the language.

Table 2. Comparison of present approach with [2]

Total Original Errors	490
Total Errors Flagged by [2]	1724
Total Errors Flagged by our approach	672
Total Common Errors Flagged by our approach and [2]	522
[2] Errors Correct	395
our approach Errors Correct	434
Overlap between [2] and our approach	360
Total(Both methods) Flagged Errors	2959
Total(Both methods) Correct Errors	483

Table 3. Comparison of overall (attachment+labeling) results of our approach with previous one [2]

Approach	Precision(%)	Recall(%)
Our approach	**64.58**	**88.57**
[2]	23.24	80.61

6   Discussion

In this paper, we presented an automatic approach to detect potential errors in an annotated treebank to provide aid in its validation task. Precision and recall of the approach are *64.58% and 88.57%* respectively. Unlike the previous approach reported in [2], our approach is language independent. Given a manually validated treebank, all one needs is to prepare a knowledge-base using parser error patterns. We expect, it will be relatively easy to adopt our approach for similar dependency treebanks.

Certain types of errors cannot be detected by our method. For categories where the parser is less confident on taking correct decision, a node was flagged as a potential error only when the annotators decision matched the parsers decision. There are chances that for a class of arcs, both, the parser and the annotators take different decisions but both the decisions are incorrect. In such a case, since the annotator's output doesn't match the parser's output, we do not flag it as an error even though it is an erroneous node.

The power of our validation approach is bounded by the performance of the parser used. If we have better parsers, we can flag potential errors with more confidence and accuracy. Also, to decide the confidence level for various classes of arcs, recall was used as a metric but there was a trade-off between precision and recall. Hence, if we use F-score as a metric in place of recall, we may achieve better results.

Acknowledgement. We would like to thank Himani Chaudhry and Riyaz Ahmad Bhat for their valuable comments which helped us to improve the quality of the paper.

The work reported in this paper is supported by the NSF grant (Award Number: CNS 0751202; CFDA Number: 47.070). Any opinions, findings, and conclusions or recommendations expressed in this material are those of the author(s) and do not necessarily reflect the views of the National Science Foundation.

References

1. Abney, S.: Parsing by Chunks. Principle-Based Parsing 44, 257–278 (1991)
2. Agarwal, R., Ambati, B., Sharma, D.: A Hybrid Approach to Error Detection in a Treebank and its Impact on Manual Validation Time. Linguistic Issues in Language Technology 7(1) (2012)
3. Ambati, B.R., Gupta, M., Husain, S., Sharma, D.M.: A High Recall Error Identification Tool for Hindi Treebank Validation. In: Proceedings of the Seventh International Conference on Language Resources and Evaluation (LREC 2010). European Language Resources Association (ELRA), Valletta (2010)
4. Ambati, B.R., Husain, S., Nivre, J., Sangal, R.: On the Role of Morphosyntactic Features in Hindi Dependency Parsing. In: Proceedings of the NAACL HLT 2010 First Workshop on Statistical Parsing of Morphologically-Rich Languages, SPMRL 2010, pp. 94–102. Association for Computational Linguistics, Stroudsburg (2010)
5. Ambati, B., Agarwal, R., Gupta, M., Husain, S., Sharma, D.: Error Detection for Treebank Validation. In: The 9th International Workshop on Asian Language Resources (ALR), Chiang Mai, Thailand (2011)
6. Begum, R., Husain, S., Dhwaj, A., Misra, D., Bai, L., Sangal, R.: Dependency Annotation Scheme for Indian Languages (2008)
7. Bharati, A., Chaitanya, V., Sangal, R., Ramakrishnamacharyulu, K.: Natural Language Processing: A Paninian Perspective. Prentice-Hall of India (1995)
8. Bhatt, R., Narasimhan, B., Palmer, M., Rambow, O., Sharma, D.M., Xia, F.: A Multi-Representational and Multi-Layered Treebank for Hindi/Urdu. In: Proceedings of the Third Linguistic Annotation Workshop, ACL-IJCNLP 2009, pp. 186–189. Association for Computational Linguistics, Stroudsburg (2009)
9. Eskin, E.: Automatic Corpus Correction with Anomaly Detection. In: North American Chapter of the Association for Computational Linguistics (2000)
10. van Halteren, H.: The Detection of Inconsistency in Manually Tagged Text. In: Proceedings of LINC 2000, Luxembourg (2000)
11. Husain, S., Agrawal, B.: Analyzing Parser Errors to Improve Parsing Accuracy and to Inform Treebanking Decisions. Linguistic Issues in Language Technology 7(1) (2012)
12. Kaljurand, K.: Checking Treebank Consistency to Find Annotation Errors (2004)
13. de Kok, D., Ma, J., van Noord, G.: A Generalized Method for Iterative Error Mining in Parsing Results. In: Proceedings of the 2009 Workshop on Grammar Engineering Across Frameworks, GEAF 2009, pp. 71–79. Association for Computational Linguistics, Stroudsburg (2009)
14. Kordoni, V.: Strategies for Annotation of Large Corpora of Multilingual Spontaneous Speech Data. In: The workshop on Multilingual Corpora: Linguistic Requirements and Technical Perspectives held at Corpus Linguistics, Citeseer (2003)
15. Nivre, J.: Incrementality in Deterministic Dependency Parsing. In: Proceedings of the Workshop on Incremental Parsing: Bringing Engineering and Cognition Together, pp. 50–57. Association for Computational Linguistics (2004)

16. Nivre, J., Hall, J.: Maltparser: A Language-Independent System for Data-Driven Dependency Parsing. In: Proc. of the Fourth Workshop on Treebanks and Linguistic Theories, pp. 13–95 (2005)
17. van Noord, G.: Error Mining for Wide-Coverage Grammar Engineering. In: Proceedings of the 42nd Annual Meeting on Association for Computational Linguistics, ACL 2004. Association for Computational Linguistics, Stroudsburg (2004)
18. Volokh, A., Neumann, G.: Automatic Detection and Correction of Errors in Dependency Tree-Banks. In: Proceedings of the 49th Annual Meeting of the Association for Computational Linguistics: Human Language Technologies: Short Papers - HLT 2011, vol. 2, pp. 346–350. Association for Computational Linguistics, Stroudsburg (2011)
19. Xia, F., Rambow, O., Bhatt, R., Palmer, M., Sharma, D.: Towards a Multi-Representational Treebank. In: The 7th International Workshop on Treebanks and Linguistic Theories, Groningen, Netherlands. Citeseer (2009)

Topic-Oriented Words as Features
for Named Entity Recognition

Ziqi Zhang, Trevor Cohn, and Fabio Ciravegna

211 Portobello, Regent Court, Department of Computer Science,
University of Sheffield, Sheffield, UK, S1 4DP
Initial.Lastname@domain.com

Abstract. Research has shown that topic-oriented words are often related to
named entities and can be used for Named Entity Recognition. Many have proposed to measure topicality of words in terms of 'informativeness' based on
global distributional characteristics of words in a corpus. However, this study
shows that there can be large discrepancy between informativeness and topicality; empirically, informativeness based features can damage learning accuracy
of NER. This paper proposes to measure words' topicality based on local distributional features specific to individual documents, and proposes methods to
transform topicality into gazetteer-like features for NER by binning. Evaluated
using five datasets from three domains, the methods have shown consistent improvement over a baseline by between 0.9 and 4.0 in F-measure, and always
outperformed methods that use informativeness measures.

1   Introduction

Named Entity Recognition (NER) identifies, and classifies text elements in a document into pre-defined categories. An important type of resource for NER is gazetteer,
which contains pre-defined candidate NEs and is often crucial for specialized domains
due to the complexity of terms [19, 24]. However, gazetteers are expensive to build.
Research has shown that NEs are often relevant to the topic of a document [5]. They
are among the most 'information dense' tokens of the text and largely define the domain of interest [12]. Thus NER can benefit from the identification of topic-oriented
words; and it has been proposed that topicality can be measured in terms of 'informativeness' [18, 10]. It is generally agreed that informative words demonstrate a
'peaked' frequency distribution over a collection of documents such that the majority
of their occurrences are found in a handful of documents [4]. In practice, informativeness measures [13, 11, 3, 18] largely depend on two kinds of word distributional
properties: document frequency and overall frequency in the corpus.

Informativeness may not always represent topicality for two reasons. First, document topics can vary largely in a collection. This is reflected by varying vocabularies
and frequency patterns of words at individual document basis. Informativeness however, studies the global distribution of words and ignores their local context. We
show that many informativeness measures can promote words that are irrelevant to a

A. Gelbukh (Ed.): CICLing 2013, Part I, LNCS 7816, pp. 304–316, 2013.

document's topic, or neglect the relevant. Second, the scores are globally uniform and specific to a collection. Therefore, ambiguous words that carry different senses in different document contexts can be mis-interpreted. For example, a fair amount of biomedical named entities can contain common English words such as 'bright' and 'cycle', which can be used to refer to protein names in some documents but also widely used as common words carrying no special senses in other documents [16]. A uniform score cannot distinguish these cases.

This paper proposes to measure topicality of words in terms of a word's relevance to a specific document – a widely adopted notion in Information Retrieval [15]. Our method begins with evaluating topicality of words using several simple relevance functions. We then hypothesize a non-linear distribution of topic-oriented words in named entities: only a few words can be highly topic-oriented in a document, as topics are often composed of a small set of keywords; however, they may be found in a large proportion of named entity mentions, since named entities that are highly relevant to the topic of a document are also likely to be repeated. To capture this nature, we use a simple equal interval binning technique to split the list of words ranked by their topicality and use their bin membership as features for statistical NER. The idea is that highly topic-oriented words will be grouped into top-ranked bins, which can contain the most indicative words of named entities. Essentially, binning creates binary features similar to gazetteers.

This approach is evaluated using five datasets from three domains, and compared against two methods based on informativeness. It consistently improves the baseline by between 0.9 and 4.0 of F-measure on all datasets, while methods that use informativeness based features have shown unstable performance. The remainder of this paper is organised as the follows. Section 2 discusses related work. Section 3 introduces the method. Section 4 presents the experiments and results. Section 5 analyses and discusses the results. Section 6 concludes this paper.

2   Related Work

NER is a field that has seen decades of research since its formal introduction in MUC6 [9]. It is often divided into two sub-tasks: NE detection or identification that identifies the boundaries of NEs, and NE classification that assigns NEs to predefined semantic types. Over the years, the mainstream techniques for NER have shifted from rule-based to statistical machine learning based approaches [17]. Typically, NEs are recognised based on the linguistic features of their component and contextual words.

There are few studies that explore the distributional characteristics of NEs. Most of these address NE detection but not classification. Silva et al. [20] argued that NEs are often Multi-Word Units (MWUs) with strong 'cohesion', i.e., the composing word units are found more often together than individually. A similar method was later introduced by Downey et al. [8] and extended to NER from the Web.

The most relevant work to this paper includes Rennie and Jaakkola [18] and Gupta and Bhattacharyya [10]. Rennie and Jaakkola used a number of informativeness measures (e.g., *idf*, *ridf*, *z*-measure) to compute a score for each unique word in the corpus based on its distributional statistics. The scores are then normalised against the mean or median score to obtain a relative score, which is then used as features in a statistical learning model for NE detection. This method proved useful in detecting restaurant names from forum posts. Gupta and Bhattacharyya proposed to create gazetteers dynamically at both training and testing phase using word informativeness measures. The core of the process is creating a lexicon by selecting the most informative words – evaluated by an informativeness measure and filtered by an arbitrary threshold – in the corpus. The list is then pruned by adding more words that are distributionally similar to highly informative words on the list and discarding noisy words based on heuristics (e.g., stopwords). The final list of words is used as gazetteers. The method was shown to be effective in NE detection for Hindi texts.

Zhang et al. [23] and Wan et al. [22] studied methods for finding the most important NEs from the output of Chinese NER tasks. They showed that the distributional statistics of NEs are strong features for this purpose. These are essentially in line with the informativeness hypothesis; however, they do not deal with NER but a post-processing task.

All these studies have only presented a partial view of the usefulness of word topicality or informativeness in NER. On the one hand, they address only a partial phase of NER – either NE detection or a post-processing task on the NER output. On the other hand, they have shown evaluations for different languages, single domains, and mostly self-created datasets that are unavailable for comparative studies. It is unclear whether the lessons can be generalised to support NER in general. In addition, this work is also related to Zhang et al. [25] who also studied NE classification using local distributional features of words. The work is extended in two ways: 1) we explore new functions for measuring topicality of words; 2) we carry out comprehensive analyses and comparative studies with state-of-the-art. This allows us to develop an explanation of the empirical findings, leading to a number of reusable conclusions.

3   Method

Given a collection of documents, we define topicality of a word – denoted by *topcat* – as a function that returns a real number for a word w found in a document d:

$$topcat(w,d) \in \mathbb{R}, \tag{1}$$

where $d \in D$ is a document in the whole collection, and (w, d) represents a word w from d. Topicality scores are used to create features for statistical machine learning.

3.1   Topcat Functions

Essentially, evaluating topicality of words in a document is similar to the task of assessing word-document relevance, a problem addressed by the IR community. Many IR-oriented word-document weighting functions can be used for this purpose. The simplest method among these is **Term Frequency**, denoted by $tf(w, d)$, which is simply the count of occurrences of w in d. Intuitively, words associated with the topics of a document are likely to be repeated and therefore have higher frequency since they are the focus of the content. Although this simple technique may be effective [7], it is also well-known for its inability to identify less frequent yet equally important words, such as those that are unique to a smaller set of documents in the whole collection. Thus the second measure is **Term Frequency - Inverse Document Frequency** which is the classic measure used in IR for evaluating the word-document relevance:

$$tfidf(w, d) = tf_{norm}(w, d)\, idf(w) \, , \tag{2}$$

where $tf_{norm}(w, d)$ is $tf(w, d)$ normalised against the total number of words in d, and $idf(w)$ is the inverse document frequency of w in the entire collection D. The intuition is that words associated with the topic of a document should be frequently used and also unique to that document. The latter can be measured by inverse document frequency – a word found only in a handful of documents is likely to be specific to the documents and bear specific meanings.

The third *topcat* function, called **Weirdness** (*wd*, [1]), compares a word's frequency in the target document against its frequency in a reference collection:

$$wd(w) = \frac{tf_{norm}(w,d)}{tf_{norm}(w,C)} \, , \tag{3}$$

where $tf_{norm}(w, C)$ is the normalised frequency of w in a reference corpus (C), indicating the probability of finding w in a different context. The idea is that words that are more likely to be found in a document than a reference collection are 'unique' to that document and more topic-oriented.

Additionally, we also introduce a function that combines both *tfidf* with *wd* to balance different views of 'uniqueness'. Given W_d the set of words found in d, we rank W_d by their *wd* and *tfidf* scores to obtain two ordered lists R_d^{wd} and R_d^{tfidf}. The final score for a word w, called **Combined Inverse Rank** (*cir*), combines both rankings:

$$cir(w,d) = \frac{1}{R_d^{wd}(w,d)R_d^{tfidf}(w,d)} \tag{4}$$

This method promotes words highly ranked by both *topcat* functions rather than any single function. For all functions, we exclude stopwords since practically they damage the learning accuracy.

3.2   Binning

The values returned by the *topcat* measures are real numbers indicating how topic-oriented a word is to a document. They are document-specific and unbounded in range, and therefore, non-generalisable for a statistical learning model. Furthermore, we hypothesize that highly topic-oriented words are rare but repeated in a large proportion of named entity mentions in a document. As the scores drop, their usefulness (i.e., being indicative of named entities) drops disproportionally faster. This implies a non-linear and long-tail distribution where a handful of highly topic-oriented words are most useful indicators of named entities. To capture this nature and also to normalise the scores to a usable range, we split the ordered list of words from a document into k intervals containing equal number of items, such that the most highly scored items are likely to be included in top ranked bines (e.g., $k=1$) and those with the lowest scores to be included in bin k. The bin membership of words is then used as a feature for NER. This is equivalent to creating k untyped, equal sized, document specific gazetteers so that the correspondence between the gazetteer membership (labelled $1~k$) and the classification decision can be automatically learned from training data.

4   Experiment and Evaluation

To thoroughly evaluate the usefulness of topic-oriented words for NER, we selected five datasets from three domains, and compared against state-of-the-art under a uniform framework.

4.1   Datasets and Domains

Five corpora are selected from the newswire, bio-medical and archaeology domains. The CoNLL03 [21] dataset is used for the news-wire domain. It contains 70% of the original English dataset used by the CoNLL2003 shared NER task. We have discarded documents that contain only tables or lists of entities, such as league tables or phone directories. This is because they do not carry a focused topic and are not suitable testbed for the hypothesis. The corpus contains articles with about 22k annotations of person names (Per, 28%), locations (Loc, 32%), organisations (Org, 21%) and miscellaneous (Misc, 19%) such as events, languages and nationalities. For the bio-medical domain, the Genia BioNLP04 (referred to as Genia) dataset is obtained from the Bio-Entity Recognition Task at BioNLP/NLPBA 2004 [14]. It contains 2,400 MEDLINE abstracts with about 56.5k annotations of protein (60%), cell type (15%), cell line (7%), DNA (16%) and RNA (2%) entities. The Bio1[1] [6] dataset contains 100 MEDLINE abstracts with about 3.2k annotations   of protein names (63%), DNA (11%), and SOURCE (26%), which include 7 sub-types such as 'cell type' and 'cell line' etc. In this study these are treated as a single type. The Yapex[2] dataset contains 200 MEDLINE abstracts with about 3.7k annotations of proteins.

While these datasets contain documents of the similar size, the Archaeo corpus consists of 38 long articles averaging about 7,000 words. These are obtained as part of the data used in [24], and contains about 17.5k annotations of three types: archaeological temporal terms (Tem, 23%), such as 'Bronze Age', and '1089AD'; location (Loc, 14%), which is UK-specific and often refers to place names of findings and events; subject of interest (Sub, 63%), which is a highly heterogeneous category

---

[1]   We ignored 'RNA' as there are less than 40 instances only
[2]   http://www.sics.se/humle/projects/prothalt

containing terms from various domains, such as architecture, warfare, and education. Each dataset is split into five equal parts for five-fold cross-validation.

4.2 Baseline

We firstly created a baseline (denoted by B) statistical learning system using a set of basic features. The topicality features generated by different methods are then added to this baseline for comparison. We used an SVM-based NER classifier, built on the SVM-light package[3] using a linear kernel with a training error and margin trade-off of 0.075. The baseline features tested include:

- The exact token string, its stem, and lemma
- The orthographic type of the token (e.g., alpha-numeric, digits only, capitalized)
- Above features applied to the previous and the following three tokens of the current token

We did not use domain or language specific features or optimise the model for each dataset because the goal of this work is to compare the effectiveness of different methods and their genericity across domains, rather than finding the highest possible learning accuracy.

4.3 Candidate Methods for Comparison

For each of the *topcat* functions introduced before, we tested $k = 10, 15, 20$. The British National Corpus (BNC) is used as the reference corpus (C) for the *wd* function.

Next, we also compare our methods against two studies that are most relevant: Rennie and Jaakkola [18] (RJ05), and Gupta and Bhattacharyya [10] (GB10). However, both originally addressed NE detection, not NER. In RJ05, several different informativeness measures are tested. Two of the most effective were found to be *idf* and *ridf* (Residual IDF [3]). It is known that these measures are biased towards document frequency. Therefore, we also added a third measure 'burstness' [4] that calculates informativeness by dividing the overall frequency of a word in a collection by its document frequency. Four settings are created:

- B, normalised (by median) *idf* (idf_{med})
- B, normalised (by mean) *ridf* ($ridf_{avg}$)
- B, normalised (by median) *burstness* (bur_{med})
- B, normalised (by mean) *burstness* (bur_{avg})

Idf scores are normalised against the median of *idf* scores; *ridf* scores are normalised against the mean of all *ridf* scores. These are chosen based on the best-performing settings reported by Rennie and Jaakkola. For the *burstness* measure, both normalisation methods are tested.

---

[3] http://svmlight.joachims.org/

The method introduced in GB10 employs fairly complex processing involving computing distributional similarity, clustering, and filtering by language and domain specific heuristics. However, the core lesson is automatically creating lexicon as gazetteers based on informativeness scores, for which they used the *burstness* measure. To compare this against *topcat* based methods, the informativeness score is calculated for each unique word extracted from a corpus using *burstness*. Then words whose scores exceed an arbitrary threshold are selected for gazetteers. Since it is impractical to manually inspect the lists and select a threshold for each dataset and each cross-fold experiment, the top *n%* of the entire list (each from the training and testing parts separately) is chosen. Empirically *n* is set to *20*. Further, for a more thorough evaluation of this approach, four other functions are tested as replacement of *burstness*. We adapt each *topcat* function to calculate a global score for each unique word in the corpus based on their statistics in the entire collection. Specifically, *tf* uses word frequency in a corpus; *tfidf* and *wd* use normalised word frequency in the entire corpus; and *cir* combines the adapted *tfidf* and *wd*. Thus for GB10 five settings are created, each using a gazetteer generated based on the *burstness* or an adapted *topcat* function.

We also excluded stopwords when using the RJ05 and GB10 methods for feature generation. Practically, this has led to better accuracy.

4.4   Results

Firstly, we ran experiments with the baseline for all five datasets, and obtained comparable results to similar implementations of state-of-the-art (see [2] [6] [14] [21] [24] for reference). Next, we apply the *topcat* based features and the RJ05 and GB10 based methods with the baseline and show the changes in Table 1 and 2 respectively.

5   Discussion and Analysis

5.1   Overview

Firstly, regardless of the choices of *topcat* measures or the number of bins *k*, this type of features has consistently improved the baseline. We note that the difference in accuracy achieved under different *k* for each method is rather trivial (0.1 ~ 0.2) in most cases, except *wd* and *cir* on the Bio1 and Yapex datasets where a maximum difference of 0.4 and 0.6 are noted. As a result, optimising the value of *k* may be insignificant. The maximum improvements for each dataset are: 1.8 for Archaeo; 1.7 for Bio1; 0.9 for CoNLL03; 1.5 for Genia; and 4.0 for Yapex. It is particularly effective for specialised domains, in which NER is harder. It is known that domain specific NER largely relies on domain specific resources such as gazetteers to boost learning accuracy. However, such resources are often unavailable in technical domains [24], expensive to build [19], and non-generalisable. Taking into account the diversity in the datasets used for this study and the consistent improvements obtained, we believe that *topcat* based features are highly generalisable and reusable across domains and can also be very effective.

In terms of individual *topcat* functions, *tfidf* and *wd* generally achieved better results than the simple *tf* function, which suggests that in addition to word frequency, the 'uniqueness' of words to the document is also an important factor in assessing topicality. Also *wd* appears to marginally outperform *tfidf* on some datasets, while combining both have led to small improvement on the Bio1 and Yapex datasets.

Table 1. Results (in terms of changes to baseline) of topcat based methods

	Archaeo			CoNLL03			Genia			Bio1			Yapex		
B	69.1			78.5			62.8			65.6			55.5		
(k)	*10*	*15*	*20*	*10*	*15*	*20*	*10*	*15*	*20*	*10*	*15*	*20*	*10*	*15*	*20*
tf	1.3	1.4	1.5	0.7	0.7	0.7	1.3	1.3	1.3	1.1	1.1	1.0	2.6	2.8	2.7
tfidf	1.6	1.6	1.6	0.7	0.8	0.8	1.3	1.3	1.3	1.0	1.1	1.1	2.9	3.2	3.2
wd	**1.8**	1.7	1.6	**0.9**	0.8	**0.9**	**1.5**	**1.5**	1.4	1.2	1.5	1.1	2.9	3.1	3.2
cir	1.7	1.7	1.6	0.8	**0.9**	**0.9**	1.4	1.4	1.4	1.5	**1.7**	1.3	3.4	3.8	**4.0**

Table 2. Results of RJ05 and GB10 based methods (in terms of changes to baseline)

	Archaeo	CoNLL03	Genia	Bio1	Yapex
			RJ05		
Idf	-4.5	-9.6	-4.8	-8.4	-19.7
Ridf	-6.6	-5.8	-4.7	-9.1	-11.8
bur_{med}	-11.6	-6.6	-5.4	-7.7	-11.6
bur_{avg}	-14.2	-6.6	-5.1	-8.3	-11.6
			GB10		
bur	0.2	0.2	-0.3	1.2	2.6
tf	-0.1	-0.5	-0.4	0.2	0.8
tfidf	0.1	-0.5	-0.4	0.7	1.2
wd	0.3	0.6	-0.3	0.9	1.2
cir	0.5	0.0	-0.3	0.3	1.8

In comparison, we have obtained contradictory results with RJ05 based methods, and the GB10 based methods are also found to be less effective than topcat based methods. For RJ05, the methods originally succeeded in detecting a single type of entity – restaurant names – in forum posts, where each thread often discusses a restaurant and is treated as a separate document. It is possible that the nature of the data (e.g., short sentences, lack of context, narrowly defined distinctive topics and focused discussion) has made informativeness scores such as *idf* a useful feature for NEs. However, in our NER experiments that involve more formally presented data, these methods have caused significant drop in accuracy for all datasets. For GB10, in most cases the gazetteer based features have contributed to small improvements over the baseline. Also there are several occasions where these features have failed and caused decreased accuracy. This is likely to be caused by inappropriate thresholds for gazetteer selection. Overall the results are rather inconsistent and empirically deriving suitable thresholds for each dataset and measure can be difficult. Although the GB10 settings are not an identical replication of the original method, we believe they provide a useful reference. Both observations have confirmed that informativeness of words calculated based on their global distributional characteristics is not always a good indicator of NEs and can make spurious features that harm learning accuracy.

5.2   Feature Analyses

In this section, we analyse whether and how topicality based features can contribute to the accuracy of a learning model. For *topcat* functions, binning transforms the topicality score returned by any *topcat* function into an integer i within the range of $[1, k]$. For a comparative analysis, we choose the RJ05 method and define $f_{RJ05}(w)$ as the feature function that normalises the informativeness score (e.g., *idf*, *ridf*) of a word into a finite range of real numbers $[x, y]$, where x and y depend on the data and specific measures.

We designed two types of analyses for each dataset using the positive and negative examples (words that are part of an NE and words that are not). The first addresses the precision perspective by computing the ratio between the number of positive and negative instances captured by each value given by a feature function. Intuitively, the higher the ratio, the more useful the feature can be to boost learning precision. This is denoted by **POSRatio**. The second analysis addresses the recall perspective by computing the fraction of all positive instances in the sample captured by each feature value. Intuitively, the higher the fraction, the more useful the feature can be to boost learning recall. This is denoted by **POSCov**. Additionally, we also study the fractions of the entire sample data represented by each feature value. We would like features that are useful to precision or recall to capture a higher volume of data. This is referred to as volume and denoted by **VOL**.

The findings: We analysed all combinations of datasets and *topcat* functions introduced in this study, as well as the informativeness measures used in RJ05 settings. In general, we found similar patterns across the combinations. Due to the space limit, we show the findings on the Genia and CoNLL03 datasets only, with the following settings: *topcat* function = *tfidf*, $k = 20$; RJ05 informativeness measure = idf_{avg}, bur_{med}.

The Genia and CoNLL03 are the most frequently used datasets for NER, thus a good representation of the problem. For informativeness measures, the numbers of unique features for these corpora are shown in Table 3. The findings are shown in Figure 1.

Table 3. Number of unique features generated on each dataset by RJ05 based methods

Corpus	*idf*	*burstness*
CoNLL03	57	242
Genia	74	363

As shown in Figure 1, both the POSRatio and POSCov curves for the *topcat* based features show a strong non-linear pattern. This suggests that highly topic-oriented words extracted based on document-level distributional features are useful to both the

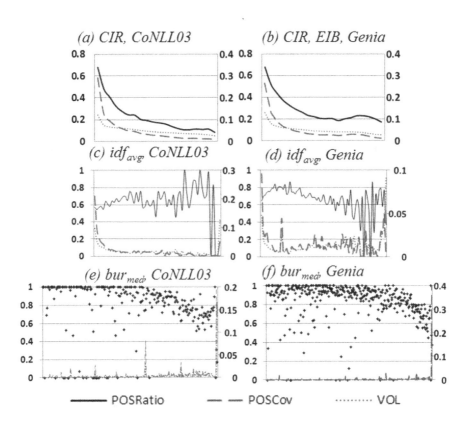

Fig. 1. Analyses of the informativeness-based and topicality-based features. The *x*-axis defines possible feature values. The highly scored features are on the left. The POSRatio curves are aligned against the left *y*-axis, and the VOL and POSCov curves are aligned against the right *y*-axis. For RJ05, due to the large number of features, we replaced curves by points.

recall and precision of the NER learner and as the *topcat* scores drop, their usefulness decreases much more rapidly. This non-linearity relation between topic-oriented words and classification decisions is well captured by the simple binning approach, which transform *topcat* scores to a handful of binary features, effectively collapsing most topic-oriented words into several groups useful for discrimination. The strongest pattern is noticed for the Yapex dataset (see supplementary data), which possibly contributed to the greatest improvement among all datasets. Furthermore, the VOL curves also show a generally consistent pattern of a fairly uniform distribution that tends to become more non-linear at the left end of the curves. Accordingly the most useful bins cover a higher proportion of data, justifying why they are useful features.

In contrast, informativeness features based on the RJ05 methods failed to show consistent patterns. POSRatio values appeared to be random with respect to the level of informativeness and higher values tend to spread towards both ends of the *x*-axis, possibly suggesting that there is no definitive correlation between the informativeness of words and NER learning precision. For the POSCov curves, non-linear patterns are noted on some datasets (Bio1, CoNLL, Yapex), while random patterns are noted on others. In all cases, there is no consistency between the POSRatio and POSCov curves. Furthermore, the VOL curves across different datasets also appear to be more random compared against those obtained with topcat based measures. For both *ridf* and *burstness*, it has been noticed that on some datasets, highly informative words seem to be useful for learning precision since these features obtained POSRatio values of 1.0 (e.g., points that cross the leftmost sections of the topmost gridlines in *e* and *f*, Figure 1), they cover a very small portion of data (less than 1%, as indicated by the corresponding points on the POSCov and VOL curves) and therefore, may be less informative to learning. Overall, this nature of randomness and inconsistency displayed by the informativeness based features make it difficult to generalise for a learner, which may explain the decreased learning accuracy.

Furthermore, we looked for concrete supporting examples for these findings. Using the Genia corpus for example, we compared manually the output of the baseline against those of the baseline with topicality based features, or the baseline with RJ05 features to identify instances labelled differently when such additional features are added. We found that, on a sample of 50 correctly labelled instances due to *topcat* features, 74% contain words that fall in bin 1 and another 17% with words in bin 2, showing a strong signal of the usefulness of the feature. The majority of these are acronyms of mixed causes (e.g., RhoA, YpkA, Rho GTPases) and long terms (e.g., cAMP-dependent protein kinase, Yersinia protein kinase A). On the contrary, for RJ05, the *idf* scores are not discriminative because 57% of words in the corpus received the highest score as they were found in only one document, while only 1% had the lowest score. Many words such as 'fifty' and 'disappointing' are highly ranked but do carry useful information regarding the topics of the document and are not part of any entity names. The globally uniform informativeness score also misled extraction of entity names containing ambiguous words such as '2alpha' – in one case it is part of protein entity names while in other cases, it refers to a type of natural prostaglandin that is irrelevant to the core topic and not as part of named entities.

6   Conclusion

This paper has introduced a method to measure topicality of words within specific document context and exploit topic-oriented words in NER. A comprehensive and comparative evaluation has shown that this method consistently improves the baseline on five datasets from three different domains. It is particularly useful to specialised domains such as archaeology and biomedicine, where NER is much harder and often requires domain specific external resources that can be expensive to build. On the contrary, methods that measure topicality by informativeness have shown unstable performance with often damaged accuracy. This has confirmed that topic-oriented words extracted specific to documents are more effective and generalisable features to NER. Further analyses have shown a highly non-linear relation between topic-oriented words and NER classification decisions, which is captured by the simple equal interval binning technique.

One limitation of this work is that the number of bins k and the range of each bin are arbitrarily defined. Although empirically, the choice of k has led to insignificant difference in learning accuracy in most cases, this has not been evaluated extensively and the question remains as to whether an optimised k and/or ranges of bins can further improve learning. Ideally, bins should be defined dynamically based on data and defines the boundaries of topic-oriented and non-topic words. Our future work will therefore explore alternatives in this direction.

Acknowledgement. Part of this research has been sponsored by the EPSRC funded project LODIE: Linked Open Data for Information Extraction, EP/J019488/1.

References

1. Ahmed, K., Gillam, L., Tostevin, L.: University of Surrey Participation in TREC8: Weirdness Indexing for Logical Document Extrapolation and Retrieval (WILDER). In: The 8th Text Retrieval Conference, TREC-8 (1999)
2. Chang, J., Schütze, H., Altman, R.: GAPSCORE: finding gene and protein names one word at a time. Bioinformatics 20(2), 216–225 (2004)
3. Church, K., Gale, W.: Inverse Document Frequency (IDF): A Measure of Deviation from Poisson. In: Proceedings of the 3rd Workshop on Very Large Corpora, Cambridge, Massachusetts, USA, pp. 121–130 (1995a)
4. Church, K., Gale, W.: Poisson mixtures. Natural Language Engineering 1(2), 163–190 (1995b)
5. Clifton, C., Cooley, R., Rennie, J.: TopCat: Data Mining for Topic Identification in a Text Corpus. In: Proceedings of the 3rd European Conference of Principles and Practice of Knowledge Discovery in Databases, pp. 949–964 (1999)
6. Collier, N., Nobata, C., Tsujii, J.: Extracting the Names of Genes and Gene Products with a Hidden Markov Model. In: Proceedings of COLING 2000, pp. 201–207 (2000)
7. Dagan, I., Church, K.: Termight: Identify-ing and Translating Technical Terminology. In: Proceedings of EACL, pp. 34–40 (1994)
8. Downey, D., Broadhead, M., Etzioni, O.: Locating Complex Named Entities in Web Text. In: Proceedings of the 20th International Joint Conference on Artificial Intelligence (2007)
9. Grishman, R., Sundheim, B.: Message Understanding Conference - 6: A brief history. In: Proceedings of the 16th International Conference on Computational Linguistics (1996)
10. Gupta, S., Bhattacharyya, P.: Think Globally, Apply Locally: Using Distributional Characteristics for Hindi Named Entity Identification. In: Proceedings of the 2010 Named Entities Workshop, ACL 2010, pp. 116–125 (2010)
11. Harter, S.: A probabilistic approach to automatic keyword indexing: Part I. On the distribution of specialty words in a technical literature. Journal of the American Society for Information Science 26(4), 197–206 (1975)
12. Hassel, M.: Exploitation of Named Entities in Automatic Text Summarization for Swedish. In: Proceedings of the 14th Nordic Conference on Computational Linguistics (2003)
13. Jones, K.: Index term weighting. Information Storage and Retrieval 9(11), 619–633 (1973)
14. Kim, J., Ohta, T., Tsuruoka, Y., Tateisi, Y.: Introduction to the Bio-Entity Recognition Task at JNLPBA. In: Proceedings of the Joint Workshop on Natural Language Processing in Biomedicine and its Applications (2004)

15. Mizzaro, S.: Relevance: The Whole History. Journal of the American Society for Information Science 48(9), 810–832 (1997)
16. Morgan, A., Hirschman, L., Yeh, A., Colosimo, M.: Gene Name Extraction Using FlyBase Resources. In: ACL 2003 Workshop on Language Processing in Biomedicine, Sapporo, Japan, pp. 1–8 (2003)
17. Nadeau, D., Sekine, S.: A survey of named entity recognition and classification. Lingvisticae Investigationes 30(1), 3–26 (2007)
18. Rennie, J., Jaakkola, T.: Using Term Informativeness for Named Entity Detection. In: Proceedings of the 28th Annual International ACM SIGIR Conference on Research and Development in Information Retrieval (2005)
19. Saha, S., Sarkar, S., Mitra, P.: Feature selection techniques for maximum entropy based biomedical named entity recognition. Journal of Biomedical Informatics 42(5), 905–911 (2009)
20. Silva, J., Kozareva, Z., Noncheva, V., Lopes, G.: Extracting Named Entities: A Statistical Approach. In: Proceeding of TALN (2004)
21. Tjong, E., Sang, K., Meulder, F.: Introduction to the CoNLL-2003 shared task: language-independent named entity recognition. In: Proceedings of the Seventh Conference on Natural Language Learning at HLT-NAACL 2003, pp. 142–147 (2003)
22. Wan, X., Zhong, L., Huang, X., Ma, T., Jia, H., Wu, Y., Xiao, J.: Named Entity Recognition in Chinese News Comments on the Web. In: Proceedings of the 5th International Joint Conference on Natural Language Processing, pp. 856–864 (2011)
23. Zhang, L., Pan, Y., Zhang, T.: Focused Named Entity Recognition using Machine Learning. In: Proceedings of the 27th Annual International ACM SIGIR Conference on Research and Development in Information Retrieval (2004)
24. Zhang, Z., Iria, J.: A Novel Approach to Automatic Gazetteer Generation using Wikipedia. In: Proceedings of the ACL 2009 Workshop on Collaboratively Constructed Semantic Resources (2009)
25. Zhang, Z., Iria, J., Ciravegna, F.: Improving Domain-specific Entity Recognition with Automatic Term Recognition and Feature Extraction. In: Proceedings of LREC 2010, Malta (May 2010)

Named Entities in Judicial Transcriptions: Extended Conditional Random Fields

Elisabetta Fersini and Enza Messina

University of Milano-Bicocca,
Viale Sarca, 336 - 20126 Milano, Italy
{fersini,messina}@disco.unimib.it

Abstract. The progressive deployment of ICT technologies in the courtroom is leading to the development of integrated multimedia folders where the entire trial contents (documents, audio and video recordings) are available for online consultation via web-based platforms. The current amount of unstructured textual data available into the judicial domain, especially related to hearing transcriptions, highlights therefore the need to automatically extract structured data from the unstructured ones for improving the efficiency of consultation processes. In this paper we address the problem of extracting structured information from the transcriptions generated automatically using an ASR (Automatic Speech Recognition) system, by integrating Conditional Random Fields with available background information. The computational experiments show promising results in structuring ASR outputs, enabling a robust and efficient document consultation.

1 Introduction

Thanks to the recent progresses in judicial proceedings management, a smart consultation of legal materials represent a key challenge [5]. In particular the introduction of audio/video recording facilities in the courtroom management systems, has originated a proliferation of multimedia evidences (audio/video tracks of the hearings) that are potentially available for consultation purposes. Although these evidences represent a valuable knowledge, judicial stakeholders envisage the necessity of effective tools for managing the complexity of such multimedia digital libraries. In this context, two main steps are fundamental for allowing end users to be tuned in a audio/video document: (1) generate a debate transcription that could be easily linked to the media source for looking at given debate portions and (2) structure the debate transcription for allowing an efficient consultation process.

Although automatic speech transcription represents a first information source for a better understanding and consultation of trial recordings, the unstructured text originated is hard to be processed by stakeholders, making therefore prohibitive the consultation phase. One of the possible solutions to this issue is

A. Gelbukh (Ed.): CICLing 2013, Part I, LNCS 7816, pp. 317–328, 2013.

represented by Information Extraction and in particular by Named Entity Recognition (NER), which is aimed at identifying and associating atomic elements in a given text to a predefined category such as names of persons, organizations, locations, dates, quantities and so on.

Several approaches have been presented in the state of the art of NER systems. Early solutions have been defined as rule-based systems with a set of fixed and manually coded rules provided by domain experts. Considering the costs, in terms of human effort, to reveal and formulate hand-crafted rules, several research communities ranging from Statistical Analysis to Natural Language Processing and Machine Learning have provided valuable contributions for automatically deriving models able to detect and categorize pre-defined entities. The first tentatives aimed at deriving these rules are based on inductive rule learner, where rules can be learnt automatically from labelled examples. Alternative approaches are represented by statistical methods, where the NER task is viewed as a decision making process aimed at assigning a sequence of labels to a set of either joint or interdependent variables. This decision making paradigm can be addressed in two different ways: (1) at segment-level, where the NER task is managed as a segmentation problem in which each segment corresponds to an entity label; (2) at token level where an entity label is assigned to each token of a sentence. For a survey see [11].

However, NER in the e-justice domain is still in an early stage. Most of the studies available in the literature relates to crime information extraction tasks, which are targeted on narrative reports coming primarily from police departments [2,8,3,9]. All of these approaches make use of data from police departments, newspaper articles and newswire in order to extract names of persons, locations, and organizations from narrative reports. In this context lexical lookup and statistical methods have been investigated in order to accomplish the IE task. However, all the sources used to accomplish a NER task comprise text that often has an explicit structure and from which typos have been removed. Narrative reports from police departments contain fixed data structures such as type of crime, date, time, and location. Newspaper articles are well organized and contain few typos and grammatical errors. Police narrative reports are often edited by police officers to remove grammatical and spelling errors to enhance readability and to protect peoples privacy. As a consequence, high precision and recall can be achieved. Concerning trial recordings at our knowledge no investigation has been carried out for structuring corresponding automatic speech transcriptions. In this paper the problem of identifying named entities in automatically generated speech transcriptions has been addressed by empowering one of the most promising probabilistic approaches, i.e. Conditional Random Fields (CRFs), by introducing available domain knowledge. The outline of the paper is the following. In section 2 the problem of identifying named-entities in judicial speech transcription is described. In section 3 the proposed extension of CRFs is detailed. In section 4 experimental results are reported and finally in section 5 conclusions are derived.

2   Structuring Judicial Speech Transcriptions

In order to address the problem of structuring text, especially in a judicial context, well-formed documents are usually assumed: sentences are short, delimited and well structured, cased words are used, misspellings are not present and punctuation marks are available and correctly placed. However, when dealing with text provided by automatic speech recognition (ASR) systems, these assumptions do not hold anymore. This is particularly true when addressing automatic transcription in judicial courtrooms. The ASR system[1] could be biased by several domain constraints/limitations, among which:

- Noisy audio streams: the audio stream recorded during a judicial trial, can be affected by different types of background environmental noises and/or by noises caused by the recording equipment (cross microphone interference and reverberation);
- Spontaneous Speech: the sentences uttered by a speaker during a trial are characterized by breaks, hesitations, and false starts. Spontaneous speech - with respect to read utterance - plays a fundamental role in ASR systems, by originating higher values of word error rate;
- Pronounce, language and lexicon heterogeneity: the actors interacting during a trial may be different for language, lexicon and pronounce types. In particular, the judicial debates could contain many words (e.g. person names, names of Institutions or Organizations, etc.) that are not included in the dictionary of the ASR, thus increasing the number of out-of-vocabulary words and, consequently, the resulting word error rate;
- Variability of the vocal signal: the word sequences can be influenced by different circumstances such as posture of the speaker, emotional state, conversation tones, and different microphone frequency responses.
- Non-native speakers: the actors involved in a proceeding can be characterized by linguistic difficulties or can be non-native speaker. These linguistic distortions negatively affect the accuracy of the generated transcriptions.

Considering so far the goal of structuring automatic transcription of judicial trials, the input text to tackle is actually characterized by long sentences that could be partially delimited and with no underlying "common structure", lower case and (some) misspelled words, and finally no punctuation. In this context, some worldly wisdom should be taken into account for accomplishing NER tasks. First of all, the lexicon used in a courtroom debate is heterogeneous and potentially huge (due to the variability of argumentation for different types of crimes). Although some rules could be automatically learned they are not able to guarantee a good coverage in detecting entities. This suggests to use probabilistic approaches, instead of fixed rules, due to their ability to adapt to uncommon situations. Despite CRFs represent the most promising of such approaches, the identification of entities using local contextual cues could be affected by training data: during inference some textual chunks could be not recognized as an entity

---

[1] For more details about the ASR systems adopted in this paper you can refer to [7].

because not available during the training phase. While the training of CRFs can be easily performed by using (previously) generated verbatim transcriptions provided by stenotypist, the inference phase is biased by the ASR output. In these cases, a gazetteer of possible entity identifiers could be useful for including additional features in the probabilistic model and therefore for representing the dependencies between a word's label and its presence in a given gazetteer. When structuring the outputs of a judicial ASR system by combining CRFs and in-domain gazetteers, it's necessary to take into account some discrepancies between the transcribed words and the elements available in the gazetteers. In the next session we propose an extension of CRFs at token level for addressing the judicial NER task.

3   Extending Conditional Random Fields

3.1   Conditional Random Fields

A CRFs is an undirected graphical model that define a single joint distribution $P(y|x)$ of the predicted labels (hidden states) $y = y_1, ..., y_N$ given the corresponding tokens (observations) $x = x_1, ..., x_N$. Formally, the definition of CRFs is given subsequently:

Definition 1 (Conditional Random Fields). *Let $G = (V,E)$ be a graph such that $Y = (Y_v)_{v \in V}$, so that Y is indexed by the vertices of G. Then (X,Y) is a Conditional Random Field, when conditioned on X, the random variables Y_v obey the Markov property with respect to the graph: $p(Y_v|X, Y_w, w \neq v) = p(Y_v|X, Y_w, w \sim v)$, where $w \sim v$ means that w and v are neighbors in G.*

According to the Hammersley-Clifford theorem [4], given $\mathcal{C}$ as the set of all cliques in the graph, a CRF defines a conditional probability distribution over G:

$$p(\boldsymbol{y}|\boldsymbol{x}) = \frac{1}{Z(\boldsymbol{x})} \prod_{C \in \mathcal{C}} \Phi_C(\boldsymbol{x}_C, \boldsymbol{y}_C) \tag{1}$$

where Φ_C represents the set of maximal cliques $C \in \mathcal{C}$, the vertices $\boldsymbol{x}_C$ and $\boldsymbol{y}_C$ of the clique C correspond to hidden states y and observations x, and $Z(\boldsymbol{x}) \in [0,1]$ is the partition function for global normalization. Formally $Z(\boldsymbol{x})$ is defined as:

$$Z(\boldsymbol{x}) = \sum_{\boldsymbol{y}_C} \prod_{C \in \mathcal{C}} \Phi_C(\boldsymbol{x}_C, \boldsymbol{y}_C) \tag{2}$$

Considering that each clique $\Phi_C(\cdot) \in \mathcal{C}$ can encode one or more potential functions (in this case we consider the log-linear ones), the probability of a sequence of label y given the sequence of observations x can be written as:

$$p(\boldsymbol{y}|\boldsymbol{x}) = \frac{1}{Z(\boldsymbol{x})} \exp \left(\sum_{t=1}^{N} \sum_{k=1}^{K} \omega_k f_k(y_t, y_{t-1}, x, t) \right) \tag{3}$$

where $f_k(y_t, y_{t-1}, x, t)$ is an arbitrary real-valued feature function over its arguments and ω_k is a learned weight that tunes the importance of each feature function. In particular when for a token x_t a given feature function f_k is active, the corresponding weight ω_k indicates how to take into account f_k: (1) if $\omega_k > 0$ it increases the probability of the tag sequence y; (2) if $\omega_k < 0$ it decreases the probability of the tag sequence y; (3) if $\omega_k = 0$ has no effect whatsoever.

The partition function $Z(x)$ is an instance-specific normalization function formally defined as:

$$Z(\boldsymbol{x}) = \sum_{\boldsymbol{y}} \left\{ \exp \sum_{t=1}^{N} \sum_{k=1}^{K} \omega_k f_k(y_t, y_{t-1}, x, t) \right\} \tag{4}$$

The conditional probability distribution can be estimated by exploiting two different kinds of feature functions such that $p(y|x)$ can be rewritten as follows:

$$p(\boldsymbol{y}|\boldsymbol{x}) = \frac{1}{Z(\boldsymbol{x})} \exp \left(\sum_{t=1}^{N} \sum_{i=1}^{|I|} \lambda_i s_i(y_t, x, t) + \sum_{j=1}^{|J|} \mu_j t_j(y_{t-1}, y_t, x, t) \right) \tag{5}$$

where I and J represent the given and fixed set of *state feature functions* $s_i(y_t, x, t)$ and *transition feature functions* $t_j(y_{t-1}, y_t, x, t)$, while λ_i and μ_j are the corresponding weights to be estimated from training data. State feature functions and transition feature functions model respectively the sequence of observations x with respect to the current state y_t and the transition from the previous state y_{t-1} to the current state y_t. The parameters λ_i and μ_j are used to weight the corresponding state and transition feature functions.

Training: Parameter Estimation

The learning problem in CRFs relates to the identification of the best feature functions $s_i(y_t, x, t)$ and $t_j(y_{t-1}, y_t, x, t)$ by estimating the corresponding weights λ_i and μ_j. Among the available training paradigms a widely used approach consists of maximizing the conditional log-likelihood of the training data. Given a training set $\mathcal{T}$ composed of training samples (x, y), with y_t ranging in the set $\mathcal{Y}$ of entity labels, the conditional (penalized) log-likelihood is defined as follows:

$$\mathcal{L}(\mathcal{T}) = \sum_{(\boldsymbol{x}, \boldsymbol{y}) \in \mathcal{T}} \log p(\boldsymbol{y}|\boldsymbol{x}, \lambda, \mu) - \sum_{i=1}^{|I|} \frac{\lambda_i^2}{2\sigma_\lambda^2} - \sum_{j=1}^{|J|} \frac{\mu_j^2}{2\sigma_\mu^2}$$

$$= \sum_{(\boldsymbol{x}, \boldsymbol{y}) \in \mathcal{T}} \sum_{t=1}^{N} \sum_{i=1}^{|I|} \lambda_i s_i(y_t, x, t) + \sum_{j=1}^{|J|} \mu_j t_j(y_{t-1}, y_t, x, t) - \log Z(x) \tag{6}$$

$$- \sum_{i=1}^{|I|} \frac{\lambda_i^2}{2\sigma_\lambda^2} - \sum_{j=1}^{|J|} \frac{\mu_j^2}{2\sigma_\mu^2}$$

where the terms $\sum_{i=1}^{|I|} \frac{\lambda_i^2}{2\sigma_\lambda^2}$ and $\sum_{j=1}^{|J|} \frac{\mu_j^2}{2\sigma_\mu^2}$ map a Gaussian prior on λ and μ used for avoiding over-fitting. The objective function $\mathcal{L}(\mathcal{T})$ is concave, and therefore both parameters λ and μ have a unique set of global optimal values. A standard approach to parameter learning computes the gradient of the objective function to be used in an optimization algorithm [12]. Among them we choose the quasi-Newton approach known as Limited-memory Broyden-Fletcher-Goldfarb-Shanno (L-BFGS) [10] due the main advantage of providing a dramatic speedups in case of huge number of feature functions.

Inference: Finding the Most Probable State Sequence

The inference problem in CRFs corresponds to find the most likely sequence of hidden state y^*, given the set of observation $x = x_1, ..., x_N$. This problem can be solved approximately or exactly by determining y^* such that:

$$y^* = \arg\max_y p(\boldsymbol{y}|\boldsymbol{x}) \qquad (7)$$

The most common approach to tackle the inference problem is represented by the Viterbi algorithm [1]. Now, consider $\delta_t(y_j|\boldsymbol{x})$ as the probability of the most likely path of generating the sequence $y^* = y_1, y_2, ..., y_j$. This path can be derived by one of the most probable paths that could have generated the subsequence $y_1, y_2, ..., y_{j-1}$. Formally, given $\delta_t(y_j|\boldsymbol{x})$ as

$$\delta_t(y_j|\boldsymbol{x}) = \max_{y_1, y_2, ..., y_{j-1}} p(y_1, y_2, ..., y_j|\boldsymbol{x}) \qquad (8)$$

we can derive the induction step as:

$$
\begin{aligned}
\delta_{t+1}(y_j|\boldsymbol{x}) &= \max_{y' \in S} \left[\delta_t(y') \Phi_{t+1}(\boldsymbol{x}, y, y') \right] = \\
&= \max_{y' \in S} \left[\delta_t(y') \exp \left(\sum_{k=1}^{K} \omega_k f_k(y_j, y', \boldsymbol{x}, t) \right) \right]
\end{aligned}
\qquad (9)
$$

The recursion terminates in $y^* = \arg\max_{y_j}[\delta_T(y_j)]$, allowing the backtrack to recover y^*.

3.2   Relational Gazetteer

In order to improve the ability of entity recognition, CRFs are supported by a relational gazetteer which includes domain knowledge related to the following entities: Court, Judge, Lawyer, Prosecutor, Victim, Defendant, Witness and Cited Subject.

The domain knowledge is used for creating views of stored instances of a given entity and subsequently for enriching the feature set of CRFs. If a text chunk of a given transcription corresponds to an instance available in the relational gazetteer, CRFs will evaluate a set of additional feature functions over their arguments. Although CRFs could gain from the available domain knowledge, we should take into account that a mismatch between the transcript and the domain knowledge could exist (due to some misspelling errors). In the next subsection

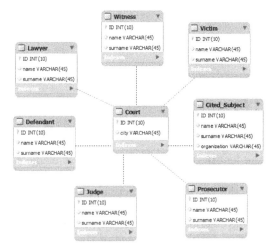

Fig. 1. Snapshot of the Judicial Relational Gazetteer

an extension of CRFs has been formalized for dealing with available knowledge in a relational form.

3.3   Extended CRF

The conditional probability distribution of traditional CRFs, as mentioned in section 3.1, can be estimated by exploiting state and transition feature functions. In order to enhance the prediction capability of CRFs, the set of feature functions has been enlarged in order to deal with the transcription generated by the ASR system and the information available in the judicial relational gazetteer. The additional set of feature functions is able to deal with the following two aspects: (1) enabling the matching with instances available in the judicial relational database in case of misspellings of transcriptions and (2) exploiting the relationships available in the judicial database for creating views that could provide more appropriate subsets of instances to be used for enhancing the probability model. In order to address the first issue, a set H of proximity feature functions has been created. Given a potentially misspelled word x_t, a proximity feature functions allows to match a given instance x_r available in the judicial gazetter R as follows:

$$s_h(y_t, x, t) = \begin{cases} 1 & \text{if } L(x_t, x_r) \leq c \\ 0 & \text{otherwise} \end{cases} \tag{10}$$

where $L(x_t, x_r)$ is a function that computes the edit distance between x_t and $x_r \in R$ (names of judges, names of lawyers, etc..), while c is the cost allowed for aligning x_t and x_r (in our investigation the cost c is set equal to 1). Formally, the edit distance between two strings x_t and x_r is given by:

$$
L(x_t, x_r) = \begin{cases}
0 & \text{if } |x_t| = |x_r| = 0 \\
|x_t| & \text{if } |x_r| = 0, |x_t| > 0 \\
|x_r| & \text{if } |x_t| = 0, |x_r| > 0 \\
\min \begin{cases}
L\Big(x_t[1...(|x_t| - 1)], x_r\Big) + 1 \\
L\Big(x_t, x_r[1...(|x_r| - 1)]\Big) + 1 \\
L\Big(x_t[1...(|x_t| - 1)], x_r[1...(|x_r| - 1)]\Big) + 1
\end{cases} & \text{otherwise}
\end{cases}
$$

(11)

where $|x|$ denotes the length of the word x. The first element in the minimum corresponds to deletion, the second to insertion and the third to mismatch. A proximity feature function $s_h(y_t, x, t)$ is active when the distance between the transcribed word x_t and the instance $x_r \in R$ is less than c (with the minimum number of substitution, deletion and insertion). An active function $s_h(y_t, x, t)$ means that there exist at least one instance of the entities available in R that could be matched by a transcribed word x_t.

Concerning the second issue about the relational gazetteer, a set of views can be created for reducing the search space when computing $Prox(x_t, x_r)$. Considering that a transcription can be easily associated to a given court, all the related instance entities could be retrieved in order to limit the search space for $s_h(y_t, x, t)$. The set H of proximity feature functions can be therefore reduced to H' according to the court location where the transcription has been required to be structured. Now, considering the set H' of proximity feature functions, the probability model of Extended Conditional Random Fields can be formally defined as:

$$
p(\boldsymbol{y}|\boldsymbol{x}) = \frac{1}{Z(\boldsymbol{x})} \exp \left(\sum_{t=1}^{N} \sum_{i=1}^{|I|} \lambda_i s_i(y_t, x, t) + \sum_{j=1}^{|J|} \mu_j t_j(y_{t-1}, y_t, x, t) + \sum_{h=1}^{|H'|} \eta_h s_h(y_t, x, t) \right)
$$

(12)

where the parameters η_h are used to weight the corresponding proximity feature function $s_h(y_t, x, t)$.

The extended CRFs allows a straightforward learning and inference as for traditional CRFs: all the parameters can be estimated by easily adapting equation (6) and, for accomplishing the inference task, proximity feature functions with the corresponding parameters will be exploited in equation (9) for determining the optimal labeling solution.

4   Experimental Results

4.1   Dataset and Performance Criteria

In order to compare the proposed model with traditional CRFs, we carried out experiments on two datasets[2]: *Italian* and *Polish*. The *Italian* dataset is a col-

---

[2] The transcriptions of the Italian and Polish datasets have been obtained from audio recordings acquired in the Court of Naples and Court of Wroclaw, with the support of the Italian and the Polish Ministry of Justice.

lection of 20 real trial transcriptions. The dataset is composed of 165.651 tokens distributed across 3.540 sentences. For the validation phase, the dataset has been divided in training set with 15 trial transcriptions (131.405 tokens across 2.325 sentences) and validation set with 5 trial transcriptions (34.246 tokens across 1.215 sentences).

Table 1. Italian and Polish label distribution

	Italian		Polish	
	Train	**Test**	**Train**	**Test**
Lawyer	420	180	40	7
Prosecutor	260	45	10	4
Victim	59	50	155	139
Defendant	976	257	464	140
Witness	586	44	235	237
Cited Subject	385	63	319	146
Cited Date	120	32	203	79
Tot.	**2822**	**671**	**1435**	**742**

The *Polish* dataset is a collection of 45 real trial transcriptions. The dataset is composed of 218.820 tokens distributed across 2580 sentences. For the validation phase, the dataset has been divided in training set with 33 trial transcriptions (166.353 tokens across 2.500 sentences) and validation set with 12 trial transcriptions (52.467 tokens across 80 sentences). Both datasets have been manually annotated by using the following entities: *Lawyer, Prosecutor, Victim, Defendant, Witness, Cited Subject (Persons, Organizations and Institutions in a deposition), Cited Date (in a deposition)*. All the tokens do not labelled by using one of the above entities have been marked *Other*. Training data have been provided as verbatim transcriptions provided by stenotypist, while test data as transcription obtained by ASR systems. The label distribution of both datasets is reported in Table 1.

The performance in terms of effectiveness have been measured by using several well known evaluation metrics, i.e. Precision, Recall and F-Measure. Given a set of labels Y we compute the Precision and Recall for each label $y_j \in Y$ as:

$$Precision(y) = \frac{\text{\# of tokens successfully predicted as } y}{\text{\# of tokens predicted as } y} \qquad (13)$$

$$Recall(y) = \frac{\text{\# of tokens successfully predicted as } y}{\text{\# of tokens effectively labelled as } y} \qquad (14)$$

The F-Measure for each class $y \in Y$ is computed as the armonic mean of Precision and Recall:

$$F(y) = \frac{2 \cdot Recall(y) \cdot Precision(y)}{Recall(y) + Precision(y)} \qquad (15)$$

The experimental investigation have been performed by using, as starting point, the CRF framework developed by Sunita Sarawagi of IIT Bombay[3].

4.2   Results and Discussion

In order to perform an experimental comparison between Traditional and Extended CRFs we used a baseline set of features, which includes the word identity, transitions among labels, start features, end features, word score features (log of the ratio of current word with the label y to the total words with label y), and features for dealing with yet unobserved words. Extended CRFs include also the proximity feature functions detailed in section 3.3. All the parameters have been estimated by minimizing the log-likelihood by L-BFGS until the convergence is achieved, i.e. when the difference between $\mathcal{L}(\mathcal{T})$ in subsequent iterations is less then $\epsilon = 0.0001$. In Tables 2 and 3 the comparisons of Precision and Recall are

Table 2. Performance comparison on the Italian dataset

	Traditional CRFs		Extended CRFs	
	Precision	Recall	Precision	Recall
Lawyer	88,30	69,01	**88,77**	**86,27**
Prosecutor	86,25	60,34	**88,90**	**84,31**
Victim	**100,00**	8,00	93,75	**45,00**
Defendant	93,08	78,51	**93,31**	**88,86**
Witness	22,88	40,30	**38,24**	**58,21**
Cited Subject	8,33	2,02	**34,40**	**43,43**
Cited Date	12,00	6,82	**100,00**	6,82

reported for the considered datasets. The effect of including proximity features when structuring a judicial transcription is significant: Extended CRFs increases the performance for most of the labels both in terms of Precision and Recall.

Table 3. Performance comparison on the Polish dataset

	Traditional CRFs		Extended CRFs	
	Precision	Recall	Precision	Recall
Lawyer	68,25	44,12	**77,15**	**56,21**
Prosecutor	69,50	42,36	**76,95**	**53,14**
Victim	0,00	0,00	**69,03**	**31,71**
Defendant	55,72	33,18	**62,45**	**74,09**
Witness	46,51	10,15	**47,78**	**57,36**
Cited Subject	20,83	4,05	**48,80**	**30,77**
Cited Date	60,70	67,51	**64,86**	**70,45**

---

[3] The CRF baseline framework is available at http://crf.sourceforge.net

The F-Measure of the proposed approach, as shown in Figure 2, considerably outperforms the traditional CRFs. The most valuable improvements are related to the entities *Victim* for both dataset (+45%), *Cited Subject* for the Italian one and *Witness* for the Polish one (+35%). A summary of the gain ensured by Extended CRFs when structuring judicial transcriptions coming from ASR systems is given by Micro-Average F-Measure, which provides an overall measure of the approaches by taking into account the dimension of each label category. Indeed, the proposed approach is able to achieve a Micro-Average F-Measure of

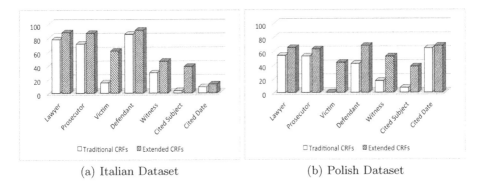

(a) Italian Dataset   (b) Polish Dataset

Fig. 2. Comparison of F-Measures

75,91% on the Italian dataset, against the 61,90% of traditional CRFs, and for the Polish one a measure of 57,76% opposed to the 23,81%.

5   Conclusion

In this paper the problem of structuring text corresponding to automatically generated speech transcriptions has been addressed by empowering one of the most promising probabilistic approach to tackle NER problems, i.e. Conditional Random Fields (CRFs), by modeling available domain knowledge. The proposed Extended Conditional Random Fields is able deal with corrupted audio transcriptions by the use of a relational gazetteer together with the proximity feature functions when computing the conditional probability distribution during training and inference. Although the performance on real judicial debate transcriptions are extremely promising, an important open question relates to the training phase. The parameters that Extended CRFs need to estimate are potentially huge, according to the dimension of the relational gazetteer available. Moreover, the proposed model should be adaptable for consecutive incoming instances (insertion or delation of lawyers, prosecutors and so on). For NER in judicial transcriptions, with consecutive incoming examples in the relational gazetteers, on-line learning could have the potential to achieve a likelihood as high as off-line learning without scanning all available training examples. Adaptation methods [6] represent a potential solution to dynamically adjust the learning rates when dealing with large scale NER problems.

Acknowledgment. This work has been partially funded by the grant "Dote Ricercatori" - FSE, Regione Lombardia.

References

1. Baum, L.E., Petrie, T.: Statistical inference for probabilistic functions of finite state markov chains. Annals of Mathematical Statistics 37(6), 1554–1563 (1966)
2. Chau, M., Xu, J.J., Chen, H.: Extracting meaningful entities from police narrative reports. In: Proc. of the National Conference for Digital Government Research, pp. 271–275 (2002)
3. Chen, H., Jie, J., Qin, Y., Chau, M.: Crime data mining: A general framework and some examples. IEEE Computer 37, 50–56 (2004)
4. Clifford, P.: Markov random fields in statistics; Disorder in Physical Systems: A Volume in Honour of John M. Hammersley. Oxford University Press (1990)
5. Fersini, E., Messina, E., Archetti, F., Cislaghi, M.: Semantics and Machine Learning: A New Generation of Court Management Systems. In: Fred, A., Dietz, J.L.G., Liu, K., Filipe, J. (eds.) IC3K 2010. CCIS, vol. 272, pp. 382–398. Springer, Heidelberg (2013)
6. Huang, H.-S., Chang, Y.-M., Hsu, C.-N.: Training conditional random fields by periodic step size adaptation for large-scale text mining. In: Proc. of the 7th IEEE International Conference on Data Mining, pp. 511–516 (2007)
7. Falavigna, L.J.D., Schlüter, R., Giuliani, D., Gretter, R., Ney, H.: Evaluation of automatic transcription systems for the judicial domain. In: Proc. of the 2010 IEEE Spoken Language Technology Workshop, pp. 206–211 (2010)
8. Ku, C., Iriberri, A., Leroy, G.: Crime information extraction from police and witness narrative reports. In: Proc. of the IEEE International Conference on Technologies for Homeland Security, pp. 193–198 (2008)
9. Ku, C.H., Iriberri, A., Leroy, G.: Natural language processing and e-government: crime information extraction from heterogeneous data sources. In: Proc. of the 2008 International Conference on Digital Government Research, pp. 162–170 (2008)
10. Malouf, R.: A comparison of algorithms for maximum entropy parameter estimation. In: Proc. of the 6th Conference on Natural Language Learning, pp. 1–7 (2002)
11. Nadeau, D., Sekine, S.: A survey of named entity recognition and classification. Linguisticae Investigationes 1, 3–26 (2007)
12. Nocedal, J., Wright, S.: Numerical Optimization. Springer (2000)

Introducing Baselines
for Russian Named Entity Recognition

Rinat Gareev[1], Maksim Tkachenko[2], Valery Solovyev[1], Andrey Simanovsky[3], and Vladimir Ivanov[4]

[1] Kazan Federal University
[2] St Petersburg State University
[3] HP Labs Russia
[4] National University of Science and Technology "MISIS"

Abstract. Current research efforts in Named Entity Recognition deal mostly with the English language. Even though the interest in multi-language Information Extraction is growing, there are only few works reporting results for the Russian language. This paper introduces quality baselines for the Russian NER task. We propose a corpus which was manually annotated with organization and person names. The main purpose of this corpus is to provide gold standard for evaluation. We implemented and evaluated two approaches to NER: knowledge-based and statistical. The first one comprises several components: dictionary matching, pattern matching and rule-based search of lexical representations of entity names within a document. We assembled a set of linguistic resources and evaluated their impact on performance. For the data-driven approach we utilized our implementation of a linear-chain CRF which uses a rich set of features. The performance of both systems is promising (62.17% and 75.05% F_1 measure), although they do not employ morphological or syntactical analysis.

1 Introduction

Named entity recognition (NER) is a process of recognizing entities like people, organizations, locations etc. in a text. It is an essential stage in various applications dealing with text analytics. NER precedes machine translation, information extraction, question answering, and a lot of other tasks. Thanks to NER, a later text analysis can make use of the knowledge about named entities mentions in the text and treat them in a special way.

The problem of NER for the English language was extensively addressed in the scientific community including such conference tracks as MUC [1] and CoNLL [2]. Spanish and German languages were also in the focus of active research. However, some other languages, in particular Russian, received far less attention. A non-sophisticated GATE [3] extension that emulated Russian NER was presented in [4]. It achieved 71% F_1-measure in recognizing and classifying entity mentions[1] on sample newswire data but it is not available anymore. NER was

---

[1] They used six classes: Date, Person, Organization, Location, Percent, and Money.

A. Gelbukh (Ed.): CICLing 2013, Part I, LNCS 7816, pp. 329–342, 2013.

also considered as a part of a machine translation task from the Russian language [5]. However, there is no NER system that is focused on Russian despite the fact that it is generally assumed that an inflective nature of the language in question requires special treatment.

In this work we explore the necessity for a Russian language NER system and introduce two baselines which we evaluate on a Russian newswire corpus that we have marked up. The first baseline implements a knowledge-based approach and is based on a set of dictionaries and hand-crafted rules. The second baseline implements a statistical approach and is a machine-learning system, namely a linear-chain conditional random field (CRF) [6] with Viterbi inference that uses full corpus statistics. We performed 5-fold cross-validation over the markup of the corpus.

Our contributions in this work are the following:

- we developed two baselines for Russian NER;
- we created and marked up an evaluation corpus for Russian NER;
- we tested our baselines on the evaluation corpus and obtained 62.17% and 75.05% F_1-measure respectively.

The rest of the paper is structured in the following way. Section 2 describes the related work. In section 3 we present the evaluation corpus that was used to measure effectiveness of the baselines. In section 4 a knowledge-based approach is introduced. Section 5 discusses a statistical approach based on CRF. Section 6 provides results of evaluation. In section 7 we summarize our work and outline future directions.

2   Related Work

NER is a topic extensively studied throughout the last one and a half decades since its introduction at Message Understanding Conference [1]. Automatic Content Extraction tasks, and Conference on Natural Language Learning [2] can be named among other conferences that addressed NER in their tasks. A NER workshop (NEWS) is regularly held at ACL conference. The above mentioned conference series dealt mostly with newswire data. Other domains were considered in such tasks as NLPBA [7] (bio-medical). Resource-specific NER, for example, NER in Twitter, has recently gained a lot of attention [8].

An overview of the main approaches to NER is given in [9]. The first effort that resulted in a number of rule-based systems was presented at MUC conferences followed by CoNLL series. Recent research mainly concentrates on machine-learning solutions and explores various unsupervised data sources. The authors of [10] introduced a perceptron-based NER system and achieved 90.8% F_1-measure on CoNLL 2003 benchmark. They also explored various approaches to modeling non-local and anaphoric dependencies between named-entity mentions, e.g. context aggregation. A new clustering technique that was proposed in [11] allowed them to achieve 90.9% F_1-measure on CoNLL data. The authors of [12] achieved 90.8% F_1-measure via combining several machine-learning algorithms. The authors of [13] considered a number of new features including phrasal clustering and DBPedia gazetteers, and achieved 91.02% F_1-measure on CoNLL benchmark and up to 87.12% on OntoNotes version 4 dataset [14]. A quite different approach to NER was proposed by Finkel and Manning in [15]. They performed NER work together with the process of parsing via incorporating named entities as special nodes into a parse tree. They obtained up to 88.18% F_1-measure on OntoNotes version 4.

Research on Russian NER is poorly represented in the literature. The main application considered in the literature is machine translation [5], for which it is usually enough to provide do-not-translate markup. The dictionary-based approach to that task was first suggested in [4]. An ongoing research on adapting an English information extraction pipeline to the Russian language is described in [16]. It is done as a part of the PULS project. There are also NER modules in Ontos[2] and RCO[3] industrial systems but their quality is unknown. We believe that introduction of an evaluation corpus and baselines can change this situation.

3   Evaluation Corpus

To the best of our knowledge, there is no available Russian corpus designed for NER evaluation purposes. Some multi-lingual corpora can be employed for NER. The work [17] proposed a Russian corpus with most of NE annotations generated automatically. They evaluated only recall (Russian had the worst results as compared to other languages). A Hungarian-Russian parallel corpus [18] has manual NE annotations. However, it does not include newswire materials and has few Organization mentions. We propose a new corpus designed for the Russian NER task.

As document sources we used ten newswires created from ten top cited "Business" feeds in Yandex "News" web directory[4]. Such domain specificity was caused by the initial context of NER components development. Ten news messages from each newswire were sampled to the corpus on a fixed date. Exact duplicates were removed.

The final number of documents in the corpus is 97. Titles were omitted; the documents in the corpus contain only the contents of message bodies. The texts were cleared from HTML tags. Source URLs are preserved as metadata. The corpus statistics is given in Table 1. The corpus is free to use for academic purposes. Currently there are some restrictions related to copyrights on the included material.[5]

Named entities of two types, Person and Organization, have been annotated in the corpus so far. This task was performed by one human annotator. The annotation conventions adhere to those described in MUC 7 Named Entity Task

---

[2] http://www.ontos.com

[3] http://www.rco.ru/eng/

[4] http://yaca.yandex.ru/yca/cat/Media/Online_Media/

[5] The corpus can be obtained by sending a request to gareev-rm@yandex.ru or to any other author of the paper.

[19]. Only explicit mentions have been annotated. There are no annotations for anaphoric references represented by pronouns or common nouns.

For cases that are not covered by [19], we follow the principle that if there is an entity mention having the same name as some organization and it is not clear whether the organization is mentioned or its metonym, this entity mention should be annotated as Organization. There are no such ambiguities for Person mentions in the corpus.

Nested entities are not annotated. Only entity mentions denoted by a whole phrase are annotated. For example, *Microsoft* is not annotated with Organization if the mention appears inside a *Microsoft Office* phrase.

Table 1. The corpus statistics

Tokens	44326
Words & Numbers	35116
Organization annotations	1317
Person annotations	486

There are several areas of improvements of the corpus: the annotation set can be extended with other generally used entity types like Location and Miscellaneous; annotating temporal and numeral expressions is also eligible. Another aspect of corpus development is to include materials of a broader stylistic variety and originating from other domains to mitigate the present bias towards business news feeds.

4   Knowledge-Based Approach

Our first approach is knowledge-based. We start with a description of our algorithmic pipeline and continue with resource building details.

4.1   Dictionary-Driven, Pattern-Based and Document-Scope Steps for NER

We use three types of knowledge resources: names dictionaries, local context indicators and patterns, and rules that define possible variations of entity mentions in a single document. These resources correspond to three main steps of our text processing pipeline[6]: dictionary matching, context pattern matching, and *docuscope* (document-wide analysis of entity names). The first two steps work on the sentence level; the third one is document-wide. The pipeline implementation is based on UIMA platform[7].

---

[6] We do not describe pre-processing steps, i.e. tokenization and sentence splitting.
[7] http://uima.apache.org/

Dictionary Matching. This step matches dictionary entity names in a text. A matched entity name can differ from the dictionary representation in flexion or case. Both stemming and case-sensitivity can be turned off if necessary. We use ConceptMapper [20] as an implementation of this step. We match only person first names and organization names on this step; person surnames are matched on the next one.

Pattern Matching. On this step named entity mentions are recognized by looking at context cues. We use preceeding noun phrases that denote the entity type, as in the following example:

Example 1. **директора по продажам** Seagate Роки Пиментель
Seagate's **director of sales** *Rocky Pimentel*
российское **издательство** "Эксмо"
Russian **publishing house** "Eksmo"

We have defined several semantic types of such indicator words and phrases and created patterns that use these types. There are two main classes of patterns. Patterns of the first class match proper names and mark them with PossibleNE annotation. Patterns of the second class assign a type to a PossibleNE using indicators. The indicators detection is implemented with ConceptMapper. TextMarker [21] language and engine are used to implement pattern matching.

Docuscope. Typically, the first mention of an entity in a document uses a full entity name and/or an indicator phrase, while the subsequent mentions use short representations. The goal of the step is to recognize entity mentions in this particular case.

First, the mentions that are candidates for resolution at this stage are collected inside a document. A candidate set consists of PossibleNE annotations that are not resolved at the pattern matching stage. The decision on which type to assign to a candidate name cn depends on information about the names recognized on the previous steps. If rn is an earlier recognized name with a short form that matches cn, rn contributes one vote to resolve cn to the same type as rn has. cn is assigned a type that has received maximum votes. Short names are produced following the options given in Table 2.

4.2 Resources Used

Several names dictionaries were prepared for the evaluation. For Organization type we have assembled four different dictionaries in Russian and one in English. The first one consists of legal body names that were obtained from web catalogs[8] providing information from the Russian government registry. Two

---

[8] For example, `http://moscowtele.ru/`

Table 2. Examples of generating shorter name variants

Entity type	Generator	Example
All	Recognized name as is	
All	Adjacent parentheses content	DensBits Technologies (DensBits) $\longrightarrow$ DensBits
Person	Second name of a full name	Pavel Filenkov $\longrightarrow$ Filenkov
Organization	Omitted designator	ОАО "РусГидро" $\longrightarrow$ "РусГидро" JSC "RusHydro" $\longrightarrow$ "RusHydro"
Organization	Acronym	World Steel Dynamics $\longrightarrow$ WSD

post-processing steps were applied to a raw set of legal names. We enriched the original set of names with their nested names using heuristics illustrated in the following example:

Example 2. ООО "АПК "Еленовское" $\longrightarrow$ АПК "Еленовское" $\longrightarrow$ Еленовское. Ltd "Agroindustrial complex "Elenovskoe" $\longrightarrow$ Agroindustrial complex "Elenovskoe" $\longrightarrow$ Elenovskoe.

The second post-processing step was to remove single-word entries that were ambiguous with common Russian words. The final dictionary (denoted **D1**) has about 2.3 million of entries.

The second dictionary (**D2**) of organizations was obtained from the Russian Wikipedia. We developed mappings for the most prominent infobox types in the category "Infoboxes:Organization" and used DBPedia extraction framework [22]. The dictionary consists of infobox labels, Wikipedia page titles (including redirect page titles). The size of the dictionary is about 40,000 entries.

Following the approach used in DBPedia Spotlight [23] system, we enriched **D2** with anchor texts of links inside Wikipedia that link to organization pages. The augmented dictionary (**D3**) contains about 87,000 entries.

The last dictionary (**D4**) of organizations was derived from a companies catalog of a specific newswire[9]. The size of the dictionary is about 2,800 entries.

English organization names were obtained from English DBPedia. The dictionary (**ED**) consists of rdfs:label and foaf:name properties of Organization class entities. It contains about 280,000 entries.

The Person names dictionary consists of approximately 10,000 male and 10,000 female first names.

We defined several indicator types for the pattern matching step. Organization indicators are noun phrases and their acronyms that denote an organization type (e.g., *ООО/Ltd.*, *фонд/foundation*) or an organization subdivision (e.g., *совет директоров/board of directors*). Most of them (about 2,700) were obtained from D1 dictionary by manual supervision of the most frequent entity name prefixes. Person indicators (50) were prepared manually.

---

[9] `http://expert.ru/dossier/companies/`

Patterns specify conditions on local contexts of named entities using TextMarker language. So far all patterns (about 30) are handcrafted. Sample patterns[10] with respective text examples are given below:

```
CW+{NOT(PARTOF({Person, Company})) -> MARK(PossibleNE)}
РАО ЭС Востока (RAO ES Vostoka).
```

```
CompanyIndicator1 PossibleNE {NOT(PARTOF(Company))
    -> MARK(Company)}
холдинг РАО ЭС Востока (holding company RAO ES Vostoka).
```

5 Statistical Approach

In the data-driven approach we tuned machine-learning-based NER implementation presented earlier for the English language [13].

5.1 Conditional Random Fields

Our approach is based on conditional random fields [6] (CRF). CRF is a graphical model of hidden and visible state variables designed for the sequence labeling task. CRF describes the behavior of a label random variable $Y = \{y1, .., y_N\}$ given a sequence random variable $X = \{x1, .., x_N\}$. CRF is defined by a graph, which vertexes are indexes of Y, and describes X and Y if the Markov property holds: $P(Y_i|X, Y_j, i \neq j) = P(Y_i|X, Y_j, i \sim j)$, where $\sim$ stands for the neighbour relationship. That is, the graph defines all possible effects that hidden variable values have on each other. A *linear-chain CRF* is a CRF for which the graph is a simple chain and each edge has a form of $(j, j + 1)$. Thus, the Markov property can be represented in the following form: $P(Y_i|X, Y_j, i \neq j) = exp(\sum_{e,k} \lambda_k f_k(e, y|_e, x) + \sum_{v,k} \mu_k g_k(v, y|_v, x))$, where k indexes features, v indexes vertexes, and e indexes adjacent pairs of vertexes in the graph; f_k and g_k stand for binary features that describe dependencies on variable values of the other label along the graph edges and dependencies on the label value in question respectively. For example, g_k might be a binary function that has the value of 1 at position i if x_i contains a dimension denoting that the corresponding i-th token starts with a capital letter and y_i label is "person-start". An f_k feature might be a binary function that has the value of 1 at position i if x_i contains a dimension denoting that the corresponding i-th token starts with a capital letter and y_i label is "person-start" *and* y_{i+1} label is "person-continued". The features need not be independent. λ_k and μ_k are considered parameters and are learned from the training data.

We use Mallet [11] implementation of the linear-chain CRF with an implementation of Viterbi inference that maximizes log-likelihood for the feature weights.

---

[10] Due to space limitations some conditions and actions in the example are omitted. 'CW' stands for 'Capitalized Word'. '+' is a regular expression quantifier. 'NOT PARTOF' condition is required to avoid overlapping with the names matched before executing the rule.

[11] http://mallet.cs.umass.edu

5.2 Feature Set

We consider the whole document as a sequence to be labeled. The tokenization that we use removes punctuation and other special characters including hyphens. Words with a hyphen are represented by two tokens. The following list describes what sub-vectors compose x_i:

- current token; x_i has lexicon-sized subspace that contains 1 at a dimension corresponding to the token at the given position in the sequence;
- number normalization; we use four regular-expression masks for a date format (DDDD) and a separate mask for other numbers;
- 5-token window centered on the current token; a subspace of x_i contains 1's at dimensions corresponding to the tokens found in the window; for example, a position corresponding to the token "ЭС" inside "акций РАО ЭС Востока выросла" will have 1's at dimensions corresponding to "акций", "РАО","ЭС","Востока", and "выросла";
- shape window (current, previous, next word shape, combined shapes of previous and current and next and current) centered on the current token; for example, a position corresponding to the token "ЭС" inside "РАО ЭС Востока" will have 1's at dimensions corresponding to XX, Xx, XX XX, and XX Xx shapes;
- current token prefixes and suffixes up to 6 characters in length; a subspace of x_i contains 1 at dimensions that correspond to the suffixes and prefixes found in the current token;
- LDA cluster labels[12] of the current token; a subspace of x_i contains 1 at dimensions that correspond to the latent Dirichlet allocation cluster [24] labels;
- 3-window for LDA cluster labels centered on the current token;
- Brown cluster [25] label[12]; a subspace of x_i contains 1's at dimensions corresponding to the paths of length of 7, 11, and 13 in the dendrogram;
- Clark cluster [26] label[12]; a subspace of x_i contains 1's at the cluster label;
- extended context (merged context of the repeats of the current token in a 401-token window centered on the current token); 200 tokens to the left and to the right of the current one are searched for a repeat of the token, the 5-token windows centered on these repeats are combined in a set of words, the set is expressed in a lexicon-sized subspace of x_i;
- extended context over cluster labels; a subspace of x_i is used to describe the set of cluster labels of the extended context.

All our features are f_k features that are characteristic functions of the values of one of the above-mentioned subspaces. That is, the features have a form like "the current 5-token window contains A, B, C, D and E, the *current* label is L_1, and the *next* label is L_2".

We use IOB2 labeling scheme [2] for the sequence labeling task and then switch to the evaluation corpus format.

---

[12] Clusterings were built by us on the unlabeled texts crawled from Russian news sites.

6   System Performances

6.1   Knowledge-Based System Performance

As a baseline we consider the plain dictionary-based recognition All configurations use the same dictionary of person first names. Person annotations were obtained via matching a first name with the subsequent capitalized word. The results for Person type are shown in Table 3.

Table 3. Performance of the plain dictionary-based recognition for Person type

P	R	F
91.44	54.94	68.64

Table 4. Performance of the plain dictionary-based recognition for Organization type

Configuration	P	R	F
ED only	41.25	12.00	18.59
(*) ED + D1, case-insensitive, with stemming	5.41	48.60	9.74
(**) previous + rejecting	24.52	48.44	32.56
(***) previous without stemming	44.10	37.21	40.36
ED + D2, case-sensitive, stemming	41.31	48.75	**44.72**
ED + D3, case-sensitive, stemming	25.57	**58.09**	35.51
ED + D4, case-sensitive, stemming	**53.42**	29.61	38.10

Table 4 shows the results for the dictionaries of organization names. In the evaluation corpus 31,8% of organization names (mostly English) use only Latin letters. The first row of the table shows the effectiveness of the ED dictionary. In all configurations ED is matched case-insensitive and without stemming. Three different configurations use D1. D1 contains case-normalized entries and, consequently, is matched case-insensitive. Low precision in row (*) is due to many D1 entries coinsiding with common words after stemming. In configuration (**) we introduce an additional filtering heuristic: all non-capitalized organization names are ignored. Row (***) contains a result with stemming switched off. D1, D2, and D3 dictionaries yield considerable recall but they are very noisy, and additional filtering of their content is required. The following list shows examples of the most frequent error types:

- ambiguous names: coinciding with geo-political entities, e.g., "Татарстан" (Tatarstan, a republic and an airline), with common words, e.g. "Эксперт" (expert, also a company name);
- incorrect stemming: "Прио" (a company name) has the same stem with "При" (a Russian preposition), "Этам" (a company name) – "Это" (this);
- partial matches: "Goldman Sachs Asset Management" was expected – only "Goldman Sachs" was matched, "ВТБ Капитал" was expected – only "ВТБ" was matched.

In Table 5 we provide the evaluation results of other components. Docuscope demonstrates considerable improvement (19% F_1) of the overall recognition quality. The best precision is achieved by configurations that do not employ any organization name dictionaries. Pattern matching is more precise than other steps because it relies upon context indicators that resolve ambiguity. The last three rows of the table show that pattern matching and Docuscope improve the recall compared to the plain dictionary matching. Pattern matching and Docuscope are not used to reject dictionary matchings, consequently, the full pipeline inherits the low precision of the first step.

Analysis of the system output discovered the following reasons of the pattern matching and Docuscope errors:

- Person false negatives are caused by incomplete dictionary of first names, incorrect stemming of first and second names and weak coverage of person indicators in pattern matching.
- Most of Person false positives are caused by ambiguous patterns, e.g., the pattern "`PersonIndicator PossibleNE`", which is intended to match cases like "вице-президент Андрей Столяров" (vice-president Andrey Stolyarov), also matches "президент Роснефти" (president of Rosneft).
- Organization false negatives are caused by the absence of indicators in the text, incompleteness of our indicators set, and incorrect detection of name boundaries, especially in case of names containing lower-case words, e.g. "Объединенная судостроительная корпорация" (United Shipbuilding Corporation).
- The main reason of Organization false positives is Organization-Product, e.g. "Yammer" (a company and a social network), Organization-Facility, Organization-Location, e.g. "Домодедово" (Domodedovo, an airport and a town) and other Organization ambiguities.

Besides further augmentation of patterns and indicators, we suppose that the following improvements would be beneficial for the knowledge-based system:

- strict filtering of output of plain dictionary matching,
- an improved stemmer or a lemmatizer for proper names,
- using an NP-chunker for better boundary detection of organization names,
- recognition of other type entities (Locations, GPE, Products) for ambiguity detection.

6.2   CRF-Based System Performance

We used an exact match evaluation protocol from the CoNLL benchmark to test our CRF-based NER system.

We evaluated our CRF-based NER system by 5-fold cross-validation; at each fold we got a roughly 80% training / 20% test split. Table 6 shows the results for the corpus for each fold. The results vary greatly from fold to fold: the mean value of the overall F_1-measure is $\mu = 75.05$ and the standard deviation is

Table 5. Performance of the KB system pipelines

Configuration	Person			Organization			Overall		
	P	R	F	P	R	F	P	R	F
Pattern matching (PM)	**79.60**	57.82	66.98	67.12	18.91	29.50	73.20	29.40	41.95
PM+Docuscope	78.16	78.81	78.48	**75.35**	40.85	52.98	**76.50**	51.08	61.26
ED+D1(***)+PM +Docuscope	78.90	76.95	77.92	50.59	58.47	54.24	57.31	63.45	60.23
ED+D2+PM+Docuscope	78.62	**79.42**	79.02	43.90	**61.43**	51.20	51.20	**66.28**	57.77
ED+D4+PM+Docuscope	79.38	79.22	**79.30**	59.04	52.32	**55.48**	65.01	59.57	**62.17**

$\sigma = 4.82$. A graph of F_1-measure against the number of training examples (as a test set we randomly chose 7 examples from our corpus) demonstrates that our corpus is not big enough to fit the parameters well and, consequently, the model is under-trained. The graph is given in Figure 1.

Table 6. 5-fold cross-validation of the CRF-based NER

Fold #	Person			Organization			Overall		
	P	R	F	P	R	F	P	R	F
1	94.59	83.33	88.61	85.77	63.27	72.82	88.46	68.66	77.31
2	95.96	81.9	88.37	87.69	70.08	77.92	90.48	73.89	81.35
3	88.89	77.11	82.58	77.92	70.59	74.07	80.53	72.19	76.13
4	87.1	78.26	82.44	85.38	55.3	67.13	85.84	60.06	70.67
5	88.16	77.01	82.21	69.78	60.19	64.63	75.19	65.1	69.78
mean	90.94	79.52	84.84	81.31	63.88	71.31	84.10	67.98	75.05
std	4.04	2.91	3.33	7.44	6.54	5.38	6.22	5.57	4.82

We assessed the errors of the system and found two main issues. First, we somewhat inconsistently tag organization mentions, e.g., in the mention of "РАО ЭС Востока" ("RAO ES Vostoka") we can tag all three tokens as well as only "РАО ЭС". This problem can be addressed with a two-stage prediction technique [27]. Second, mixed language and Latin-spelled entities as well as transliterated entites are not recognized (e.g. "агентство Moody's" / "Moody's agency", "Джорджу Осборну" / "George Osborn") or are labeled incorrectly (e.g. "Dow Jones" is tagged as an Organization). Gazetteers and an English language NER can be applied to tackle this problem. Minor contributions to the errors come from (1) recognizing only acronyms but not the full names of the organizations when they occur together, e.g. in "Федеральная антимонопольная служба, ФАС" / "Federal Anti-monopoly Service, FAS" we tag only the last token as an Organization, (2) our ignorance of punctuation, e.g. we do not recognize "S.P." as an Organization and we recognize "Газпрома, Роснефти" / "Gazprom, Rosneft" as one Organization mention while it is a list of two. Mentions of multiple-word entities, e.g. "Бюро по вопросам охраны окружающей среды и экологического законодательства" / "Environment Protection and Ecological Legislation Bureau", are often tagged incorrectly. Some of the problems described above can be addressed if we use more data to build clusterings that we utilize to create features. Our analysis shows that some of the present clusters fail to separate semantic groups well.

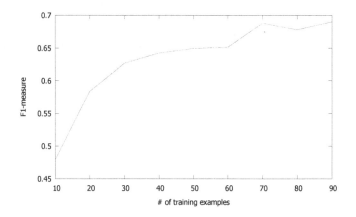

Fig. 1. F_1-measure against the number of training examples

7 Conclusions

In this work we proposed two baselines (knowledge-based and statistical) for Russian language NER. We created an evaluation corpus and evaluated these baselines on the corpus. Our early results of 62.17% and 75.05% F_1-measure are very promising, given that neither of our baselines employs morphological or syntactical analysis.

We plan to develop our evaluation corpus in three directions: consider new document sources, label more documents, add other classes of named entities (Location etc.). We also see our future work in converting our baselines to high-quality NER systems for the Russian language.

References

1. Grishman, R., Sundheim, B.: Message understanding conference-6: a brief history. In: Proceedings of the 16th Conference on Computational Linguistics, vol. 1, pp. 466–471. ACL, Stroudsburg (1996)
2. Tjong Kim Sang, E.F., De Meulder, F.: Introduction to the CoNLL-2003 shared task: language-independent named entity recognition. In: Proceedings of the Seventh Conference on Natural Language Learning at HLT-NAACL 2003, pp. 142–147. ACL, Morristown (2003)
3. Cunningham, H., Wilks, Y., Gaizauskas, R.J.: GATE: a general architecture for text engineering. In: Proceedings of the 16th Conference on Computational Linguistics, vol. 2, pp. 1057–1060. ACL, Stroudsburg (1996)

4. Popov, B., Kirilov, A., Maynard, D., Manov, D.: Creation of reusable components and language resources for named entity recognition in Russian. In: Proceedings of the Fourth International Conference on Language Resources and Evaluation. European Language Resources Association (2004)
5. Babych, B., Hartley, A.: Improving machine translation quality with automatic named entity recognition. In: Proceedings of the 7th International EAMT Workshop, pp. 1–8. ACL, Stroudsburg (2003)
6. Lafferty, J.D., McCallum, A., Pereira, F.C.N.: Conditional random fields: Probabilistic models for segmenting and labeling sequence data. In: Proceedings of the Eighteenth International Conference on Machine Learning, pp. 282–289. Morgan Kaufmann Publishers Inc., San Francisco (2001)
7. Kim, J.D., Ohta, T., Tsuruoka, Y., Tateisi, Y., Collier, N.: Introduction to the bio-entity recognition task at JNLPBA. In: Proceedings of the International Joint Workshop on Natural Language Processing in Biomedicine and its Applications, pp. 70–75. ACL, Stroudsburg (2004)
8. Ritter, A., Clark, S., Mausam, E.O.: Named entity recognition in tweets: an experimental study. In: Proceedings of the Conference on Empirical Methods in Natural Language Processing, pp. 1524–1534. ACL, Stroudsburg (2011)
9. Nadeau, D., Sekine, S.: A survey of named entity recognition and classification. Lingvisticae Investigationes 30(1), 3–26 (2007)
10. Ratinov, L., Roth, D.: Design challenges and misconceptions in named entity recognition. In: Proceedings of the 13th Conference on Computational Natural Language Learning, pp. 147–155. ACL, Stroudsburg (2009)
11. Lin, D., Wu, X.: Phrase clustering for discriminative learning. In: Proceedings of the Joint Conference of the 47th Annual Meeting of the ACL and the 4th International Joint Conference on Natural Language Processing of the AFNLP, pp. 1030–1038. ACL, Suntec (2009)
12. Ciaramita, M., Altun, Y.: Named-entity recognition in novel domains with external lexical knowledge. In: Proceedings of the NIPS Workshop on Advances in Structured Learning for Text and Speech Processing (2005)
13. Tkachenko, M., Simanovsky, A.: Named entity recognition: Exploring features. In: Jancsary, J. (ed.) Proceedings of KONVENS 2012, ÖGAI, pp. 118–127 (2012)
14. Hovy, E., Marcus, M., Palmer, M., Ramshaw, L., Weischedel, R.: OntoNotes: the 90% solution. In: Proceedings of the Human Language Technology Conference of the NAACL, Companion Volume: Short Papers, pp. 57–60. Association for Computational Linguistics, Stroudsburg (2006)
15. Finkel, J.R., Manning, C.D.: Joint parsing and named entity recognition. In: Proceedings of Human Language Technologies: The 2009 Annual Conference of the North American Chapter of the Association for Computational Linguistics, pp. 326–334. ACL, Stroudsburg (2009)
16. Du, M., von Etter, P., Kopotev, M., Novikov, M., Tarbeeva, N., Yangarber, R.: Building Support Tools for Russian-Language Information Extraction. In: Habernal, I., Matoušek, V. (eds.) TSD 2011. LNCS, vol. 6836, pp. 380–387. Springer, Heidelberg (2011)
17. Ehrmann, M., Turchi, M., Steinberger, R.: Building a multilingual named entity-annotated corpus using annotation projection. In: Proceedings of the International Conference Recent Advances in Natural Language Processing, RANLP 2011 Organising Committee, Hissar, Bulgaria, pp. 118–124 (2011)

18. Szabó, M.K., Vincze, V., Nagy T., I.: HunOr: A Hungarian-Russian parallel corpus. In: Calzolari, N., Choukri, K., Declerck, T., Dogan, M.U., Maegaard, B., Mariani, J., Odijk, J., Piperidis, S. (eds.) Proceedings of the 8th International Conference on Language Resources and Evaluation (LREC 2012). European Language Resources Association, Istanbul (2012)

19. Chinchor, N.A.: MUC-7 named entity task definition. In: Proceedings of the 7th Message Understanding Conference (MUC-7), Fairfax, VA, USA (1998)

20. Tanenblatt, M., Coden, A., Sominsky, I.: The ConceptMapper approach to named entity recognition. In: Calzolari, N., Choukri, K., Maegaard, B., Mariani, J., Odijk, J., Piperidis, S., Rosner, M., Tapias, D. (eds.) Proceedings of the 7th International Conference on Language Resources and Evaluation (LREC 2010), European Language Resources Association, Valletta (2010)

21. Kluegl, P., Atzmueller, M., Puppe, F.: TextMarker: A tool for rule-based information extraction. In: Proceedings of the 2nd UIMA@GSCL Workshop, 2009 Conference of the GSCL (Gesellschaft fur Sprachtechnologie und Computerlinguistik) (2009)

22. Bizer, C., Lehmann, J., Kobilarov, G., Auer, S., Becker, C., Cyganiak, R., Hellmann, S.: DBpedia - a crystallization point for the Web of Data. Web Semant. 7(3), 154–165 (2009)

23. Mendes, P.N., Jakob, M., García-Silva, A., Bizer, C.: DBpedia Spotlight: shedding light on the web of documents. In: Proceedings of the 7th International Conference on Semantic Systems, pp. 1–8. ACM, New York (2011)

24. Chrupala, G.: Efficient induction of probabilistic word classes with LDA. In: Proceedings of the 5th International Joint Conference on Natural Language Processing, Chiang Mai, Thailand, pp. 363–372. Asian Federation of Natural Language Processing (2011)

25. Brown, P.F., Della Pietra, V.J., Desouza, P.V., Lai, J.C., Mercer, R.L.: Class-based n-gram models of natural language. Comp. Linguistics 18(4), 467–479 (1992)

26. Clark, A.: Combining distributional and morphological information for part of speech induction. In: Proceedings of the 10th Conference on European Chapter of the Association for Computational Linguistics, vol. 1, pp. 59–66. ACL, Stroudsburg (2003)

27. Krishnan, V., Manning, C.D.: An effective two-stage model for exploiting non-local dependencies in named entity recognition. In: Proceedings of the 21st International Conference on Computational Linguistics and the 44th Annual Meeting of the Association for Computational Linguistics, pp. 1121–1128. ACL, Stroudsburg (2006)

Five Languages Are Better Than One: An Attempt to Bypass the Data Acquisition Bottleneck for WSD

Els Lefever[2,3], Véronique Hoste[1,2], and Martine De Cock[3]

[1] Department of Linguistics, Ghent University
[2] LT3, Language and Translation Technology Team, University College Ghent
[3] Department of Applied Mathematics and Computer Science, Ghent University

Abstract. This paper presents a multilingual classification-based approach to Word Sense Disambiguation that directly incorporates translational evidence from four other languages. The need of a large predefined monolingual sense inventory (such as WordNet) is avoided by taking a language-independent approach where the word senses are derived automatically from word alignments on a parallel corpus. As a consequence, the task is turned into a cross-lingual WSD task, that consists in selecting the contextually correct translation of an ambiguous target word.

In order to evaluate the viability of cross-lingual Word Sense Disambiguation, we built five classifiers with English as an input language and translations in the five supported languages (viz. French, Dutch, Italian, Spanish and German) as classification output. The feature vectors incorporate both local context features as well as translation features that are extracted from the aligned translations. The experimental results confirm the validity of our approach: the classifiers that employ translational evidence outperform the classifiers that only exploit local context information. Furthermore, a comparison with state-of-the-art systems for the same task revealed that our system outperforms all other systems for all five target languages.

Keywords: cross-lingual word sense disambiguation, classification, multilingual, parallel corpora.

1 Introduction

Word Sense Disambiguation (WSD), the task that consists in selecting the correct sense of an ambiguous word in a given context, is a well-researched NLP problem. For a complete overview of the field, we refer to [1] and [2].

Supervised approaches to WSD, which are trained on manually sense-tagged corpora, invariably yielded the best results for various WSD tasks at the different SemEval competitions [1]. These supervised approaches face a number of challenges, though, that prevent them from being integrated in efficient and reliable WSD systems for general-purpose applications. Firstly, training an accurate all-words supervised WSD system requires a huge training corpus that

A. Gelbukh (Ed.): CICLing 2013, Part I, LNCS 7816, pp. 343–354, 2013.

contains enough labeled examples per ambiguous focus word. Although sense-tagged corpora are available (E.g. SemCor [3], SemEval benchmark data sets[1]), these often contain too few training examples to cover all senses of all ambiguous words and hardly exist for languages other than English. Secondly, the sense divisions in most dictionaries and sense inventories such as WordNet [4] are often based on subjective decisions and too fine-grained to be useful for real world applications. Furthermore, there is a desperate lack of electronic sense inventories for languages other than English, even though initiatives have been launched for other languages, amongst which the EuroWordNet database[2]. Since it is unlikely that large quantities of hand-annotated text and robust sense inventories will be available any time soon for a large variety of languages, unsupervised approaches have been introduced as an alternative.

Translation-based unsupervised methods use a word-aligned parallel corpus in order to extract cross-lingual evidence for Word Sense Disambiguation. These methods start from the hypothesis that the different senses of a word might result in different translations in a given target language. [5] showed with a set of empirical studies of translingually-based sense inventories that sense distinctions can indeed be captured by translations into second languages. The authors also revealed that multiple languages are needed to cover all different sense distinctions, as some senses are lexicalized in one particular language but not in the other language. Using word-aligned parallel corpora thus enables one to extract the sense inventory in an automatic way, based on the various translations of a particular word in these corpora. In addition, a sense-tagged corpus based on the induced senses - or translations - can be automatically created and used to train a more traditional supervised classifier.

Several studies have already underlined the validity of using parallel corpora for **bilingual** word sense disambiguation. [6] use the idea that different meanings of the word are lexicalized differently across languages to automatically create a sense-tagged dataset for supervised WSD. They first apply word alignment to identify all translation candidates of an ambiguous word in the French-English parallel corpus. In a next step, a system is trained to predict a correct translation of the ambiguous focus word by using source-language local context information. [7] follow a similar approach. They try to identify valid translations of the ambiguous words by using association measures such as Mutual Information. The English training sentences are tagged with the corresponding French translations and the resulting corpus is used to train a WSD classifier. [8] presents an unsupervised bootstrapping approach to WSD, which expects a word-aligned parallel corpus as input and clusters all source words which translate into the same target word, which are then used to apply WSD using a similarity measure. [9] propose a hybrid approach that employs an inductive logic programming algorithm to learn disambiguation rules based on corpus-based evidence and several

---

[1] www.senseval.org

[2] EuroWordNet is a multilingual database with wordnets for several European languages (Dutch, Italian, Spanish, German, French, Czech and Estonian). The wordnets are structured in the same way as the American wordnet for English.

knowledge sources. The system aims to provide a correct Portuguese translation for a set of highly ambiguous English verbs.

Other WSD systems exploit the **multilingual** sense alignments in multilingual wordnets. [10] combine earlier research on word clustering to automatically extract translational equivalents with aligned wordnets. Given two aligned words in a parallel corpus, these words receive the sense label of the synsets of the two words which are mapped by EuroWordNet's interlingual index. This way, the method can be used to automatically sense-tag corpora in several languages at once.

For the WSD system proposed in this paper, we elaborate on two well-known research ideas:

1. the possibility to use parallel corpora to extract translation labels and features in an automated way.
2. the assumption that incorporating evidence from multiple languages into the feature vector will be more informative than a more restricted set of monolingual or bilingual features. This assumption is based on the idea of [11] who argued that "two languages are better than one" to select the right sense of an ambiguous focus word. They showed that it was possible to use the differences between languages in order to obtain a certain leverage on word meanings.

Our system is different from the bilingual WSD approaches mentioned above ([6,9,12]) in that it takes a truly multilingual approach incorporating evidence from five different languages. Unlike other multilingual approaches as presented by, for instance, [10], we do not use aligned WordNets in the languages under consideration to enhance the WSD classification results.

We believe, based on the following assumptions, that our Cross-lingual WSD (CLWSD) algorithm proposes an integrated solution for the most important remaining WSD issues. Firstly, using multilingual unlabeled parallel corpora goes some way towards clearing the data acquisition bottleneck for WSD, because using translations as sense labels excludes the need for manually created sense-tagged corpora and sense inventories such as WordNet or EuroWordNet. Moreover, as there is fairly little linguistic knowledge involved, the framework can be easily deployed for a variety of different languages. Secondly, this approach also deals with the sense granularity problem; finer sense distinctions are only relevant as far as they get lexicalized in different translations of the word. At the same time, the subjectivity problem is tackled that arises when lexicographers have to construct a fixed set of senses for a particular word that should fit all possible domains and applications. In our approach, the use of domain-specific corpora allows to derive sense inventories that are tailored towards a specific target domain or application and to train a dedicated CLWSD system using these particular sense inventories. Thirdly, working immediately with translations instead of more abstract sense labels allows to bypass the need to map (WordNet) senses to corresponding translations. This makes it easier to integrate a dedicated WSD module into real multilingual applications such as machine translation or information retrieval. Finally, including evidence from multiple languages might help to further refine the obtained sense distinctions [5].

2 Information Sources

Instead of using a predefined monolingual sense inventory, such as WordNet, we use a language-independent framework where the word senses are derived automatically from word alignments on a parallel corpus. We based our data set on the Europarl parallel corpus[3], which is extracted from the proceedings of the European Parliament [13]. We experimented with 6 languages from the Europarl corpus, viz. English (our source language), Dutch, French, German, Italian and Spanish.

We selected a set of 20 polysemous English nouns[4] and then linguistically preprocessed (tokenized, lemmatized, part-of-speech tagged and chunked) all Europarl sentences containing these nouns and their aligned translations in the other five languages. For the feature vector construction, we combined a set of local context features that were extracted from the English sentence and a set of bag-of-words features that were extracted from the aligned translations in the four other languages that are not the target language of the classifier. We created two flavors of the translation features: a set of binary bag-of-words features and a set of *latent semantic* translation features that resulted from applying Latent Semantic Analysis [14] to the content words of the aligned translations.

The idea to enrich the more commonly used local context features with multilingual translational evidence starts from the assumption that incorporating evidence from multiple languages into the feature vector will be more informative than only using monolingual or bilingual features. The working hypothesis we adopted for all experiments is that the differences between the different languages that are integrated in the feature vector will enable us to refine the obtained sense distinctions and that adding more languages will improve the classification results accordingly.

2.1 Local Context Features

We extracted the same set of local context features from both the English training and test instances. The linguistically preprocessed English instances were used as input to build a set of commonly used WSD features [1]:

- features related to the **focus word itself** being the word form of the focus word, the lemma, part-of-speech and chunk information.
- **local context features** related to a window of three words preceding and following the focus word containing for each of these words their full form, lemma, part-of-speech and chunk information.

---

[3] http://www.statmt.org/europarl/

[4] coach, education, execution, figure, job, letter, match, mission, mood, paper, post, pot, range, rest, ring, scene, side, soil, strain, test.

The motivation to incorporate local context information for a seven-word window containing the ambiguous focus word is twofold. Firstly, we assume that the immediate context of the ambiguous focus word will be more efficient in capturing compound and collocation information for the ambiguous focus word. Secondly, previous research has shown that a classifier using this specific set of features performs very well for WSD [15].

2.2 Translation Features

In addition to the commonly deployed local context features, we also extracted a set of binary bag-of-words (BOW) features from the aligned translations that are not the target language of the classifier. For the French classifier for instance, we extract bag-of-words features from the Italian, Spanish, Dutch and German aligned translations. Per ambiguous focus word, a list of all content words (nouns, adjectives, adverbs and verbs) that occurred in the linguistically preprocessed aligned translations of the English sentences containing this word, were extracted. Each content word then corresponds to exactly one binary feature per language. For the construction of the translation features for the training set, we used the Europarl aligned translations.

As we do not dispose of similar aligned translations for the test instances for which we only have the English test sentences at our disposal, we had to adopt a different strategy. We decided to use the Google Translate API[5] to automatically generate translations for all English test instances in the five target languages. This automatic translation process can be done using whatever machine translation tool, but we chose the Google API because of its easy integration into our programming code. Once the automatic translations were generated, we preprocessed them in the same way as we did for the aligned training translations. Subsequently, we selected all content words from these aligned translations and constructed the binary bag-of-words features per language.

2.3 Latent Semantic Translation Features

Manual inspection of the binary translation features revealed two potential issues with this type of bag-of-words features:

1. The translation features result in very sparse feature vectors where only a small amount of the binary translation features has a positive value per instance.
2. It is often the case that synonyms occur in instances denoting the same meaning of the ambiguous focus word. These synonyms, however, are considered as two completely different words, as only exact overlap of lexical units is taken into account when measuring the similarity of these binary features.

---

[5] http://code.google.com/apis/language/

In order to tackle both issues, we made an alternative version of the translations features by applying Latent Semantic Analysis (LSA) on the set of bag-of-words translation features. LSA [14] once again starts from the distributional hypothesis that words that are close in meaning will occur in similar contexts. In order to compare distributions of different words, first a term-document matrix is created where the rows represent unique words, the columns stand for the documents and the cells contain the word counts per document. As is done by [16], we build one matrix per ambiguous focus word and use instances instead of full documents. In order to normalize the term frequencies in our feature-by-instance matrix, we applied TF-IDF (term frequency – inverse document frequency). In a next step, the singular value decomposition (SVD) technique is applied to construct a condensed representation of the feature space by reducing its dimensionality, which makes it possible to infer much deeper (*latent semantic*) relations between features.

SVD decomposes a given $m \times n$ term-by-document – or in our case feature-by-instance – matrix A into the product of three new matrices:

$$A = USV^T$$

where

- U is the $m \times r$ matrix whose columns are orthogonal eigenvectors of $A^T A$
- S is a diagonal $r \times r$ matrix whose diagonal elements are the r singular values of A (which are represented in descending order)
- V is the $r \times n$ matrix whose columns are orthogonal eigenvectors of AA^T, also called the right singular vectors.

When computing the SVD of a matrix, it is desirable to reduce its dimensions by keeping its first k singular values, that are supposed to represent *latent semantic dimensions* or the most important concepts related to the instances and terms. We selected the first 50 dimensions ($k = 50$) and mapped our training and test data in the newly constructed latent space.

2.4   Selection of the Classification Label

In our cross-lingual WSD approach, the classification label for a given training instance corresponds to the translation of the ambiguous focus word in the aligned translation in the target language. In order to detect all relevant translations for the twenty ambiguous focus words, we ran the unsupervised statistical word alignment method GIZA++ [17] with its default settings on the selected English Europarl sentences and their aligned translations. The obtained word alignment output for the ambiguous word was then considered to be the classification label for the training instances for a given classifier (e.g. the French translation resulting from the word alignment is the label that is used to train the French classifier). We created two experimental setups. The first training set incorporates the automatically generated word alignments as labels. The second training set uses manually verified word alignments as labels for the training instances. This second setup is then to be considered as the upper bound on the current experimental setup.

3   Classifier

In our CLWSD system, we used the memory-based learning (MBL) algorithms implemented in TIMBL [18], which has successfully been deployed in previous WSD classification tasks [15]. During the training phase, this method stores all training examples – together with their sense label – in memory (hence called memory-based learning). At classification time, a previously unseen test example is presented to the system and the algorithm looks for the k most similar examples or *nearest neighbors* in memory and performs an "average" of their senses to predict a class label. In order to measure the distance between the new occurrence and the examples in memory, a similarity metric is used. As is the case for most machine learning algorithms, TIMBL offers a large range of parameter settings (distance metrics, weighting methods, k parameter, class weights, etc.) that can all have a major impact on the classification results. This is why we decided to run optimization experiments on our training data by means of a Genetic Algorithm [19].

Genetic algorithms are search methods that are inspired by Darwin's theory of evolution – and its central idea of natural selection – and by genetics in biological systems. They start with a population of candidate solutions, called *individuals*, for a given search problem and explore different areas of the search space in parallel. Based on the Darwinian principle of "survival of the fittest", these candidate solutions are then combined to find the optimal, or better, solutions. In the case of parameter optimization, the individuals contain possible values for the classifier parameter settings. An essential part of the genetic algorithm is the *fitness function* that judges the quality of the obtained solutions and decides which individual will survive into the next generation. In our setup, the fitness function evaluates the selected parameter settings with respect to the classification accuracy. In a next step, new individuals are combined using procedures of mutation and crossover.

4   Evaluation

To evaluate our five classifiers, we used the sense inventory and test set of the SemEval "Cross-Lingual Word Sense Disambiguation" task [20]. The sense inventory was built up on the basis of the Europarl corpus: all retrieved translations of a polysemous word were manually grouped into clusters, which constitute different senses of that given word. Native speakers labeled 50 test sentences per ambiguous target word: they were asked to provide three contextually correct translations from the predefined clusters of Europarl translations. As evaluation metric, we use a straightforward accuracy measure, that divides the number of correct answers by the total amount of test instances. In order to define the baseline scores, the most frequent lemmatized translation was selected that resulted from the automated word alignment (GIZA++).

Table 1 gives an overview of all experimental results when combining different flavors of the system (binary versus latent semantic translation features, TIMBL applied with default versus optimized parameter settings). The figures clearly show that the optimized system incorporating local context and binary translation features achieves the best results and constantly beats the most frequent translation (MFT) baseline. Applying LSA on the translation features does not seem to improve the classification results. From a qualitative point of view, however, we discovered possible benefits of using the latent semantic translation features. A detailed analysis of the Dutch system output showed that for the system incorporating binary translation features, in average 71.2% of the sentences receive the MFT label, with on average 2.3 different translations per test word. The system that uses the latent semantic translation features only assigns the MFT label to 53% of the instances, which is comparable to the number of MFT labels in the gold standard (49.7%). This results in an average of 7.85 different translation labels per ambiguous target word. The system using the latent semantic translation features is less biased towards predicting the MFT label, which results in a more varied set of more precise translations for the ambiguous words under consideration. However, since the MFT is such a strong baseline, the latter system generates more varied but also more incorrect translation labels.

Table 1. Overall classification results expressed in accuracy scores averaged over all twenty test words

	French	Italian	Spanish	Dutch	German
Most Frequent Translation Baseline					
Baseline	0.658	0.526	0.601	0.594	0.523
CLWSD Classification results					
TIMBL Baseline local context + latent semantic features	0.588	0.532	0.603	0.455	0.449
TIMBL Baseline local context + binary translation features	0.691	0.593	0.621	0.560	0.524
Optimized TIMBL local context + latent semantic features	0.704	**0.634**	0.696	0.644	0.613
Optimized TIMBL local context + binary translation features	**0.752**	0.624	**0.705**	**0.684**	**0.668**
SemEval competition results					
T3-COLEUR	0.669	0.507	0.598	0.400	0.542
UvT-WSD	NA	NA	0.702	0.641	NA

Comparison with State-of-the-art Systems. We also compared the results with the two winning SemEval systems for the Cross-Lingual Word Sense Disambiguation task, UvT-WSD (that only participated for Dutch and Spanish) and T3-COLEUR. The UvT-WSD system [21], which also uses a k Nearest Neighbor classifier and a variety of local and global context features, yielded the best scores for Spanish and Dutch in the SemEval CLWSD competition. Although we also used a memory-based learner, our method is different from this system in the way the feature vectors are constructed. Alongside similar local context features, we also included translational evidence from multiple languages in our feature vector. For French, Italian and German, the T3-COLEUR system [22] outperformed the other systems in the SemEval competition. This system adopts a different approach: during the training phase a monolingual WSD system processes the English input sentence and a word alignment module is used to extract the aligned translation. The English senses, together with their aligned translations (and probability scores), are then stored in a word sense translation table, in which look-ups are performed during the testing phase. This system also differs from the other systems in that the word senses are derived from WordNet, whereas our system as well as the UvT-WSD system do not use any external resources. The classification scores show that the CLWSD system that incorporates the binary translation features clearly outperforms the winning SemEval systems for all five languages. These results confirm the potential advantages of using a multilingual approach to solving the cross-lingual WSD task.

Contribution of the Translational Evidence. We also ran a set of additional experiments to examine the contribution of the translation features to the WSD performance. To this goal, we built classifiers incorporating a different number of languages and different language combinations. If we take for instance French, we built four classifiers containing translation features from one particular language (German, Dutch, Italian, Spanish), four classifiers containing translation features from three different language combinations and six classifiers containing translation features from two different languages. We then averaged the score over all different language combinations in order to measure the average score according to the number of languages that is used in the feature vector.

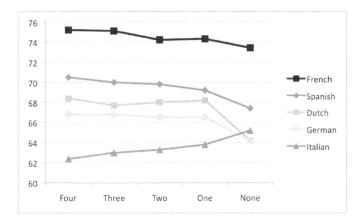

Fig. 1. Relation between the (averaged) classification performance and the number of languages incorporated in the feature vector

As is shown in Figure 1, the classifiers using translational evidence constantly achieve better results than the ones that merely use English local context features for four target languages, viz. French, Spanish, German and Dutch. For Italian, however, adding translation features does not result in better classification scores. For French and Spanish, and German to a minor extent, we even notice a clear correspondence between the number of languages that is incorporated in the feature vector and the performance of the WSD classifier.

Impact of Word Alignment Errors. One major advantage of the presented cross-lingual WSD approach is that all steps to create the system can be run automatically. As a consequence, the approach might be sensitive to error percolation between the different steps that are run automatically for the creation of the training and test feature vectors. Especially automatic word alignment is not error-prone yet, which might lead to erroneous translation labels in the training corpus. In order to measure the performance decrease caused by those errors that were introduced by the statistical word alignment module, we built a version of the system which contains manually-validated translation labels, and compared it to the system which contains the automatically-generated translation labels.

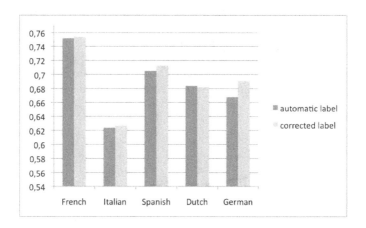

Fig. 2. Accuracy scores for the CLWSD system using the automatically generated translation labels and for the system using the manually corrected translation labels

Figure 2 clearly shows that the classification scores decrease only slightly when the automatically-generated word alignments are used; only the German classifier considerably improves (+ 2.3%) when using the corrected translation labels. These results confirm the viability of our setup as manual interventions in the data seem to result in very modest performance gains. As a consequence, our system can be developed fully automatically, which makes it very flexible and language-independent.

5   Conclusion

We presented a real multilingual classification-based machine learning approach to cross-lingual word sense disambiguation. Unlike other multilingual approaches to WSD, our CLWSD system does not use any manually created lexical resources, but relies only on parallel corpora for the automatic creation of the sense inventory and the construction of the training data set. The experimental results clearly confirm the validity of our multilingual approach to word sense disambiguation: the classifiers that employ translational evidence outperform the classifiers that merely use English local context features for four out of five target languages. Furthermore, a detailed comparison with all systems that participated in the SemEval cross-lingual word sense disambiguation task reveals that the proposed system obtains state-of-the-art results for the task in hand.

In future research, we would like to improve and extend the cross-lingual WSD system in different ways. Firstly, we would like to add more distant languages to the feature vector in order to measure whether languages from other language families contribute more/less to the classification results. As different languages tend to lexicalize different meaning of the word, one would expect that more distant languages will enable the system to distinguish between sense distinctions that are possibly not made by languages from the same language family. Secondly, we would like to test our approach for other Part-of-Speech categories. We have validated the CLWSD approach for ambiguous nouns, but the scope could be expanded to verbs or adjectives. We expect more noise due to erroneous word alignment for both PoS categories, but our system revealed to be quite robust against word alignment mistakes. Thirdly, we would like to apply a decompounding module to the training data and test the impact on the word alignment performance and classification accuracy.

To conclude, we strongly believe that our multilingual classification-based approach offers a very flexible, efficient and language-independent solution for the word sense disambiguation task. As all steps are performed automatically, and we only use a parallel corpus, we believe our approach proposes a valid answer to the knowledge-acquisition bottleneck for WSD.

References

1. Agirre, E., Edmonds, P.: Word Sense Disambiguation. Algorithms and Applications. Text, Speech and Language Technology. Springer (2006)
2. Navigli, R.: Word Sense Disambiguation: a Survey. ACM Computing Surveys 41(2), 1–69 (2009)
3. Landes, S., Leacock, C., Tengi, R.: Building Semantic Concordances, ch. 8, pp. 199–216. MIT Press, Cambridge (1998)
4. Fellbaum, C.: WordNet: An Electronic Lexical Database. MIT Press (1998)
5. Resnik, P., Yarowsky, D.: Distinguishing systems and distinguishing senses: New evaluation methods for word sense disambiguation. Natural Language Engineering 5(3), 113–133 (2000)

6. Brown, P., Pietra, S., Pietra, V., Mercer, R.: Word-sense disambiguation using statistical methods. In: Proceedings of the 29th Annual Meeting of the Association for Computational Linguistics, Berkeley, California, pp. 264–270 (1991)
7. Gale, W., Church, K., Yarowsky, D.: A method for disambiguating word senses in a large corpus. Computers and the Humanities 26, 415–439 (1992)
8. Diab, M.: Word Sense Disambiguation within a Multilingual Framework. Phd, University of Maryland, USA (2004)
9. Specia, L., Nunes, M., Stevenson, M.: Learning Expressive Models for Word Sense Disambiguation. In: Proceedings of the 45th Annual Meeting of the Association of Computational Linguistics, Prague, Czech Republic, pp. 41–48 (2007)
10. Tufiş, D., Ion, R., Ide, N.: Fine-Grained Word Sense Disambiguation Based on Parallel Corpora, Word Alignment, Word Clustering and Aligned Wordnets. In: Proceedings of the 20th International Conference on Computational Linguistics (COLING 2004), Geneva, Switzerland, pp. 1312–1318. Association for Computational Linguistics (August 2004)
11. Dagan, I., Itai, A., Schwall, U.: Two Languages are more Informative than One. In: Proceedings of the 29th Annual Meeting of the Association for Computational Linguistics, pp. 130–137 (1991)
12. Ng, H., Wang, B., Chan, Y.: Exploiting parallel texts for word sense disambiguation: An empirical study. In: 41st Annual Meeting of the Association for Computational Linguistics (ACL), Sapporo, Japan, pp. 455–462 (2003)
13. Koehn, P.: Europarl: a parallel corpus for statistical machine translation. In: Tenth Machine Translation Summit, Phuket, Thailand, pp. 79–86 (2005)
14. Landauer, T., Foltz, P., Laham, D.: An introduction to latent semantic analysis. Discourse Processes 25, 259–284 (1998)
15. Hoste, V., Hendrickx, I., Daelemans, W., van den Bosch, A.: Parameter Optimization for Machine-Learning of Word Sense Disambiguation. Natural Language Engineering, Special Issue on Word Sense Disambiguation Systems 8, 311–325 (2002)
16. Lopez de Lacalle, O.: Domain-Specific Word Sense Disambiguation. Phd, Lengoiaia eta Sistema Informatikoak Saila (UPV-EHU). Donostia 2009ko Abenduaren 14ean (2009)
17. Och, F.J., Ney, H.: A systematic comparison of various statistical alignment models. Computational Linguistics 29(1), 19–51 (2003)
18. Daelemans, W., van den Bosch, A.: Memory-based Language Processing. Cambridge University Press (2005)
19. Holland, J.: Adaptation in natural and artificial Systems. MIT Press (1975)
20. Lefever, E., Hoste, V.: Construction of a Benchmark Data Set for Cross-Lingual Word Sense Disambiguation. In: Calzolari, N., Choukri, K., Maegaard, B., Mariani, J., Odijk, J., Piperidis, S., Tapias, D. (eds.) Proceedings of the Seventh International Conference on Language Resources and Evaluation (LREC 2010), Valletta, Malta. European Language Resources Association, ELRA (May 2010)
21. van Gompel, M.: UvT-WSD1: A Cross-Lingual Word Sense Disambiguation System. In: Proceedings of the 5th International Workshop on Semantic Evaluation (SemEval 2010), Uppsala, Sweden, pp. 238–241. Association for Computational Linguistics (2010)
22. Guo, W., Diab, M.: COLEPL and COLSLM: An Unsupervised WSD Approach to Multilingual Lexical Substitution, Tasks 2 and 3 SemEval 2010. In: Proceedings of the 5th International Workshop on Semantic Evaluation, Uppsala, Sweden, pp. 129–133. Association for Computational Linguistics (2010)

Analyzing the Sense Distribution of Concordances Obtained by Web as Corpus Approach

Xabier Saralegi[1] and Pablo Gamallo[2]

[1] Elhuyar Foundation,
Osinalde Industrialdea 3, 20160 Usurbil, Spain
x.saralegi@elhuyar.com
[2] Centro de Investigação em Tecnologias da Informação (CITIUS)
Universidade de Santiago de Compostela, Galiza, Spain
pablo.gamallo@usc.es

Abstract. In corpus-based lexicography and natural language processing fields some authors have proposed using the Internet as a source of corpora for obtaining concordances of words. Most techniques implemented with this method are based on information retrieval-oriented web searchers. However, rankings of concordances obtained by these search engines are not built according to linguistic criteria but to topic similarity or navigational oriented criteria, such as page-rank. It follows that examples or concordances could not be linguistically representative, and so, linguistic knowledge mined by these methods might not be very useful. This work analyzes the linguistic representativeness of concordances obtained by different relevance criteria based web search engines (web, blog and news search engines). The analysis consists of comparing web concordances and *SemCor* (the reference) with regard to the distribution of word senses. Results showed that sense distributions in concordances obtained by web search engines are, in general, quite different from those obtained from the reference corpus. Among the search engines, those that were found to be the most similar to the reference were the informational oriented engines (news and blog search engines).

Keywords: Web As Corpus, Word Sense Disambiguation.

1   Introduction

Most statistical approaches to solving tasks related to Natural Language Processing (NLP) as well as lexicographic works use corpora as a resource of evidences. However, one of the biggest problems encountered by these approaches is to obtain an amount of data that could be large enough for statistical and linguistic analysis. Taking into account the rapid growth of the Internet and the quantity of texts included in it, some researchers have proposed using the Web as a source for building corpora [1]. Two strategies have been proposed for exploiting the web with that objective in mind:

A. Gelbukh (Ed.): CICLing 2013, Part I, LNCS 7816, pp. 355–367, 2013.
© Springer-Verlag Berlin Heidelberg 2013

- *Web As Corpus*: The web is accessed directly as a corpus. This access is usually performed by means of commercial web search engines (*Bing, Yahoo, Google...*), which are used to retrieve concordance lines showing the context in which the user's search term occurs. *WebCorp* [2] is a representative linguistic tool based on this strategy.
- *Web For Corpus*: This strategy consists of compiling a corpus from the web to be accessed later. This compilation process can be performed by crawling the web and also by using commercial web search engines. This latter approach consists of sending a set of queries including seed terms corresponding to a certain topic or domain, and then retrieving the pages returned by the engine.

The two strategies usually rely on web search engines (SEs) in order to retrieve pages with text and in this way build word concordances or text corpora. Using APIs provided by SEs offers several advantages. It makes the treatment of spam and other low-quality, undesired contents easier. Besides, these APIs provide a high coverage of the web.

The Web as Corpus approach is more suitable than Web for Corpus for those tasks requiring an acceptable quantity of examples of concordances for any word (e.g., Distributional Similarity, Information Extraction, Word Sense Disambiguation, etc...). However, some problems can arise from using SEs for concordance compilation. For example, Aguirre et al. [4] found that the great number of erotic web pages strongly influenced their experiments on WSD. The set of pages retrieved by the web SE is dependent on ranking criteria1, which are not specified according to linguistic features such as frequency of use of each sense. The users of commercial SEs have other needs than those focused on obtaining specific pieces of text information. Broder [5] states that the target of search queries is often non-informational (more than 50%), since it might be navigational when the queries are looking for websites, or transactional when the objective is to shop, download a file, or find a map. Thus, criteria related to these needs are also reflected in the above mentioned ranking factors. The main page-ranking factors mainly rely on popularity, anchor text analysis and trusted domains, but not on content.

Our work is based on two main assumptions:

- *Assumption 1. SemCor* is a sense-tagged corpus [6], which can be regarded as a gold-standard reference of the "real" distribution of word senses in an open domain. According to Martinez and Agirre [7], corpora with similar distribution to that of *SemCor* get the best results in the task of WSD for an open domain. Word Sense Disambiguation (WSD) systems with the best performance in Senseval-2 were trained with it. We are aware that the concept of reference corpora is arguable as some author point out [8].
- *Assumption 2*. The Web is, in terms of variety of genres, topics and lexicons, close to the most traditional open domain "balanced" corpora, such as the Brown or BNC [3].

---

[1]   An example of factors used for search engine rankings are listed here:
   http://www.seomoz.org/article/search-ranking-factors

On the basis of these two assumptions, we aim to validate the following hypothesis: the ranked results obtained by SEs are not a representative sample in terms of sense distribution, since they follow ranking criteria focused on non-linguistic relevance and only first results are usually compiled. In other words, the linguistic examples or concordances extracted by web SEs are biased by non-linguistic criteria. A high bias would indicate that web-based concordances compilation is not useful at all in some NLP and lexicography tasks. For example, linguistic information obtained by an SE used for knowledge extraction such as cross-lingual distributional similarity [9], semantic-class learning [10], or consultation (e.g., lexicographic, translators, writers or language learners) might not be reliable. In the remaining sections, we will attempt to confirm or reject such a hypothesis.

2   Related Work

There are few works which deal with linguistic adequacy of concordances obtained by SEs. Chen et al. [11] describe an experiment to manually evaluate concordances included in web documents retrieved from an SE for 10 test words. They annotated by hand about 1,000 to 3,000 instances for each test word. In particular, the authors evaluate two pieces of information: the quality of the web documents returned by the SE, and sense distributions. Concerning sense distribution, they concluded that, on the one hand, the most frequent senses from web-based corpora are similar to *SemCor* and, on the other, web-based corpora may provide more diverse senses than *SemCor*. However they do not perform any correlation analysis to draw that conclusion. As we will show later in Section 5, a correlation analysis performed over their results shows a low correlation in terms of sense distribution between web-based corpus and *SemCor*.

Other works analyze some aspects related to linguistic adequacy but for different purposes. Diversity in web-rankings is a topic closely related to word-sense distribution analysis. Santamaría et al. [12] propose a method to promote diversity of web search results for small (one-word) queries. Their experiments showed that, on average, 63% of the pages in search results belong to the most frequent sense of the query word. This suggests that "diversity does not play a major role in the current *Google* ranking algorithm". As we will show in Section 5, the degree of diversity of the concordances we have retrieved is still lower: in our experiments 72% of the concordances belong to the most frequent sense.

3   Methodology for Analyzing the Adequacy of *Web as Corpus*

Our objective is to verify whether the distribution of senses in the rankings obtained from SEs are representative in linguistic terms for an open domain. Our analysis relies on *SemCor* as evidence, since it is a well-known, manually annotated open domain reference corpus for word senses according to *Wordnet*. In addition, we also measure two further properties of the concordances retrieved by SEs, namely sense diversity and linguistic coherence (i.e., typos, spelling errors, etc...).

3.1   How to Obtain Concordances of a Word from the Web?

Several SEs have been used in the literature in order to collect examples of concordances from the web. Most authors use SEs directly by collecting the retrieved snippets. Web As Corpus tools such as *WebCorp* are more linguistically-motivated tools. In that sense they offer parameters to post-process SE rankings (case sensitive searches to avoid named entities, no searches over links to avoid a somehow navigational bias...). Anyway they still depend on SE rankings. So they raise the same problems mentioned in Section 1. Other emergent SEs are those focused on specific domains and genres such as news or blogs. These SEs are interesting for linguistic purposes because bias produced by factors related to navigational and transactional is avoided. In addition, their text sources are not domain restricted. In fact, newswire-based corpora are often built for open domain knowledge extraction purposes. However, some authors [3] point out that, in terms of variety of genres and topics, Web is closer to traditional "balanced" corpora such as the BNC. Blog is a new genre not present in traditional written sources but similar to them in terms of distribution of topics. In order to analyze and compare the influence of these characteristics, the following engines have been evaluated: *WebCorp*, *Google News Archive* (*GNews*), and *Google Blog Search* (*GBlog*). See Table 1 for more details.

Table 1. Characteristics of SEs

SE	Domain	Genre	Query	Ranking
WebCorp	Open	Open	inform.	topic
			navig.	popularity
			transac.	... (See first note)
GBlog	Open	Blogs	inform.	topic
GNews	Open	News	inform.	topic

In order to guarantee a minimum linguistic cohesion of the concordances, the following parametrizations were used for each SE. English is selected as the target language in all of them. In *WebCorp2*, Bing has been selected as the API because provides the best coverage. Case-sensitive searches were performed. The search over links option was disabled in order to mitigate navigational bias. Span of ± 5 words for concordances was established. *GNews* and *GBlog* do not offer choice for case-sensitive searches. So, case-sensitive treatment was done after retrieving the snippets. In the cases of *GNews* and *GBlog*, searches were performed only on the body of documents and not on the titles (*allintext* operator was used).

3.2   Selecting Test Words

10 test words (see Table 2.) are randomly selected from the *SemCor* 1.6 corpus, a corpus where all words are tagged with their corresponding sense according to

---

[2]   *WebCorp* has included recently *Gnews* and *Gblog* APIs.

WordNet 1.6. Due to the small size of the sample several conditions were established in order to guarantee the representativeness of the test words and the corresponding contexts:

- Nouns are selected because they are the largest grammatical group.
- More than 1 sense in *SemCor* because we want to focus on ambiguous words.
- Minimum frequency of 50 on *SemCor* corpus. As McCarthy [13] pointed out, *SemCor* comprises a relatively small sample of words. Consequently, there are words where the first sense in *WordNet* is counter-intuitive. For example, the first sense of *"tiger"* according to *SemCor* is an audacious person, whereas one might expect carnivorous animal to be a more common usage.

Table 2. Selected test words from *SemCor* and their sense distribution

Word	Sense distribution
church	1=0.47,2=0.45,3=0.08
particle	1=0.63,2=0.35,3=0.02
procedure	1=0.73,2=0.27
relationship	1=0.60,2=0.21,3=0.19
element	1=0.71,2=0.21,3=0.06,4=0.02
function	1=0.58,2=0.32,3=0.09
trial	1=0.45,2=0.03,4=0.52
production	1=0.64,2=0.21,3=0.11,4=0.04
newspaper	1=0.66,3=0.02,2=0.29,2;1=0.02
energy	1=0.74,2=0.10,3=0.12,4=0.05

3.3   Annotation of Web Based Concordances

Each test word is submitted to the three different SEs. The number of retrieved snippets, i.e., word concordances, may change depending on both the query word and the SE. So, in order to obtain more comparable samples, the first 250 concordances are retrieved for each case. As the number of test examples is still too much to analyze by hand, to save work without missing the rank information, only an interpolated sample of 50 concordances was analyzed. The hand analysis involves manually tagging the sense of the test words according to *WordNet* 1.6.

3.4   Measuring the Adequacy of Concordances

The main objective here is to measure differences in terms of sense distribution between *SemCor* and the concordances retrieved by the SE. However, besides sense distribution, our aim is also to measure both sense diversity and linguistic coherence. Let us describe first how we measure sense diversity, then linguistic coherence, and finally sense distribution.

3.4.1   Sense Diversity

We associate the term *"sense diversity"* with text corpora whose word occurrences cover a great variety of senses. It is possible to know to a certain extent the degree of diversity of a corpus by observing the senses of a sample of words. In particular, diversity can be measured by comparing the number of possible senses of the test words (e.g., their *WordNet* senses) with the number of different senses that are actually found in the corpus, i.e., in our collections of concordances. The higher the number of senses covered by the concordances, the greater their degree of diversity. Concordances with much sense diversity tend to be open to many domains.

3.4.2   Linguistic Coherence

The quality and linguistic coherence of the retrieved concordances can vary from totally nonsensical expressions to high quality texts. So, coherence, or more precisely "level of coherence", is also taken into account in our evaluation protocol. To do this, the annotators can assign four possible coherence values to each retrieved concordance:

- Score 0. The concordance has no problems.
- Score 1. The concordance has serious typographical errors or morphosyntactic problems, but it can be understood.
- Score 2. The query word is part of a Named Entity, e.g., *"town"* in *"Luton Town FC Club"*.
- Score 3. The concordance is totally nonsensical and cannot be understood at all.

The range of values is from 0 (coherent) to 3 (totally incoherent or nonsensical). It should be borne in mind that values 1 and 2 could be unified since named entities written in lower-case seem to be typographical errors. However, we preferred to keep the two coherence levels because value 1 still allows us to assign a *WordNet* sense to the key word, but it is not the case when the coherence level is 2. On the basis of the notion of level of coherence, we define "degree of incoherence", which is associated with a concordance collection. The degree of incoherence of a concordance collection, noted ϕ, is computed as follows:

$$\phi = \left(\sum_{i}^{n} L(c_i) \right) / 3n \tag{1}$$

where $L(c_i)$ stands for the level of coherence of concordance c_i, and n is the number of concordances in the collection. Let us suppose that we have a collection of 4 concordances, with the following levels of coherence for each concordance: 0, 1, 0, 3. The degree of incoherence of the total collection is then 4/12 = 0.33. The values of this function are ranged from 0 (fully coherent) to 1 (totally incoherent).

3.4.3   Sense Distribution

The distributions of senses found in the three SE concordance collections are compared with those extracted from *SemCor* by analyzing the Pearson correlation between them. The correlation between two sets of concordances is computed by considering the relative frequencies of those senses of the word that have been found in, at least, one of the two sets.

Besides the strict correlation, we are also interested in verifying properties concerning sense dominance. For instance, two collections may share (or not share) the same dominant sense. When their dominant senses are not the same, and one of them is domain-specific (scientific, technical, etc.), then the two collections should be considered very different in terms of sense distribution, regardless of their specific Pearson correlation. On the other hand, when they share the same dominant sense, it is also important to observe whether there are differences in terms of degree of dominance. If the main sense is very dominant in one collection and not so dominant in the other one, we may infer that there are significant differences in sense distribution. This is true even if the Pearson correlation is actually very high. Let us see an example. Word *"production"* has 4 senses with the following two sense distributions, in *SemCor* and *GNews*, respectively:

- *SemCor*: 0.64 0.21 0.11 0.04
- *Gnews*: 0.98 0.02 0.0 0.0

The Pearson correlation between *SemCor* and *GoogleNews* is very high: > 0.97. However, from a linguistic perspective, the distributions are very different. While the sense distribution in *SemCor* may be considered as an evidence for content heterogeneity (there are three senses with more than 10% occurrences), sense distribution in *GoogleBlogs* shows that concordances are content homogeneous. As in *GoogleBlogs* only one sense covers more than 98% of the word occurrences, it means that concordances are retrieved from a domain-specific source. By contrast, concordances of *SemCor* seem to represent more open and balanced text domains. Here, we also should take into account the conclusions to which Kilgarriff came in [14], where he argued that the expected dominance of the commonest sense rises with the number of corpus instances. It follows that the dominance of one sense is higher in GoogleNews because of the corpus size.

The rank of retrieved concordances for each SE is also analyzed. We are interested in observing the order of appearance of the senses among the web ranking. An adequate order will be that one in which concordances are ordered according to the probability of senses of the search word. Thus, concordances including the most probable ones should be on the top of the rank. For linguistic consultations, for instance, it is better to show concordances including the most common senses of the search word at the top of the ranking. In addition, those strategies that only retrieve the first concordances of the SE could also perform better if top concordances corresponded to the most common senses. Once again, we use *SemCor* to prepare a reference rank according to sense probabilities calculated from *SemCor*. In order to measure the adequacy of the rank of web-concordances, we compute the Spearman correlation between the web concordances rank and the *SemCor* based reference rank.

4   Results

The results concerning sense diversity, linguistic coherence, and sense distribution are shown and analyzed as follows:

4.1   Sense Diversity

In total, the 10 test words have 49 different *WordNet* senses. Among these 49 senses, the collection of concordances from *WebCorp* contains instances of 34 senses, two more (32) than the senses found in *SemCor* for the same 10 words. The concordances of *Gblog* contain a different 31 senses while those of *Gnews* only 27. In Table 3, we show the percentage of different senses we found in each corpus with regard to the total number of *Wordnet* senses attributed to the 10 test words (first column), as well as to the number of senses these words have in *SemCor* (second column). In the last column, we show the number of senses appearing in each collection of concordances that do not appear in *SemCor*. It follows that the *WebCorp* corpus may provide more diverse senses and, therefore, more domain diversity, than the two corpora built from the Google engines. In addition, we may also infer that the journalism articles seem to be more restricted in terms of domain diversity than the posts of blogs. However, we have to take into account that high diversity does not imply balanced sense distribution.

Table 3. Sense diversity

	% senses of test words	% senses in *SemCor*	#new-senses
WebCorp	69%	81%	8
Gblog	63%	78%	6
Gnews	55%	72%	5

4.2   Linguistic Coherence

Table 4 shows information on levels of incoherence associated with the three web-based concordance collections. The three first columns show the total values of 3 levels of coherence (level 0 is not shown but can be inferred). The forth column measures the degree of incoherence for each collection, according to formula 1 (see above). In the last column, we show the percentage of concordances having some positive incoherence value (i.e., 1, 2, or 3) for each collection.

We can observe that *WebCorp* is the SE that provides more incoherent concordances at the 3 levels. In addition, in *WebCorp* almost 1 context out of 4 has some problems of coherence. This is probably due to the fact that *WebCorp* covers the whole Web, containing many not very confident text sources. The degree of incoherence in *Gblog* is also relatively high (0.12), against only 0.04 of *Gnews*, which is then the most reliable source of textual data in our experiments So, the linguistic quality of the corpora built with Google engines is clearly better than that of *WebCorp*.

Table 4. Sense diversity

	level 1	level 2	level 3	φ	incoherence (%)
WebCorp	68	6	48	0.15	24%
Gblog	34	1	25	0.07	12%
Gnews	32	0	9	0.04	8%

4.3   Sense Distribution

4.3.1   Pearson Correlation

The senses found in the web-based concordances are compared with those extracted from *SemCor* by analyzing the Pearson correlation between them (see Table 5). This table is organized as follows. The test word is in the first column. The following columns show the Pearson correlation between the *SemCor* and sense distributions corresponding to the different web-based concordances (*WebCorp*, *Gnews*, or *Gblog*).

Table 5. Pearson correlation of sense distribution regarding to *SemCor*

	WebCorp	*GBlog*	*GNews*
church	0.78	0.85	0.70
particle	0.10	0.53	0.20
procedure	0.62	0.88	0.89
relationship	-0.28	-0.41	0.50
element	0.28	0.29	0.95
function	0.50	-0.03	0.03
trial	0.60	0.63	0.7
production	0.61	0.89	0.97
newspaper	0.93	0.99	0.66
energy	0.97	0.97	0.98
average	0.51	0.56	0.66

As far as the Pearson coefficient is concerned, the average correlation of WebCorp and *SemCor* is 0.51, which is the lowest correlation. The average correlation between *Gblog* and *SemCor* is 0.56, and the one between *GNews* and *SemCor* is the highest:

0.66. As the correlation values between 0.51 and 0.79 are interpreted as being "low", we may consider that there is always a low correlation between *SemCor* and our three web-based concordance collections. If we conduct a more detailed analysis word by word and compute the correlations of each test word, we can observe that there are three words ("*newspaper*", "*production*", and "*procedure*") with moderate correlations (between 0.80 and 0.86), four words ("*energy*", "*church*", "*trial*", and "*element*") with low correlations (between 0.51 and 0.79), and three ("*particle*", "*function*", and "*relationship*") without any correlation at all, since their values are lower than the significance level (0.35, p<.01) established for tests with 50 pairs.

It should be noticed that these results seem to be not in accordance with those experiments reported in [10], where the web-based corpus is "quite similar" to *SemCor* in terms of sense distribution. In that work, the notion of "quite similar" must be considered as a non-technical and naive intuition, since no correlation measure was computed. Considering the sense frequencies reported in that work, we computed the Pearson correlation between *SemCor* and their concordances for the 9 ambiguous test words used in their experiments. Table 6 shows that the average correlation is low (0.59), very close to our results (total average of the three collections: 0.58).

Table 6. Pearson correlation between concordances reported in Chen et al. [11] and *SemCor*

	Chen et al. [11] concordances
author	1
back	0.94
cart	-1
case	0.85
center	0.15
core	0.35
mind	1
sequence	1
toast	0.99
average	0.59

4.3.2   Analysis on Dominant Senses

Besides the statistical test, it is also important to verify further qualitative aspects, in particular whether concordances are comparable in terms of sense dominance. More precisely, given two collections of concordances, we checked both whether they share the same dominant senses and whether the same dominant senses have a similar degree of dominance.

We observed that *SemCor* and each web-based concordances do not share the same dominant sense in several cases. In addition, for most of these words, their dominant senses in the web-based concordances are domain-specific senses, e.g., physics for "*particle*", computer science for "*function*", show business for "*production*", or gossip news for "*relationship*". It should be noticed that these specific senses are not dominant in *SemCor* and they do not correspond to the first sense in *WordNet*. The high relevance given by non-linguistic ranking criteria to webs dealing with scientific, business or gossip topics could explain the large number of domain-specific senses in the concordances.

On the other hand, for many cases where the test word shares the same dominant sense in both *SemCor* and the web-based concordances, we observe that there are significant differences in terms of degree of dominance. In general, the sense distribution of *SemCor* seems to be more balanced than that of web-based concordances. Six words had same dominant sense in both *SemCor* and *GNews*, but in all cases the shared sense is clearly more dominant in *GNews*. The average degree of dominance in *SemCor* is 62% against 72% in the web-based concordances. As we showed above in 4.3.1, these differences may not be very significant for the Pearson correlation, but from a linguistic point of view, they are very significant since they denote that the web-based concordances are more homogeneous in terms of linguistic content. Once again, the ranking criteria of the SE could give more relevance to a very restricted subset of topics among all of those we can find in the open domain web.

In addition, we can find further qualitative differences between *SemCor* and web-based concordances. On the one hand, it should be noted that web-based concordances introduce new technical or domain-specific senses that are not in *SemCor*. On the other hand, we can find cases where the transactional function of some webs may influence sense distribution. For instance, the second sense of "*trial*" (very marginal in *SemCor*) is very important in *WebCorp* and *GBlog* because of the high number of commercial pages with "free trial" software.

4.3.3   Spearman Correlation

Finally, the order of the senses appearing in the web concordances ranking is also analyzed. We check whether web-concordances of test words are sorted by the frequency of use of the included sense. For this purpose, the reference ranking for each test word and SE is prepared by sorting all the collected concordances according to sense probabilities mined from *SemCor*.

For example, suppose we collect the following concordances ranking from the web for a test word including 4 senses: $\{(context_1, sen_1), (context_2, sen_2), (context_3, sen_1), (context_4, sen_1)\}$. The *SemCor*-based ranking (the reference) is built by sorting all concordances according to sense probability estimated from *SemCor* ($sen_2 = 0.8$, $sen_1 = 0.2$): $\{(context_2, sen_2), (context_1, sen_1), (context_3, sen_1), (context_4, sen_1)\}$. Contexts with the same sense keep the original order. Then, the Spearman correlation between original concordances ranking and *SemCor*-based reference ranking is calculated (see Table 7).

Notice that if all concordances of the ranking are analyzed, we observe that only *GBlog* concordances are correlated. Other SEs provide some correlation only if the first 50 concordances are selected. So, it seems that top of rankings are more adequate in terms of sense probability.

Table 7. Spearman correlation of concordance ranking with respect to SemCor

	WebCorp		GBlog		GNews	
	All	top50	All	top50	All	top50
church	0.41	0.98	0.58	0.68	0.43	-0.25
particle	0.63	1	0.57	1	0.58	0.65
procedure	-0.59	-0.76	0.46	1	0.11	1
relationship	0.61	0.53	0.49	-0.25	0.70	0.37
element	0.53	1	0.53	0.99	0.38	0.95
function	-0.42	0.94	-0.08	-0.59	-0.75	-0.75
trial	0.13	0.58	0.80	0.99	0.55	0.82
production	-0.11	-0.01	1	1	0.19	0.82
newspaper	0.80	0.79	0.30	0.2	0.17	0.31
energy	0.42	0.66	0.68	1	0.51	0.4
average	0.24	0.57	0.53	0.6	0.30	0.43

5   Conclusions

We have proposed an experimental method to verify whether the distribution of senses in the rankings obtained from SEs are balanced, representative, and coherent in linguistic terms. Taking *SemCor* as a balanced reference, we observed that the concordances retrieved by different SEs have low correlation with *SemCor* with regard to sense distribution. If we consider that the diversity of topics and domains in the web is close to that of most traditional open domain balanced corpora, we may infer from our experiments that the sense distribution bias is due to the fact that web engines rank their pages using non-linguistic criteria. It should be noted that the best correlation was achieved with SEs that only cover a part of the web (news and blogs), whose text sources are thus rather far, in terms of topic and genre distribution, from a traditional balanced corpus. By contrast, the worse correlation was achieved by the search engine (*WebCorp*) using the entire web as text source. We can surmise that ranking factors related to popularity or navigational queries introduce some non-linguistic bias in the concordances retrieved by general-purpose SEs. Furthermore, some SEs may retrieve concordances with serious problems concerning linguistic coherence (24% of concordances of *WebCorp* display problems of linguistic coherence). All these observations lead us to conclude that word sense information obtained by SEs used for knowledge extraction, word sense disambiguation, or lexicographic consultation might not be totally reliable. However, this final conclusion should be corroborated by carrying out further experiments over larger samples.

References

1. Kilgarriff, A., Grefenstette, G.: Introduction to the special issue on the web as corpus. Comput. Linguist. 29, 333–347 (2003)
2. Morley, B.: WebCorp: A tool for Online Linguistic Information Retrieval and Analysis. In: Renouf, A., Kehoe, A. (arg.) The Changing Face of Corpus Linguistics. Rodopi (2006)

3. Baroni, M., Bernardini, S.: WaCky!: Working Papers on the Web as Corpus. Gedit (2006)
4. Agirre, E., Ansa, O., Hovy, E., Martínez, D.: Enriching very large ontologies using the WWW. In: Proceedings of the ECAI 2000 Workshop «OntologyLearning» (2000)
5. Broder, A.: A taxonomy of web search. SIGIR Forum 36, 3–10 (2002)
6. Mihalcea, R.: Semcor semantically tagged corpus (1998),
   `http://www.cse.unt.edu/rada/downloads.html`
7. Martinez, D., Agirre, E.: The effect of bias on an automatically-built word sense corpus. In: Proceedings of the Fourth International Conference on Language Resources and Evaluations (LREC 2004), Lisbon, Portugal (2004)
8. Kilgarriff, A.: Getting to Know Your Corpus. In: Sojka, P., Horák, A., Kopeček, I., Pala, K. (eds.) TSD 2012. LNCS, vol. 7499, pp. 3–15. Springer, Heidelberg (2012)
9. Nakov, P., Nakov, S.: Improved word alignments using the web as a corpus. In: Proceedings of RANLP 2007 (2007)
10. Kozareva, Z., Riloff, E., Hovy, E.: Semantic class learning from the web with hyponym pattern linkage graphs. In: Proceedings of ACL 2008: HLT, pp. 1048–1056 (2008)
11. Chen, P., Brown, D., Tran, A., Ozoka, N., Ortiz, R.: Word sense distribution in a web corpus. In: 2010 9th IEEE International Conference on Cognitive Informatics (ICCI), pp. 449–453 (2010)
12. Santamaría, C., Gonzalo, J., Artiles, J.: Wikipedia as sense inventory to improve diversity in Web search results. In: Proceedings of the 48th Annual Meeting of the Association for Computational Linguistics, pp. 1357–1366. Association for Computational Linguistics, Stroudsburg (2010)
13. McCarthy, D., Koeling, R., Weeds, J., Carroll, J.: Finding predominant word senses in untagged text. In: Proceedings of the 42nd Annual Meeting on Association for Computational Linguistics. Association for Computational Linguistics, Stroudsburg (2004)
14. Kilgarriff, A.: How Dominant Is the Commonest Sense of a Word? In: Sojka, P., Kopeček, I., Pala, K. (eds.) TSD 2004. LNCS (LNAI), vol. 3206, pp. 103–111. Springer, Heidelberg (2004)

MaxMax:
A Graph-Based Soft Clustering Algorithm
Applied to Word Sense Induction

David Hope and Bill Keller

University of Sussex
Cognitive and Language Processing Systems Group
Brighton, Sussex, UK
davehope@gmail.com, billk@sussex.ac.uk

Abstract. This paper introduces a linear time graph-based soft clustering algorithm. The algorithm applies a simple idea: given a graph, vertex pairs are assigned to the same cluster if either vertex has maximal affinity to the other. Clusters of varying size, shape, and density are found automatically making the algorithm suited to tasks such Word Sense Induction (WSI), where the number of classes is unknown and where class distributions may be skewed. The algorithm is applied to two WSI tasks, obtaining results comparable with those of systems adopting existing, state-of-the-art methods.

1 Introduction

A Natural Language Processing (NLP) task may require a set of words to be grouped or clustered into subsets, where each subset represents a distinct lexicological class. For example, Word Sense Induction (WSI), the task of automatically determining word senses from text, is approached in this paper by clustering words associated with a polysemous target word into subsets of semantically related words. The words in each subset are then taken to define a different sense of the target word. For example, if *orange* is a target word associated with the set of words {*red, apple, yellow, banana, green, pear*}, assignment of the words in this set to two subsets {{*red, yellow, green*},{*apple, banana, pear*}} defines two senses of *orange*: the first representing the colour sense of *orange*, the second, its fruit sense. In practice of course, the use of very large corpora in NLP means that there is a need to cluster much larger sets of words than illustrated here. A computationally efficient clustering process is therefore needed. In addition, as words associated with a target word may themselves be polysemous, clustering should also be able to assign words to two or more senses of the target word.

A WSI system such as that outlined above has the potential to alleviate the lexicographer's task of manually identifying, collating, and exemplifying word senses: an enormous undertaking, given both the number of existing senses and the rate at which new senses are introduced into language. In principle, WSI

A. Gelbukh (Ed.): CICLing 2013, Part I, LNCS 7816, pp. 368–381, 2013.

avoids reliance on a pre-defined sense inventory[1], as required in Word Sense Disambiguation (WSD). WSD systems assign pre-defined senses to words on the basis of context. In contrast, WSI systems follow the dictum that *"The meaning of a word is its use in the language."* [3] to discover senses through examination of context of use in large text corpora. As a consequence, rare, fine-grained and domain specific senses not defined in existing inventories can be induced [4].

2   A Graph-Based Approach to Word Sense Induction

WSI is approached in this paper using a graph-based model of word co occurrence. A graph $G = (V, E)$ consists of a set of vertices V and a set of edges $E \subseteq V \times V$. In the present approach, each vertex $v \in V$ represents a word. An edge $(u, v) \in E$ is a pair of vertices. An edge represents a symmetrical relationship between words u and v; here, that u and v co-occur in the contexts of a target word. An edge-weighted graph assigns to each edge a weight $w(u, v)$. In the present work edge-weights can be understood as quantifying the strength or significance of word co-occurrence relationships.

Figure 1 shows an edge-weighted graph G_{orange} in which the target word vertex is *orange* and the set of words associated with *orange* are represented as adjacent vertices (i.e. words found to co-occur in *orange*'s contexts of use).

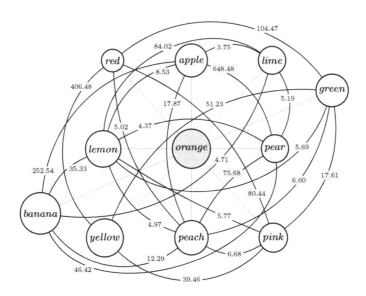

Fig. 1. G_{orange}

---

[1] In practice, evaluation of a WSI approach requires the use of a gold standard sense inventory, for example WordNet [1] or OntoNotes [2].

In principle, 'contexts of use' might be interpreted as sentences, paragraphs, or context windows containing the target word. In the work described, context words are nouns occurring in co-ordination patterns [5][2]. Edge weights are scores provided by the Log Likelihood Ratio (LLR) [6], a measure of how significant it is that two words u and v co-occur [7].

Senses may be induced by applying a clustering algorithm to identify subgraphs of G_{orange}, as illustrated in Fig. 2. Each subgraph (cluster) can then be assigned a sense of the target word, either by mapping the cluster to a sense given in an inventory (e.g. WordNet [8, 9]) or to a gold standard class [10–12].

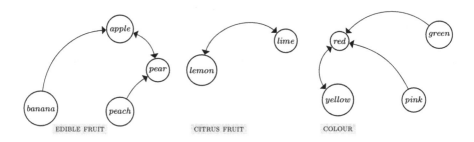

Fig. 2. $C_{G_{orange}}$, a clustering solution for the graph G_{orange} shown in Fig. 1. Sense labels are obtained using the cluster to sense mapping algorithm proposed in [8]

The word co-occurrence model outlined here is similar to models previously applied in WSI, notably to those presented in [5, 10]. A key difference however is the use of a novel clustering algorithm, MaxMax.

3   MaxMax

MaxMax is a non-parameterised, soft-clustering algorithm applicable to edge-weighted graphs. A notion of *maximal affinity* is used, where affinity between vertex pairs u and v is quantified by edge weights $w(u, v)$. A vertex u is said to have maximal affinity to a vertex v if the edge weight $w(u, v)$ is maximal amongst the weights of all edges incident on u. In this case, v is said to be a *maximal vertex* of u (v need not be unique). Two principles are applied: 1) vertex pairs u,v are assigned to the same cluster if either vertex is a maximal vertex of the other; and 2) maximal affinity implies a *directed* relationship: if v is a maximal vertex of u then there is a directed relationship from v to u.

---

[2] Nouns are extracted from the British National Corpus (BNC) using the regular expression NP(, NP)*,?(CJC NP)+ where CJC = (and|or|nor) and NP, a noun phrase, = AT?(CRD)*(ADJ)*(NOUN)+.

Algorithm 1. MaxMax

1: **procedure** MAXMAX($G = (V, E)$)
2:     construct a directed graph $G' = (V, E')$ where:
3:         $(v, u) \in E'$ iff $(u, v) \in E$ and v is a maximal vertex for u
4:     mark all vertices of G' initially as *root*
5:     **for** each vertex v of G' **do**
6:         **if** v is marked *root* **then**
7:             mark any descendant u of v ($u \neq v$) as $\neg root$
8:         **end if**
9:     **end for**
10: **end procedure**

MaxMax consists of two discrete stages:

Stage 1. Graph Transformation. In stage 1 (lines 2 and 3 of Algorithm 1) MaxMax takes a weighted graph G and transforms it to an unweighted, directed graph (*digraph*) G'. The maximal affinity relationships between vertices of G are used to determine the direction of the edges in G'. An example of the way in which a weighted undirected graph is transformed to an unweighted, directed graph is shown in Fig. 3.

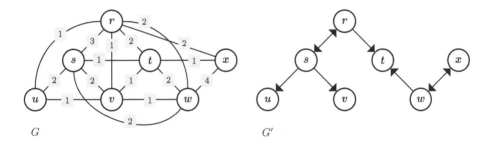

Fig. 3. G and its transformation to an unweighted directed graph G'

Stage 2. Identifying Clusters. In a digraph G', a vertex v is said to be a *descendant* of a vertex u if there is a directed path from u to v. For example, in Fig. 3 vertex v is a descendant of vertices s and r. In stage 2, clusters are found by tracing directed paths in G' to identify rooted subgraphs of a particular type (lines 4 to 9 of Algorithm 1). The vertices of each subgraph define a distinct cluster. This is made precise as follows.

A directed graph is said to be *quasi-strongly connected* (QSC) if for any vertices v_i and v_j, there is a vertex v_k (not necessarily distinct from v_i and v_j) such that there is a directed path from v_k to v_i and a directed path from v_k to v_j.

It is not hard to show that a QSC digraph must contain at least one vertex v_r which is a *root* in the sense that every other vertex can be reached by following a directed path from v_r. Given a directed graph G', a subgraph of G' is a *maximal QSC subgraph* if it is a QSC digraph and it is not possible to add any further vertices or edges from G' without rendering the subgraph non-QSC.

Clusters are identified by finding the *root* vertices of maximal QSC subgraphs of G'. This is achieved simply by marking all descendants of a given vertex as ¬*root*. For example, consider vertex s in the directed graph G' of Fig 4, which is initially marked as a *root*. The descendant vertices of s are u and v thus marked as ¬*root*. In turn, s, as a descendant of r, is marked ¬*root*[3]. At the end of stage two, vertices that are still marked as *root* vertices uniquely identify clusters, since they correspond to the roots of maximal QSC subgraphs of G'.

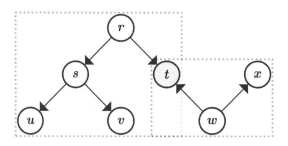

Fig. 4. Two clusters in G'

As Fig. 4 shows, this process allows vertices to be soft clustered to more than one cluster. In this particular example, vertex t is soft clustered to cluster $\{r, s, t, u, v\}$ and cluster $\{w, t, x\}$.

3.1   Time Complexity

It can be shown that for a connected graph $G = (V, E)$, MaxMax runs in time $O(|E|)$, that is, linear in the number of edges of G. The transformation of an edge weighted graph G to an unweighted directed graph G' in the first stage can be computed in $O(|E|)$. In constructing G' it is necessary to find maximal vertices of each vertex in G. For a given vertex u, the set of maximal vertices can be identified by scanning each of the edges from u to a vertex adjacent to u in order to determine those of maximal weight. This is done for each vertex of G, with each edge in G inspected just once[4]. Consequently, G' can be constructed in time linear in the number of edges of G.

---

[3] In Fig. 3 vertex r or vertex s is a permissible *root* of the cluster $\{r, s, t, u, v\}$; similarly, either x or w may be the *root* of the cluster $\{t, w, x\}$.

[4] Connections u to v and v to u are considered to be two separate edges in undirected graphs [13].

In the second stage, vertices are initially marked as *root* (line 4 of Algorithm 1), taking $O(|V|)$ time. The for loop (lines 5 to 9) iterates over vertices to identify descendant vertices (line 7) that should be marked $\neg root$. Naively tracing all of the descendants of each vertex in turn could in the worst case entail visiting $O(|V|)$ vertices on each pass through the loop and so, result in an overall complexity of $O(|V|^2)$. However, it is easy to show that once a vertex has been marked $\neg root$ then none of its descendants needs to be visited again. Equivalently, no directed edge needs to be traversed more than once thus, overall complexity of the for loop is linear in the number of edges of G' (and hence linear in the number of edges of G). This yields an overall time complexity of $O(|E|)$.

4   WSI Tasks

In this section MaxMax is applied to two different WSI tasks. Results show that approaches based on MaxMax are comparable with those of state-of-the-art WSI systems.

4.1   Task 1: The SemEval 2010 WSI Task

The SemEval 2010 Word Sense Induction and Disambiguation task [12] provides a formal evaluation framework, enabling participants to compare systems. Systems may be evaluated on a supervised WSD task or alternatively on an unsupervised WSI task. The WSI task is considered here.

Participants are required to induce the senses of 100 target words: 50 verbs and 50 nouns. The test set consists of 8,915 instances (sentences or paragraphs) containing a target word: 5,285 for nouns; 3,630 for verbs. Instances are tagged with OntoNotes senses [2]. Participants are required to tag each instance in the test set with a sense of the target word, the sense being derived by the participant's WSI system.

The SNN$_{swf}$ System. The WSI system applied in this evaluation, SNN$_{swf}$, adapts the Shared Nearest Neighbours (SNN) algorithm [14] to fit the task. SNN$_{swf}$ first extracts unordered and ordered pairs of words from test instances. For example, given an instance $[w_1 \; tw \; w_2 \; w_3]$, where tw represents the target word and w_1, w_2, w_3 represent context words, the following information is extracted -

1. unordered pairs: $\{w_1, w_2\}, \{w_1, w_3\}, \{w_2, w_3\}$.
2. ordered pairs: $(w_1, w_2), (w_1, w_3), (w_2, w_3)$.

Each context word in a target word instance is associated with a set of *word features*. Thus, for the instance above, w_2 is associated with the word features

w_1 and w_3 and the relative word order features (extracted from the ordered pairs) $w_{1\,Left2}, w_{3\,Right1}$. The rationale for the inclusion of the word order features is that these may function as proxies for dependency relations between context words [15].

Word features are filtered using the Log Likelihood Ratio (LLR) measure [6]. In this evaluation, a LLR threshold is set at 10.83^5. Thus, if the LLR score between w_2 and w_3 is greater than or equal to 10.83, then w_3 is taken to be a *significant word feature* of w_2. The threshold filters out features shared by many words. Features passing the LLR threshold should provide strong indicators for the senses of a target word.

Similarity between instance pairs is then calculated as the number of shared significant word features (for all context words in both instances). This approach allows both first order and second order similarity to be computed. Thus, even if two instances have no words in common, the words themselves may share many features, indicating a degree of semantic relatedness between two instance pairs.

A target word graph is constructed using similarity between instance (vertex) pairs as edge weights. MaxMax is then applied to identify a set of sense clusters. A perfect clustering solution would thus assign test instances of each sense of the target word to a separate cluster.

Evaluation Measures. Two evaluation measures are used to assess system performance: the V-Measure [16] and the Paired F-Score [17]. Both measures purport to reflect alignment between a hypothesis K, the clusters returned by a system, and a reference C, the set of gold standard classes in the test set. The V-Measure is defined as the harmonic mean of homogeneity and completeness, where homogeneity is the degree to which each cluster in K consists of instances belonging to a single gold standard class in C and completeness is the degree to which each cluster in K consists of all instances of a single gold standard class in C. The Paired F-Score pairs instances in gold standard classes C and instances in clusters K, then measures the extent to which pairs in C and K overlap.

Results. Tables 1 and 2 report results for SNN$_{swf}$ along with the best, worst, and average score returned by participating systems. The baselines, provided by the organisers of the task, are: *1CPI*, one cluster per instance; *MFS*, most frequent senses (all instances in one cluster), and *Random*, which randomly assigns instances to one of four clusters. Table 1 shows that SNN$_{swf}$ is the best performing system, by some margin, according to the V-measure. However, Table 2 shows that SNN$_{swf}$ is the worst performing system if the Paired F-Score is applied.

---

[5] Lower LLR thresholds were applied (3.84, 6.63, 7.9), returning worse results. Given the number of possible word, feature pairs in the test set, a threshold set higher than 10.83 would be statistically invalid.

Table 1. V-Measure results

System	Verbs	Nouns	|Clusters|
1CPI	25.6	35.8	89.15
SNN$_\mathbf{swf}$	**24.6**	**32.8**	32.31
Hermit	**15.6**	16.7	10.78
UoY	8.5	**20.6**	11.54
Average	6.37	7.73	4.07
Random	4.6	4.2	4.00
Duluth-WSI-SVD-Gap	0.1	0.0	1.02
MFS	0.0	0.0	1.00

Table 2. Paired F-Score results

System	Verbs	Nouns	|Clusters|
MFS	72.7	57.0	1.00
Duluth-WSI-SVD-Gap	**72.4**	**57.0**	1.02
Average	52.3	42.8	4.07
Random	34.1	30.4	4.00
Hermit	30.1	24.4	10.78
SNN$_\mathbf{swf}$	14.4	13.2	32.31
1CPI	0.08	0.11	89.15

It can be observed that Paired F-Score is biased towards clustering solutions returning large clusters: each instance in a cluster of size n pairs with $n-1$ other instances, and so punishes misclassification in small clusters disproportionately [18, 19]. This penalises the MaxMax system, which tends to generate relatively high numbers of fine-grained senses. Such senses may very well have high standard Precision and Recall [19]. The V-Measure on the other hand favours clustering solutions returning numerous small clusters [18, 20, 19, 21]. In this case the bias is due to the normalisation applied in the completeness term of the measure which monotonically increases with the number of induced clusters [20].

4.2   Task 2: Inducing WordNet Senses

The aim of this task is to induce the senses, as defined in WordNet 3.0, of the 27,071 nouns found in co-ordination patterns extracted from the British National Corpus (BNC). The evaluation methodology follows that in [5], which reports results comparable with those reported in [9] – the best results reported to date for this task. In [5], an unweighted graph G_{tw} is constructed, where each vertex represents a noun found in co-ordination patterns and each edge represents noun co-occurrence. The vertex cohesion measure of *curvature* [22, 23] is applied to partition the graph into a set of clusters. The graph theoretical concept of *percolation* [24, 25] is used to find a suitable curvature threshold to apply. Nouns in each cluster are taken to represent a candidate WordNet sense.

A problem observed with this approach is that many semantically unrelated words may be assigned to the same cluster [19]. Consequently, an alternative approach is adopted here. The weighted graph G_{tw} is first transformed to a weighted graph G_{tw}^{T}. G_{tw} consists of a target word (a noun in co-ordination patterns) and its adjacent neighbours. Edge weights $w(u, v)$ between vertices u and v are values returned by an association measure for two nouns co-occurring in patterns. G_{tw}^{T} is derived by deleting edges in G_{tw} with edge weight $\leq$ a predefined threshold. MaxMax is then applied to G_{tw}^{T}, returning a set of candidate sense clusters for the target word.

Candidate sense clusters are mapped to WordNet senses using the method proposed in [9]. This method returns a similarity score between a cluster and the sense of the target word the cluster maximises. If the similarity score exceeds a predefined threshold, the cluster is taken to be a sense of the target word. Validity of cluster to sense mappings is measured using Precision, Recall and F-Score [26]. Precision for a target word tw is defined as:

$$Precision(tw) = \frac{|\{c_i \in C_{tw} \mid \exists\, s_j \in S_{tw} : similarity(c_i, s_j) \geq \sigma\}|}{|C_{tw}|}, \quad (1)$$

and Recall as:

$$Recall(tw) = \frac{|\{s_i \in S_{tw} \mid \exists\, c_j \in C_{tw} : similarity(c_j, s_i) \geq \sigma\}|}{|S_{tw}|}. \quad (2)$$

In (1) and (2) C_{tw} denotes the set of clusters returned by MaxMax given G_{tw}^{T}, and S_{tw} is the set of WordNet senses of tw. σ is the cluster-sense similarity threshold applied[6].

Precision is defined in (1) as a *many to one* mapping; that is, many clusters may map to a single sense of the target word tw[7]. Arguably, this is a fairer measure for evaluating WSI approaches than that of standard Precision [26] as a sense of a target word may be distributed across a number of clusters. Note that each cluster mapped to a sense of tw must pass the similarity threshold σ thus, each cluster counted in the numerator of (1) is, according to the definition given in [9], a valid sense of the target word.

Results. Table 3 reports results, where |Words| is the number of words that can be evaluated by a particular measure, LLR is the Log Likelihood Ratio [6] threshold used to transform G_{tw} to G_{tw}^{T}, and $Counts$ is a word co-occurrence model using raw co-occurrence counts as edge weights between noun pairs in coordination patterns (counts > 1). Results show that the graph transformation approach outperforms the curvature approach. Coverage of words is also far higher. It is interesting to note that the best results are returned by a simple graph model of word co-occurrence ($Counts$). Coverage here, at 9101 words, is

---

[6] σ is set to 0.25, the threshold applied in [9] and [5].
[7] Defined as accuracy in [14].

Table 3. Results for the curvature and graph transformation approaches

System	Precision	Recall	F-Score	|Words|
$MaxMax_{LLR=15.13}$	72.1	53.2	61.2	11138
$MaxMax_{LLR=10.83}$	65.5	53.1	58.7	16850
$MaxMax_{LLR=6.63}$	59.6	55.8	57.6	21716
$MaxMax_{LLR=3.84}$	56.6	**59.1**	57.8	22899
$MaxMax_{Counts}$	**74.2**	58.6	**65.5**	9101
$Curvature$	60.9	40.7	48.8	3906

relatively low yet, is still over twice the number of words that can be evaluated using the curvature approach. This result suggests that comparatively complex measures of word association may not be required to induce word senses.

5   Discussion

Graph-based models (GBM) have been previously applied to WSI, using words as vertices in [27, 28, 10, 29–31] and word collocations as vertices in [32–35]. Alternative approaches include the use of a vector space model (VSM) in [36, 37, 9], Latent Semantic Analysis (LSA) in [38, 39], and a Bayesian approach in [40]. Surveys of WSI approaches are provided in [41–43].

As noted in [28, 44], a VSM approach using context words can conflate senses, as each vector merges the senses context words take. A GBM clearly delineates the uses of context words. LSA, a dimensionality reduction technique, aims to remove information that is irrelevant to the problem space however, this can lead to information pertinent to finding rarer senses being discarded. In contrast, a GBM retains all information. A GBM using collocations as vertices is based on Yarowsky's tenet: 'one sense per collocation' [45]. The argument given [32–35] is that collocation vertices are less sense conflating than single word vertices. Arguably though, collocation vertices are not required if the set of word vertices that define target word senses is filtered using a significance threshold set on edge weights.

Existing soft clustering algorithms such as Fuzzy c-Means [46] and Expectation Maximization [47] require the number of clusters to be pre-defined. Parameter tuning is therefore necessary in order to find a good clustering solution [14]. Consequently, these algorithms are not well-suited to WSI, as the number of senses target words take is often undefined.

MaxMax bears some resemblance to single-link Hierarchical Agglomerative Clustering (HAC) [14] in that leaf vertices (clusters consisting of one vertex) in the first iteration of HAC are clustered using maximal affinity between vertex pairs. However, whereas HAC hard clusters vertices in $O(|V|^3)$, MaxMax hard/soft clusters vertices in linear time. MaxMax also shares some properties with Chinese Whispers [48], a non-parameterised, graph-based clustering algorithm shown to have utility in NLP [4, 49–52, 30]. Both algorithms use affinity within the local neighbourhood of vertices to generate clusters and both have linear run times. However, there are key differences. MaxMax is deterministic, whilst Chinese Whispers may return different solutions for the same graph. In addition, MaxMax is able to soft cluster vertices, whilst Chinese Whispers cannot. Thus, given the input graph G in Fig. 5, Chinese Whispers randomly assigns vertex c to cluster 1 or 2 whereas MaxMax returns the clustering solution C_G.

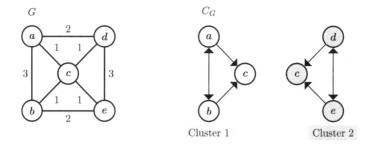

Fig. 5. Soft clustering example

6   Conclusions

This paper introduced MaxMax, a novel non-parameterised soft clustering algorithm that finds the number of clusters in a graph automatically by identifying root vertices of maximal quasi strongly connected subgraphs, a process shown to be computable in linear time. Examples showed that descendant vertices of more than one root vertex can be soft clustered, with a descendant vertex assigned to each cluster containing a vertex to which it has maximal affinity: a straightforward process that, in comparison to existing soft clustering algorithms, is both fast and transparent. As a non-parameterised clustering algorithm, MaxMax is well-suited to WSI or, indeed, to any task in which the number of clusters is not known in advance. To test its utility for WSI, MaxMax was incorporated into two induction systems. Results in two tasks showed the systems to return scores comparable with, if not better than, those of existing state-of-the-art systems. However, further tests are required thus, future research plans to apply the algorithm in the forthcoming SemEval 2013 WSI evaluations and to carry out a comparative analysis against the recently introduced SquaT++ and B-MST clustering algorithms [30]. Additionally, as the WSI tasks in this paper have no special requirement for soft clustering, future research also plans to apply the algorithm to networks in which soft clustering may be of use. For example, social networks [53] and contagion networks [54] typically have many vertices with ties to more than one maximal vertex therefore, MaxMax may be particularly suited to studying these types of networks.

Acknowledgements. The authors wish to thank the anonymous reviewers for their feedback on a previous version of this paper.

References

1. Miller, G., Beckwith, R., Fellbaum, C., Gross, D., Miller, K.: Introduction to Word-Net: An On-line Lexical Database. International Journal of Lexicography 3, 235 (1990)
2. Hovy, E., Marcus, M., Palmer, M., Ramshaw, L., Weischedel, R.: OntoNotes: the 90% Solution. In: Proceedings of the Human Language Technology Conference of the NAACL, pp. 57–60. Association for Computational Linguistics (2006)
3. Wittgenstein, L.: Philosophical Investigations. Blackwell (1953)
4. Klapaftis, I., Manandhar, S.: Word Sense Induction Using Graphs of Collocations. In: Proceeding of the 2008 Conference on ECAI, pp. 298–302 (2008)
5. Dorow, B.: A Graph Model for Words and their Meanings. PhD thesis, Institut für Maschinelle Sprachverarbeitung der Universität Stuttgart (2007)
6. Dunning, T.: Accurate Methods for the Statistics of Surprise and Coincidence. Computational Linguistics 19, 61–74 (1993)
7. Evert, S.: The Statistics of Word Cooccurrences: Word Pairs and Collocations. PhD thesis, Institut für Maschinelle Sprachverarbeitung der Universität Stuttgart (2005)
8. Widdows, D.: Geometry and Meaning. CSLI Lecture Notes. CSLI Publications, Center for the Study of Language and Information (2004)
9. Pantel, P., Lin, D.: Discovering Word Senses from Text. In: Proceedings of the Eighth ACM SIGKDD International Conference on Knowledge Discovery and Data Mining, pp. 613–619. ACM (2002)
10. Biemann, C.: Unsupervised and Knowledge-Free Natural Language Processing in the Structure Discovery Paradigm. PhD thesis, University of Leipzig (2007)
11. Agirre, E., Soroa, A.: SemEval-2007 Task 02: Evaluating Word Sense Induction and Discrimination Systems. In: Proceedings of the 4th International Workshop on Semantic Evaluations, pp. 7–12. Association for Computational Linguistics (2007)
12. Manandhar, S., Klapaftis, I., Dligach, D., Pradhan, S.: SemEval-2010 Task 14: Word Sense Induction and Disambiguation. In: Proceedings of the 5th International Workshop on Semantic Evaluation, pp. 63–68. Association for Computational Linguistics (2010)
13. Dasgupta, S., Papadimitriou, C., Vazirani, U.: Algorithms. McGraw-Hill (2006)
14. Tan, P., Steinbach, M., Kumar, V.: Introduction to Data Mining. Pearson Addison Wesley (2006)
15. Yuret, D.: Discovery of Linguistic Relations Using Lexical Attraction. PhD thesis, Massachusetts Institute of Technology (1998)
16. Rosenberg, A., Hirschberg, J.: V-Measure: A Conditional Entropy-Based External Cluster Evaluation Measure. In: Proceedings of the 2007 Joint Conference on Empirical Methods in Natural Language Processing and Computational Natural Language Learning, pp. 410–420 (2007)
17. Artiles, J., Amigó, E., Gonzalo, J.: The Role of Named Entities in Web People Search. In: Proceedings of the 2009 Conference on Empirical Methods in Natural Language Processing, vol. 2, pp. 534–542 (2009)
18. Pedersen, T.: Duluth-WSI: SenseClusters Applied to the Sense Induction Task of SemEval-2. In: Proceedings of the 5th International Workshop on Semantic Evaluation, pp. 363–366. Association for Computational Linguistics (2010)

19. Hope, D.: Graph-Based Approaches to Word Sense Induction. PhD thesis, University of Sussex (2012) (unpublished)
20. Reichart, R., Rappoport, A.: The NVI Clustering Evaluation Measure. In: Proceedings of the Thirteenth Conference on Computational Natural Language Learning, CoNLL 2009, pp. 165–173. Association for Computational Linguistics (2009)
21. Amigó, E., Gonzalo, J., Artiles, J., Verdejo, F.: A Comparison of Extrinsic Clustering Evaluation Metrics Based on Formal Constraints. Information Retrieval 12, 461–486 (2009)
22. Watts, D., Strogatz, S.: Collective Dynamics of 'Small-World' Networks. Nature 393, 440–442 (1998)
23. Eckmann, J., Moses, E.: Curvature of Co-Links Uncovers Hidden Thematic Layers in the World Wide Web. Proceedings of the National Academy of Sciences 99, 5825 (2002)
24. Erdős, P., Rényi, A.: On the Evolution of Random Graphs. Akad. Kiadó (1960)
25. Bollobás, B., Riordan, O.: Percolation. Cambridge University Press (2006)
26. Van Rijsbergen, C.: Information Retrieval. Butterworths (1979)
27. Dorow, B., Widdows, D.: Discovering Corpus-Specific Word Senses. In: Proceedings of the Tenth Conference, European Chapter of the Association for Computational Linguistics, vol. 2, pp. 79–82. Association for Computational Linguistics (2003)
28. Véronis, J.: Hyperlex: Lexical Cartography for Information Retrieval. Computer Speech & Language 18, 223–252 (2004)
29. Agirre, E., Martínez, D., de Lacalle, O., Soroa, A.: Two Graph-Based Algorithms for State-of-the-Art WSD. In: Proceedings of the 2006 Conference on Empirical Methods in Natural Language Processing, pp. 585–593. Association for Computational Linguistics (2006)
30. Di Marco, A., Navigli, R.: Clustering and Diversifying Web Search Results with Graph-Based Word Sense Induction. Computational Linguistics 39(4) (2013)
31. Navigli, R., Crisafulli, G.: Inducing Word Senses to Improve Web Search Result Clustering. In: Proceedings of the 2010 Conference on Empirical Methods in Natural Language Processing, pp. 116–126. Association for Computational Linguistics (2010)
32. Dorow, B., Widdows, D., Ling, K., Eckmann, J., Sergi, D., Moses, E.: Using Curvature and Markov Clustering in Graphs for Lexical Acquisition and Word Sense Discrimination. In: 2nd Workshop Organized by the MEANING Project (2005)
33. Klapaftis, I., Manandhar, S.: Word Sense Induction Using Graphs of Collocations. In: Proceeding of the 2008 Conference on ECAI 2008: 18th European Conference on Artificial Intelligence, pp. 298–302. IOS Press (2008)
34. Klapaftis, I., Manandhar, S.: Word Sense Induction and Disambiguation Using Hierarchical Random Graphs. In: Proceedings of the 2010 Conference on Empirical Methods in Natural Language Processing, pp. 745–755. Association for Computational Linguistics (2010)
35. Bordag, S.: Word Sense Induction: Triplet-Based Clustering and Automatic Evaluation. In: Proceedings of EACL 2006, Trento (2006)
36. Schütze, H.: Automatic Word Sense Discrimination. Computational Linguistics 24, 97–123 (1998)
37. Purandare, A., Pedersen, T.: Word Sense Discrimination by Clustering Contexts in Vector and Similarity Spaces. In: Proceedings of the Conference on Computational Natural Language Learning, pp. 41–48 (2004)

38. Van de Cruys, T.: Using Three Way Data for Word Sense Discrimination. In: Proceedings of the 22nd International Conference on Computational Linguistics (Coling 2008), pp. 929–936 (2008)
39. Van de Cruys, T., Apidianaki, M.: Latent Semantic Word Sense Induction and Disambiguation. In: Proceedings of the 49th Annual Meeting of the Association for Computational Linguistics: Human Language Technologies (ACL/HLT), pp. 1476–1485 (2011)
40. Brody, S., Lapata, M.: Bayesian Word Sense Induction. In: Proceedings of the 12th Conference of the European Chapter of the Association for Computational Linguistics, pp. 103–111. Association for Computational Linguistics (2009)
41. Navigli, R.: Word Sense Disambiguation: A Survey. ACM Computing Surveys (CSUR) 41, 10 (2009)
42. Apidianaki, M., Van de Cruys, T.: A Quantitative Evaluation of Global Word Sense Induction. In: Gelbukh, A.F. (ed.) CICLing 2011, Part I. LNCS, vol. 6608, pp. 253–264. Springer, Heidelberg (2011)
43. Navigli, R.: A Quick Tour of Word Sense Disambiguation, Induction and Related Approaches. In: Bieliková, M., Friedrich, G., Gottlob, G., Katzenbeisser, S., Turán, G. (eds.) SOFSEM 2012. LNCS, vol. 7147, pp. 115–129. Springer, Heidelberg (2012)
44. Klapaftis, I.: Unsupervised Concept Hierarchy Induction: Learning the Semantics of Words. PhD thesis, University of York (2008)
45. Yarowsky, D.: Unsupervised Word Sense Disambiguation Rivaling Supervised Methods. In: Proceedings of the 33rd Annual Meeting on Association for Computational Linguistics, pp. 189–196. Association for Computational Linguistics (1995)
46. Bezdek, J.: Pattern Recognition with Fuzzy Objective Function Algorithms. Kluwer Academic Publishers (1981)
47. Dempster, A., Laird, N., Rubin, D.: Maximum Likelihood from Incomplete Data via the EM Algorithm. Journal of the Royal Statistical Society, Series B (Methodological), 1–38 (1977)
48. Biemann, C.: Chinese Whispers - an Efficient Graph Clustering Algorithm and its Application to Natural Language Processing Problems. In: Proceedings of the HLT-NAACL 2006 Workshop on Textgraphs 2006 (2006)
49. Korkontzelos, I., Manandhar, S.: Detecting Compositionality in Multi-Word Expressions. In: Proceedings of the ACL-IJCNLP 2009 Conference Short Papers, pp. 65–68. Association for Computational Linguistics (2009)
50. Zhang, Z., Sun, L.: Improving Word Sense Induction by Exploiting Semantic Relevance. In: Proceedings of the 5th International Joint Conference on Natural Language Processing, pp. 1387–1391 (2011)
51. Jurgens, D.: An Evaluation of Graded Sense Disambiguation Using Word Sense Induction. In: Proceedings of *SEM First Joint Conference on Lexical and Computational Semantics. ACL (2012)
52. Fountain, T., Lapata, M.: Taxonomy Induction Using Hierarchical Random Graphs. In: 2012 Conference of the North American Chapter of the Association for Computational Linguistics: Human Language Technologies, pp. 446–476 (2012)
53. Wasserman, S., Faust, K.: Social Network Analysis: Methods and Applications. Cambridge University Press (1994)
54. Newman, M., Barabási, A.L., Watts, D.J.: The Structure and Dynamics of Networks. Princeton University Press (2006)

A Model of Word Similarity Based on Structural Alignment of Subject-Verb-Object Triples

Dervla O'Keeffe and Fintan Costello

School of Computer Science and Informatics,
University College Dublin,
Belfield, Dublin 4, Ireland
dervla.okeeffe@ucdconnect.ie, fintan.costello@ucd.ie

Abstract. In this paper we propose a new model of word semantics and similarity that is based on the structural alignment of $\langle Subject\ Verb\ Object \rangle$ triples extracted from a corpus. The model gives transparent and meaningful representations of word semantics in terms of the predicates asserted of those words in a corpus. The model goes beyond current corpus-based approaches to word similarity in that it reflects the current psychological understanding of similarity as based on structural comparison and alignment. In an assessment comparing the model's similarity scores with those provided by people for 350 word pairs, the model closely matches people's similarity judgments and gives a significantly better fit to people's judgments than that provided by a standard measure of semantic similarity.

1   Introduction

Models of word similarity based on the statistics of word co-occurrence in large corpora have been used successfully in many natural language processing tasks such as word sense discrimination [13], text segmentation [2], thesaurus construction [9] and information retrieval [16]. All such models gather co-occurrence data at the level of single words; the atomic element of analysis in these models is a single word that is related to the target word either by simple closeness in text or by some form of grammatical dependency. We present a model in which the elements of co-occurence do not consist of single words, but instead consist of $\langle Verb\ Object \rangle$ predicates that co-occur syntactically with the target word. We represent the semantics of a given target word by gathering from a corpus of text all $\langle Verb\ Object \rangle$ predicates with that target word as a subject. We compute the similarity between two words by structurally aligning the $\langle Verb\ Object \rangle$ predicates for those two words.

This approach has two advantages over models based on the co-occurence of single words with the target. The first is that a $\langle Verb\ Object \rangle$ predicate that occurs with the target word as a subject represents a complete semantic feature of that target. For example, the sentence

Linguists love grammar.

asserts that the predicate or semantic feature $\langle loves\ grammar \rangle$ is true of the subject *linguists*. If, from a given corpus, we gather a series of such predicates that have the

A. Gelbukh (Ed.): CICLing 2013, Part I, LNCS 7816, pp. 382–393, 2013.

target word *linguists* as their subject, then we have a representation of the semantic meaning of *linguists* in terms of the semantic features that are asserted of linguists in that corpus. By contrast, single words that co-occur with the target (either by simple proximity or by taking part in some dependency relation with that target) are not predicates and typically do not describe semantic features of that target. A model that represents word meaning in terms of co-occurence with $\langle Verb\ Object \rangle$ predicates thus provides a more direct and meaningful representation of semantics than that provided by standard models.

The second advantage of this approach is that it allows for a measure of similarity that is closer to actual psychological similarity. The model of word similarity used in standard approaches to corpus semantics is based on shared features: the two target words being compared are each represented by a list of features (words that the targets co-occur with), and the similarity of those words is computed as a function of the number of features the words have in common. The current psychological understanding of similarity is more complex. In this view, people judge the similarity between two objects by comparing and aligning the higher-order structure of those two objects. Our model of semantics based on $\langle Verb\ Object \rangle$ predicates allows this form of 'structural similarity' to be judged between two words, and so goes beyond the featural approach typically used in corpus semantics.

In testing our model we gathered a large list of $\langle Subject\ Verb\ Object \rangle$ triples from the British National Corpus (BNC), accessed via Sketch Engine [7]. We calculated similarity between pairs of nouns by structurally aligning the sets of $\langle Verb\ Object \rangle$ predicates derived from those triples. We compared the model's similarity scores for 350 pairs of nouns with similarity scores produced by people for those nouns. The model's correlation with the average of people's similarity scores was significantly higher than that obtained from Wu & Palmer's WordNet-based model of word similarity [19], which has become somewhat of a standard for measuring similarity in lexical taxonomies [6]. The model's correlation with people's similarity scores was also close to the average inter-person correlation, showing that the model's similarity scores were a relatively close reflection of people's similarity judgments.

2 Background: Current Approaches to Corpus Semantics

Researchers have proposed a range of different models in which word meaning is represented in terms of co-occurrence statistics. The basic idea in these models is that the context in which a given word occurs provides important information about the meaning of that word, and so two words which co-occur in a similar set of contexts in a given corpus are likely to have the same or similar meanings.

These models take a standard 'vector space' approach to word meaning, where the meaning of a word is represented by a vector (a list) of boxes with each box labelled with some linguistic element (such as a word) and each box containing a count of the number of times that element co-occured with the target word. Target words with similar vectors tend to occur in the same contexts and so are expected to be semantically similar, while target words with dissimilar vectors do not occur in the same contexts and so are semantically different (see [17] for an overview).

There are three central aspects to these models: how they define co-occurrence, what they take to be the elements of co-occurence (that is, what they use as box labels in the vectors representing word meaning), and how they compute semantic similarity from these vectors of co-occurence data.

2.1 Distance-Based Models

Many vector-space models of meaning take single words to be the elements of co-occurence. In these models target words are represented by vectors with each element or box labelled by some other word and containing a count of the number of times that word occurs in the corpus in some "context" around the target. Perhaps the best known model of this type is Latent Semantic Analysis (LSA) [8], in which context is defined to be the entire document in which the target word occurs; other models define context to be the paragraph in which target occurs or the sentence in which the target word occurs. A general form of this approach defines context to be the set of W words on either side of an occurence of the target word. The parameter W is often referred to as the window size. Models taking this approach have used various window sizes, typically testing various window sizes to find the best size for the task at hand [10]. To take an example, consider a corpus

$$Linguists\ love\ grammar.\ Linguists\ parse\ sentences.$$

consisting of only two sentences. With a window size of 2, a distance-based model would represent the target word $linguists$ by the vector

$$[love = 2, grammar = 2, parse = 1, sentences = 1]$$

($love$ and $grammar$ occuring twice within 2 words of the target, $parse$ and $sentences$ occuring once within 2 words).

2.2 Dependency-Based Models

Another group of models retain single words as the elements of co-occurence but define co-occurence in terms of syntactic dependency relationships between the target word and other words, rather than in terms of the distance between words in text [9]. A word occurs in a dependency relationship with the target if it has a direct grammatical link to that target: if, for example, it modifies that target, has that target as a subject or object, is determiner for that target, and so on. In models taking this approach, other words do not have to occur within a fixed window of the target word to be counted as co-occuring with the target. Instead, these models represent the meaning of a target word by a vector of values containing, for every other word, a count of the number of times that other word occurs as a subject of the target word, as a modifier of the target word, as an object of the target word, as a noun modified by the target word, and so on (see [15] for an overview and a general framework for such dependency-based models). Given our example corpus, a dependency-based model would represent the target word $linguists$ by the vector

$$[love_v = 1, parse_v = 1, grammar_o = 1, sentences_o = 1]$$

(*love* and *parse* co-occuring with *linguists* once as a verb, *grammar* and *sentences* co-occuring with *linguists* once as an object). Note that, while these models make use of the ⟨*S V O*⟩ structure to gather co-occurence information for a given target word, the elements of co-occurence are still single words, as in distance-based models.

2.3   The Opacity of Word Co-occurence

In all of these models, whether distance-based or dependency-based, the meaning of a given target word consists of a list or vector giving the number of times each other word co-occurred with the target word, either within the given window distance, or with some dependency relation. The semantic similarity between two target words then corresponds to the similarity between the two lists representing those words.

These lists tell us little about the meaning of the target words, because they tell us little about the relationship between the co-occuring words and the target (other than that they occurred near each other in the text). Dependency-based models address this problem of opacity to some extent because these models represent the relationship between the target and co-occuring words. However these representations still suffer from a degree of opacity, especially in the case of transitive verbs. In contexts where a target word occurs as a subject of a transitive verb, for example, the dependency-based approach counts the verb as co-occuring with the target word, and the object as co-occuring with the target, but does not record the fact that those two words occured together as a predicate describing the target word. This means that an important part of the target word's meaning in that context (the object of the verb to which it is subject) is not included in the meaning representation.

To return again to our example, the dependency-based co-occurence vector given above tells us that the *linguists* occurs as a subject of *love* and *parse* (that is, it tells us that linguists love something and parse something) and as the subject of some verb with *grammar* as an object and some verb with *sentences* as an object (that is, it tells us that linguists do something to grammar and do something to sentences). The vector doesn't tell us whether linguists love grammar or sentences, or whether they parse sentences or grammar. These important parts of the target word's meaning are missing from the dependency-based representation.

3   A Model Based on ⟨*verb object*⟩ Predicate Co-occurrence

Amost all models of word meaning based on the statistics of co-occurrence take single words, or single words plus some syntactic marker, to be the elements of co-occurence[10,9,14,18]. Even the most sophisticated dependency-based models, such as that given by Padó [15], assume that the elements of co-occurence have this 'single word plus syntactic marker' form. Models that are explicitly based on ⟨*S V O*⟩ triples (such as Burek et al.'s ⟨*S V O*⟩ model of textual entailment) also take single words plus syntactic markers as the elements of co-occurence: 'entries to the [co-occurence] matrices are the frequency of the co-occurence of each labeled stem with itself or with another labeled stem' (quote from Burek et al.'s paper [1]; 'stem' here represents a stemmed word).

Table 1. An example of structural alignment

	comparison	linguists	physicists
common	aligned commonality	*linguists write papers*	*physicists write papers*
structure	aligned difference	*linguists love grammar*	*physicists love equations*
	nonalignable difference	*linguists parse sentences*	
	nonalignable difference		*physicists perform experiments*

Our model uses more complex elements of co-occurence: $\langle verb\ object \rangle$ predicates, which consist of two words each with its syntactic marker. For the example corpus given above, this model would produce a co-occurence vector

$$[\langle love\ grammar \rangle = 1, \langle parse\ sentences \rangle = 1]$$

consisting of the two predicates asserted of the target word *linguists* in the corpus. Using such predicates addresses the problem of opacity because it represents meanings in terms of semantic features asserted of the target noun.

3.1 Structural Alignment and Similarity

Equally importantly, this approach allows for a more psychologically realistic and structured measure of similarity. In the current psychological view of similarity, people judge the similarity between two objects by comparing and aligning the higher-order structure of those two objects [3,4,5,11,12]. After the two objects have been aligned (that is, after the common structure shared between the objects has been identified), commonalities and differences between the objects are identified relative to that shared structure. For example, consider a comparison between linguists and physicists. We can give (extremely simplified) predicate representations of this comparison as in Table 1. In this table the common or aligned structure between these two subjects is represented by the verbs *write* and *love*. In the structural view of similarity when people compare linguists and physicists, they first align common structures like these. They then use this structural alignment to find commonalities and differences between the subjects being compared. A commonality between the two subjects is identified when the values in the aligned structures are the same: the first row in Table 1. This structural alignment generates two types of differences. Alignable differences occur when a common structure contains different values for the two subjects: 2nd row in Table 1. Non-alignable differences, on the other hand, occur when there is no common or aligned structure between the two subjects: 3rd and 4th row in Table 1.

In the structural alignment view of similarity, different components of alignment make different contributions to the similarity between the subjects being compared. Most important is the degree of structural similarity between the two subjects. Two subjects with a large degree of aligned structure will be judged similar, even if they do not have many specific features in common. According to this view the main sources of similarity between linguists and physicists is the fact that they share a relational structure: that both are subjects of the verbs *write* and *love*. Next most important in

$S(A, B)$ = number of verbs V such that, for some object O, the triples $\langle A\ V\ O \rangle$ and $\langle B\ V\ O \rangle$ both occur in the corpus (Aligned Structure).

$C(A, B)$ = number of predicates $\langle V\ O \rangle$ such that triples $\langle A\ V\ O \rangle$ and $\langle B\ V\ O \rangle$ both occur in the corpus (Structural Commonalities).

$AD(A, B)$ = number of predicates $\langle V\ O \rangle$ such that $\langle A\ V\ O \rangle$ and $\langle B\ V\ X \rangle$ occur but $\langle B\ V\ O \rangle$ does not occur, or $\langle B\ V\ O \rangle$ and $\langle A\ V\ X \rangle$ occur but $\langle A\ V\ O \rangle$ does not occur (Aligned Differences)

$ND(A, B)$ = (number of predicates $\langle V\ O \rangle$ such that $\langle A\ V\ O \rangle$ occurs but no triple $\langle B\ V\ X \rangle$ occurs, or $\langle B\ V\ O \rangle$ occurs but no triple $\langle A\ V\ X \rangle$ occurs (Nonalignable Differences).

$$sim(A, B) = \frac{W_1 \times S(A, B) + C(A, B)}{W_1 \times S(A, B) + C(A, B) + W_2 \times AD(A, B) + ND(A, B)}$$

Fig. 1. Similarity factors and similarity expression. A and B are target words being compared, W_1 and W_2 are free weighting parameters.

the judgment of similarity is the number of aligned commonalities between the subjects being compared. In our example, the aligned commonality between linguists and physicists is that both write papers (both verb and object are common across the two subjects). Subjects with a large degree of aligned structure and many aligned commonalities in that structure will be judged more similar than those with the same degree of aligned structure but fewer commonalities.

After these two factors come aligned and nonalignable differences. In our example, the aligned difference between linguists and physicists is that both *love* something, but linguists love grammar while physicists love equations (the verb is common but the object is not). Finally, the nonalignable differences are that *linguists parse sentences* and *physicists perform equations* (each verb occurs for one subject but not the other). Some researchers argue that the process of aligning the structures of two subjects draws attention to alignable differences, and conclude that alignable differences will thus contribute more to judgements of similarity. Others argue, however, that nonalignable differences are 'more different' than alignable differences (because alignable differences involve a component of shared similarity due to their aligned structure, while non-alignable differences do not have any such shared structure), and conclude that non-alignable differences will thus contribute more to similarity judgments. The status of these two dimensions is equivocal, with some results showing an advantage for alignable differences and some an advantage for non-alignable differences.

3.2   The Model

Our model of similarity judgment makes use of these four factors (see Figure 1). For a given corpus and a given pair of nouns A and B, we define four measures representing the four factors in structural similarity. In these measures, the degree of aligned structure between nouns A and B is represented by the number of verbs common to both A and B. The degree of commonality is represented by predicates $\langle V\ O \rangle$ that occur for both A and B. The degree of aligned difference is represented by the number of predicates $\langle V\ O \rangle$ where V occurs for both A and B but O only occurs for one of A or B. Finally, the degree of nonalignable difference is represented by the number of predicates $\langle V\ O \rangle$ where V only occurs for one of A or B.

We combine these four factors to give an overall expression of similarity. This expression gives a weighted sum of the structural similarities and commonalities between A and B, as a proportion of a weighted sum of the number of points of comparison between A and B. We use two free parameters as weights in this expression, to control the relative importance of aligned structure (W_1) and aligned differences (W_2) in similarity.

Note that in measuring the aligned structure between two nouns A and B, we do not simply count the number of verbs to which both A and B are subject. Given the high level of semantic ambiguity of many frequent verbs, this would incorrectly count many semantically different verb subcases as the same relation. To avoid this problem and to ensure that the verbs we count as elements of aligned structure actually represent the same semantic relation for both A and B, we only count a verb V as occuring for both A and B if there exists some object O such that both $\langle A\ V\ O \rangle$ and $\langle B\ V\ O \rangle$ occur in the corpus. This requirement is intended to ensure a verb is counted as occuring for both A and B ony when the same sense of that verb occurs for both.

4   Dataset

To apply this model of similarity we need a set of $\langle S\ V\ O \rangle$ triples derived from some corpus. We extracted a large set of such triples from the British National Corpus (BNC) via Sketch Engine [7][1], an online Corpus Query System that allows macro queries on dependency relations such as `object/object\_of`.

These macros expand into regular expressions in a Corpus Query Language that uses tags as in the BNC tagset. Figure 2 gives the query produced by expanding one such macro. This query returns nouns taking part in the `object/object\_of` relation with an lexical verb. Similarly, the query in Figure 3 returns nouns taking part in the `subject/subject\_of` relation with a lexical verb. In these queries the label `:2` represents the target noun of the query; in these cases the label `:1` identifies the related verb, and the other tags represent other other parts of speech (articles, determiners, possessives, and so on) and may occur in sentences matching the query.

To retrieve a set of $\langle S\ V\ O \rangle$ triples from the BNC we combined these two queries as in Figure 4. This combined query returns all sentences in the BNC containing subjects, objects and verbs matching the given expression. We run this query and for each returned sentence extract the verb, a single subject noun, and a single object noun, to give the required $\langle S\ V\ O \rangle$ triple. Just over $600,000$ results matched this query, giving approximately $600,000$ triples.

---

[1] http://www.sketchengine.co.uk

```
Macro object/object\_of
 = lex_verb adv_string long_np

Expansion of macro
  1:"VV." [tag="AV0"|tag="XX."] {0,3}
  [tag="AT."|tag="DT."|tag="DPS"|tag="[OC]RD"|tag="AV0"|
  tag="AJ."|tag="POS"]{0,3} "N.."{0,2} 2:"N.."
  [tag!="N.." \& tag != "POS"]
```

Fig. 2. Corpus query for nouns taking part in one form of *Object_of* dependency relation with some verb

```
Macro subject/subject\_of
 = any_noun rel_start? adv_aux_string_incl_be aux_have
   adv_string past_part

Expansion of macro
2:"N.." [tag="DTQ"|tag="PNQ"|tag="CJT"]?
  [tag="AV0"|tag="XX."|tag="V[MHB]."]{0,6}
  "VH." [tag="AV0"|tag="XX."]{0,3} 1:"VVN"
```

Fig. 3. Corpus query for nouns taking part in one form of *Subject_of* dependency relation with some verb

```
Combined corpus query representing Subject-Verb-Object triple
     1:"N.."
[tag="DTQ"|tag="PNQ"|tag="CJT"]?
[tag="AV0"|tag="XX."|tag="V[MHB]."]{0,6}
"VH." [tag="AV0"|tag="XX."]{0,3}
2:"VVN" [tag="AV0"|tag="XX."]{0,3}
[tag="AT."|tag="DT."|tag="DPS"|
     tag="[OC]RD"|tag="AV0"|tag="AJ."|tag="POS"]
{0,3} "N.."{0,2} 3:"N.." [tag!="N.." & tag != "POS"]
```

Fig. 4. Corpus query for one form of ⟨*S V O*⟩ triple, produced by combining the queries in Figures 2 and 3. Tags from the BNC tagset, where AV represents adverbs, XX negative particles, AT articles, DT determiners, DPS possessive determiners and so on (see http://www.natcorp.ox.ac.uk/docs/c5spec.html).

5   Assessing the Model

We assess our model in two ways. We first consider the transparency or meaningfulness of the representations produced by this model by presenting the most frequent triples from the BNC analysis for two example subject words, *bank* and *company*

Table 2. Most frequent triples with subjects *company* and *bank*. Note that there were 9 triples with subject *bank* and frequency of 2: for space reasons, not all of these are shown here.

triple	frequency	triple	frequency
⟨*company employ people*⟩	16	⟨*bank lend money*⟩	3
⟨*company issue report*⟩	13	⟨*bank make loan*⟩	3
⟨*company bear name*⟩	11	⟨*bank bear name*⟩	2
⟨*company make profit*⟩	11	⟨*bank make refund*⟩	2
⟨*company report figures*⟩	11	⟨*bank pay interest*⟩	2
⟨*company hold position*⟩	11	⟨*bank charge fees*⟩	2
		⟨*bank break law*⟩	2
		⟨*bank demand cash*⟩	2

(Table 2). As this table shows, the triples that are part of the generated representation for these target words are all semantically transparent and their contribution to the meaning of the target word is clear. Indeed in most cases these triples represent semantic features that are characteristic or defining of those words (e.g. *banks lend money*, *banks charge fees*, *companies employ people*, *companies make profits*).

We can also use the BNC triples to illustrate the structural alignment and comparison process in the model. For this we selected *man* and *woman* as subject words and considered triples with frequency above 4 (for *woman*) and above 6 (for *men*), giving a roughly equal number of triples for both words (see Table 3).

As we can see from Table 3, the commonalities, aligned differences and non-alignable differences obtained are easily understood and provide a transparent justification for any similarity score computed from these points of comparison. Calculating values for the factors in our model from the data in this table we get $S(man, woman) =$

Table 3. An example of alignment and comparison with triples from the BNC dataset

	comparison	man	woman
common structure {	aligned commonality	⟨*man open door*⟩	⟨*woman open door*⟩
	aligned commonality	⟨*man shake head*⟩	⟨*woman shake head*⟩
	aligned difference	⟨*wan wear hat*⟩	⟨*woman wear dress*⟩
	non-alignable difference	⟨*man walk dog*⟩	
	non-alignable difference	⟨*man drive car*⟩	
	non-alignable difference	⟨*man make love*⟩	
	non-alignable difference		⟨*woman give birth*⟩
	non-alignable difference		⟨*woman hold baby*⟩

$3, C(man, woman) = 2, AD(man, woman) = 1$ and $ND(man, woman) = 5$. Taking $W_1 = 1$ and $W_2 = 1$ and so giving no differential weighting to aligned structure or aligned difference we get

$$sim(man, woman) = \frac{3 + 2}{3 + 2 + 1 + 5} = 0.46$$

While this example is purely illustrative and is simply intended to demonstrate the basic workings of the model, the result obtained is relatively close to the average similarity score of 0.59 (5.9 out of 10) given by people for the $man, woman$ pair in our similarity judgment task (see below).

5.1 Comparing the Model with People's Similarity Judgments

To provide a quantative measure of the model's performance, we gathered people's similarity judgments for a set of 350 word pairs. We selected these pairs to have varying numbers of common verbs ranging from from no common verbs up to 35 common verbs. We selected pairs at random, subject to this requirement.

We asked three people to rate the similarity of the words in each of those 350 pairs, on a scale from 1 to 10. We also asked three people to rate the difference between words in each of those 350 pairs, again on a 1 to 10 scale. We inverted each participants's difference judgments to convert them to similarity judgments by subtracting each difference judgments from the maximum of 10. We used both similarity and difference judgments because we were interested in any distinctions between those responses. The results showed that similarity judgments tended to be less consistent than difference judgments, with average pairwise correlation between participant's responses of $r = 0.56$ for similarities, versus average pairwise correlation of $r = 0.73$ for differences ($n = 350$ for all comparisons). Apart from that, however, there was little difference between the two sets of judgments: there was a high correlation between average similarity judgments and average inverted difference judgments across the 350 pairs ($r = 0.76, p < 10^{-61}$). For this reason we combine similarity judgments and inverted diference judgments to produce an overall average similarity score for each pair of words. The average pairwise correlation between all participants' responses in the full set was $r = 0.61$. This value represents an upper bound on the degree of correlation between model-derived similarity values and those given by participants.

For each pair A, B in the set of 350 we used the BNC triples to calculate values for the model's similarity factors and overall similarity expression, as in Figure 1. There was a highly significant correlation between participants' similarity scores and the aligned structure measure $S(A, B)$ ($r = 0.50, p < 10^{-21}$), and between participants' scores and the aligned commonalities measure $C(A, B)$ ($r = 0.47, p < 10^{-18}$). Correlation with the aligned difference measure $AD(A, B)$ was lower but still highly significant ($r = -0.32, p < 10^{-8}$), while correlation with the nonalignable difference measure $ND(A, D)$ was lower still ($r = -0.15, p < 0.01$; these two negative values for r indicate that similarity decreases as the number of differences increases, as you would expect). These results agree with the structural alignment view of similarity, with degree of aligned structure being the strongest predictor of participant's similarity judgments.

For each pair A, B in the set of 350 we also calculated the value of our combined similarity expression $sim(A, B)$ for a range of different values of the parameters W_1 and W_2. Taking $W_1 = 1$ and $W_2 = 1$ and so giving no differential weighting to aligned structure or aligned difference we get a correlation between participants' judgments and $sim(A, B)$ of $r = 0.54$ ($p < 10^{-26}, n = 350$), showing that the $sim(A, B)$ expression gives a closer fit to participants' similarity judgments than any of its component factors. Finally, we found highest fit occurred with parameter values $W_1 = 10$ and $W_2 = 0.1$, giving a correlation of $r = 0.56$ ($p < 10^{-29}, n = 350$). Note that these values are not far from the upper bound of $r = 0.61$.

Finally, we compared our similarity measure against Wu & Palmer's WordNet-based similarity measure [19], which has become somewhat of a standard for similarity in lexical taxonomies [6]. Wu & Palmer's measure computes the similarity of a pair of words by taking the inverse of their distance in the WordNet hierarchy and applying a normalization for relative hierarchy depth: the closer two words are in the hierarchy, the higher their similarity score. This measure had a correlation of $r = 0.38$ ($p < 10^{-15}$) with participants' similarity judgments, significantly lower than obtained from the $C(A, B)$, $S(A, B)$ and $sim(A, B)$ measures in our model.

6 Conclusions and Further Work

Our model of similarity based on structural alignment of $\langle S\ V\ O \rangle$ triples has a number of advantages over other approaches to word similarity. First, it provides transparent and meaningful reprsentations of word semantics in terms of the predicates asserted of words in a corpus. Second, it is closer to the current psychological understanding of similarity and so provides a more realistic reflection of how people actually judge similarity. Third, it closely matches people's similarity judgments, giving a significantly better fit to people's judgments than that provided by a standard measure of semantic similarity (at least for the word pairs used in our assessment).

The main disadvantage to the model is coverage: the model can only represent the meaning of a word if that word occurs in an $\langle S\ V\ O \rangle$ triple in the corpus being used, and so even for a large corpus like the BNC many words will not be represented. An important aim for future work is to extend the coverage of our model by constructing queries that capture other $\langle S\ V\ O \rangle$ forms beyond those identified by the query in Figure 4. We would also like to extend the model to syntactic structures beyond $\langle S\ V\ O \rangle$, such as intransitive verbs (which take no object) and ditransitive verbs (which take a subject, a direct object, and an indirect object). With such extensions this model should provide an automatic process for constructing semantic representations that match those used by people, both in content and in structure.

References

1. Burek, G., Pietsch, C., De Roeck, A.: SVO triple based latent semantic analysis for recognising textual entailment. In: Proceedings of the ACK-PASCAL Workshop on Textual Entailment and Paraphrasing, WTEP (2007)

2. Choi, F.Y.Y., Wiemer-Hastings, P., Moore, J.: Latent semantic analysis for text segmentation. In: Proceedings of EMNLP, pp. 109–117 (2001)
3. Gentner, D.: Structure-mapping: A theoretical framework for analogy. Cognitive Science 7 (1983)
4. Gentner, D., Markman, A.B.: Structural alignment in comparison: No difference without similarity. Psychological Science 5(3), 152–158 (1994)
5. Gentner, D., Markman, A.B.: Structure mapping in analogy and similarity. American Psychologist 52(1), 45 (1997)
6. Heylen, K., Peirsman, Y., Geeraerts, D.: Automatic synonymy extraction. In: A Comparison of Syntactic Context Models. LOT Computational Linguistics in the Netherlands, pp. 101–116 (2008)
7. Kilgarriff, A., Rychly, P., Smrz, P., Tugwell, D.: The sketch engine. In: Proceedings of EURALEX (2004)
8. Landauer, T.K., Dutnais, S.T.: A solution to plato's problem: The latent semantic analysis theory of acquisition, induction, and representation of knowledge. Psychological Review, 211–240 (1997)
9. Lin, D.: Automatic retrieval and clustering of similar words. In: COLING-ACL, pp. 768–774 (1998)
10. Lund, K., Burgess, C.: Producing high-dimensional semantic spaces from lexical co-occurrence. Behavior Research Methods, Instruments, and Computers 28(5), 203–208 (1996)
11. Markman, A.B., Gentner, D.: Splitting the differences: A structural alignment view of similarity. Journal of Memory and Language 32, 517–517 (1993)
12. Markman, A.B., Gentner, D.: Commonalities and differences in similarity comparisons. Memory & Cognition 24(2), 235–249 (1996)
13. McCarthy, D., Koeling, R., Weeds, J., Carroll, J.A.: Finding predominant word senses in untagged text. In: ACL, pp. 279–286 (2004)
14. McDonald, S., Brew, C.: A distributional model of semantic context effects in lexical processing. In: Proceedings of the 42nd Annual Meeting on Association for Computational Linguistics, p. 17. Association for Computational Linguistics (2004)
15. Padó, S., Lapata, M.: Dependency-based construction of semantic space models. Computational Linguistics 33(2), 161–199 (2007)
16. Salton, G., McGill, M.: Introduction to Modern Information Retrieval. McGraw-Hill Book Company (1984)
17. Turney, P.D., Pantel, P.: From frequency to meaning: Vector space models of semantics. J. Artif. Intell. Res. (JAIR) 37, 141–188 (2010)
18. Widdows, D.: Unsupervised methods for developing taxonomies by combining syntactic and statistical information. In: Proceedings of the 2003 Conference of the North American Chapter of the Association for Computational Linguistics on Human Language Technology, vol. 1, pp. 197–204. Association for Computational Linguistics (2003)
19. Wu, Z., Palmer, M.: Verb semantics and lexical selection. In: 32nd Annual Meeting of the Association for Computational Linguistics (1994)

Coreference Annotation Schema for an Inflectional Language*

Maciej Ogrodniczuk[1], Magdalena Zawisławska[2],
Katarzyna Głowińska[3], and Agata Savary[4]

[1] Institute of Computer Science, Polish Academy of Sciences
[2] Institute of Polish Language, Warsaw University
[3] Lingventa
[4] François Rabelais University Tours, Laboratoire d'informatique

Abstract. Creating a coreference corpus for an inflectional and free-word-order language is a challenging task due to specific syntactic features largely ignored by existing annotation guidelines, such as the absence of definite/indefinite articles (making quasi-anaphoricity very common), frequent use of zero subjects or discrepancies between syntactic and semantic heads. This paper comments on the experience gained in preparation of such a resource for an ongoing project (CORE), aiming at creating tools for coreference resolution.

Starting with a clarification of the relation between noun groups and mentions, through definition of the annotation scope and strategies, up to actual decisions for borderline cases, we present the process of building the first, to our best knowledge, corpus of general coreference of Polish.

1   Introduction

Although the notion of coreference is no longer a subject of much controversy and there are many more or less ready-to-use annotation guidelines available, in a case where a "new" language is being investigated — which has not yet received any formalized coreference description — they usually need to be supplemented with details specific to this language, and the task of creating a coreference corpus requires establishing detailed rules concerning annotation scope, strategies and typology of coreferential constructs.

This paper comments on the experience gained in the process of creating the first substantial Polish corpus of general coreference (500K words and 160K mentions are intended), which is currently being completed. We hope our analysis can provide a valuable source of information for creators of new coreference corpora for other inflectional and free-word-order languages. We believe that they could particularly benefit from studying our assumptions based on such specific properties as the absence of definite/indefinite articles (introducing quasi-anaphoricity), frequent use of zero subjects or discrepancies between

---

* The work reported here was carried out within the *Computer-based methods for coreference resolution in Polish texts (CORE)* project financed by the Polish National Science Centre (contract number 6505/B/T02/2011/40).

A. Gelbukh (Ed.): CICLing 2013, Part I, LNCS 7816, pp. 394–407, 2013.

syntactic and semantic heads. These phenomena are fundamental for building computational coreference resolvers.

Construction of a large high-quality corpus is of great importance in the context of further tasks in the ongoing CORE project, whose central aim is the creation of an efficient coreference resolver for Polish. We wish to surpass the previous early attempts, both rule-based [1] and statistical [2], which yielded tools trained and evaluated on a very limited amount of data. We believe that a more efficient tool can boost the development of higher-level Polish NLP applications, on which coreference resolution has a crucial impact [3]. Such applications include: 1) machine translation (when translating into Polish, coreferential relations are needed to deduce the proper gender of pronouns), 2) information extraction (coreference relations help with merging partial data about the same entities, entity relationships, and events described at different discourse positions), 3) text summarization, 4) cross-document summarization, and 5) question answering.

2   Reference, Anaphora and Coreference

In order to define the scope of coreference annotation we must bring back the underlying concept of *reference* to discourse-world objects, leading to an important limitation: only nominal groups (NGs), including pronouns, can be referencing expressions.

Recall that coreference annotation is usually performed (and evaluated) in two steps: (i) identifying *mentions* (or *markables*), i.e. phrases denoting entities in the discourse world, (ii) clustering mentions which denote the same referent. Consequently, the definition of a mention, and of the difference between a mention and a NG in particular, is of crucial importance to the whole process. We, unlike e.g. [4], consider this difference too controversial to be reliably decided in a general case.

For instance, multi-word expressions (MWEs) show opaque semantics, thus the NGs they include might be seen as non-referential. However, most MWEs do inherit some part of the semantics of their components, and might be coreferential in some stylistically marked cases, as in (1)[1]. Defining a clear-cut frontier between non-referential and referential NGs in these cases seems very hard.

(1)   *Nie wahał się włożyć kij w mrowisko.*
       Mrowisko to, czyli cały senat uniwersytecki, pozostawało zwykle niewzru-szone.
       'He didn't hesitate to put a stick into an anthill (i.e. to provoke a disturbance).
       This anthill, i.e. the whole university senate, usually didn't care.'

---

[1] Henceforth, we will mark coreferent NGs with (possibly multiple) underlining, and non-coreferent NGs with dashed underlining.

Thus, our annotation process consists in retaining – as mentions – all NGs (whether referential or not), and establishing coreference chains among them wherever appropriate. In other words, we do not distinguish non-referential NGs from referential, but non-coreferential, NGs (e.g. singleton mentions). This decision obviously has a big influence on coreference resolution quality measures which take singleton mentions into account.

We also consider that the reference is context-dependent, not surface-form dependent, cf.

(2)   *Spotkałam nową dyrektorkę. Osoba ta zrobiła na mnie dobre wrażenie.*
        'I met the new manager. This person made a good impression on me.'
(3)   *Nasza nowa dyrektorka to młoda kobieta.*
        'Our new manager is a young woman.'
(4)   *Nasza dyrektorka, młoda kobieta, przyszła na spotkanie.*
        'Our manager, a young woman, came to the meeting.'
(5)   *Młoda kobieta, która przejęła funkcję dyrektora, zrobiła na mnie dobre wrażenie.*
        'The young woman who overtook the manager's duties made a good impression on me.'

In example (2) the NG *osoba ta* ('this person') has a defined referent, i.e. a concrete human being the speaker refers to. In (3)–(4), the nominal group *młoda kobieta* does not carry reference, but is used predicatively — assigns certain properties to the subject of the sentence. Our understanding of nominal coreference is therefore strictly limited to direct nominal constructs; expressions that do not denote the object directly are not included in coreference chains.

There is an additional, operational, criterion that we admit, contrary to many common coreference annotation and resolution approaches, e.g. [5]. If semantic identity relations between NGs are directly expressed by the syntax, we see no point in including them in coreferential chains. Typical cases here are predicates, as in (3), relative clauses, as in (5), and appositions, as in (4), where we see one, not two, mentions in the NG *Nasza dyrektorka, młoda kobieta* ('Our manager, a young woman').

Such definition of reference creates links between the text and discourse world and is of different nature than *anaphora* — an inter-textual reference to previously mentioned objects. Even if, in most cases, anaphora and coreference co-occur, it is not necessarily the case. In example (6), the underlined NGs are anaphoric but not coreferential, cf. [3]. Conversely, NGs in separate texts can be coreferential, but not anaphoric.

(6)   *Człowiek, który dał piękne kwiaty swojej żonie, wydał mi się sympaty-czniejszy niż człowiek, który odmówił kupienia ich swojej.*
        'The man who gave beautiful flowers to his wife seemed nicer to me than the one who refused buying them for his (wife).'

3   Scope of Annotation

3.1   Mentions

As it was said in the previous section, all NGs (both referential and non-referential) are marked as mentions, while coreference chains can only concern referential NGs (mentions). In particular, some types of nominal pronouns, which seem non-referential by nature, are marked as mentions (since they are NGs) but never included in coreference chains: (i) indefinite pronouns (*ktoś* 'somebody'), (ii) negative pronouns (*nic* 'nothing'), (iii) interrogative pronouns (*kto* 'who')[2]. Note also that some Polish lexemes designated traditionally as pronouns behave morphosyntactically like other parts of speech. Namely, demonstrative pronouns introducing subordinates other than relative clauses (*o tym, że* 'of-this-that = of the fact that') are in fact parts of correlates. The reflexive pronoun (*się* 'oneself') is a particle. Finally, possessive pronouns (*mój* 'mine') behave like adjectives. Consequently, these three types of pronouns are never considered as NGs, i.e. they are never marked as mentions.

Finally, coreference relations between phrases other than nominal ones (e.g. *tam* 'there') are obviously never marked, since only NGs are considered as mentions.

3.2   Types of Relations

The major goal of coreference annotation is to determine the type of relation holding among discourse-world entities referred to by two or more mentions. We are essentially interested in identity relations. We also consider, experimentally, the notion of *near-identity* proposed by [6]. Due to the pioneering (wrt. Polish) nature of our project, all other types of relations (whether among entities or among mentions) have been explicitly ruled out, including non-identity, indirect anaphora, bound anaphora, ellipses (with the exception of zero anaphora), predicative relations, and identity of sense.

Identity. Textual techniques used in Polish to signal the identity of referred entities are manifold:

- lexical and grammatical (personal and demonstrative pronouns),
- stylistic, such as *synonymy*,
- lexical and grammatical anaphora and cataphora between nominal groups,
- "quasi-anaphora" – when a group with syntactic-functional properties of anaphora introduces new information, e.g.

  (7)   *Duszą towarzystwa był zięć Kowalskich. Młody prawnik właśnie wrócił ze Stanów.*

  'Kowalski's son-in-law was the life and soul of the party. The young lawyer had just returned from the US.'

---

[2] Surprisingly enough, recent experiences show that such pronouns may be referential in stylistically marked cases such as: *Ktoś ukradł łopatę. Ten sam ktoś zniszczył ogrodzenie.* 'Someone stole the spade. The same someone broke the fence.'. We wish to review these cases in the final annotation stage.

– zero-anaphora, very frequent in Polish – a personal pronoun may be omitted whenever the subject's person and gender are recognizable from the verb's agreement; therefore the annotation denoting the missing referential NG is most naturally attached to the verb, as in example (8).[3]

(8)   _Maria_ wróciła już z Francji. _ØSpędziła tam miesiąc._
        '_Maria_ came back from France. _ØHad$_{singular:feminine}$_ spent a month there.'

Note that some approaches introduce a typology of coreference links which takes the above techniques into account. We, conversely, think that these types of linguistic data should be documented either at other annotation levels or in external linguistic resources. One – formal and practical – reason is that we see coreference chains as clusters, i.e. results of splitting the set of all mentions via a (unique and uniform) equivalence relation. If subtypes of this relation were to be used, clustering would no longer be possible and each pair of coreferent mentions would have to be marked explicitly. Such a methodology might not only have a prohibitive cost in some types of texts but would also be hard to evaluate by classical quality measures.

Near-Identity. [7] define the notion of _near-identity_, taking place in two contexts called refocusing and neutralization. Our understanding of these phenomena involves the following:

– Refocusing – two mentions refer to the _same_ entity but the text suggests the opposite. This stylistic technique is often used to account for a temporal or spatial change of an object as in[4]:

(9)   _Warszawa przedwojenna i ta z początku XXI wieku_
        '_Pre-war Warsaw and the one at the beginning of the 21st century_'

– Neutralization – two mentions refer to _different_ entities but the text suggests the opposite. This situation is typical for metonymy, as in example (10), where a container and its contents are merged, and unlike (11), which is a case of a classical identity:

(10)   _Wziął wino z lodówki i wypił je._
        'He took _the wine_ from the fridge and drank _it_.'

(11)   _Wziął wino z lodówki i włożył je do torby._
        'He took _the wine_ from the fridge and put _it_ into the the bag'.

[6] put forward a detailed typology of near-identity relations. However, in the experimental annotation stage of our project, the annotators marked very few examples of near-identity, most of them concerning, in fact, more typical semantic relations, like homonymy, meronymy, metonymy, element of a set or — sometimes — hypernymy, e.g.:

---

[3] Elliptical constructions concerning functions other than the subject, as in _Czytałeś książki Lema? Czytałem Ø._ 'Did you read _Lem's books_? I read _Ø_.' are not annotated in our model.

[4] Henceforth, near-identity-related mentions will be marked by a wavy underline.

(12) Cała Warszawa była właściwie jednym wielkim cmentarzem. Ginęli ludzie,
       mnóstwo ludzi! Na podwórku, już tak po 15 sierpnia, praktycznie codzien-
       nie był pogrzeb przed kapliczką. Warszawa była bardzo pobożna...

       'The whole Warsaw was in fact one big graveyard. People were dying, plenty of people!

       After the 15th of August there were funerals in the courtyard, in front of the chapel,

       almost every day. Warsaw was very pious...[5]

That experience made us think that near-identity is either too infrequent to
deserve a rich typology, or too hard to capture and classify reliably by annotators.
That is why we mark near-identity links in our corpus, but we assign no type
labels to them. Once the annotation has been completed, we plan to compare
our examples of near-identity more thoroughly with the types proposed in [6].

3.3 Dominant Expressions

Despite the fact that all mentions within a cluster are (mathematically speaking)
equivalent, we enrich each cluster with a pointer towards the *dominant expression*, i.e. the one that carries the richest semantics. For instance in the following
chain the last element is dominant: *stworzenie* 'creature' → *zwierzę* 'animal' → *pies*
'dog' → *jamnik* 'dachshund'.

   In many cases, pointing at the dominant expression helps the annotators sort
out a large set of pronouns denoting various persons (e.g. in fragments of plays or
novels). We think that it might also facilitate linking mentions within different
texts, and creating a semantics frame containing different descriptions of the
same object.

4 Annotation Strategies

4.1 Mention Boundaries

In order to encompass the wide range of mentions, we set the boundaries of
nominal groups as broadly as possible. Therefore, an extended set of elements
is allowed within NG contents, i.e., 1) adjectives as well as adjectival participles
in agreement (with respect to case, gender and number) with superior noun,
2) subordinate noun in the genitive case, 3) nouns in case and number agreement with superior nouns (i.e. nouns in apposition); but also 4) prepositional-
nominal phrase that is a subordinate element of a noun (e.g. *koncert na skrzypce i
fortepian* 'a concerto for violin and piano')[6]; 5) relative clause (e.g., *dziewczyna,
o której rozmawiamy* 'the girl that we talk about'). Moreover, the following
phrases are treated as nominal groups: 1) numeral groups (e.g., *trzy rowery*

---

[5] *The whole Warsaw* refers to the place, while *Warsaw* is a metonymy and refers to
    people who lived in the city.
[6] Such cases should be distinguished from situations where a prepositional-nominal
    phrase is a subordinate element of a verb, e.g. *Kupił mieszkanie z garażem.* 'He
    bought a flat with a garage.'

'three bicycles'), 2) adjectival phrases with elided nouns (e.g., *Zrób bukiet z tych czerwonych kwiatów i z tych niebieskich.* 'Make a bouquet of these red flowers and these blue ones.'), 3) date/time expressions of various syntactic structures, 4) coordinated nominal phrases, including conjoining commas (*krzesło, stół i fotel* 'a chair, a table, and an armchair').

For each phrase, the semantic head is selected, being the most relevant word of the group in terms of meaning. The semantic head of a nominal group is usually the same element as the syntactic head, but there are some exceptions, e.g., in numeral groups, the numeral is the syntactic head, and the noun is the semantic head.

4.2 Mention Structure

The deep structure of noun phrases, i.e. all embedded phrases not containing finite verb forms having semantic heads other than those of the superior phrase (which reference different entities), is subject to annotation, therefore the fragment *dyrektor departamentu firmy* 'manager of a company department' contains 3 nominal phrases, referencing *dyrektora departamentu firmy* ('manager of a company department'), *departamentu firmy* ('a company department') and *firmy* ('the company') alone.

This assumption is also valid for coordination — we annotate both the individual constituents and the resulting compound, because they can be both referred to:

(13)   *Asia i Basia mnie lubią. One są naprawdę ładne, szczególnie Asia.*
       '*Asia and Basia like me. They are really pretty, particularly Asia.*'

Discontinuous phrases and compounds are also marked:

(14)   *To był delikatny, że tak powiem, temat.* '*It was a touchy, so to speak, subject.*'

5   Task Organization

Texts for annotation were randomly selected from the National Corpus of Polish [8]. Similarly to this resource, we aimed at creating a 500-thousand-word balanced subcorpus. It was divided into over 1700 samples between 250 and 350 segments each. These samples were automatically pre-processed with a shallow parser detecting nominal groups and their semantic heads[7], and a baseline coreference resolution tool marking potential mentions and identity clusters.

---

[7] More precisely, nominal groups were precomputed from parse trees produced by a shallow parser Spejd [9] supported with the Pantera tagger [10] and a named entity recognizer Nerf [11]. The scope of each NG was heuristically determined in that the longest NG was retained among all potential NGs sharing the same head, e.g. *dyrektor departamentu firmy* 'the director of the department of the company' was retained rather than *dyrektor departamentu.* Nested NGs were then marked within each retained maximal NG, e.g. *[dyrektor [departamentu [firmy]]].*

The manual revision of this automatically performed pre-annotation is being carried out in the MMAX2 tool [12] adapted to our needs. In particular, the annotators can correct the pre-annotation results (i.e., remove or change NG marking, change the semantic heads, and modify the content of identity clusters). They can also add mentions and clusters that were not detected by the tool.

Each fragment of the corpus is prepared by one annotator (entitled to change the pre-annotation in all respects) and then checked by the supervising annotator. Note that although the best practice in other annotation tasks [8] is parallel annotation, in which two independent annotators work on each text, and an adjudicator reviews cases of disagreement, we find this practice hard to apply in coreference annotation.

Nevertheless, a part of the corpus, namely 210 texts, has been annotated independently by two people, and then adjudicated by the supervising annotator, in order to check the inter-annotator agreement. Statistics were calculated for each level of annotation separately, i.e., 1) NG scope: $F1 = 85,55\%$, 2) semantic heads: 97%, 3) identity clusters: see below, 4) near-identity links: 22,20%, and 5) dominant expressions: 63,04%.

The agreement in identity clusters annotation was calculated using κ coefficient (taking agreement by chance into account) for the decision about each mention, whether it is a singleton or not. This method is similar to the agreement computation in the An-Cora corpus [4]. We achieved κ of 0.7424. Note also the particularly low agreement in near-identity links which indicates the hardness of this task.

6   Difficult Cases

Recall that the annotators' main task in the project is to indicate identity of reference, i.e. that two or more linguistic elements point to the same extralinguistic referent in the text. The task does not sound very difficult, but in practice, things turn out to be different. There are relatively many cases when the recipient cannot decide if the NGs are coreferential or not. The mistakes can occur on the three main levels: lexical, grammatical and conceptual.

6.1   Lexical Level

Frequent occurrences of annotator's "false friends" are due to polysemy and homonymy. In the first sentence of example (15), the noun *misja* 'mission' means "a responsible task someone is entrusted with", while in the second it is "a representation of a country/organization with special assignment". Such cases of graphically identical but non-coreferential NGs may be hard to detect.

(15)   *Misja rozpoczęła się 8 kwietnia. W skład misji weszły 24 osoby.*

   'The mission started on April 8th. 24 people were members of the mission.'

In some extreme cases, two NGs may be both graphically and semantically identical, and still remain non-coreferential as in example (16). The speaker clearly assigns here different characteristics to both expressions, i.e. he means that there are many different types of mothers, e.g. good ones and bad ones. Detecting this particular type of repetition might be useful in a future automatic coreference resolver.

(16)   *Są matki i matki.* 'There are mothers and (then there are) mothers.'

Another, perhaps the most problematic, lexical issue involves the so-called *co-extension*. It occurs when two or more NGs refer to objects which belong to the same conceptual field. The referents of these NGs can be linked by various semantic relations, e.g. hypo-/hypernymy, meronymy, antonymy, etc. Such relations very often make it difficult to decide if the NGs are coreferential or not. In example (17), the annotator made an excessive cluster in which s/he placed phrases *mity* 'myths' and *mitologia* 'mythology', while a myth is a meronym of a mythology, and a mythology is a holonym of a myth.

(17)   *[...] mity są niezastąpionym narzędziem dla psychologa, usiłującego prze-śledzić wzorce ludzkich zachowań. Wysiłki archeologów, religioznawców, antropologów doprowadziły z jednej strony do porzucenia eurocentrycznego spojrzenia na mitologię...*
       '[...] myths are irreplaceable tools for a psychologist, who is trying to follow through the standards of human behaviour. Efforts of archaeologists, specialists in religious studies and anthropologists resulted, on the one hand, in giving up the Eurocentric point of view on the mythology...'

Similarly, in example (18), the annotator had a problem with deciding on an identity connection between *okupacja* 'occupation' and *wojna* 'war'. Obviously, those two words have something in common (occupation is a result of war, therefore the WordNet *entailment* relation would be relevant here), but they are not coreferential.

(18)   *Od czasu okupacji (...) — Ale tu je masz z powrotem, w metryce — powiedział dyrektor. — Kiedy to jest stara metryka, którą mi odtworzono zaraz po wojnie.*
       'After occupation (..) — But here you have it back, in your birth certificate — said the headmaster. — But it is an old birth certificate, which was reconstructed after the war.'

6.2   Grammar Level

We omit the most obvious cases, e.g. a speaker's grammar mistakes, and concentrate on the less typical examples. There are no articles in Polish, therefore the most difficult task for the annotators was to distinguish definite and indefinite objects. In example (19), an annotator wrongly created one cluster in which s/he placed all forms of the word *asystent* ('*assistant*'), e.g.:

(19)   *Każdy szanujący się poseł ma <u>asystenta</u>. <u>Asystentami</u> są z reguły ludzie młodzi,*
        ale nie brakuje również szczerze zaangażowanych emerytów. Pracują jako
        wolontariusze tak jak Marek Hajbos, <u>asystent</u> Zyty Gilowskiej.

    'Every decent Member of Parliament has <u>an assistant</u>. <u>Assistants</u> are usually young peo-
    ple, but there are also genuinely involved senior citizens. They work as volunteers like
    Marek Hajbos, <u>the assistant</u> of Zyta Gilowska.'

6.3   Conceptual Level

A crucial problem in establishing the identity relations between NGs is the lack
of annotator's competence in some fields. In example (20) the annotator was
unaware that the players of the Silesian football team Ruch Chorzów wear blue
shirts.

(20)   *W trzecim kwartale 2010 roku <u>Ruch Chorzów</u> zarobił na czysto aż 5.5 mln*
        zł. Wiadomość o zysku <u>Niebieskich</u> na pewno ucieszy jego kibiców.

    'In the third quarter of 2010 year <u>Ruch Chorzów</u> earned 5.5 million złoty net. The news
    about the profit of <u>the Blues</u> will please their supporters for sure.'

7   Related Work

In this section we review some of the coreference annotation schemes admitted
in previous efforts for several languages. While an exhaustive state-of-the-art
contrastive study is beyond the scope of our paper, we are particularly interested in languages that show coreference-relevant morphosyntactic similarities
with Polish. Slavic languages are obviously of highest importance, but Spanish
is also relevant, in particular due to its frequent zero subjects. Finally, for obvious dominance reasons in NLP, we also address one of the most recent studies
dedicated to English.

[13] describes BulTreeBank, a syntactically annotated corpus of Bulgarian based
on an HPSG model. Coreferential chains link nodes in HPSG trees. Each noun
phrase is linked to an (extra-linguistic) index representing, roughly, the discourse-
word entity. Coreference is expressed by linking several phrases to the same index. In principle, only coreferential relations which cannot be inferred from the
syntactic structure are annotated explicitly, however, some inferable ones are annotated too (it is unclear which ones). Zero subjects and other elliptical elements
(e.g. headwords missing due to coordination) are represented whenever they belong to coreference chains. Syntactic trees may help represent split mentions but it
is uncertain if they do. Possessive pronouns are considered as mentions. Three relations are encoded: identity, member-of, and subset-of. Discourse deixis is probably
taken into account. It seems that the annotation concerns coreference occurring
within one sentence only. No inter-annotator agreement results are given.

[14] presents annotation efforts for a 94,000-word English corpus. Special attention is paid to two difficult phenomena: discourse-inherent coreference ambiguity and discourse deixis. The former yields an annotation scheme in which

coreference is not an equivalence relation (one mention can appear in several chains). All nominal groups are considered mentions, but some are later marked as non-coreferential. A limited set of bridging relations is taken into account. Problems related to zero subjects, nested, split and attributive NGs, as well as semantic heads, are not discussed and are probably not addressed in the annotation scheme.

[15] extends the coreference annotation in the Prague Dependency Treebank of Czech, a language rather close to Polish. It builds on previously constructed annotation layers including the so-called tectogrammatical layer, which provides ready mention candidates and (probably) their semantic heads. Mentions include nominal phrases and coreferential clauses (discourse deixis). Nested groups are delimited except in named entities (where only embedded groups which are NEs themselves are marked). Attributive phrases are not considered uniformly: appositions are marked as mentions even if they are never included in coreference chains, while predicate nominals are not considered at all. The notable contribution of this approach is addressing a wide range of bridging relations between nominals. The relatively low scores of the inter-annotator agreement might be an evidence that coreference annotation is particularly difficult in Slavic languages.

[16] describes coreference annotation in AnCora-CO, a 400K-word corpus of Spanish and Catalan, for which, like in [15] and [13], other annotation layers had previously been provided, including syntax. Thus, possible candidates for mentions had already been delimited. The annotation schema is rather complete. Three types of relations are considered: identity, predicative link and discourse deixis. Zero subjects are marked, clitic pronouns which get attached to the verb are delimited, embedded and discontinuous phrases are taken into account, and referential NGs are distinguished from attributive ones. Bridging references are not considered.

[17] addresses the anaphoric relations in a parallel, 5-language Copenhagen Dependency Treebank, in which unified annotation of morphology, syntax, discourse and anaphora is being performed. It consists of a 100,000-word Danish corpus with its translations into English, German, Italian and Spanish. Possible specificities of mention detection are not addressed, however, relation typology is extensively discussed. Both coreference and bridging relations (called *associative anaphora*) are considered. The former are split into 6 categories, according to linguistic techniques used to express the coreference, including discourse deixis. The latter count as many as 12 types. The inter-annotator agreement (expressed in percentages, i.e. not accounting for agreement by chance) varies highly among relation types.

Table 1 shows a contrastive study of some coreference annotation schemata and of our approach. In view of this analysis, our approach shows some novelty. It seems to be the first one to experiment with near-identity (introduced by [7]) on a large scale. It is the only one to focus on pointing at dominant expressions and semantic heads. Along with [13], it belongs to two out of three approaches dedicated to Slavic languages which introduce zero subjects in coreference chains.

Table 1. Contrastive analysis of coreference annotation schemes and tools

| Reference Language | Mention scope | | | | Discourse Deixis | Semantic Head Markup | Relations | Dominating Expression Markup | Inter-Annotator Agreement |
	Zero Subject	Nested NGs	Split NGs	Attributive NGs					
[13] Bulgarian	✓	✓	?	?	✓		identity, part-of, subset-of		unknown
[14] English				✓	✓		identity, predicative link, discourse deixis bridging relations		$\alpha = 0.6 - 0.7$
[15] Czech		partly	✓	partly	✓	?	identity, predicative link, discourse deixis with several sentences, bridging relations		F_1-measure= $0.39 - 0.80$
[16] Spanish, Catalan	✓	✓	✓	delimited, predicative-linked	✓		identity, predicative link, discourse deixis		$\alpha = 0.85 - 0.89$
[17] Danish, English, German, Italian, Spanish					✓		identity, discourse deixis bridging relations		$25 - 100\%$
Our approach Polish	✓	✓	✓	delimited but never linked		✓	identity, near-identity	✓	$\kappa = 0.7424$

8   Conclusions and Further Work

We believe that the reported notes on coreference annotation could prove valuable for other coreference corpora creators and the underlying resource is in itself an important step towards general-purpose coreference resolution for Polish.

We also hope that the current work could help harmonize efforts aimed at creating similar corpora for a group of related languages and, in turn, testing new cross-lingual concepts such as coreference projection (see, e.g., [18]), which require stable and consistent annotation model for all languages involved.

References

1. Ogrodniczuk, M., Kopeć, M.: End-to-end coreference resolution baseline system for Polish. In: Vetulani, Z. (ed.) Proceedings of the Fifth Language & Technology Conference: Human Language Technologies as a Challenge for Computer Science and Linguistics, Poznań, Poland, pp. 167–171 (2011)
2. Kopeć, M., Ogrodniczuk, M.: Creating a Coreference Resolution System for Polish. In: Proceedings of the Eighth International Conference on Language Resources and Evaluation, LREC 2012, pp. 192–195. ELRA, Istanbul (2012)
3. Mitkov, R.: Anaphora Resolution. In: Mitkov, R. (ed.) The Oxford Handbook of Computational Linguistics. Oxford University Press (2003)
4. Recasens, M.: Coreference: Theory, Annotation, Resolution and Evaluation. PhD thesis, Department of Linguistics, University of Barcelona, Barcelona, Spain (2010)
5. Haghighi, A., Klein, D.: Simple coreference resolution with rich syntactic and semantic features. In: Proceedings of the 2009 Conference on Empirical Methods in Natural Language Processing, EMNLP 2009, vol. 3, pp. 1152–1161. Association for Computational Linguistics, Stroudsburg (2009)
6. Recasens, M., Hovy, E., Martí, M.A.: A Typology of Near-Identity Relations for Coreference (NIDENT). In: Proceedings of the Seventh International Conference on Language Resources and Evaluation (LREC 2010), pp. 149–156 (2010)
7. Recasens, M., Hovy, E., Martí, M.A.: Identity, non-identity, and near-identity: Addressing the complexity of coreference. Lingua 121(6) (2011)
8. Przepiórkowski, A., Bańko, M., Górski, R.L., Lewandowska-Tomaszczyk, B. (eds.): Narodowy Korpus Języka Polskiego (Eng.: National Corpus of Polish). Wydawnictwo Naukowe PWN, Warsaw (2012)
9. Przepiórkowski, A., Buczyński, A.: Spejd: Shallow Parsing and Disambiguation Engine. In: Vetulani, Z. (ed.) Proceedings of the 3rd Language & Technology Conference, Poznań, Poland, pp. 340–344 (2007)
10. Acedański, S.: A Morphosyntactic Brill Tagger for Inflectional Languages. In: Loftsson, H., Rögnvaldsson, E., Helgadóttir, S. (eds.) IceTAL 2010. LNCS, vol. 6233, pp. 3–14. Springer, Heidelberg (2010)
11. Waszczuk, J., Głowińska, K., Savary, A., Przepiórkowski, A., Lenart, M.: Annotation Tools for Syntax and Named Entities in the National Corpus of Polish. International Journal of Data Mining, Modelling and Management (to appear)
12. Müller, C., Strube, M.: Multi-level annotation of linguistic data with MMAX2. In: Braun, S., Kohn, K., Mukherjee, J. (eds.) Corpus Technology and Language Pedagogy: New Resources, New Tools, New Methods, pp. 197–214. Peter Lang, Frankfurt a.M., Germany (2006)

13. Osenova, P., Simov, K.: BTB-TR05: BulTreeBank Stylebook. BulTreeBank Version 1.0. Technical Report BTB-TR05, Linguistic Modelling Laboratory, Bulgarian Academy of Sciences, Sofia, Bulgaria (2004)
14. Poesio, M., Artstein, R.: Anaphoric Annotation in the ARRAU Corpus. In: Proceedings of the International Conference on Language Resources and Evaluation (LREC 2008). European Language Resources Association, Marrakech (2008)
15. Nedoluzhko, A., Mírovský, J., Ocelák, R., Pergler, J.: Extended Coreferential Relations and Bridging Anaphora in the Prague Dependency Treebank. In: Proceedings of the 7th Discourse Anaphora and Anaphor Resolution Colloquium (DAARC 2009), Goa, India. AU-KBC Research Centre, Anna University, Chennai, pp. 1–16 (2009)
16. Recasens, M., Martí, M.A.: AnCora-CO: Coreferentially annotated corpora for Spanish and Catalan. Language Resources and Evaluation 44(4), 315–345 (2010)
17. Korzen, I., Buch-Kromann, M.: Anaphoric relations in the Copenhagen Dependency Treebanks. In: Proceedings of DGfS Workshop, Göttingen, Germany, pp. 83–98 (2011)
18. Rahman, A., Ng, V.: Translation-Based Projection for Multilingual Coreference Resolution. In: Proceedings of the 2012 Conference of the North American Chapter of the Association for Computational Linguistics: Human Language Technologies, pp. 720–730. Association for Computational Linguistics, Montréal (2012)

Exploring Coreference Uncertainty of Generically Extracted Event Mentions

Goran Glavaš and Jan Šnajder

University of Zagreb, Faculty of Electrical Engineering and Computing
Unska 3, 10000 Zagreb, Croatia
{goran.glavas,jan.snajder}@fer.hr

Abstract. Because event mentions in text may be referentially ambiguous, event coreferentiality often involves uncertainty. In this paper we consider event coreference uncertainty and explore how it is affected by the context. We develop a supervised event coreference resolution model based on the comparison of generically extracted event mentions. We analyse event coreference uncertainty in both human annotations and predictions of the model, and in both within-document and cross-document setting. We frame event coreference as a classification task when full context is available and no uncertainty is involved, and a regression task in a limited context setting that involves uncertainty. We show how a rich set of features based on argument comparison can be utilized in both settings. Experimental results on English data suggest that our approach is especially suitable for resolving cross-document event coreference. Results also suggest that modelling human coreference uncertainty in the case of limited context is feasible.

1 Introduction

Events are often defined as situations that *happen* or *occur* [13]. A real-world event necessarily instantiates a spatiotemporal region and therefore, as noted in [14], real-world events may be individuated by virtue of spatiotemporal co-extensiveness. However, due to ambiguity and vagueness of natural language, the linguistic representations of real-world events – the *event mentions* – may be referentially ambiguous. Therefore, in some cases it may be impossible to uniquely identify a real-world event denoted by a mention and definitively determine whether two event mentions corefer. Thus, from an epistemological point of view, event coreferentiality may involve uncertainty. We are able to resolve much of this ambiguity by relying on a wider context and common-sense knowledge. However, in some cases a wider context required for determining event coreference is not available. A case in point is the event-oriented IR, in which a query typically lacks a wider context (e.g., *"Bush visits Haiti"*). On the other hand, exploiting information from a wider context (e.g., full document), even if such a context is available, is rather challenging from a computational linguistics point of view. Because of this, coreference resolution is often effectively limited to comparing only event mentions (i.e., it does not make use of a wider context), which is more feasible computationally.

A. Gelbukh (Ed.): CICLing 2013, Part I, LNCS 7816, pp. 408–422, 2013.

An event mention consists of an *event anchor*, which is a single-word (or, less often, a multi-word) expression that bears the core meaning of the real-world event, and one or more *event arguments*. Each event argument is assigned a semantic type (role) that describes its relation with the event anchor. In this paper we address event coreference resolution by robustly extracting and comparing four main semantic types of arguments (agent, target, time, and location), which may be assigned to the vast majority of real-world events. From a conceptual point of view, we argue that these arguments bear the most information about an event, as they provide the answers to the main wh-questions: *"who did what to whom, where and when"*. Consequently, these four argument types contribute the most to reducing referential ambiguity of event mentions. For example, *"Sebastian Vettel won the Formula 1 World Championship in São Paulo on Sunday"* is less referentially ambiguous than *"Sebastian Vettel won the Formula 1 World Championship"* (as Vettel won the World Championship three times so far), which is in turn less referentially ambiguous than *"Sebastian Vettel won"* (as he could have won a race, a championship, or something else). On the other hand, from a computational point of view, extracting arguments of generic types can be performed more robustly, thus overcoming the problem of limited coverage that is common to semantic role labelling approaches [16].

The contribution of this work is twofold: we (1) develop a supervised event coreference resolution model based on the comparison of generically extracted event mentions and (2) explore the uncertainty of event coreference in both human annotations and the coreference resolution model, and in both within-document and cross-document setting. For the latter, we (2a) investigate whether the use of a limited context (consisting of event anchors and their arguments) suffices to make a confident coreference decision, (2b) examine how revealing the full document context affects event coreference uncertainty, and (2c) model human uncertainty on event coreference in a limited context setting. The modelling of human coreference uncertainty can be useful in applications where event mentions lack a wider context (e.g., event-oriented IR, sentence retrieval, semantic similarity of short texts). For example, in event-oriented IR, documents relevant to the query will most likely be those containing event mentions that are coreferring with the event mention expressed in the query. We show that the supervised model based on features comparing mention anchors and arguments is especially suitable for resolving cross-document event mention coreference. We also show that modelling human coreference uncertainty in the case of limited context is feasible. We conduct our experiments on English data; the methods may be applied to other languages provided that the required language tools and resources exist (e.g., a dependency parser, a named entity recognizer, WordNet, etc.).

The remainder of the paper is organized as follows. In the next section we briefly outline the related work on event coreference resolution. In Section 3 we describe the event coreference uncertainty in more detail, while in Section 4 we analyse human event coreference confidence annotations. In Section 5 we describe the event coreference models. We present and discuss experimental results in Section 6. Section 7 concludes the paper and outlines future work.

2 Related Work

The task of event coreference resolution has first been addressed by Humphreys et al. [11]. Mentions from text were arranged in a hierarchical semantic network according to an event ontology determined by the specific information extraction template (e.g., in a *succession* template, *hire* is a sub-event of *incoming* and *retire* is a subevent of *outgoing*). Coreference was determined pairwise, by computing coreference score based on the distance of events in the template event network and comparing event attributes (e.g., animate, time). The major shortcomings of this approach are that the comparison is limited to event anchors and the need for a predetermined information extraction template.

Subsequent work of Bagga et al. [3] dealt specifically with cross-document event coreference, considering only texts from narrow domains such as *resignations*, *elections*, and *espionage*. They first build summaries for each event anchor in the document by selecting all sentences in which the anchor or one of its nominalizations appear. To determine whether a pair of cross-document event mentions is coreferent, they compare the named entities and temporal expressions overlap in the events' summaries. Unfortunately, this approach assumes that matching of event anchors implies within-document coreference, which in general is not the case, especially not for frequently occurring event anchors (e.g., reporting event anchors such as *told* or *said*, often used in news domain).

Event coreference resolution was addressed as a separate task in the ACE evaluation campaign, where Ahn [2] used a maximum entropy classifier to resolve coreference of paired event mentions. The probabilistic output of the classifier was used as input to a linking algorithm to build coreference chains. Unlike previous approaches, Ahn utilizes event argument comparison features along with the anchor comparison features. However, event argument types being extracted and compared are very fine-grained (35 different types) and domain specific (e.g., *attacker* or *vehicle*). Moreover, only the within-document event coreference is considered.

Bejan and Harabagiu [5] propose a clustering approach based on nonparametric Bayesian models for both within- and cross-document event coreference resolution. Their models are based on a rich set of linguistic features, including features comparing event arguments. However, they extract arguments according to semantic frames defined in PropBank and FrameNet, effectively limiting the approach to the predicates covered by those resources. Chen et al. [8] propose a framework that integrates five different event coreference resolvers, arguing that coreferent pairs with different parts-of-speech exhibit different behaviour (e.g., coreference of verb-verb event pairs differs from coreference of verb-noun event pairs). They train binary classifiers and use a rich set of linguistic features including argument matching features. Finally, they apply a spectral graph partitioning algorithm to cluster the mentions into event chains. Their argument extraction approach is limited to the extraction of arguments from anchor pre-modifiers and prepositional attachments.

Most of the aforementioned approaches ([2,5,8]) address event coreference as a classification task, using a rich set of features based on comparison of event mentions. In contrast, we focus on exploring event coreference uncertainty, as there are situations in which the full contexts of event mentions are not available (e.g., queries in event-oriented IR). We analyse and model human coreference uncertainty in the setting where only event mentions are available and also demonstrate how this uncertainty manifests in a full-context setting. Furthermore, we employ a robust argument extraction approach extracting only four main semantic types of arguments (agent, target, time, and location), assignable to the vast majority of real-world events, making the approach applicable across domains. While some approaches ([2,8]) intuitively utilize classifier confidence (usually to build coreference chains), we justify this by showing that confidence of a binary classifier correlates with human uncertainty scores when provided limited context.

3   Event Coreference Uncertainty

Our investigation of event coreference uncertainty began with the observation that in many situations we cannot reach a definite coreference decision when comparing only the event mentions, even after the four main semantic types of arguments have been correctly identified. Uncertainty of event coreference is a direct consequence of the referential ambiguity of event mentions. The referential ambiguity is, in turn, usually caused by one or more of the following:

1. Implicit arguments – some of the arguments may be omitted, usually because they are implied by the discourse structure [15]. Consider the following example: *"US Secretary of State addressed the U.N. on gay rights on Saturday. Clinton's speech in Geneva was a another confirmation of a strong support..."*. When comparing only event mentions (anchors and arguments), the coreference of events *"addressed"* and *"speech"* is uncertain, because arguments *"the U.N."* and *"on Saturday"* are implicit for anchor *"speech"*, while argument *"in Geneva"* is implicit for anchor *"addressed"*. This type of coreference uncertainty is more characteristic for within-document, anaphoric event mention pairs;
2. Missing information – some information relevant for establishing event coreference may be omitted in the text, resulting in missing event arguments. This is usually the case when some aspect of the event (e.g., time or location) is perceived as not being particularly important for the intended discourse goal (e.g., *"The Spurs won their ninth game in a row"*);
3. Argument ambiguity/vagueness – in some cases the coreference cannot be established with certainty due to ambiguity of arguments or vagueness of spatiotemporal location of the event. Argument ambiguity is typical for named entities; e.g., *"the celebration in Paris"* may refer to an event that occurred

in any of twenty cities around the world bearing that name. Similarly, the coreference of *"the bomb exploded in Baghdad on Saturday"* and *"In Baghdad the explosion left 4 dead on Saturday morning"* is uncertain because the first explosion might have occurred on Saturday evening or on some other Saturday (temporal vagueness) or the explosions might have happened in different parts of Baghdad (spatial vagueness). This type of coreference uncertainty is more characteristic for cross-document mention pairs;

4. Limited contextual information – other contextual information, such as relation to other events (e.g., temporal or causal relations) may resolve referential ambiguity of a mention. E.g., in *"Vettel won the Formula 1 Championship after finishing 6th in São Paulo"*, the existence of a temporal relation between the mentions (*"Vettel"*, *"won"*, *"Formula 1 Championship"*) and (*"Vettel"*, *"finishing"*, *"6th"*, *"in São Paulo"*) eliminates the referential ambiguity of the first mention, as only one of three different Vettel's Championship victories was achieved after finishing 6th in São Paulo. In the lack of the context that expresses the temporal relation between the two mentions (*"... after finishing 6th in São Paulo"*), the first mention becomes referentially ambiguous.

Intuitively, allowing for more context reduces the uncertainty, as we potentially explicitate and uncover some of the missing information (notice, however, that widening the context may, in principle, also increase the ambiguity of events, but we feel that this phenomenon is less prominent). However, as previously noted, utilizing more context is not a straightforward task, and moreover a wider context may not always be available. We can remedy these limitations by explicit modelling of event coreference uncertainty.

4 Human Coreference Annotations

Our initial observations on event coreference uncertainty led to annotating event coreference confidence given only event mention pairs, but also given the full context available – the full document text in within-document case and the full text of both documents in the cross-document case.

4.1 Dataset Preparation

The fraction of coreferential mention pairs is negligible when compared to all possible event mention pairs for both within- and cross-document setting. In order to ensure a sufficient amount of coreferring event mention pairs, we decided to draw the events from a limited number of topically related groups of documents. For this purpose we used a set of newspaper documents obtained through NewsBrief service of the European Media Monitor.[1] NewsBrief groups the documents describing the same seminal events. We selected 70 seminal event groups and 2–3 documents from each group. From the chosen documents, we

---

[1] http://emm.newsbrief.eu/NewsBrief/clusteredition/en/latest_en.html

extracted the event anchors (using an in-house anchor extraction tool performing at F-score of 81%) and main event arguments (cf. Section 5.1). In the next step, we employed a heuristic for selecting the event pairs in order to increase the chance of selecting a sufficient number of coreferential pairs. The heuristic searches for event mentions (both within- and cross-document) that have similar anchors and similar arguments. The lexico-semantic similarity of event anchors (and argument heads) is measured on WordNet using Wu and Palmer's algorithm [18]. Our heuristic obviously introduces a bias in the distribution of the dataset; to account for this, we use the same heuristic as the baseline model in our evaluation. The final dataset consists of 1006 event mention pairs (437 within-document pairs and 569 cross-document pairs).[2]

4.2 Guidelines and Annotations

Each mention pair is annotated with two scores. The first score represents the annotator's level of certainty that the event mentions corefer when considering only the event mentions, i.e., event anchors and arguments for both mentions. We refer to this as the *mentions-only* setting. The second score represents the annotator's level of certainty that event mentions corefer when considering the full context, i.e., the full text of the documents from which the mentions were extracted. We refer to this as the *full-context* setting. The annotators were allowed to assign the second score only after they have assigned the first score. The annotators were instructed to assign the scores on a five-level scale, as follows: -2 (confident that mentions do not corefer), -1 (it is more likely that mentions do not corefer), 0 (cannot tell); 1 (it is more likely that mentions corefer); 2 (confident that mentions corefer).

Two annotators independently annotated a subset of 300 mention pairs (150 within- and 150 cross-document pairs). The disagreements were resolved by consensus, after which one of the annotators annotated the remaining 700 pairs. Table 1 shows the inter-annotator agreement measured in terms of the mean absolute error (MAE), Pearson correlation, and the coefficient of determination (R^2). Besides overall agreement, we also report agreement on within-document mention pairs as well as agreement on cross-document mention pairs. The agreement is reported for both *mentions-only* scores and *full-context* scores.

Expectedly, the inter-annotator agreement was significantly higher in the *full-context* setting, as uncertainty decreases when more context is available. Deeper inspection of annotator disagreements in the *mentions-only* setting revealed that most of the disagreements occurred between adjacent labels (in particular between 0 and 1 and between -1 and -2). On the other hand, *full-context* disagreements (-2 vs. 2) were caused by obvious annotation mistakes in all cases (e.g., when one of the annotators did not read the full text carefully enough).

Revealing the full context of event mentions in most cases had an expected influence on the confidence score: what has been judged as likely in the *mentions-only* setting usually gets confirmed after inspection of the full context

---

[2] Available online at http://takelab.fer.hr/evcoref

Table 1. Inter-annotator agreement for *mentions-only* (limited context) and *full-context* setting

Mention pairs	Mentions-only			Full-context		
	MAE	Pearson	R^2	MAE	Pearson	R^2
All	0.51	0.78	0.61	0.18	0.88	0.74
Within-document	0.55	0.70	0.47	0.19	0.82	0.61
Cross-document	0.48	0.83	0.67	0.16	0.90	0.79

(i.e., 1 to 2 and -1 to -2). Mention pairs for which the coreference was indecisive (those labelled 0 in the *mentions-only* setting) get resolved as coreferring (0 to 2) and not coreferring (0 to -2) similarly often. Particularly interesting are the pairs for which the introduction of the full context switched the polarity of coreference uncertainty (i.e., a change from -1 and -2 to 2, and from 1 and 2 to -2). As an example, consider the event mentions (*"the man"*, *"set"*, *"himself"*, *"on fire"*) and (*"the man"*, *"started"*, *"the fire"*). Intuitively, it is more likely that these mentions do not corefer. However, a wider context reveals that they in fact do corefer: *"...Police said the man doused himself in a flammable liquid and set himself on fire..."* vs. *"...Police operations leader Finn Belle told The Associated Press that the man doused himself in a flammable liquid and started the fire..."*. While such examples are individually interesting, they do not seem to have much in common.

5   Event Coreference Resolution Model

In this section we describe the event coreference model and the rich set of features used for modelling event coreference in the *full-context* setting and limited context (*mentions-only*) setting. We first briefly describe a rule-based approach for extraction of main event arguments (agent, target, time, and location) and then describe in more detail the employed features and models.

5.1   Argument Extraction

Because most of our features are based on argument comparison, for the sake of completeness we briefly discuss the robust rule-based approach for event argument extraction. The robust rule-based approach for argument extraction is based on a rich set of dependency-based patterns [9].

We use two types of syntactic patterns: *extractive patterns* and *distributive patterns*. *Extractive patterns* recognize the arguments of interest (agent, target, time, and location) and assign them to the event satisfying the pattern constraint. Some of the many extractive patterns are listed in Table 2 for illustration (the argument word is shown in bold and the event anchor is underlined). *Distributive patterns* serve to distribute arguments shared by two or more conjoined events (e.g., *"Woody Allen produced and directed the movie"*) or to assign another argument to the event, conjoined with the previously extracted argument (e.g., *"Billy Crystal and Woody Allen directed the movie"*).

Table 2. Some of the patterns for argument extraction

Name	Example	Dependency relations	Arg. type
Nominal subject	*"John saw a monkey"*	*nsubj(saw, John)*	Agent
Direct object	*"Chelsea bought Torres"*; *"John visited Paris"*	*dobj(bought, Torres)*; *dobj(visited, Paris)*	Target Location
Prepositional object	*"Torres scored on Sunday"*; *"Torres scored on Stamford Bridge"*; *"Torres scored against ManUtd"*	*prep(scored, on)* and *pobj(on, Sunday)*; *prep(scored, on)* and *pobj(on, Bridge)*; *prep(scored, against)* and *pobj(against, ManUtd)*	Time Location Target
Participial modifier	*"The boy carrying flowers"*; *"The soldier killed in the attack"*	*partmod(boy, carrying)*; *partmod(soldier, killed)*	Agent Target
Noun compound	*"Mumbai conference"*; *"Sunday protests"*; *"UN initiative"*	*nn(conference, Mumbai)*; *nn(protests, Sunday)*; *nn(initiative, UN)*	Location Time Agent

Some of the extraction patterns (e.g., *prepositional object* pattern) require further semantic processing in order to determine the argument type. We employ named entity recognition [10], temporal expression extraction [6], and WordNet-based lexico-semantic similarity [18] in order to disambiguate argument types in such cases. The performance of the rule-based argument extraction method, evaluated on a held-out set, is as follows (F-score): agent – 88.0%, target – 83.1%, time – 82.3%, location – 67.5%.

5.2   Features

As one of our goals was to model event coreference uncertainty, we focused on designing the *mentions-only* features – features computable only from event mentions, i.e., event anchors and the extracted arguments. *Mentions-only* features can be divided into bag-of-words (BoW) features, named entity (NE) features, and anchor and argument comparison features. Additionally, we compiled a set of simple context-based features to examine whether such features increase classification performance in a *full-context* setting.

Bag-of-words Features. Bag-of-words (BoW) features are computed based on the bag-of-words representations of event mentions. A BoW representation of an event mention is a set of lemmas obtained from event anchor and all event arguments. These features, inspired by work on the semantic textual similarity [1,17], measure the semantic similarity between events at the surface level, utilizing the overlap in lexical information. All lexico-semantic similarities were measured on WordNet using the algorithm of Wu and Palmer [18].

Greedy bag-of-words overlap is computed by greedily pairing lemmas from the BoW of the first event with lemmas from the BoW of the second event. In each iteration we select the pair (l_1, l_2), where l_1 is the lemma from the BoW set of the first mention and l_2 is the lemma from the BoW set of the second mention. Each lemma from each BoW set can form at most one pair (i.e., once selected, it cannot be selected again). Let P be the set of all the pairs constructed following this greedy iterative procedure. The final value for this feature is computed as the weighted normalized sum of semantic similarities (l_1, l_2) of all pairs in P:

$$grOv(e_1, e_2) = \frac{\sum_{(l_1,l_2) \in P} w(l_1, l_2) \cdot wupalmer(l_1, l_2)}{\max(|bow(e_1)|, |bow(e_2)|)}$$

where $w(l_1, l_2)$ is the weighting factor based on the *information content* of lemmas l_1 and l_2. Information content, a measure based on lemma frequencies from a corpus, allows to put more importance on lemmas bearing more information. Information content of the lemma l in the corpus C is computed as follows:

$$ic(l) = \ln \frac{\sum_{l' \in C} freq(l')}{freq(l')} .$$

Weighted lemma overlap computes the sum of information content of the lemmas found in the overlap of BoW of events and normalizes that sum with the sums of information content of lemmas of each BoW individually. *Weighted lemma coverage* of the first mention over the second is computed as:

$$wlc(e_1, e_2) = \frac{\sum_{l \in bow(e_1) \cap bow(e_2)} ic(l)}{\sum_{l' \in bow(e_2)} ic(l')} .$$

The final *weighted lemma overlap* score is computed as the harmonic mean between $wlc(e_1, e_2)$ and $wlc(e_2, e_1)$.

Content ngram overlap counts ngrams (unigrams, bigrams, and trigrams) consisting of "content lemmas" (nouns, verbs, adjectives, and adverbs) from BoWs of both events. Bigrams and trigrams in a BoW are all combinations of two and three lemmas from the set. Let $bow(e, n)$ be the set of all different ngrams of size n obtained from the BoW set of the mention e. The *content ngram coverage* for ngrams of size n, measuring the coverage of the first mention over the second is computed as:

$$cngc(e_1, e_2, n) = \frac{|bow(e_1, n) \cap bow(e_2, n)|}{|bow(e_2, n)|}$$

and the *content ngram overlap* score is computed as the harmonic mean of $cngc(e_1, e_2, n)$ and $cngc(e_2, e_1, n)$.

Named Entity Features. Named entity features measure the overlap in named entity occurrences between the two events, looking for named entity matches regardless of the argument roles those named entities occupy.

Capital letter overlap feature measures the overlap of capitalized words (as one form of rudimentary NER) between the two events. This measure counts the number of capitalized words occurring within arguments of both mentions and normalizes it with the total number of capitalized words in the first and second event mentions, respectively. The harmonic mean of these two normalized scores is the final feature value.

Person, location, organization, and temporal expression overlap features measure the overlap of NEs of type person, location, and organization and temporal expressions, respectively. The number of overlapping NEs of the corresponding type (e.g., person) is normalized by the number of NEs of the same type within the first and second event mention, respectively. Final feature score for each NE type is the harmonic mean of the normalized overlap scores. For each NE type overlap score is accompanied with a feature indicating whether comparison is in place (i.e., at least one NE of the type in question is present within at least one of the events).

Numbers overlap feature measures the overlap in numbers between the two events. Numbers often occur as part of dates, temporal expressions, money, or percentages. If differing between the two events, numbers often strongly indicate that events do not corefer (e.g., *"Australians <u>won</u> three gold medals"* vs. *"Australians <u>won</u> seven gold medals"*). An additional feature counting all the numbers within both mentions (i.e., $|num(e_1)| + |num(e_2)|$) indicates whether measuring the overlap is in place.

Anchor and Argument Comparison Features. An event anchor bears the core meaning of the event. Hence, the intuition is that coreferent mentions have semantically closely related anchors. The same observation holds for arguments of matching types. Bag-of-words and NE overlap imply close semantic relatedness, but not necessarily coreference. Consider, for example, events *"The dog <u>bit</u> a boy"* and *"The boy <u>bit</u> a dog"*. These events are identical if observed only through lexical overlap features, but obviously do not corefer.

We use one feature to measure semantic similarity of anchors of the two mentions on WordNet. For each of the four main argument types (*agent, target, location, time*) we use three different features: *argument head word similarity*, *argument chunk similarity*, and a binary feature indicating if the comparison is in place (i.e., at least one of the events has at least one argument of the type in question). *Argument head word similarity* is also computed using the algorithm of Wu and Palmer on WordNet. *Argument chunk similarity* feature measures the similarity between the argument chunks instead of only argument words. We use four other features reflecting the matches between event arguments: the difference in argument number between the mentions, the total number of matching arguments, and the number of unmatched arguments for each mention.

Information content value of the mention anchor may be indicative of the referential ambiguity of the mention. Anchors that generally occur more frequently introduce more referential ambiguity (e.g., *"said"* vs. *"invaded"* in news domain). To account for this, we use a feature indicating *average information content* of mentions' anchors. The *difference in information content* between mentions' anchors is also used as a feature as it may indicate semantic relatedness for pairs for which one or both of the mentions cannot be found in WordNet.

Context Features. We employ several simple features that utilize the context of the event mentions in order to examine whether such features increase coreference classification performance in the *full-context* setting.

Context similarity computes the cosine similarity between the bag-of-word vectors of the contexts of event mentions. We define the context of the mention to be a three-sentence window including the sentence of the mention and one sentence before and after the sentence of the mention. *Context overlap* computes the overlap in information content of the two contexts. It is computed the same way as *weighted lemma overlap* feature, only on context BoW vectors instead of mention BoW vectors. Some additional context information for within-document mention pairs are also added: the distance between the mentions expressed in number of tokens, a binary feature indicating if the mentions are in the same sentence, and a binary feature indicating if the mentions are from adjacent sentences. A feature indicating if the mentions originate from the same document (i.e., within- or cross-document) is also included.

5.3   Models

We trained two classification models and one regression model. The first classification model attempts to recognize coreferent mention pairs using *mentions-only* features (BoW, NE, and anchor and arguments comparison features). The second classification model uses context features in addition to features used by the first model, allowing us to inspect how the use of simple context features affects the classification accuracy. The regression model attempts to parallel human uncertainty when only event mentions are available. We used support vector machines (SVM) with radial basis kernel for classification and support vector regression (SVR) for regression. We used the LIBSVM implementation of the SVM algorithm [7].

6   Experiments

In this section we describe both the experiments on binary event coreference resolution and event coreference uncertainty modelling. We used the *full-context* human annotations for the binary coreference classification task and the *mentions-only* coreference confidence scores for modelling coreference uncertainty.

6.1 Event Coreference

The first model (*mentions-only* model) was trained using only mention-based features in order to assess how well we can predict event coreference based only on event mentions without additional context. In the second model (*mention+context* model) we additionally use the context features in order to determine whether such features improve the classification performance.

The performance of the two models is given in Table 3. Besides analysing the overall performance, we separately analyse the performance on the within-document and cross-document pairs. All scores are obtained by averaging over 10 cross-validation folds. As the baseline, we use the same heuristic that was used in dataset preparation to increase the odds of selecting positive examples. The heuristic considers a pair of event mentions as coreferent if (1) the semantic similarity of their anchors (measured on WordNet) is above a given threshold (set to 0.5 in our experiments) and (2) event mentions have at least one matching argument.

Table 3. Event coreference pairwise classification performance

Model	All			Within-document			Cross-document		
	P	R	F	P	R	F	P	R	F
Baseline	35.1	88.8	50.3	14.6	78.9	24.7	50.4	91.2	64.9
Mention-only	85.2	73.4	78.9	53.4	56.4	54.9	95.1	77.5	85.4
Mention+context	88.5	76.2	**81.9**	68.3	50.9	**58.3**	92.6	82.4	**87.2**

Both models (with and without context features) significantly outperform the baseline. The model with simple context features consistently outperforms the model without context features, although not by a very wide margin. This implies that context features have a positive impact on event coreference resolution.

The gap in performance on within-document and cross-document examples is striking. Such a significant difference suggests that there exists a systematic difference in feature values between coreferring within-document pairs and coreferring cross-document pairs. To investigate this, we assumed that each numeric feature (of both within-document and cross-document positive examples) is normally distributed. For each feature we calculate the maximum-likelihood estimate of the mean, separately for within-document and cross-document examples. We then compared the estimated means and identified the features for which the within-document and cross-document mean values differ the most; the top three features with the largest differences are listed in Table 4.

Features that differ the most all relate to the number of arguments exhibited by the mentions. In within-document coreferent pairs it is more often the case

Table 4. Differences in feature value means between positive within- and cross-document examples

	Mean value		
Feature	Within-document	Cross-document	Difference
No. matched arguments	1.18	1.67	0.49
Diff. in no. of arguments	0.54	0.89	0.35
No. unmatched arg. first ev.	1.00	0.72	0.28

that one of the mentions has fewer arguments realized (or even none), as they are being implied by the discourse structure of the document (e.g., introduced by some coreferent mention earlier in the text). This explains why our models, primarily based on features comparing main event arguments, are less successful in resolving within-document event coreference.

6.2   Coreference Uncertainty

Our regression model for predicting coreference confidence (or, equivalently, coreference uncertainty) was trained on human coreference annotations in the *mentions-only* setting. The results are shown in the first row of the Table 5. We use the same performance measures as for the inter-annotator agreement and average the scores over 10 cross-validation folds. The performance of the regression model suggests that modelling human understanding of event coreference uncertainty is feasible.

One interesting question is whether there is a correlation between the coreference confidence scores predicted by the *mentions-only* classification model and human confidence scores. To investigate this, we mapped the classification confidence scores of the SVM classifier into a $[-2, +2]$ interval (e.g., classification confidence of 60% maps to confidence score of 0.4). The correlation between classifier and human coreference confidence scores is given in the second row of Table 5. The result suggests that the binary model and humans are similarly uncertain when there is no context available other than the event mentions themselves.

Table 5. The performance of the regression model and the correlation of classifier's confidence

	MAE	Pearson	R^2
Regression model predictions	0.62	0.71	0.49
Classification model confidence	0.72	0.71	0.47

7   Conclusion and Future Work

We explored the concept of event coreference uncertainty and how it manifests depending on the provided context of event mentions. Human annotations confirmed the intuition that there is no uncertainty when full context of both event mentions is available. In some circumstances, however, a full context of an event mention might not exist (e.g., queries in event-based IR). We addressed the uncertainty of event corefentiality in such a setting and described a regression-based model for estimating event coreference uncertainty. We demonstrated that we can parallel human level of confidence by comparing event mentions comprised of event anchors and four main event argument types (agent, target, time, and location), which we can robustly extract using a rule-based extraction method. Experimental results support our hypothesis that modelling human understanding of event coreference uncertainty is feasible.

We showed that the same set of mention comparing features are suitable for resolving cross-document event coreference in the full-context setting. However, a set of more sophisticated discourse-level features may be required for resolving within-document coreference. We plan to evaluate the proposed approach on other event coreference datasets such as OntoNotes [12] or EventCorefBank [4].

Further improvements and extensions can be achieved both conceptually (e.g., considering pronominal event anaphora as well) and technically (e.g., separate models for within-document and cross-document coreference resolution). There are notions of similarity and relatedness between event mentions other than coreference as well as other interesting event relations such as causality or subordination. We intend to investigate whether and how comparison of event arguments can be used to measure such similarities or detect such relations.

Acknowledgments. We thank the anonymous reviewers for their useful comments and suggestions. This work has been supported by the Ministry of Science, Education and Sports, Republic of Croatia under Grant 036-1300646-1986.

References

1. Agirre, E., Cer, D., Diab, M., Gonzalez-Agirre, A.: SemEval-2012 task 6: A pilot on semantic textual similarity. In: Proceedings of the 6th International Workshop on Semantic Evaluation (SemEval 2012), in Conjunction with the First Joint Conference on Lexical and Computational Semantics (*SEM 2012) (2012)
2. Ahn, D.: The stages of event extraction. In: Proceedings of the Workshop on Annotating and Reasoning about Time and Events, pp. 1–8. Association for Computational Linguistics (2006)
3. Bagga, A., Baldwin, B.: Cross-document event coreference: Annotations, experiments, and observations. In: Proceedings of the Workshop on Coreference and its Applications, pp. 1–8. Association for Computational Linguistics (1999)
4. Bejan, C., Harabagiu, S.: A linguistic resource for discovering event structures and resolving event coreference. In: Proceedings of the 6th International Conference on Language Resources and Evaluation, Marrakech, Morocco (2008)

5. Bejan, C., Harabagiu, S.: Unsupervised event coreference resolution with rich linguistic features. In: Proceedings of the 48th Annual Meeting of the Association for Computational Linguistics, pp. 1412–1422. Association for Computational Linguistics (2010)

6. Chang, A., Manning, C.: SUTime: a library for recognizing and normalizing time expressions. Language Resources and Evaluation (2012)

7. Chang, C.C., Lin, C.J.: LIBSVM: a library for support vector machines. ACM Transactions on Intelligent Systems and Technology 2, 27:1–27:27 (2011)

8. Chen, B., Su, J., Pan, S., Tan, C.: A unified event coreference resolution by integrating multiple resolvers. In: Proceedings of 5th International Joint Conference on Natural Language Processing (2011)

9. De Marneffe, M., Manning, C.: Stanford typed dependencies manual (2008), http://nlp.stanford.edu/software/dependencies_manual.pdf

10. Finkel, J., Grenager, T., Manning, C.: Incorporating non-local information into information extraction systems by gibbs sampling. In: Proceedings of the 43rd Annual Meeting on Association for Computational Linguistics, pp. 363–370. Association for Computational Linguistics (2005)

11. Humphreys, K., Gaizauskas, R., Azzam, S.: Event coreference for information extraction. In: Proceedings of a Workshop on Operational Factors in Practical, Robust Anaphora Resolution for Unrestricted Texts, pp. 75–81. Association for Computational Linguistics (1997)

12. Pradhan, S., Ramshaw, L., Weischedel, R., MacBride, J., Micciulla, L.: Unrestricted coreference: Identifying entities and events in OntoNotes. In: International Conference on Semantic Computing, ICSC 2007, pp. 446–453. IEEE (2007)

13. Pustejovsky, J., Castano, J., Ingria, R., Sauri, R., Gaizauskas, R., Setzer, A., Katz, G., Radev, D.: TimeML: Robust specification of event and temporal expressions in text. In: New Directions in Question Answering 2003, pp. 28–34 (2003)

14. Quine, W.: Events and reification. In: Actions and Events, pp. 162–171 (1985)

15. Roth, M., Frank, A.: Aligning predicate argument structures in monolingual comparable texts: A new corpus for a new task. In: Proceedings of the First Joint Conference on Lexical and Computational Semantics, Montreal, Canada (June 2012)

16. Ruppenhofer, J., Ellsworth, M., Petruck, M., Johnson, C., Scheffczyk, J.: FrameNet II: Extended theory and practice. International Computer Science Institute (2006)

17. Šarić, F., Glavaš, G., Karan, M., Šnajder, J., Bašić, B.: TakeLab: Systems for measuring semantic text similarity. In: Proceedings of the 6th International Workshop on Semantic Evaluation (SemEval 2012), in Conjunction with the First Joint Conference on Lexical and Computational Semantics (*SEM 2012) (2012)

18. Wu, Z., Palmer, M.: Verbs semantics and lexical selection. In: Proceedings of the 32nd Annual Meeting of the Association for Computational Linguistics, pp. 133–138. Association for Computational Linguistics (1994)

LIAR$_C$: Labeling Implicit ARguments in Spanish Deverbal Nominalizations

Aina Peris[1], Mariona Taulé[1], Horacio Rodríguez[2], and Manuel Bertran Ibarz[1]

[1] University of Barcelona
CLiC, Centre de Llenguatge i Computació
Gran Via de les Corts Catalanes 585, 08007 Barcelona
[2] Technical University of Catalonia TALP Research Center
Jordi Girona Salgado 1-3, 08034 Barcelona
{aina.peris,mtaule}@ub.edu, {horacio,mbertran}@lsi.upc.edu

Abstract. This paper deals with the automatic identification and annotation of the implicit arguments of deverbal nominalizations in Spanish. We present the first version of the LIAR system focusing on its classifier component. We have built a supervised Machine Learning feature based model that uses a subset of AnCora-Es as a training corpus. We have built four different models and the overall F-Measure is 89.9%, which means an increase F-Measure performance approximately 35 points over the baseline (55%). However, a detailed analysis of the feature performance is still needed. Future work will focus on using LIAR to automatically annotate the implicit arguments in the whole AnCora-Es.

1 Introduction

In the absence of wide-coverage semantic interpreters able to process unrestricted domain texts in depth, the growing need for semantic knowledge for many Natural Language Processing (NLP) tasks is covered by partial semantic taggers, especially by Semantic Role Labelers (SRL). A SRL tries to assign to each predicate its corresponding semantic roles (both argument positions and thematic roles) that appear within the sentence in the case of verbs, or within the Noun Phrase (NP) in the case of nouns. In a nutshell, they focus on the identification of explicit arguments. The success of applying a SRL component to NLP tasks has resulted in a growing interest in extending its coverage, going beyond explicitly occurring role fillers. The work presented here is placed in this scenario. We deal with predicates expressed by deverbal nominalizations (nouns morphologically derived from verbs) and with arguments not explicitly assigned to these predicates, i.e. with implicit arguments. This is especially important in the case of deverbal nominalizations since the degree of optionality of their explicit arguments is higher than for verbs.

This work presents the first version of LIAR (Labeling Implicit ARguments in Spanish deverbal nominalizations), a system that aims to automatically identify

A. Gelbukh (Ed.): CICLing 2013, Part I, LNCS 7816, pp. 423–434, 2013.

implicit arguments and assign an argument position –iarg0[1], iarg1, etc.– and a thematic role (agent, patient, cause, etc.) to them, using Machine Learning (ML) techniques based on linguistically informed features. These arguments can be recovered if a wider discursive context is taken into account and their identification is important for achieving a deep semantic representation of sentences and, therefore, a deep understanding of texts. Our system allows for the annotation of the implicit arguments of Spanish deverbal nominalizations in the AnCora-Es corpus[2]. This work is part of the *IARG-AnCora* project, the aim of which is to enrich the Spanish AnCora corpus by annotating the implicit arguments of deverbal nominalizations. The goal of our research is four-fold: 1) To perform an empirical study of the knowledge involved in the relationship between a predicate and its implicit arguments; 2) To explore to what extent the features valid for English are also applicable to Spanish; 3) To build a classifier for allowing the automatic detection and tagging of the implicit arguments of the nominalizations in Spanish texts; 4) To apply the classifier to the AnCora-Es corpus in order to semiautomatize the implicit argument annotation process. Only the three first goals are presented in this paper.

The paper is organized as follows. First, in section 2, we define what we understand by implicit argument and set out the annotation scheme used. Then, in section 3, we briefly discuss the related work. In section 4, we present the $LIAR_C$, its architecture, its components and the resources involved in its building. Section 5 describes the results obtained. Finally, the main conclusions and final remarks are given in section 6.

2   Defining an Implicit Argument

An implicit argument, or a null instantiation (NI) in terms of the FrameNet framework [1], [2], is an argument syntactically not realized in the local context of the predicates (verbs, adjectives or nouns) whose semantic interpretation depends on the linguistic or extralinguistic context. The implicit argument can be either a core argument or an adjunct argument. In FrameNet, the NIs are further classified as definite null instantiation (DNI) -anaphorically bound within the discourse- or indefinite null instantiation (INI) -existentially bound within the discourse. In this work, as in [3,4], we do not take into account the distinction between INI and DNI. We are only able to detect and classify the implicit core arguments of the deverbal nominalizations whose semantic interpretation depends on the linguistic context. If the semantic interpretation depends on the extralinguistic context it cannot be recovered from the surrounding discourse. In contrast, if it depends on the linguistic context it can be recovered and linked to an entity. Concretely, they can be recovered from the sentence containing the nominalization (1) or from the previous or following sentences (2).

---

[1] The letter 'i' at the beginning of the argument position label stands for implicit argument. We annotate the implicit arguments as iarg$n$-r, where n stands for an argument position and r for a thematic role. See more in Section 2.1.
[2] AnCora corpora are freely available at: `http://clic.ub.edu/corpus/ancora`

(1)   [Las escuelas de samba de Sao Paulo]$_{iarg1\text{-}pat}$ han conseguido [el **apoyo**[de la empresa privada]$_{arg0\text{-}agt}$ para mejorar las fiestas de carnaval]$_{NP}$.
*[Samba schools in Sao Paulo]$_{iarg1\text{-}pat}$ have obtained [the **support** [of private industry]$_{arg0\text{-}agt}$ to improve Carnival celebrations]$_{NP}$.*

(2)   [El carnaval de Sao Paulo es feo]$_{iarg1\text{-}pat}$, dijo hoy [el alcalde de Río de Janeiro]$_{iarg0\text{-}agt}$ en una conversación informal con periodistas cariocas, y encendió la polémica. [. . .] [Esa **opinión**[3]]$_{NP}$ fue respaldada por el gobernador de Río de Janeiro [...]".
*[The Carnival of Sao Paulo is ugly]$_{iarg1\text{-}pat}$, said [the mayor of Rio de Janeiro]$_{iarg0\text{-}agt}$ in an informal conversation with Carioca journalists, and ignited the controversy. [. . .] [This **opinion**]$_{NP}$ was supported by the governor of Rio de Janeiro [...]".*

Example (1) shows the deverbal nominalization 'apoyo', *support*, with the agent argument ('de la empresa privada', *of private industry*) realized inside the NP, whereas the patient argument ('las escuelas de samba de Sao Paulo', *Samba schools in Sao Paulo*) is realized in the same sentence but outside the NP. In (2), the nominalization 'opinión', *opinion*, appears without any explicit argument in the NP. However, the agent argument ('el alcalde de Río de Janeiro', *the mayor of Rio de Janeiro*) as well as the patient argument ('el carnaval de Sao Paulo es feo', *the carnival of Sao Paulo is ugly*) are realized implicitly (iarg0-agt and iarg1-pat, respectively) in the previous sentence. Currently, the AnCora corpus is only annotated with explicit arguments, therefore 'opinión' *opinion* has no associated argument and 'apoyo' *support* only has the *agent* argument annotated.

2.1   Annotation Scheme

We use the same annotation scheme as the one followed to annotate the explicit arguments of deverbal nouns [5], and the argument structure of verbs in AnCora [6], which was in turn based on PropBank [7] and NomBank [8]. In this way, it is ensured the consistency of the annotation of arguments of different predicates -nouns and verbs-, as well as the compatibility of Spanish and English resources.

We use the iarg$n$ tag to identify implicit arguments and to differentiate them from explicit arguments (arg$n$ tag) [3,4]. The list of thematic roles includes 20 different labels based on VerbNet [9] proposals[4]. The combination of the six argument position labels (iarg0, iarg1, iarg2, iarg3, iarg4, iargM) with the different thematic roles results in a total of 36[5] possible semantic tags (iarg0-cau, iarg1-agt, iarg0-agt, iarg2-loc, etc.).

---

[3] In Spanish, the noun 'opinión', *opinion*, is derived from the verb 'opinar', *to express an opinion*.

[4] agt (agent), cau (cause), exp (experiencer), scr (source), pat (patient), tem (theme), cot (cotheme), atr (attribute), ben (beneficiary), ext (extension), ins (instrument), loc (locative), tmp (time), mnr (manner), ori (origin), des (goal), fin (purpose), ein (initial state), efi (final state), and adv (adverbial).

[5] Not all the combinations of argument positions and thematic roles are valid semantic tags.

3   Related Work

In this section we present related work on implicit arguments focusing on the systems developed to automatically detect and classify them. One of the first automatic methods to retrieve extrasentential arguments was proposed by [10]. They approach the detection of implicit arguments using knowledge about predicates and coreference chains in sentences belonging to the same topic domain. However, since this method is applied to a specific domain (computer equipment maintenance reports), it is not clear how to extend this method to unrestricted domains. [11] also describes another early attempt at solving the implicit arguments of verbs by means of an algorithm that uses metrics information about thematic roles from a previous corpus based study. In this case, the domain is also very specific (a subset of the TRAINS corpus [12]) and the approach, based on specific algorithms manually written for each relevant thematic role, seems difficult to scale up.

Recently developed systems can be divided in two groups. On the one hand, those systems related to the SemEval-2010 Task 10 [13], and concretely, those that tackled the NI resolution subtask. We include in this group the two participating systems, Semafor [14] and Venses++ [15], and those systems that use the same data set and evaluation measures used in this subtask [16,17,18,19]. All these systems identify implicit arguments for different English predicates (verbs and nouns), following the typology of implicit arguments proposed in [1] and [2] and the FrameNet annotation scheme [20]. The systems use different approaches to solve the binding of NIs. [16] approach the problem as a coreference resolution task; [19], Semafor, Venses++ as an extension of semantic role labeling systems, while [17] and [18] adopt a mixed approach to carry out the task combining both strategies. On the other hand, these systems can use supervised ML techniques such as [16] and Semafor, or they can be based in hand written algorithms that use different type of information ([17], [18], [19], [15]). It is also worth noting that all systems except [14] (which works in parallel) deal with the problem sequentially, that is, breaking down the task into different subtasks. On the other hand, [4] developed a parallel supervised ML feature-based model for detecting the core implicit arguments of the ten most frequent unambiguous nouns in the standard training, development and testing sections of the Penn Treebank [21]. This was done in order to avoid the problem of sparseness present in the SemEval-2010 task 10 data set. The annotation scheme follows PropBank [7] and NomBank [8] proposals.

LIAR is the first system to deal with the implicit arguments of deverbal nominalizations in Spanish. Like [3] and [4], we have built a supervised ML feature-based model for detecting core implicit arguments that uses a subset of the AnCora-Es as a training corpus. In fact, in this first version of LIAR we replicate the experiments carried out by these authors and propose a number of variations and improvements.

4   LIAR

Taking into account our goals, we selected the approach used by G&C [4] as the initial model for our system instead of SemEval based systems since is more likely to scale up and does not suffer from the data-sparseness problem found in the SemEval training corpus. As in G&C, we selected the eight[6] most frequent and unambiguous (with only one role set associated) deverbal nominalization lemmas in AnCora-Es for the building of LIAR. This set of eight nominalizations corresponds to a total of 469 instances shown in the first two columns of Table 1. The implicit arguments of these 469 instances were manually annotated for building the training corpus, AnCora-Es-IARG. The remaining columns in Table 1 present statistics on the annotations. The number of explicit and implicit core arguments are shown for each predicate. It is worth noting that the gain in role coverage is, on average, superior when taking into account adjunct arguments (317% on average) rather than core arguments alone (277% on average), possibly due to the fact that core arguments are those selected when the noun expresses the arguments explicitly. These figures are much higher than those reported by G&C for English. Although these figures are not directly comparable, the difference is due to both the lower number of explicit arguments for Spanish nominalizations (0.6 on average) compared to English (1.1 in G&C) and to the higher number of implicit arguments (1.3 *vs.* 0.8). The average number of implicit arguments per predicate instance is 1.3 with a standard deviation of 0.41 (0.8 in G&C)[7].

Table 1. Statistics for implicit arguments

Nominalization	Instances	Nr. iargs	Avg. iargs	Nr. args	Avg. args
Actuación, *actuation*	67	35	0.5	29	0.4
Comunicado, *communication*	63	116	1.8	16	0.2
Daño, *harm*	42	67	1.6	12	0.3
Empate, *draw*	41	54	1.3	12	0.2
Negocio, *business*	50	59	1.2	16	0.3
Opción, *option*	48	56	1.2	34	0.7
Propuesta, *proposal*	104	177	1.7	78	0.8
Viaje, *travel*	54	66	1.2	30	0.5
TOTAL	**469**	**630**	**1.3**	**227**	**0.5**

4.1   LIAR Architecture

The LIAR$_C$ classifier is a component of a larger system able to detect all the valid fillers for all the deverbal nominalizations occurring in AnCora-Es. The basic

---

[6] We selected eight predicates and not ten like in G&C because there is a severe drop in frequency from the ninth predicate in AnCora-Es.

[7] A third explanation could be the use of different criteria in the annotation of both explicit and implicit arguments.

scenario of this system is the following: For each AnCora-Es file d the system looks for all the occurrences of all the nominalizations p whose lemma occurs in the AnCora-Nom lexicon [22]. From this lexicon, it obtains all the valid frames of p, and from each frame the set of possible arguments (argument positions and thematic roles), ARG. From ARG, the arguments annotated as explicit arguments of p, i.e. already realized, are removed. From d, we also obtain the set of mentions of entities (regular entities and singletons) annotated as explicit arguments of a predicate. This set, M, is the search space in which the fillers of the implicit arguments of p will be searched. The process of classification is now undertaken. $LIAR_C$ is applied to all the triples $<p, iarg, e_i>$, for each nominalization p, each $iarg$ in ARG and each mention e_i in M. Each triple is classified as supporting or not the assignment of e_i as the implicit argument $iarg$ of p. Four possibilities arise:

1. No tuple has been classified positively, so there is no filler for the $iarg$ of p.
2. Just one mention e_i has been classified positively, so the filler for the $iarg$ of p is the entity corresponding to this mention.
3. More than one mention e_i has been classified positively, and all of them belong to the same regular entity. This entity is chosen as the result.
4. More than one mention e_i has been classified positively, but they belong to several entities. The entity corresponding to the mention e_i closest to p is chosen.

The key point of the algorithm is the performance of $LIAR_C$ classifier, all the other components are very simple. This paper only describes the classifier component. Two weaknesses can be detected:

(1) The rather straightforward way of assigning the filler in the last case;
(2) $LIAR_C$ has been learned from a corpus of monosemous nominalizations and is also applied to polysemous ones.

These issues require further study.

4.2   Learning $LIAR_C$

In Ancora-Es, for each document d, the implicit arguments of a nominalization instance are annotated at entity level, i.e. given a predicate instance p, and an implicit argument tag iarg$n$-r, the iarg$n$-r filler, if existing, is annotated with the identifier of an entity $e \in E$, being E the set of entities occurring in d. Entities in E can be regular ones, occurring in coreference chains, or those appearing just once (singletons). We have observed that 75% of core implicit arguments are retrieved from regular entities while the remaining 25% correspond to singletons. For a regular entity e, the set of mentions (the coreference chain) is noted as $e' = \{e_1, \ldots, e_i, \ldots, e_n\}$. At least one mention e_i, in the coreference chain e', is often an explicit argument of a predicate. When e is a regular entity, mentions in e' tend to be NPs, while in the case of singletons other possibilities exist (prepositional, adjectival, adverbial phrases, possessives or subordinated clauses). Several possible scenarios can be considered for defining the instances that can be used in the learning process of the classifier. Basically, an instance is defined as a triple $<p, iarg, filler>$, when the p is fixed and the $iarg$ can be expressed in terms of argument positions (iarg$n$) or argument positions plus thematic roles (iarg$n$-r). In the case of the *filler*, the following possibilities are considered:

1. $<p,iarg,e>$, $e \in$ E, i.e. the filler is an entity.
2. $<p,iarg,e_i>$, $e_i \in e'$, i.e. the filler is not an entity but one of its mentions.
3. $<p,iarg,e_i>$, $e_i \in e^F$, when e^F is the result of applying a filtering process F for obtaining a subset e^F of e', i.e. the filler is a mention of the subset e^F. Many options for implementing this filtering, F, can be considered.

For a closer comparison with G&C, we have choose the third possibility, using as a filtering mechanism one that discards mentions that are not explicit arguments of a predicate. In this way, a few positive examples are lost but the search space is heavily pruned. We formulate our classification problem as a binary classification, i.e. given a predicate occurrence p, for each *iarg* tag and each mention $e_i \in e^F$, the classifier has to decide whether or not the entity corresponding to the mention is a valid *filler* for the *iarg*.

4.3   Training Corpus

Currently, the AnCora corpus is only annotated with the explicit arguments of deverbal nouns, that is, with those overtly realized. AnCora-Es is a Spanish corpus of 500,000 words, annotated at different linguistic levels. We built a manually annotated training corpus, AnCora-Es-IARG, which consists of the sentences containing the 469 deverbal noun occurrences in AnCora-Es. In order to ensure the quality and the consistency of the annotated data an inter-annotator agreement test was conducted on a subsample of 200 occurrences. The average pairwise result obtained between the three pairs of annotators was 81% of observed agreement (58.3% kappa *vs.* 64.3% in G&C). The features for the classification model are inferred from this training corpus.

4.4   Learning Setting

Our learning setting tries to tackle two incompatible constraints: 1) to follow G&C's proposal as closely as possible due to the highly accurate performance of its features and the possibility of comparing their results with ours; and 2) to learn LIAR$_C$ for the semi-automatic annotation of the whole AnCora-Es, the fourth of our goals, as described in section 1. These constraints are incompatible because most of the best features used by G&C are highly lexicalized. They contain, for instance, the specific predicate involved and, frequently, the specific words, lemmas, synsets and predicates surrounding the candidate filler. In order to exploit the LIAR$_C$ in the whole AnCora-Es scenario, these lexicalized features have to be avoided. Therefore, the classifier learned four different models: 1)lexicalized-specific; 2)lexicalized-generalized; 3) non-lexicalized-specific; 4) non-lexicalized-generalized.

The first model is reasonably comparable with G&C, while the latter can be used for automatically annotating AnCora-Es. Non-lexicalized models are built by simply replacing the name of the predicate p with a null string. Generalization is performed by synsets occurring in the features by the Top Concept Ontology (TCO) labels[8] [23] attached to them[9], in the case of nouns, and, in the case of verbs, their lemmas are replaced by the semantic classes of AnCora-Verb [24].

For learning, we used the Weka toolbox [25]. Due to the highly lexicalized nature of most of the features, and their precision oriented type, we choose Adaboost as classifier model. Due to the highly unbalanced distribution of positive and negative examples, we weight the positive ones by a factor of 69, leading to a more balanced distribution. As in G&C, we use feature templates. Each template is expanded into a set of actual features. For instance, the feature template in row 1 of Table 2, $<pe_i, arge_i, p, iargn >$has to be instantiated for all the possible values of p (8 values), for all the possible values of $iargn$ (up to 36 argument positions + thematic roles), for all the possible values of $arge_i$ (up to 6 argument positions) and for all the possible values of pe_i. Depending on our settings, these values can change. For instance, when reducing our search to core arguments argM and iargM are removed, when not using thematic roles the number of $iargn$ is reduced to 6. The actual number of instances of this template is the reduced to the combination occuring in the training corpus (500 in our setting). An instantiation of this template is 'afirmar-arg0-actuación-iarg0:agt'. The following sections present some details of the features used and the results of the learning process.

4.5   Features for Learning

Many of the features and feature templates we use for learning the $LIAR_C$ were replicated from G&C. These authors rank their features by accuracy, we took the most accurate features of them. In some cases, our features basically reproduce theirs, while, in others, our features are often simply inspired by theirs, sometimes because we use different knowledge sources such as AnCora-Es, AnCora-Nom, AnCora-Verb or the TCO, and sometimes because the differences in language suggest using different schemas.

In Table 2, we list the most important features used for training our four models. In the third column, the external resources used in each feature are shown: the concepts of WordNet TCO; AnCora-Verb lexicon and AnCora-Es

---

[8] TCO aims to provide WordNet synsets with a neutral ontological assignment. The ontology contains 63 features organized as 1st order entities (physical things), 2nd order entities (situations) and 3rd order entities (unobservable things). 1st order features are further classified as form, composition, origin and function.

[9] As AnCora-Es mentions are annotated with correct synsets, no Word Sense Disambiguation is needed.

corpus. In the fourth column, '+' and '-' mark those features that are used or not by G&C respectively, and '≈' marks those that are similar to theirs. Features marked with an asterisk (*) are template features discussed in section 4.4.

Table 2. Features used for training and classifying

Nr.	Feature Description	Resources	G&C
*1	For every e_i, $<p\ e_i,\ arg\ e_i,\ p,\ iargn >$		+
2	Sentence distance from e_i to p		+
*3	For every e_i, $<$head word of e_i, the verbal form of $p, iargn>$		+
*4	Same as 1 except generalizing pe_i and p to their TCO	TCO	≈
*5	Same as 3 except generalizing e_i to its TCO	TCO	≈
6	Whether or not any $arge_i$ and iarg$n$ have the same integer argument position		+
7	Whether or not the verb forms of pe_i and p are in the same AnCora-Verb class and $arge_i$ and $iargn$ have the same thematic role	AnCora-Verb	≈
8	PMI between $arge_i$ of pe_i and iarg$n$ of p.	AnCora-Es	≈
9	$<p$, the number of p's complements $>$	AnCora-Es	+
*10	Same as 5 except generalizing the verbal form of p to the semantic verbal class of AnCora-Verb	AnCora-Verb	-

5   Results

The classifier performance of the four models was evaluated by a five fold cross-validation method. We performed the learning process for the four models described above. We compared the results obtained with a baseline that consists of using just the 500 features corresponding to the most accurate template in G&C (first row in Table 2). The baseline obtains an F-Measure of 55%. The results are not significantly different for the four models and the overall F-Measure is 89.9%. Therefore, our models increase F-Measure performance approximately 35 points over the baseline. The lack of difference between lexicalized and non-lexicalized models is rather surprising. In fact, we expected a serious decrease on F-Measure in the non-lexicalized models. In order to analyze this unexpected outcome, we performed a feature analysis and we observed that none of the lexicalized features occurred among the top ranked ones (the first lexicalized feature occurs in the nineteenth position). Therefore, the process of delexicalizing shows a very limited effect on the figures obtained. These results strongly contrast with the results reported by G&C for English. The higher F-Measure in Spanish can be explained by the fact that the explicit arguments in the training corpus are manually annotated, while G&C used an automatic SRL system to obtain this information. The best performing features of our models use this semantic information.

6   Conclusions

This paper presents the first version of LIAR, a system that aims to automatically identify the implicit arguments of Spanish deverbal nominalizations. In a previous empirical linguistic study, we observed that the annotation of implicit arguments involves an important gain in role coverage (277% on average, taking into account core arguments alone). This figure is due to the high degree of optionality of explicit arguments in Spanish deverbal nominalizations and, therefore, it is important to detect implicit arguments to achieve a deep semantic representation of these predicates.

In this paper, we focus on the $LIAR_C$ classifier component, which was built using ML techniques based on linguistically informed features. We experimented four models depending on whether the features were lexicalized or not, and whether the features were specific or generalized. The overall F-Measure for all the models is, on average, 89.9%, showing that there are no significant differences between lexicalized and non-lexicalized models. We note that our models do not use lexicalized features, when they are available, among the first ones applied. A further study on this issue is needed. The results obtained are better than those reported by G&C for English. This could be explained by the fact that the explicit arguments in G&C are automatically obtained while they are manually annotated in AnCora-Es-IARG. However, a deeper analysis of the differences in performance of the similar features for English and Spanish is required. An application of the current learned model of LIAR presented in this paper is being carried out to automatically annotate the implicit arguments in the whole AnCora-Es (including both monosemic and polisemic nominalizations).

Acknowledgments. We are grateful to David Bridgewater for the proof-reading of English. We would also like to express our gratitude to the two anonymous reviewers for their comments and suggestions to improve this article. This work was partly supported by the projects IARG-ANCORA (FFI2011-13737-E), Know2 (TIN2009-14715-C04-04) and TEXT-MESS 2.0 (TIN2009-13391-C04-04) from the Spanish Ministry of Economy and Competitiveness.

References

1. Fillmore, C.J.: Pragmatically Controlled Zero Anaphora. Technical report, Department of Linguistics. University of California (1986)
2. Fillmore, C.J., Baker, C.F.: Frame semantics for text understanding. In: Proceedings of the Workshop on WordNet and Other Lexical Resources, NAACL. Association for Computational Linguistics, Pittsburgh (2001)
3. Gerber, M., Chai, J.Y.: Beyond NomBank: a study of implicit arguments for nominal predicates. In: Proceedings of the 48th Annual Meeting of the Association for Computational Linguistics, ACL 2010, pp. 1583–1592. Association for Computational Linguistics, Stroudsburg (2010)
4. Gerber, M., Chai, J.Y.: Semantic Role Labeling of Implicit Arguments for Nominal Predicates. Computational Linguistics 38, 755–798 (2012)

5. Peris, A., Taulé, M.: Annotating the argument structure of deverbal nominalizations in Spanish. In: Language Resources and Evaluation (2011), doi:10.1007/s10579-011-9172-x
6. Taulé, M., Martí, M., Recasens, M.: AnCora: Multilevel Annotated Corpora for Catalan and Spanish. In: Proceedings of the Sixth International Language Resources and Evaluation (LREC 2008), pp. 96–101. European Language Resources Association (ELRA), Marrakech (2008)
7. Palmer, M., Kingsbury, P., Gildea, D.: The Proposition Bank: An Annotated Corpus of Semantic Roles. Computational Linguistics 31, 76–105 (2005)
8. Meyers, A.: Annotation Guidelines for NomBank Noun Argument Structure for PropBank. Technical report, University of New York (2007)
9. Kipper, K.: VerbNet: A broad-coverage, comprehensive verb lexicon. PhD thesis, Computer and Information Science Dept., University of Pennsylvania, PA (2005)
10. Palmer, M.S., Dahl, D.A., Schiffman, R.J., Hirschman, L., Linebarger, M., Dowding, J.: Recovering Implicit information. In: Proceedings of the 24th Annual Meeting of the Association for Computational Linguistics, pp. 10–19. Association for Computational Linguistics, New York (1986)
11. Tetreaul, J.R.: Implicit Role Reference. In: International Symposium on Reference Resolution for Natural Language Processing, pp. 109–115 (2002)
12. Heeman, P., Allen, J.: Trains 93 dialogues. Technical report, University of Rochester (1995)
13. Ruppenhofer, J., Sporleder, C., Morante, R., Baker, C., Palmer, M.: SemEval-2010 Task 10: Linking Events and Their Participants in Discourse. In: Proceedings of the 5th International Workshop on Semantic Evaluation, pp. 296–299. Association for Computational Linguistics, Uppsala (2010)
14. Chen, D., Schneider, N., Das, D., Smith, N.A.: SEMAFOR: Frame argument resolution with log-linear models. In: Proceedings of the 5th International Workshop on Semantic Evaluation, SemEval 2010, pp. 264–267. Association for Computational Linguistics, Stroudsburg (2010)
15. Tonelli, S., Delmonte, R.: VENSES++: Adapting a deep semantic processing system to the identification of null instantiations. In: Proceedings of the 5th International Workshop on Semantic Evaluation, SemEval 2010, pp. 296–299. Association for Computational Linguistics, Stroudsburg (2010)
16. Silberer, C., Frank, A.: Casting Implicit Role Linking as an Anaphora Resolution Task. In: *SEM 2012: The First Joint Conference on Lexical and Computational Semantics – Volume 1: Proceedings of the Main Conference and the Shared Task, and Volume 2: Proceedings of the Sixth International Workshop on Semantic Evaluation (SemEval 2012), pp. 1–10. Association for Computational Linguistics, Montréal (2012)
17. Laparra, E., Rigau, G.: Exploiting Explicit Annotations and Semantic Types for Implicit Argument Resolution. In: ICSC, pp. 75–78 (2012)
18. Ruppenhofer, J., Gorinski, P., Sporleder, C.: In search of missing arguments: A linguistic approach. In: Proceedings of the International Conference Recent Advances in Natural Language Processing, RANLP 2011 Organising Committee, Hissar, Bulgaria, pp. 331–338 (2011)
19. Tonelli, S., Delmonte, R.: Desperately seeking implicit arguments in text. In: Proceedings of the ACL 2011 Workshop on Relational Models of Semantics, pp. 54–62. Association for Computational Linguistics, Stroudsburg (2011)

20. Baker, C.F., Fillmore, C.J., Lowe, J.B.: The Berkeley FrameNet Project. In: Proceedings of the 36th Annual Meeting of the Association for Computational Linguistics and 17th International Conference on Computational Linguistics, ACL 1998, vol. 1, pp. 86–90. Association for Computational Linguistics, Stroudsburg (1998)
21. Marcus, M.P., Santorini, B., Marcinkiewicz, M.A.: Building a large annotated corpus of english: The penn treebank. Computational Linguistics 19, 313–330 (1993)
22. Peris, A., Taulé, M.: AnCora-Nom: A Spanish Lexicon of Deverbal Nominalizations. Procesamiento del Lenguaje Natural 46, 11–19 (2011)
23. Alvez, J., Atserias, J., Carrera, J., Climent, S., Oliver, A., Rigau, G.: Consistent annotation of eurowordnet with the top concept ontology. In: Proceedings of Fourth International WordNet Conference (GWC 2008). Association for Computational Linguistics (2008)
24. Aparicio, J., Taulé, M., Martí, M.: AnCora-Verb: A Lexical Resource for the Semantic Annotation of Corpora. In: Proceedings of the Sixth International Language Resources and Evaluation (LREC 2008), pp. 797–802. European Language Resources Association (ELRA), Marrakech (2008)
25. Witten, I.H., Frank, E.: Data Mining: Practical Machine Learning Tools and Techniques, 2nd edn. Morgan Kaufmann, San Francisco (2005)

Automatic Detection of Idiomatic Clauses

Anna Feldman[1,2] and Jing Peng[1]

[1] Department of Computer Science
[2] Department of Linguistics
Montclair State University, Montclair, NJ 07043, USA
{feldmana,pengj}@mail.montclair.edu

Abstract. We describe several experiments whose goal is to automatically identify idiomatic expressions in written text. We explore two approaches for the task: 1) idiom recognition as outlier detection; and 2) supervised classification of sentences. We apply principal component analysis for outlier detection. Detecting idioms as lexical outliers does not exploit class label information. So, in the following experiments, we use linear discriminant analysis to obtain a discriminant subspace and later use the three nearest neighbor classifier to obtain accuracy. We discuss pros and cons of each approach. All the approaches are more general than the previous algorithms for idiom detection – neither do they rely on target idiom types, lexicons, or large manually annotated corpora, nor do they limit the search space by a particular type of linguistic construction.

1   Introduction

Idioms are conventionalized expressions that have figurative meanings that cannot be derived from the literal meaning of the phrase. The prototypical examples of idioms are expressions like *I'll eat my hat, He put his foot in his mouth, Cut it out, I'm going to rake him over the coals, a blessing in disguise, a chip on your shoulder,* or *kick the bucket.* Researchers have not come up with a single agreed-upon definition of idioms that covers all members of this class (Glucksberg, 1993; Cacciari, 1993; Nunberg et al., 1994; Sag et al., 2002; Villavicencio et al., 2004; Fellbaum et al., 2006). The common property ascribed to the idiom is that it is an expression whose meaning is different from its simple compositional meaning.

Some idioms become frozen in usage, and they resist change in syntactic structure, while others do allow some variability in expression (Fellbaum, 2007; Fazly et al., 2009). In addition, Fazly et al. (2009) have argued that in many situations, a Natural Language Processing (NLP) system will need to distinguish a usage of a potentially-idiomatic expression as either idiomatic or literal in order to handle a given sequence of words appropriately. We discuss previous approaches to automatic idiom detection in section 3.1.

2   Our Approach

Following Degand and Bestgen (2003), we have identified three important properties of idioms. (1) A sequence with literal meaning has many neighbors, whereas

A. Gelbukh (Ed.): CICLing 2013, Part I, LNCS 7816, pp. 435–446, 2013.

a figurative one has few. (2) Idiomatic expressions should demonstrate low semantic proximity between the words composing them. (3) Idiomatic expressions should demonstrate low semantic proximity between the expression and the preceding and subsequent segments.

Based on the properties of idioms outlined above, we have experimented with two ideas: (1) The problem of idiom recognition be reduced to the problem of identifying a *semantic outlier*. By an *outlier* we mean an observation which appears to be inconsistent with the remainder of a set of data. We apply principal component analysis (PCA) (Jolliffe, 1986; Shyu et al., 2003) for outlier detection. (2) We view the process of idiom detection as a binary classification of sentences (idiomatic vs literal sentences). We use linear discriminant analysis (LDA) (Fukunaga, 1990) to obtain a discriminant subspace and later use the three nearest neighbor classifier to obtain accuracy. We first provide a few words on previous work.

3   Related Work

3.1   Automatic Idiom Detection

Previous approaches to automatic idiom detection can be classified into two major groups: 1) Type-based extraction, i.e., detecting idioms at the type level; 2) token-based detection, i.e., detecting idioms in context.

Type-based extraction is based on the idea that idiomatic expressions exhibit certain linguistic properties that can distinguish them from literal expressions. Sag et al. (2002); Fazly et al. (2009), among many others, discuss various properties of idioms. Some examples of such properties include 1) lexical fixedness: e.g., neither 'shoot the wind' nor 'hit the breeze' are valid variations of the idiom *shoot the breeze.* and 2) syntactic fixedness: e.g., *The guy kicked the bucket* is idiomatic whereas *The bucket was kicked* is not idiomatic anymore; and of course, 3) non-compositionality. While many idioms do have these properties, many idioms fall on the continuum from being compositional to being partly unanalyzable to completely non-compositional (Cook et al., 2008). Fazly et al. (2009); Li and Sporleder (2010), among others, notice that type-based approaches do not work on expressions that can be interpreted idiomatically or literally depending on the context and thus, an approach that considers tokens in context is more appropriate for the task of idiom recognition. A number of token-based approaches has been discussed in the literature, both supervised (Katz and Giesbrecht, 2006), weakly supervised (Birke and Sarkar, 2006) and unsupervised (Fazly et al., 2009; Sporleder and Li, 2009).

Li and Sporleder (2009); Sporleder and Li (2009) propose a graph-based model for representing the lexical cohesion of a discourse. Nodes correspond to tokens in the discourse, which are connected by edges whose value is determined by a semantic relatedness function. They experiment with two different approaches to semantic relatedness: 1) Dependency vectors, as described in Pado and Lapata (2007); 2) Normalized Google Distance (NGD) (Cilibrasi and Vitányi, 2007).

Li and Sporleder (2009) show that this method works better for larger contexts (greater than five paragraphs).

Li and Sporleder (2010) assume that literal and figurative data are generated by two different Gaussians, literal and non-literal, and the detection is done by comparing which Gaussian model has a higher probability to generate a specific instance. The approach assumes that the target expressions are already known and the goal is to determine whether this expression is literal or figurative in a particular context. The important insight of this method is that figurative language in general exhibits fewer semantic cohesive ties with the context than literal language. Their results are inconclusive, unfortunately, due to the small size of the test corpus. In addition, it relies heavily on target expressions. One has to have a list of potential idioms a priori to model semantic relatedness.

4 Idea 1: Idioms as Outliers

We first explore the hypothesis that the problem of automatic idiom detection can be formulated as the problem of identifying an outlier in a dataset. We investigate principal component analysis (PCA) (Jolliffe, 1986; Shyu et al., 2003) for outlier detection.

4.1 Idiom Detection Based on Principal Component Analysis

The approach we are taking for idiom detection is based on principal component analysis (PCA) (Jolliffe, 1986; Shyu et al., 2003). PCA has several advantages in outlier detection. First, it does not make any assumptions regarding data distributions. Many statistical detection methods assume a Gaussian distribution of normal data, which is far from reality. Second, by using a few principal modes to describe data, PCA provides a compact representation of the data, resulting in increased computational efficiency and real time performance.

PCA computes a set of mathematical features, called principal components, to explain the variance in the data. These principal components are linear combinations of the original variables describing the data and are orthogonal to each other. The first principal component corresponds to the direction along which the data vary the most. The second principal component corresponds to the direction along which the data vary the second most, and so on. Furthermore, total variance in all the principal components explains total variance in the data.

Let $\mathbf{z} = \{\mathbf{x}_i\}_{i=1}^m$ be a set of data points. Each $\mathbf{x}_i = (x_i^1, \cdots, x_i^q)^t$, where t denotes the transpose operator. That is, each data point is described by q attributes or variables. PCA computes a set of eigenvalue and eigenvector pairs $\{(\lambda_1, e_1, \cdots, (\lambda_q, e_q)\}$ with $\lambda_1 \geq \lambda_2 \geq \cdots \geq \lambda_q$ by performing singular value decomposition of the covariance matrix of the data: $\Sigma = \sum_{i=1}^m (\mathbf{x}_i - \bar{\mathbf{x}})(\mathbf{x}_i - \bar{\mathbf{x}})^t$, where $\bar{x} = 1/m \sum_{i=1}^m$. Then the ith principal component of an observation $\mathbf{x}$ is given by $y_i = e_i^t(\mathbf{x} - \bar{\mathbf{x}})$.

Note that the major components correspond strongly to the attributes having relatively large variance and covariance. Consequently, after projecting the data onto the principal component space, idioms that are outliers with respect to the major components usually correspond to outliers on one or more of the original attributes. On the other hand, minor (last few) components represent a linear combination of the original attributes with minimal variance. Thus, the minor components are sensitive to observations that are inconsistent with the variance structure of the data but are not considered to be outliers with respect to the original attributes (Jobson, 1992). Therefore, a large value along the minor components strongly indicates a potential outlier that otherwise may not be detected based solely on large values of the original attributes.

Our technique (we call it "principal minor component analysis": PMC) computes two functions for a given input $\mathbf{x}$. The first one is computed along major components: $f(x) = \sum_{i=1}^{p} \frac{y_i^2}{\lambda_i}$. The second one is computed along minor components: $g(x) = \sum_{i=q-r+1}^{q} \frac{y_i^2}{\lambda_i}$. Here y_i are projections along each component. p represents the number of major components and captures sufficient variance in the data, while r denotes the number of minor components. Both p and r can be determined through cross-validation.

It can be seen from our earlier discussion that $f(\mathbf{x})$ captures extreme observations with large values along some original attributes. On the other hand, $g(\mathbf{x})$ measures observations that are outside of the normal variance structure in the data, as measured by minor components. Thus, the strength of our approach is that it detects an outlier that is either extremely valued along the major components, or does not confirm to the same variance structure along the minor components in the data.

Our technique then decides an input $\mathbf{x}$ as outlier if $f(\mathbf{x}) \geq T_f$ or $g(\mathbf{x}) \geq T_g$, where T_f and T_g are outlier thresholds that are associated with the false positive rate α (Kendall et al., 2009). Suppose that the data follow the normal distribution. Define $\alpha_f = \Pr\{\sum_{i=1}^{p} \frac{y_i^2}{\lambda_i} > T_f | \mathbf{x} \text{ is normal}\}$, and $\alpha_g = \Pr\{\sum_{i=q-r+1}^{q} \frac{y_i^2}{\lambda_i} > T_g | \mathbf{x} \text{ is normal}\}$. Then $\alpha = \alpha_f + \alpha_g - \alpha_f \alpha_g$. The false positive rate has the following bound (Kendall et al., 2009) $\alpha_f + \alpha_g - \sqrt{\alpha_f \alpha_g} \leq \alpha \leq \alpha_f + \alpha_g$. Different types of outliers can be detected based on the values of α_f and α_g. If $\alpha_f = \alpha_g$, α can be determined by solving a simple quadratic equation. For example, if we want a 2% false positive rate (i.e., $\alpha = 0.02$), we obtain $\alpha_f = \alpha_g = 0.0101$.

Note that the above calculation is based on the assumption that our data follow the normal distribution. This assumption however is unlikely to be true in practice. We therefore determine α_f and α_g values based on the empirical distributions of $\sum_{i=1}^{p} y_i^2/\lambda_i$ and $\sum_{i=q-r+1}^{q} y_i^2/\lambda_i$ in the training data. That is, for a false positive rate of 2%, T_f and T_g represent the 0.9899 quantile of the empirical distributions of $\sum_{i=1}^{p} y_i^2/\lambda_i$ and $\sum_{i=q-r+1}^{q} y_i^2/\lambda_i$, respectively.

5   Idea 2: Supervised Classification for Idiom Detection

Another way to look at the problem of idiom detection is one of the supervised classification.

Our idiom detection algorithm is based on linear discriminant analysis (LDA). To obtain a discriminant subspace, we train our model on a small number of randomly selected idiomatic and non-idiomatic sentences. We then project both the training and the test data on the chosen subspace and use the three nearest neighbor (3NN) classifier to obtain accuracy. The proposed approach is more general than the previous algorithms for idiom detection — neither does it rely on target idiom types, lexicons, or large manually annotated corpora, nor does it limit the search space by a particular type of linguistic construction. The following sections describe the algorithm, the data and the experiments in more detail.

5.1 Idiom Detection Based on Discriminant Analysis

A similar approach has been discussed in Peng et al. (2010). LDA is a class of methods used in machine learning to find the linear combination of features that best separate two classes of events. LDA is closely related to principal component analysis (PCA) that concentrates on finding a linear combination of features that best explains the data. Discriminant analysis explicitly exploits class information in the data, while PCA does not.

Idiom detection based on discriminant analysis has several advantages. First, as previously mentioned, it does not make any assumptions regarding data distributions. Many statistical detection methods assume a Gaussian distribution of normal data, which is far from reality. Second, by using a few discriminants to describe data, discriminant analysis provides a compact representation of the data, resulting in increased computational efficiency and real time performance.

6   Datasets

6.1 Dataset 1: Outlier Detection

For the outlier detection, it is important that the training corpus is free of any idiomatic or metaphoric expressions. Otherwise, idiomatic or metaphoric expressions as outliers can distort the variance-covariance structure of the semantics of the corpus.

Our training set consists of 1,200 sentences (22,028 tokens) randomly extracted from the British National Corpus (BNC, http://www.natcorp.ox.ac.uk/). The first half of the data comes from the social science domain [1] and the other is defined in BNC as "imaginative"[2]. Our annotators were asked to identify clauses containing (any kind of) metaphors and idioms and paraphrase them literally. We used this paraphrased corpus for training.[3] The training data contains 139 paraphrased sentences.

---

[1] This is an excerpt from *Cities and Plans: The Shaping of Urban Britain in the Nineteenth and Twentieth Centuries* by Gordon Emanuel Cherry.

[2] This is an excerpt taken from *Heathen*, a thriller novel written by Shaun Hutson.

[3] We understand that this task is highly subjective, but the inter-annotator agreement was relatively high for this task (Cohen's kappa: 75%).

Our test data are 99 sentences extracted from the BNC social science (non-fiction) section, annotated as either literal or figurative and additionally labeled with the information about the figures of speech they contain (idioms (I), dead metaphors (DM), and living metaphors (LM)). The annotator has identified 12 idioms, 22 dead metaphors, and 2 living metaphors in that text.

6.2   Dataset 2: LDA

In the LDA experiments, we used the dataset described by Fazly et al. (2009). This is a dataset of verb-noun combinations extracted from the British National Corpus (BNC, Burnard (2000)). The VNC tokens are annotated as either literal, idiomatic, or unknown. The list contains only those VNCs whose frequency in BNC was greater than 20, and that occurred at least in one of two idiom dictionaries (Cowie et al., 1983; Seaton and Macaulay, 2002). The dataset consists of 2,984 VNC tokens[4].

Since our task is framed as sentence classification rather than MWE extraction and filtering, we had to translate this data into our format. Basically, our dataset has to contain sentences with the following tags: I (=idiomatic sentence), L (=literal), and Q (=unknown). Translating the VNC data into our format is not trivial. A sentence that contains a VNC idiomatic construction can be unquestionably marked as I (=idiomatic); however, a sentence that contains a non-idiomatic occurrence of VNC cannot be marked as L since these sentences could have contained other types of idiomatic expressions (e.g., prepositional phrases) or even other figures of speech. So, by automatically marking all sentences that contain non-idiomatic usages of VNCs, we create an extremely noisy dataset of literal sentences. The dataset consists of 2,550 sentences, of which 2,013 are idiomatic sentences and the remaining 537 are literal sentences.

7   Experiments

7.1   Detecting Outliers

We compare the proposed technique (PMC) against a random baseline approach. The baseline approach flips a fair coin. If the outcome is head, it classifies a given sentence as outlier (idiom, dead metaphor or living metaphor). If the outcome is tail, it classifies a given sentence as a regular sentence. The outlier thresholds T_f and T_g at a given false positive rate are determined from the training data by setting $\alpha_f = \alpha_g$.

In this experiment, we treat each sentence as a document. We created a bag-of-word model for the data set, i.e., we use TF-IDF to represent the data. Single value decomposition is then applied to the bag of words and the number of principal modes for representing the latent semantics space is calculated that capture 100% variance in the data.

---

[4] To read more about this dataset, the reader is referred to Cook et al. (2008).

Experimental Results: PCA. The following table shows the detection rates of the two competing methods at a given false positive rate. The results reported here were based on 10% of major components (p in fuction f) and 0.1% minor components (r in function g). It turns out that the technique is not sensitive to p values, while r represents a trade-off between detection and precision.

Table 1. The detection rates of the two competing methods at a given false positive rate: Second column: idioms and metaphors; Third column: idioms only; Fourth column: metaphors (dead and living); Fifth column: dead metaphors only; and Sixth column: living metaphors only

False Positive Rate	PMC					Baseline
1%	47%	44%	45%	43%	100%	50%
2%	53%	44%	55%	52%	100%	50%
4%	63%	56%	63%	62%	100%	50%
6%	70%	67%	73%	71%	100%	50%
8%	73%	77%	73%	71%	100%	50%
10%	87%	89%	86%	86%	100%	50%

7.2 LDA and Supervised Classification

We first apply the bag-of-words model to create a term-by-sentence representation of the 2,550 sentences in a 6,844 dimensional term space. The Google stop list is used to remove stop words.

We randomly choose 300 literal sentences and 300 idiomatic sentences as training and randomly choose 100 literals and 100 idioms from the remaining sentences as testing. Thus the training dataset consists of 600 examples, while the test dataset consists of 200 examples. We train our model on the training data and obtain one discriminant subspace. We then project both training and test data on the chosen subspace. Note that for the two-class case (literal vs. idiom), one dimensional subspace is sufficient. In the reduced subspace, we compare three classifiers: the three nearest neighbor (3NN) classifier, the quadratics classifier that fits multivariate normal densities with covariance estimates stratified by classes (Krzanowski, 2000), and support vector machines (SVMs) with the Gaussian kernel (Cristianini and Shawe-Taylor, 2000). The kernel parameter was chosen through 10-fold cross-validation. We repeat the experiment 20 times to obtain the average accuracy rates registered by the three methods. We also include the performance by a random baseline approach. The baseline approach (BL) flips a fair coin. If the outcome is head, it classifies a given sentence as idiomatic. If the outcome is tail, it classifies a given sentence as a regular sentence. The results are the following–*3NN*: 0.802, *Quadratic*: 0.769, *SVMs*: 0.789, and *BL*: 0.50.

Table 2 shows the precision, recall and accuracy for the nearest neighbor performance as a function of false positive rates. For the nearest neighbor method, the false positive rates are achieved by varying the number of nearest neighbors

Table 2. LDA performance in more detail

False Positive Rate	recall	precision	accuracy
15%	76%	83%	81%
16%	77%	83%	81%
17%	77%	82%	80%
20%	76%	79%	78%

in predicting idioms. By varying the number of nearest neighbors from 1 to 39, the false positive rates from 15% to 20% are obtained.

Even though we used Fazly et al. (2009)'s dataset for these experiments, the direct comparison with their methods is impossible here because our tasks are formulated differently. Fazly et al. (2009)'s unsupervised model that relies on the so-called canonical forms (CForm) gives 72.4% (macro-)accuracy on the extraction of idiomatic tokens when evaluated on their test data.

8   Comparison of the Two Methods

Unfortunately, the direct comparison of the two methods discussed above is not possible because they are not using the same datasets. To make the comparison fair, we decided to run the PMC outlier detection algorithm on the data used for LDA. In section 6.2 we already described how the data was created. Unlike the dataset used for the PCA outlier detection algorithm, where human annotators were given the task to make the training data as literal as possible and eliminate any possible figurative expressions, the data used by the LDA alogorithm was noisy. It was crudely translated by labeling all sentences that contained a VNC idiom from Cook et al. (2008) as idiomatic; the rest were labeled as literal. Unfortunately, this approach creates an extremely noisy dataset of literal sentences that can potentially contain other types of figurative expressions, other types of idioms etc. Another observation that our human annotators made when preparing the dataset of paraphrases used by PMC is that it was really difficult to avoid figurative language when paraphrasing and that it was also difficult to notice figurative expressions because some of them had become conventional.

So, having taken the issues discussed above into consideration, we decided to reverse our PMC approach and hypothesize that most sentences contain some figurative language and true outliers are in fact the literal sentences. So, we train the PMC algorithm on 1,800 sentences containing idioms (used by the LDA approach) rather than on literals. We test PMC on randomly selected 150 idiomatic and 150 literal sentences. We repeat the experiment 20 times to obtain the average accuracy rates. The numbers are reported in Table 3.

The performance of the LDA approach is better than that of PMC. However, the LDA method is supervised and requires training data, i.e., sentences marked as idiomatic or literal, while PMC does not use class label information. However, PMC's detection rates (recall) are higher than the detection rates of the LDA algorithm. Depending on an application, the higher recall might be preferred.

Table 3. Performance of the PMC algorithm on the LDA dataset

False Positive Rate	PMC recall	PMC precision	PMC accuracy
1%	93.3%	51.3%	52.3%
2%	91.9%	51.5%	52.6%
4%	91.0%	51.8%	53.2%
6%	89.5%	52.1%	53.6%
8%	88.1%	52.1%	53.6%
10%	81.2%	52.8%	54.3%

For example, if a researcher looks for sentences containing figurative speech for the purpose of linguistic analysis, after some postediting, s/he might wind up with more interesting examples.

9   Qualitative Analysis of the Results

Our error analysis of the two methods reveals that many cases are fuzzy and clear literal/idiomatic demarcation is difficult.

In examining our false positives (i.e., non-idiomatic expressions that were marked as idiomatic by the model), it becomes apparent that the classification of cases is not clear-cut. The expression *words of the sixties/seventies/ eighties/nineties* is not idiomatic; however, it is not entirely literal either. It is metonymic – these decades could not literally produce words. Another false positive contains the expression *take the square root*. While seemingly similar to the idiom *take root* in *plans for the new park began to take root*, the expression *take the square root* is not idiomatic. It does not mean "to take hold like roots in soil." Like the previous false positive, we believe *take the square root* is figurative to some extent. A person cannot literally take the square root of a number like s/he can literally take milk out of the fridge.

When it comes to classifying expressions as idiomatic or literal, our false negatives (i.e., idiomatic expressions that were marked as non-idiomatic by the model) reveal that human judgments can be misleading. For example, *It therefore has a long-term future* was marked as idiomatic in the test corpus. While our human annotators may have thought that an object could not literally have (or hold) a long-term future, this expression does not appear to be truly idiomatic. We do not consider it to be as figurative as a true positive like *lose our temper*. Another false negative contains a case of metonymy *Italy will pay a reciprocal visit* and the verbal phrase *take part*. In this case, our model correctly predicted that the expression is non-idiomatic. Properties of metonymy are different from those of idioms, and the verbal phrase *take part* has a meaning separate from that of the idiomatic expression *take someone's part*.

9.1   Inter-annotator Agreement

To gain insights into the performance of the second approach, we created a dataset that is manually annotated to avoid noise in the literal dataset. We asked three human subjects to annotate 200 sentences from the VNC dataset as idiomatic, non-idiomatic or unknown. 100 of these sentences contained idiomatic expressions from the VNC data. We then merged the result of the annotation by the majority vote.

We also measured the inter-annotator agreement (the Cohen kappa k, Cohen (1960); Carletta (1996)) on the task. Interestingly, the Cohen kappa was much higher for the idiomatic data than for the so-called literal data: k (idioms) = 0.91; k (literal) = 0.66. There are several explanations of this performance. First, the idiomatic data is much more homogeneous since we selected sentences that already contained VNC idiomatic expressions. The rest of the sentences might have contained metaphors or other figures of speech and thus it was more difficult to make the judgments. Second, humans easily identify idioms, but the decision whether a sentence is literal or figurative is much more challenging. The notion of "figurativeness" is not a binary property (as might be suggested by the labels that were available to the annotators). "Figurativeness" falls on a continuum from completely transparent (= literal) to entirely opaque (=figurative)[5] Third, the human annotators had to select the label, literal or idiomatic, without having access to a larger, extra-sentential context, which might have affected their judgements. Although the boundary between idiomatic and literal expressions is not entirely clear (expressions do seem to fall on a continuum in terms of idiomaticity), some expressions are clearly idiomatic and others clearly literal based on the overall agreement of our annotators. By classifying sentences as either idiomatic or literal, we believe that this additional sentential context could be used to further investigate how speakers go about making these distinctions.

10   Conclusion

The binary classification approach, offered in this paper, has multiple practical applications. We also feel that identifying idioms at the sentence level may provide new insights into the kinds of contexts that idioms are situated in. These findings could further highlight properties that are unique to specific idioms if not idioms in general. Our current work is concerned with improving the detection rates, by incorporating textual cohesion and compositionality measures into our models.

Acknowledgements. This material is based in part upon work supported by the National Science Foundation under Grant Numbers 0916280, 1033275, and 1048406. The findings and opinions expressed in this material are those of the authors and do not reflect the views of the NSF.

References

Birke, J., Sarkar, A.: A clustering approach to the nearly unsupervised recognition of nonliteral language. In: Proceedings of the 11th Conference of the European Chapter of the Association for Computational Linguistics (EACL 2006), Trento, Italy, pp. 329–336 (2006)

---

[5] A similar observation is made by Cook et al. (2008) with respect to idioms.

Burnard, L.: The British National Corpus Users Reference Guide. Oxford University Computing Services (2000)

Cacciari, C.: The Place of Idioms in a Literal and Metaphorical World. In: Cacciari, C., Tabossi, P. (eds.) Idioms: Processing, Structure, and Interpretation, pp. 27–53. Lawrence Erlbaum Associates (1993)

Carletta, J.: Assessing Agreement on Classification Tasks: The Kappa Statistic. Computational Linguistics 22(2), 249–254 (1996)

Cilibrasi, R., Vitányi, P.M.B.: The google similarity distance. IEEE Trans. Knowl. Data Eng. 19(3), 370–383 (2007)

Cohen, J.: A Coefficient of Agreement for Nominal Scales. Education and Psychological Measurement (20), 37–46 (1960)

Cook, P., Fazly, A., Stevenson, S.: The VNC-Tokens Dataset. In: Proceedings of the LREC Workshop: Towards a Shared Task for Multiword Expressions (MWE 2008), Marrakech, Morocco (June 2008)

Cowie, A.P., Mackin, R., McCaig, I.R.: Oxford Dictionary of Current Idiomatic English, vol. 2. Oxford University Press (1983)

Cristianini, N., Shawe-Taylor, J.: An Introduction to Support Vector Machines and other kernel-based learning methods. Cambridge University Press, Cambridge (2000)

Degand, L., Bestgen, Y.: Towards Automatic Retrieval of Idioms in French Newspaper Corpora. Literary and Linguistic Computing 18(3), 249–259 (2003)

Fazly, A., Cook, P., Stevenson, S.: Unsupervised Type and Token Identification of Idiomatic Expressions. Computational Linguistics 35(1), 61–103 (2009)

Fellbaum, C.: The Ontological Loneliness of Idioms. In: Schalley, A., Zaefferer, D. (eds.) Ontolinguistics. Mouton de Gruyter (2007)

Fellbaum, C., Geyken, A., Herold, A., Koerner, F., Neumann, G.: Corpus-based Studies of German Idioms and Light Verbs. International Journal of Lexicography 19(4), 349–360 (2006)

Fukunaga, K.: Introduction to statistical pattern recognition. Academic Press (1990)

Glucksberg, S.: Idiom Meanings and Allusional Content. In: Cacciari, C., Tabossi, P. (eds.) Idioms: Processing, Structure, and Interpretation, pp. 3–26. Lawrence Erlbaum Associates (1993)

Jobson, J.: Applied Multivariate Data Analysis, vol. II: Categorical and Multivariate Methods. Springer (1992)

Jolliffe, I.: Principal Component Analysis. Springer, New York (1986)

Katz, G., Giesbrecht, E.: Automatic identification of non-compositional multi-word expressions using latent semantic analysis. In: Proceedings of the ACL 2006 Workshop on Multiword Expressions: Identifying and Exploiting Underlying Properties, Sydney, Australia, pp. 12–19 (2006)

Kendall, M., Stuart, A., Ord, J.: Kendall's Advanced Theory of Statistics, vol. 1: Distribution Theory. John Wiley and Sons (2009)

Krzanowski, W.J.: Principles of Multivariate Analysis. Oxford University Press (2000)

Li, L., Sporleder, C.: A Cohesion Graph Based Approach for Unsupervised Recognition of Literal and Non-literal Use of Multiword Expresssions. In: Proceedings of the 2009 Workshop on Graph-based Methods for Natural Language Processing (ACL-IJCNLP), Singapore, pp. 75–83 (2009)

Li, L., Sporleder, C.: Using Gaussian Mixture Models to Detect Figurative Language in Context. In: Proceedings of NAACL/HLT 2010 (2010)

Nunberg, G., Sag, I.A., Wasow, T.: Idioms. Language 70(3), 491–538 (1994)

Pado, S., Lapata, M.: Dependency-based construction of semantic space models. Computational Linguistics 33(2), 161–199 (2007)

Peng, J., Feldman, A., Street, L.: Computing linear discriminants for idiomatic sentence detection. Research in Computing Science, Special Issue: Natural Language Processing and its Applications 46, 17–28 (2010)

Sag, I.A., Baldwin, T., Bond, F., Copestake, A., Flickinger, D.: Multiword Expressions: A Pain in the Neck for NLP. In: Gelbukh, A. (ed.) CICLing 2002. LNCS, vol. 2276, pp. 1–15. Springer, Heidelberg (2002)

Seaton, M., Macaulay, A. (eds.): Collins COBUILD Idioms Dictionary, 2nd edn. HarperCollins Publishers (2002)

Shyu, M., Chen, S., Sarinnapakorn, K., Chang, L.: A novel anomaly detection scheme based on principal component classifier. In: Proceedings of IEEE International Conference on Data Mining (2003)

Sporleder, C., Li, L.: Unsupervised Recognition of Literal and Non-literal Use of Idiomatic Expressions. In: EACL 2009: Proceedings of the 12th Conference of the European Chapter of the Association for Computational Linguistics, pp. 754–762. Association for Computational Linguistics, Morristown (2009)

Villavicencio, A., Copestake, A., Waldron, B., Lambeau, F.: Lexical Encoding of MWEs. In: Proceedings of the Second ACL Workshop on Multiword Expressions: Integrating Processing, Barcelona, Spain, pp. 80–87 (2004)

Evaluating the Results of Methods
for Computing Semantic Relatedness

Felice Ferrara and Carlo Tasso

Artificial Intelligence Lab.
Department of Mathematics and Computer Science
University of Udine, Italy
{felice.ferrara,carlo.tasso}@uniud.it

Abstract. The semantic relatedness between two concepts is a measure that quantifies the extent to which two concepts are semantically related. Due to the growing interest of researchers in areas such as Semantic Web, Information Retrieval and NLP, various approaches have been proposed in the literature for automatically computing the semantic relatedness. However, despite the growing number of proposed approaches, there are still significant criticalities in evaluating the results returned by different semantic relatedness methods. The limitations of the state of the art evaluation mechanisms prevent an effective evaluation and several works in the literature emphasize that the exploited approaches are rather inconsistent. In this paper we describe the limitations of the mechanisms used for evaluating the results of semantic relatedness methods. By taking into account these limitations, we propose a new methodology and new resources for comparing in an effective way different semantic relatedness approaches.

1 Introduction

The terms *semantic similarity* and *semantic relatedness* (on which we focus in this paper) have often been used as synonyms in the areas of Natural Language Processing, Information Retrieval and Semantic Web, but some researchers highlighted significant differences between these two concepts. The concept of semantic relatedness is defined in the literature as the extent to which two concepts are related by semantic relations [18]. On the other hand, a possible definition of semantic similarity describes it as the measure which quantifies the extent to which two concepts can be used in an interchangeable way. According to this definition two semantically similar entities are also semantically related, but two semantically related concepts may be semantically dissimilar [3]. For example, the concepts of bank and trust-company are semantically similar and their similarity implies that they are also semantically related, but two concepts related by an antonymic[1] relation (such as the adjectives *bad* and *good*) are semantically

---

[1] Antonymy is the semantic relation which connects concepts with an opposite meanings.

A. Gelbukh (Ed.): CICLing 2013, Part I, LNCS 7816, pp. 447–458, 2013.

related and semantically dissimilar. According to [21], semantic similarity is a more strict relation since it takes into account a focused set of semantic relations which are often stored in lexical ontologies such as Wordnet. In Wordnet, for example, synonyms[2] are grouped in synsets and a hierarchical structure connects hyponyms and hypernyms[3]. On the other hand, the semantic relatedness between two concepts depends on all the possible relations involving them. For example, in order to compute the semantic relatedness between two Wordnet concepts, we should use all the available semantic connections by including, for example, meronomy[4] and antonymy. However, two concepts can be related by more complex semantic relations which can/are usually not explicitly stored in lexical ontologies. For example, from "Albert Einstein received the Nobel prize" we would be able to infer the existence of a relation between *Albert Einstein* and *Nobel prize*. This relation is not explicitly defined in Wordnet as well as all the other possible relations which can be entailed between concepts which are not directly related by standard relations. Moreover, it has to be noticed that humans organize their knowledge according to complex schemas by connecting concepts according to their background knowledge and experience [7]. The reasoning task where units of meaning are processed by the human mind in order to identify connections between concepts is referred in literature as *evocation* [2], which can be also defined as the degree to which a concept brings to mind another one. Evocation adds cross-part-of-speech links among nouns, verbs, and adjectives [14]. Since the human mind works under the influence of personal experience, the evocation process builds relations which may be not true in an absolute way (for instance the relations between emotions and objects/animals) and that is why these relations cannot be available in knowledge bases such as Wordnet.

Obviously, all these aspects must be considered when we have to plan the evaluation of methods aimed at automatically quantifying the semantic relatedness (*SR methods*) or the semantic similarity (*SS methods*). Thesaurus-like resources, such as the Roget dataset or the TOEFL Synonym Question dataset, can be effectively used for evaluating the precision of SS methods: they connect terms by RT (Related-Term) links and by UF (Used-For) links, however such links are just a few for each term, whereas many others could be entailed. For this reason, the feedback of people about the relatedness between pairs of terms is commonly used in order to evaluate the precision of SR methods. However, the methodology currently used for both collecting this feedback and evaluating the precision of SR methods is widely criticized by the same researchers who use it to analyze their results. In this paper we address these limitations and we propose an original approach for evaluating the precision of SR methods. More specifically, we focus on the idea of evaluating the results of SR methods which calculate the relatedness between the concepts included in Wikipedia. This choice is mainly due to the growing interest of the research community on the usage of Wikipedia as knowledge source for computing semantic

---

[2] Two terms are synonyms if they have the identical or very similar meaning.

[3] A hyponym shares a *type-of* relationship with its hypernym.

[4] The meronomy denotes a *part of* relation.

relatedness. In fact, the large coverage of concepts and the support to multilinguism makes Wikipedia very attractive for developing SR methods. Moreover, other researches point out that the refinements of the Wikipedia articles do not significantly influences the results of SR methods [20] while new concepts can be easily introduced and connected to the existing ones.

The paper is organized as follows: in Section 2 we describe the state of the art mechanisms used for evaluating the precision of SR methods while the drawbacks of these approaches are the object of Section 3; in Section 4 we describe our proposal which is evaluated in Section 5; final considerations conclude the paper in Section 6.

2   Evaluating SR Methods: State of the Art

As reported in [3], three main approaches have been proposed in the literature for evaluating the precision of SR methods.

A possible approach, utilized for example in [11], evaluates SR methods according to a set of qualitative heuristics. The simplest heuristic takes into account if the evaluated measure is a metric, but in [8] the authors report a list of suitable features for SR methods such as domain independence, independence from specific languages, coverage of included words, and coverage of the meanings of each word. The heuristic-based strategy is the simplest one but it also does not provide very significant results since it cannot quantify the accuracy of results. For this reason, even if this strategy is a useful tool for designing new SR methods, it cannot be used to have a significant comparison of state of the art mechanisms [3].

More concrete results can be obtained by integrating the SR methods in other systems such as metonymy resolution mechanisms [10], recommender systems [5], and approaches to text similarity [1]. In these cases, different SR methods are compared and evaluated according to the improvement produced by the integration of a specific SR method. However, it is quite clear that this strategy increases the difficulties in performing an extensive comparison of SR methods since: (i) different works face different tasks and use different datasets preventing, in this way, the repeatability of experimentations and (ii) the computed precision can be influenced by other components embedded in the adopted system.

In order to overcome these drawbacks, a more direct strategy can be implemented by comparing the feedback of a set of humans with the results produced by SR approaches. The feedback of volunteers has been collected in order to create datasets which have been used in the majority of the works where the precision of SR methods has been evaluated. The first experiments aimed at creating this kind of datasets was exploited by Rubenstein and Goodenough [17]. In their experiments they exploited a deck of 65 cards where on each card there was a pair of nouns written in English. The researchers asked to 51 judges both to order the 65 pairs of words (from the most related pair to the most unrelated one) and to assign a score in [0.0,4.0] for quantifying the relatedness of each pair of terms. This experiment was also replicated by other researchers in different settings: Miller and Charles used 30 pairs selected from the Rubenstein and Goodenough's deck of cards by using a larger set of judges [12]; Resnik used the feedback of 10 human evaluators for executing his experiments [16]. The idea of ordering and assigning a value to pairs of nouns was also replied for languages different from English. In particular, Gurevych replicated the experiment of Rubenstein and Goodenough by translating the 65 pairs into German [9]. The 65 pairs of nouns were a reference point for many studies and for this reason also Finkelstein et al. decided to start from these pairs for creating a larger dataset (referred in this paper as Related353 dataset) constituted by 353 word pairs [6]. In this case, the pairs were annotated with an integer in $[0, 10]$ by two sets of evaluators (composed by 13 and 16 judges respectively). Other works focused on the task of defining similar datasets for specific domains. In the biomedical field, Pedersen et al. collected the feedback of medics and physicians in order to evaluate SR methods in that specific domain [15]. Other researchers also worked on the task of generating larger sets of pairs of terms in a more automatic way. In [19], for instance, a corpus of document is analyzed in order to extract pairs of semantically related terms by following the idea that pairs of terms which appear frequently in the same document are probably semantically related.

The numeric scores acquired in these experiments have been extensively used for evaluating the precision of SR methods. In order to reach this aim the Pearson product-moment and the Spearman rank order correlation coefficient have been used. The Pearson product-moment is a statistical tool used to check if the results of a SR method resemble human judgments. On the other hand, the comparison of two rankings of the pairs (i.e. the ranking where pairs are ordered according to the feedback of the humans and (ii) the ranking where pairs are ordered according to the results of a SR method) can be executed by the Spearman coefficient. These two coefficients are in $[-1, +1]$ where -1 corresponds to completely uncorrelated rankings (low precision) and, conversely, $+1$ corresponds to a perfect correlation (high precision).

3   Drawbacks of the State of the Art

The experiments proposed in the literature mainly use datasets constituted by pairs of terms annotated by a group of humans. However, this approach has many criticalities which are emphasized even by the same researchers who adopted it. In this section we report these limitations by organizing the discussion in two parts: in Section 3.1, we focus on the characteristics of the collections of pairs of terms and, in Section 3.2, we describe the features of both the human feedback and the procedures exploited for computing the precision of SR methods.

3.1   Characteristics of the Pairs of Terms

The quality of the feedback collected in the experiments described in Section 2 strongly depends on the task submitted to the volunteers. The following points summarize the main limitations:

Shortage. The dataset proposed by Rubenstein and Goodenough is constituted by only 65 pairs of nouns which cannot be used to exploit an extensive analysis for generalizing the findings. This limitation is partially faced by the Related353 dataset which is constituted by 353 pairs.

Terms Instead of Concepts. The datasets are build up by terms which do not identify concepts. On the other hand, SR methods compute the semantic relatedness among concepts such as the synsets of Wordnet or the pages of Wikipedia. The proliferation of senses in knowledge bases such as Wordnet and Wikipedia makes hard the task of manually associating a sense to each term included in a dataset [18]. For example, the term *love* can be associated to 6 synsets of Wordnet and, on the other hand, in Wikipedia the term *love* identifies an emotion as well as people, songs, fictional characters, and movies. For tackling this problem, it is possible to manually associate some of the terms of the considered dataset to the Wikipedia concept that most probably was adopted by the evaluators. On the other hand, in order to avoid the need for manual disambiguation of terms, the semantic relatedness between all the possible senses of the two terms can be identified and fixed in the following way: the pair of senses with the highest semantic relatedness computed by the evaluated SR method is considered for assigning two specific senses to the two terms. Both these approaches are questionable since the judges were not conscious of the meanings of the words when they annotated the pairs.

Uncovered Topics and Semantic Relations. The datasets created by Rubenstein and Goodenough as well as the Related353 dataset were defined with the main goal of covering many possible degree of similarity. Following this idea, the authors used very general terms without taking into account the idea of choosing terms in different topics. Moreover there are not details about the semantic relations which involve the terms in the dataset. These limitations do not allow to generalize the computed results.

3.2 Characteristics of the Feedback and Evaluation Procedure

The agreement among the evaluators is used in the literature for estimating the quality of the collected feedback by following the idea that higher is the agreement more reliable is the collected feedback. According to the works in literature, the agreement of the available datasets is sufficient to evaluate the precision of SR methods. Actually, there is not a threshold for the required agreement between the judges and this is also true for domain-dependent datasets. However, also other features of the feedback collected from humans may greatly influence negatively the quality of the evaluation. We point out specifically the following points:

Pairs with Low Agreement. Different works use different strategies to manage the pairs with low agreement among judges. An example of these pairs is *(monk, oracle)* in the Related353 dataset which was annotated by 13 evaluators who returned the following votes $(7, 8, 3, 4, 4, 6, 5, 8, 6, 3, 4, 6, 1)$. In the majority of the works available in the literature these pairs are threaten exactly like the others, but in [15] the authors proposed to discard the pairs with a very low agreement in order to have more significant results. Obviously, this idea can be applied only when the dataset is constituted by a large set of pairs. This is a very important issue since, as noticed in [3], the freely available datasets show a significant agreement only when the existence of the semantic relation is very clear (for instance the terms are synonyms or they are completely unrelated).

The Choice of the Scale. The choice of the scale for collecting the feedback is a controversial point and has a strong impact on the agreement among the judges. By adopting a very fine-grained scale the judges have many possible choices and they can provide more accurate responses. This was the motivating idea of the approach proposed by Rubenstein and Goodenough who also asked people to order the pairs in order to have more coherent responses. In fact, by ordering the pairs each judge could assign a decreasing list of values to quantify the semantic relatedness. However, this mechanism does not scale up to a large set of pairs since it would require a huge load of work for ordering many pairs of terms. For this reason, the collection of the feedback for larger datasets like the Related353 dataset did not require the evaluators to order the pairs of terms. In this case, the humans could not rely on the order imposed to the pairs for assigning a vote and, consequently, it was harder for a judge to be coherent with his previous votes. For this reason, when the judges only annotate pairs of terms with a number it is better to renounce to a very fine-grained scale in order to have more significant responses.

Bias Introduced by Specific Communities. Different communities of evaluators may evaluate the semantic relatedness between two concepts according to different perspectives. This is clearly reported in [15] where the author show that physicians and medics judged differently the semantic relatedness between terms in the field of biology. On the other hand, it makes sense to evaluate SR methods only on pairs where the feedback is not biased by the perspective of a community of people.

Metric Robustness. The Pearson coefficient is a statistical tool used to catch the strength of the linear correlation between the human judgments and the score computed by a SR method. However, the correlation between the votes of humans and the results of a SR method can be nonlinear. Moreover, the Pearson correlation is based on the assumption that the two compared random variables are normally distributed, whereas the actual distribution of the relatedness values is at the moment unknown [3]. On the other hand, the Spearman coefficient, which, as we said, does not directly compare the human votes with the results of a SR method, seems to be more robust. This shows again that by allowing the evaluators to order concepts according to the degree of relatedness it is possible to acquire a more reliable feedback.

4   New Feedback for Evaluating Semantic Relatedness

In order to face the limitations described in the previous section we propose a new approach: collecting a different kind of feedback and, consequently, adopting a new way for evaluating the precision of SR methods. In our work we follow the idea that the source of the drawbacks of the state of the art datasets is primarily caused by the task assigned to the judges. In fact, as we said, other researchers showed that humans can judge the semantic relatedness by using a number only if the answer is quite obvious (for example, the terms are synonyms or the terms are completely unrelated). Our hypothesis is that humans can perceive the semantic relatedness, but they are not used to quantify it by using a number. Starting from this hypothesis we define a new approach for developing a new dataset (described in Section 4.1) which can be effectively used to evaluate SR methods (as described in Section 4.2).

4.1   Creating the Dataset

We decided to change the task used to collect the feedback by avoiding both expensive workload, such as the task of ordering a long sequence of pairs of terms, and tricky/noisy tasks, like the task of choosing a number to quantify the semantic relatedness among two terms. We ask to the judges to select the concept (from a set of proposed concepts) which is more related to a fixed/given concept, where each concept is associated to a specific Wikipedia page. By following this idea we defined the questions for the judges as triples of concepts. In particular, we defined a set of triples $T = (t_1, \ldots, t_m)$ where the triple $t_i = (target_i, c_{i1}, c_{i2})$ is constituted by a reference/fixed concept and two other concepts c_{i1}, c_{i2}. For each triple t_i, the judges have to decide which one among c_{i1} and c_{i2} is more related to $target_i$. For example, given the triple $t=(Musician, Watch, Trumpet)$, the evaluator can select *Watch* or *Trumpet* as more related to *Musician*. By selecting the concept semantically more related to the target concept the judge orders the three concepts according to the relatedness to the target. By following the previous example, if a judge chooses *Trumpet* then he implicitly defines the ordered list of concepts (*Musician, Trumpet, Watch*) since *Trumpet* is semantically more related to *Musician* then *Watch*. We also believe that there are some cases where humans cannot provide a response due to:

– Lack of knowledge. The judge may be not familiar with a concept or even a topic. In this case the judge may prefer to skip the question.
– Other possible ambiguities. In some cases two concepts may be (more or less) equally semantically related to the Target concept.

We decided to manage these two issues by allowing the judges to skip the evaluation of a triple, since we want to be able to identify the responses for which the judges are sufficiently confident.

By adopting this task we can better face some of the limitations presented
in Section 3. First of all, by associating each term to a Wikipedia page we can
overcome the limitation of having datasets composed by just terms. In our case
the meaning of a term is specified by a specific page in Wikipedia. By associating
the terms to Wikipedia pages we obtain two advantages: judges can take into
account the real meaning of the concepts when they produce their responses and,
moreover, also the evaluated SR method can exploit the Wikipedia page associated to the concept for computing the semantic relatedness. Another advantage
of the proposed task is that it does not use a specific scale for collecting the
feedback and this also simplifies the work of the judges who have to select only
the most related concept. By simplifying the task, we can (at least partially)
face the shortage problem since we require lower efforts the evaluators.

Obviously, the approach used to build the triples has a significant impact on
the results. As we said in Section 3, one of the main drawbacks of the datasets
described in the literature depends on the number of domains and of different semantic relations included in the dataset. In order to face this issue, we
have defined a specific set of templates for the triples, such as $(\langle TARGET \rangle,$
$\langle Emotion_1 \rangle, \langle Emotion_2 \rangle)$ and $(\langle TARGET \rangle, \langle Work_1 \rangle, \langle Work_2 \rangle)$. Then, we create some triples by creating instantiating each template. For example, from the
template $(\langle TARGET \rangle, \langle Emotion_1 \rangle, \langle Emotion_2 \rangle)$, we can build the triple (*Love,*
Graditude, Jelausy), the triple (*Clown, Humor, Fear*) and so on. We also include
other triples by picking concepts from systems such as Delicious and Open Directory. In particular, tags, categories, and other terms are extracted from these
systems in order to create new triples. By using stacks of Delicious and categories
in Open Directory we also select concepts (that must be concepts of Wikipedia)
belonging to different domains. In this way we face (at least partially) the problem of covering semantic relations in different domains. This choice has an impact
on the number of the possible covered relations since by selecting concepts in different topics it is more likely to pick concepts linked by many different semantic
relations.

From a practical point of view, we used a Web application for collecting the
feedback of 10 judges who had to provide their feedback for 420 triples in a
month. By using our application the judges were allowed for each triple t_i to
both take a look at the description of the concepts included in t_i (the gloss
available in the corresponding Wikipedia page) and to select one among three
possible responses: c_{i1}, c_{i2} and the *I DON'T KNOW* option (for skipping the
response).

4.2   Evaluating the Precision of SR Methods

By following the idea that potential ambiguities can be discovered by taking
into account the agreement among the judges, we defined a filter in order to
throw out from our evaluation ambiguous triples. In order to reach this aim, for
each triple t_i, we compute: an *agreement score* $A(t_i)$ and *indecision score* $I(t_i)$.
In particular, $A(t_i)$ is equal to the maximum between (i) $A(t_i, c_{i1})$ which is the
ratio between the number of judges who voted the concept c_{i1} and the total

number of judges and (ii) $A(t_i, c_{i2})$ which is ratio between the number of judges who voted the concept c_{i2} and the total number of judges. In the rest of the paper we will refer the concept which maximizes the agreement score $A(t_i)$ as c_{max_i}. On the other hand, $I(t_i)$ is computed as the ratio between the number of judges who skipped the triple (by choosing the *I DON'T KNOW* option) and the number of judges. We used these these two values in order to throw out from our analysis possible ambiguities. In particular, we use only the set of triples $FT = (t_1, .., t_n)$ characterized by: an agreement score higher or equal to 0.7 (i.e. we require that at least 7 of the 10 judges provided the same response); an indecision score lower or equal to 0.2 (i.e. we require that at maximum 2 judges skipped the question). In this way we throw out the triples where the responses of the judges are more or less equally divided between the two concepts as well as the triples for which the judges could not identify the most related concept. By following this strategy we removed only 27 triples and two examples of these triples are (*Mammal, Dolphin, Lion*) and (*Lifeguard, Holiday, Work*).

We use the triples in FT for evaluating the precision of the SR methods. Our metric, named *Order Count*, aims at checking if the evaluated SR methods order the concepts in the triples exactly as the judges did. In order to reach this aim, for each triple t_i, we use the evaluated SR method for computing the semantic relatedness between $target_i$ and c_{i1} (named $SR(target_i, c_{i1})$) and the semantic relatedness between $target_i$ and c_{i2} (named $SR(target_i, c_{i1})$)). In particular, our metric counts the number of times that the evaluated SR method computes a higher value for $SR(target_i, C_{max_i})$ of each triple t_i in FT.

5   Evaluation

In this section we focus on assessing if the proposed approach is useful to evaluate the precision of SR methods. In order to reach this aim we both implemented and modified some state of the art SR methods (as described in Section 5.1) in order to have a pool of SR methods. We compare our results to the state of the art approaches by estimating the significance of our results in Section 5.2.

5.1   The Evaluated SR Methods

Two methods, referred in this paper as *COUT* and *GDIN*, are proposed in [13]. The COUT metric describes each concept as a weighted vector of Wikipedia pages: given a concept of Wikipedia, the pages linked by the concept describe it and the weight of each page is equal to $log(|W| / |P|)$ where W is the set of pages in Wikipedia and P is the number of articles linked by the page. Given such representation of concepts, the COUT metric computes the semantic relatedness between two concepts as the cosine similarity between the two corresponding vectors. We modified the COUT approach by defining the *CIN* metric which represents a concept by means of the Wikipedia pages with a link to the concept. In this case, the weight of a page in the vector is equal to $log(|W| / |P|)$ where W is the set of pages in Wikipedia and P is the number of articles linked by the page and the cosine similarity is still used to compute the semantic relatedness.

The GDIN metric, on the other hand, adapts the Google Distance measure [4] in order to compute the semantic relatedness among the concepts a and b of Wikipedia as

$$GDIN = 1 - \frac{log\,(max\,(|A|,|B|)) - log\,(|A \cap B|)}{log\,(|W|) - log\,(min\,(|A|,|B|))}$$

where A is the set of pages with a link to concept a, B is the set of pages with a link to concept b, and W is the set of pages available in Wikipedia. We modified also the GDIN metric in order to produce the GDOUT method where A is the set of pages linked by the concept a, B is the set of pages linked by the concept b, and W is still the set of pages available in Wikipedia.

5.2   Results

The agreement among the judges is used in the literature in order to assess if datasets are reliable. We have specifically exploited the Fleiss' kappa for estimating the agreement among the judges. This analysis showed a very significant agreement among the evaluators (kappa=0.783) and that is why the pruning step (executed in order to identify ambiguities) removed only 27 triples from the initial set of 420 triples T.

We use also the Related353 dataset for computing the precision of the SR methods described in the Section 5.1. Since this dataset contains terms we cannot compute the semantic relatedness between fixed Wikipedia pages and to face this issue we adopt a strategy utilized in the literature. In particular, given a SR method and a pair of terms, we compute the semantic relatedness between all the Wikipedia pages associated to the two terms in the pair. The highest computed score is taken for the pair in order to compute both the Pearson and the Spearman coefficients. We use these two coefficients for ranking the SR methods and we show these rankings as well as the coefficients (reported in the parenthesis) in Table 1. In the same table we also report the results produced by using our approach/dataset.

Table 1. The results of the compared metrics

	Pearson	Spearman	Order Count
1	GDIN (0.555)	GOUT (0.5)	CIN (0.87)
2	GOUT (0.473)	COUT (0.497)	GDIN (0.85)
3	CIN (0.472)	GDIN (0.49)	COUT (0.83)
4	COUT (0.479)	CIN (0.48)	GDOUT (0.8)

It is interesting to observe that the Pearson and the Spearman correlations produce completely different results also if the coefficients are very near. However, the ranking produced by our approach is still different from the ones produced by the Pearson and the Spearman metrics.

In order to evaluate if our approach makes sense, we integrated the 4 SR methods in a Collaborative Filtering (CF) recommender system as described in [5]. We computed the *Mean Average Error (MAE)* of the predictions of the CF system by using the MovieLens dataset with a 5-cross fold validation evaluation. According to this analysis, the CF recommender system produces the most accurate results when the CIN method is integrated in the recommender (MAE=0.735). This confirms the result predicted by our approach by showing that our approach has sense. Moreover, it seems to show that the methods which represent a Wikipedia concept by using the incoming links are more precise than others.

6   Conclusion

In this paper we surveyed the main limitations of the state of the art mechanisms aimed at evaluating the precision of SR methods. Starting from this analysis we proposed a new approach and a new dataset for evaluating the accuracy of SR methods which compute the semantic relatedness between concepts of Wikipedia. Our results differ from the outcomes produced by the approaches described in the literature. However, we showed that our results makes sense and for this reason both the built dataset and the evaluation approach can be useful tools for evaluating SR methods. Future works will use crowdsourcing systems for collecting feedback from a larger set of judges in order to better evaluate the significance of the proposed approach. We are also interested in associating concepts defined in other knowledge sources (such as Wordnet) in order to extend our evaluation to mechanisms which use different knowledge sources.

References

1. Agirre, E., Cer, D., Diab, M., Gonzalez-Agirre, A.: Semeval-2012 task 6: A pilot on semantic textual similarity. In: *SEM 2012: The First Joint Conference on Lexical and Computational Semantics – Proceedings of the Sixth International Workshop on Semantic Evaluation, June 7-8, pp. 385–393. Association for Computational Linguistics, Montréal (2012)
2. Boyd-graber, J., Fellbaum, C., Osherson, D., Schapire, R.: Adding dense, weighted connections to wordnet. In: Proceedings of the Third International WordNet Conference (2006)
3. Budanitsky, A., Hirst, G.: Evaluating wordnet-based measures of lexical semantic relatedness. Comput. Linguist. 32(1), 13–47 (2006)
4. Cilibrasi, R.L., Vitanyi, P.M.B.: The google similarity distance. IEEE Trans. on Knowl. and Data Eng. 19(3), 370–383 (2007)
5. Ferrara, F., Tasso, C.: Integrating semantic relatedness in a collaborative filtering system. In: Proceedings of the 19th Int. Workshop on Personalization and Recommendation on the Web and Beyond, pp. 75–82 (2012)
6. Finkelstein, L., Gabrilovich, E., Matias, Y., Rivlin, E., Solan, Z., Wolfman, G., Ruppin, E.: Placing search in context: the concept revisited. ACM Trans. Inf. Syst. 20(1), 116–131 (2002)

7. Gabrilovich, E., Markovitch, S.: Computing semantic relatedness using wikipedia-based explicit semantic analysis. In: Proceedings of the 20th International Joint Conference on Artifical Intelligence, IJCAI 2007, pp. 1606–1611. Morgan Kaufmann Publishers Inc., San Francisco (2007)
8. Gracia, J.L., Mena, E.: Web-Based Measure of Semantic Relatedness. In: Bailey, J., Maier, D., Schewe, K.-D., Thalheim, B., Wang, X.S. (eds.) WISE 2008. LNCS, vol. 5175, pp. 136–150. Springer, Heidelberg (2008)
9. Gurevych, I.: Using the Structure of a Conceptual Network in Computing Semantic Relatedness. In: Dale, R., Wong, K.-F., Su, J., Kwong, O.Y. (eds.) IJCNLP 2005. LNCS (LNAI), vol. 3651, pp. 767–778. Springer, Heidelberg (2005)
10. Hayes, J., Veale, T., Seco, N.: Enriching wordnet via generative metonymy and creative polysemy. In: Proceedings of the Fourth International Conference on Language Resources and Evaluation, pp. 149–152. European Language Resources Association (2004)
11. Lin, D.: Automatic retrieval and clustering of similar words. In: Proceedings of the 36th Annual Meeting of the Association for Computational Linguistics and 17th International Conference on Computational Linguistics, ACL 1998, vol. 2, pp. 768–774. Association for Computational Linguistics, Stroudsburg (1998)
12. Miller, G.A., Charles, W.G.: Contextual correlates of semantic similarity. Language and Cognitive Processes 6(1), 1–28 (1991)
13. Milne, D., Witten, I.H.: An effective, low-cost measure of semantic relatedness obtained from wikipedia links. In: Proceeding of AAAI Workshop on Wikipedia and Artificial Intelligence: an Evolving Synergy, pp. 25–30. AAAI Press (2008)
14. Nikolova, S., Boyd-Graber, J., Fellbaum, C.: Collecting Semantic Similarity Ratings to Connect Concepts in Assistive Communication Tools. In: Mehler, A., Kühnberger, K.-U., Lobin, H., Lüngen, H., Storrer, A., Witt, A. (eds.) Modeling, Learning, and Proc. of Text-Tech. Data Struct. SCI, vol. 370, pp. 81–93. Springer, Heidelberg (2011)
15. Pedersen, T., Pakhomov, S.V.S., Patwardhan, S., Chute, C.G.: Measures of semantic similarity and relatedness in the biomedical domain. Journal of Biomedical Informatics 40(3), 288–299 (2007)
16. Resnik, P.: Using information content to evaluate semantic similarity in a taxonomy. In: Proceedings of the 14th International Joint Conference on Artificial Intelligence, IJCAI 1995, vol. 1, pp. 448–453. Morgan Kaufmann Publishers Inc., San Francisco (1995)
17. Rubenstein, H., Goodenough, J.B.: Contextual correlates of synonymy. Commun. ACM 8(10) (October 1965)
18. Strube, M., Ponzetto, S.P.: Wikirelate! computing semantic relatedness using wikipedia. In: Proceedings of the 21st National Conference on Artificial Intelligence, AAAI 2006, vol. 2, pp. 1419–1424. AAAI Press (2006)
19. Zesch, T., Gurevych, I.: Automatically creating datasets for measures of semantic relatedness. In: Proceedings of the Workshop on Linguistic Distances, LD 2006, pp. 16–24. Association for Computational Linguistics, Stroudsburg (2006)
20. Zesch, T., Gurevych, I.: The more the better? assessing the influence of wikipedia's growth on semantic relatedness measures. In: Chair, N.C.C., Choukri, K., Maegaard, B., Mariani, J., Odijk, J., Piperidis, S., Rosner, M., Tapias, D. (eds.) Proceedings of the Seventh International Conference on Language Resources and Evaluation. European Language Resources Association, Valletta (2010)
21. Zesch, T., Gurevych, I.: Wisdom of crowds versus wisdom of linguists; measuring the semantic relatedness of words. Nat. Lang. Eng. 16(1), 25–59 (2010)

Similarity Measures Based on Latent Dirichlet Allocation

Vasile Rus, Nobal Niraula, and Rajendra Banjade

Department of Computer Science, The University of Memphis
Memphis, TN, 38152, USA
{vrus,nobal}@memphis.edu

Abstract. We present in this paper the results of our investigation on semantic similarity measures at word- and sentence-level based on two fully-automated approaches to deriving meaning from large corpora: Latent Dirichlet Allocation, a probabilistic approach, and Latent Semantic Analysis, an algebraic approach. The focus is on similarity measures based on Latent Dirichlet Allocation, due to its novelty aspects, while the Latent Semantic Analysis measures are used for comparison purposes. We explore two types of measures based on Latent Dirichlet Allocation: measures based on distances between probability distribution that can be applied directly to larger texts such as sentences and a word-to-word similarity measure that is then expanded to work at sentence-level. We present results using paraphrase identification data in the Microsoft Research Paraphrase corpus.

Keywords: semantic similarity, paraphrase identification, unsupervised meaning derivation.

1   Introduction

We address in this paper the problem of semantic similarity between two texts. The task of semantic similarity is about making a judgment with respect to how semantically similar two texts are. The judgment can be quantitative, e.g. a normalized score, or qualitative, e.g. one text is (or is not) a paraphrase of the other.

The problem of semantic similarity is a central topic in Natural Language Processing as it is important to a number of applications such as providing evidence for the correctness of answers in Question Answering [1], increase diversity of generated text in Natural Language Generation [2], assessing the correctness of student responses in Intelligent Tutoring Systems [3, 4], or identifying duplicate bug reports in Software Testing [5].

We focus in this paper on the problem of semantic similarity at word- and sentence- level. At word-level, the task is about judging how similar two words are, e.g. *procedure* and *technique*. As an example of sentence-to-sentence similarity [6, 7], we show below a pair of sentences from the Microsoft Research Paraphrase (MSRP; [8]). The shown example constitutes a positive instance of a paraphrase in MSRP (to be precise, it is instance #51 in MSRP test data).

A. Gelbukh (Ed.): CICLing 2013, Part I, LNCS 7816, pp. 459–470, 2013.
© Springer-Verlag Berlin Heidelberg 2013

Text A: *The procedure is generally performed in the second or third trimester.*

Text B: *The technique is used during the second and, occasionally, third trimester of pregnancy.*

We propose novel solutions to the task of semantic similarity both at word and sentence level. We rely on probabilistic and algebraic methods that can automatically derive word and sentence meaning representations from large collection of texts in the form of latent topics or concepts. The probabilistic method we use is Latent Dirichlet Allocation (LDA, [9]). LDA models documents as topic distributions and topics as distributions over words in the vocabulary. Each word has a certain contribution to a topic. Based on these distributions and contributions we define both word-to-word semantic similarity measures and text-to-text semantic similarity measures. The LDA-based word-to-word semantic similarity measure is further used in conjunction with a greedy and optimal matching method to measure similarity between larger texts such as sentences. The text-to-text measures are directly used to compute the similarity of texts such as between two sentences, the focus of our work.

The proposed semantic similarity solutions based on LDA are compared with solutions based on Latent Semantic Analysis [10]. Like LDA, LSA is fully automated. LSA starts with a term-document matrix that represents the distribution of words in documents and the distribution of documents over the words. The term vectors (as well as the document vectors) in the original term-document matrix are mapped using the mathematical procedure of Singular Value Decomposition (SVD) into a reduced dimensionality space (300-500 dimensions). Words are represented as vectors in this LSA semantic space whose dimensions form latent semantic concepts. Documents are also represented as vectors in the reduced space. Similarity of individual words and texts are computed based on vector algebra. For instance, the similarity between two words is computed as the cosine (normalized dot-product) between the corresponding word vectors.

Given that both LDA and LSA require the specification of a desired number of latent topics or concepts a priori, an interesting question relates to which of these methods best capture the semantics of words and texts for the same number of topics or concepts. The broader question would be which of these two methods can best capture the meaning of words and texts and in what conditions. This paper is one step in that direction of elucidating the strengths and weaknesses of these methodologies for meaning inference in the context of the paraphrase identification.

We have experimented with the above methods using the Microsoft Research Paraphrase corpus [8]. We provide experimental results on this data set using both a greedy method and an optimal matching method based on the job assignment problem, a famous combinatorial optimization problem.

The rest of the paper is organized as in the followings. The next section provides an overview of related work. Then, we describe the semantic similarity measures based on LDA. The Experiments and Results section describes our experimental setup and the results obtained. We conclude the paper with Discussion and Conclusions.

2   Related Work

The task of semantic similarity can be formulated at different levels of granularity ranging from word-to-word similarity to sentence-to-sentence similarity to document-to-document similarity or a combination of these such as word-to-sentence or sentence-to-document similarity. We will first review word-level semantic similarity measures followed by sentence-level measures. It should be noted that some approaches, such as LSA, are directly applicable at both at word- and sentence-level. Standard word-to-word similarity can be expanded to larger texts through some additional mechanism [7, 8]. It is important to add that the review of related work that follows is by no means an exhaustive one.

Research on semantic similarity of texts focused initially on measuring similarity between individual words. A group of word-to-word similarity measures were defined that use lexico-semantic information in WordNet [11]. WordNet is a lexical database of English that groups together words that have the same meaning, i.e. synonyms, into synsets (synonymous sets). Synsets are also referred to as concepts. For instance, the synset of {affectionate, fond, lovesome, tender, warm} corresponds to the concept of (having or displaying warmth or affection), which is the definition of the concept in WordNet.

There are nearly a dozen WordNet-based similarity measures available [12]. These measures are usually divided into two groups: similarity measures and relatedness measures. The similarity measures are limited to within-category concepts and usually they work only for the nouns and verbs categories. The text relatedness measures on the other hand can be used to compute similarity among words belonging to different categories, e.g. between a noun and an adjective.

Examples of word relatedness measures (implemented in the WordNet::Similarity package [12]) are: HSO [13], LESK [14], and VECTOR [15]. Given two WordNet nodes, i.e. concepts, these measures provide a real value indicating how semantically related the two concepts are. The HSO measure is path based, i.e. uses the relations between concepts, and assigns direction to relations in WordNet. For example, is-a relation is upwards, while has-part relation is horizontal. The LESK and VECTOR measures are gloss-based. That is, they use the text of the gloss as the source of meaning for the underlying concept.

One challenge with the above word-to-word relatedness measures is that they cannot be directly applied to compute similarity of larger texts such as sentences. Researchers have proposed methods to extend the word-to-word (W2W) relatedness measures to text-to-text (T2T) relatedness measures [7, 8]. Another challenge with the WordNet similarity measures is the fact that texts express meaning using words and not concepts. To be able to use the WordNet-based word-to-word similarity measures, the words must be mapped to concepts in WordNet, i.e. the word sense disambiguation (WSD) problem must be solved. Other solutions can be adopted such as selecting the most frequent sense for each word or even trying all possible senses [16]. Our proposed word-to-word similarity measure based on Latent Dirichlet Allocation or LSA-based word-level measures do not have this latter challenge.

Given its importance to many semantic tasks such as paraphrase identification [8] or textual entailment [17], the semantic similarity problem at sentence level has been addressed using various solutions that range from simple word overlap to greedy methods that rely on word-to-word similarity measures [7] to algebraic methods [18] to machine learning based solutions [19].

The most relevant work to ours, [18], investigated the role of Latent Semantic Analysis [10] in solving the paraphrase identification task. LSA is a vectorial representation in which a word is represented as a vector in a reduced dimensionality space, where each dimension is believed to be representative of an abstract/latent semantic concept. Computing the similarity between two words is equivalent to computing the cosine, i.e. normalized dot product, between the corresponding word vectors. The challenge with such vectorial representations is the derivation of the semantic space, i.e. discovering the latent dimensions or concepts of the LSA space. In our work, we experimented with an LSA space computed from the TASA corpus (compiled by Touchstone Applied Science Associates), a balanced collection of representative texts from various genres (science, language arts, health, economics, social studies, business, and others).

Two different ways to compute semantic similarity between two texts based on LSA were proposed in [18]. First, they used LSA to compute a word-to-word similarity measure which then they combined with a greedy-matching method at sentence level. For instance, each word in one sentence was greedily paired with a word in the other sentence. An average of these maximum word-to-word similarities was then assigned as the semantic similarity score of the two sentences. Second, LSA was used to directly compute the similarity of two sentences by applying the cosine (normalized dot product) of the LSA vectors of the sentences. The LSA vector of a sentence was computed using vector algebra, i.e. by adding up the vectors of the individual words in a sentence. We present later results with these methods as well as with a method based on optimal matching.

LDA was rarely used for semantic similarity. To the best of our knowledge LDA has not been used so far for addressing the task of paraphrase identification, which we address here. The closest use of LDA for a semantic task was for ranking answers to a question in Question Answering (QA; [20]). Given a question, they ranked candidate answers based on how similar these answers were to the target question. That is, for each question-answer pair they generated an LDA model which they then used to compute a degree of similarity (DES) that consists of the product of two measures: sim1 and sim2. Sim1 captures the word-level similarities of the topics present in an answer and the question. Sim2 measures the similarities between the topic distributions in an answer and the question. The LDA model was generated based solely on each question and candidate answers. As opposed to our task in which we compute the similarity between sentences, the answers to questions in [20] are longer, consisting of more than one sentence. For LDA, this particular difference is important when it comes to semantic similarity as the shorter the texts the sparser the distributions, e.g. the distribution over topics in the text, based on which the similarity is computed.

Another use of LDA for computing similarity between blogs relied on a very simple measure of computing the dot product of topic vectors as opposed to a similarity of distributions [21].

Similar to [20], we define several semantic similarity measures based on various distributions used in the LDA model. We do use Information Radius as [20] and, in addition, propose similarity measures based on Hellinger and Manhattan distances.

Furthermore, we use LDA for measuring word-to-word similarities and use these values in a greedy and optimal matching method at sentence-level. Finally, we compare the results with LSA-based results. The comparison is so designed to be as informative as possible, e.g. the number of topics in LDA matches the number of latent concepts/dimensions in LSA.

It should be noted that LDA has a conceptual advantage over LSA. LDA models explicitly different meaning of words, i.e. it handles polysemy. In LDA, each topic is a set of words that together define a meaning or concept, i.e. a word sense. A word belongs to different topics, which can be regarded as its various senses. On the other hand, LSA has a unique representation for a word. That is, all meanings of a word are represented by the same LSA vector making difficult to infer which meaning is being represented. Some argue that the LSA vector represents that dominant meaning of a word while others believe the LSA vector represents an average meaning of all meanings of the word.

3   Similarity Measures Based on Latent Dirichlet Allocation

As we already mentioned, LDA is a probabilistic generative model in which documents are viewed as distributions over a set of topics and each word in a document is generated based on a distribution over words that is specific to each topic. We denote by θ the distributions over topics and by φ distributions over words. Full details about the LDA model can be found in [9].

A first semantic similarity measure among words would then be defined as a dot-product between the corresponding vectors representing the contributions of each word to a topic ($\varphi_t(w)$ – represents the contribution of word w to topic t). It should be noted that the contributions of each word to the topics does not constitute a distribution, i.e. the sum of contributions does not add up to 1. Assuming the number of topics T, a word-to-word measure is defined by the formula below.

$$LDA - w2w(w,v) = \sum_{t=1}^{T} \varphi_t(w)\varphi_t(v)$$

More global text-to-text similarity measures could be defined based on the distributions over topics (θ) and distributions over words (φ) defined by LDA. Because a document is a distribution over topics, the similarity of two texts needs to be computed in terms of similarity of distributions. The Kullback-Leibler (KL) divergence defines a distance, or how dissimilar, two distributions p and q are as in the formula below.

$$KL(p,q) = \sum_{i=1}^{T} p_i \log \frac{p_i}{q_i}$$

If we replace p with θ_d (text/document d's distribution over topics) and q with θ_c (text/document c's distribution over topics) we obtain the KL distance between two documents (documents d and c in our example). Furthermore, KL can be used to compute the distance between two topics using their distributions over words (φ_{t1} and

φ_{t2}). The KL distance has two major problems. In case q_i is zero KL is not defined. Furthermore, KL is not symmetric which does not fit well with semantic similarity measures which in general are symmetric. That is, if text A is a paraphrase of text B that text B is a paraphrase of text A. The Information Radius measure solves these problems by considering the average of p_i and q_i as below.

$$IR(p,q) = \sum_{i=1}^{T} p_i \log \frac{2 \times p_i}{p_i + q_i} + \sum_{i=1}^{T} q_i \log \frac{2 \times q_i}{p_i + q_i}$$

The IR can be transformed into a similarity measure as in the following (Dagan, Lee, & Pereira, 1997):

$$SIM(p,q) = 10^{-\delta R(c,d)}$$

All our results reported in this paper for LDA similarity measures between two documents c and d are computed by multiplying the similarities between the distribution over topics (θ_d and θ_c) and distribution over words (φ_{t1} and φ_{t2}). That is, two texts would be similar if the texts have similar topic distributions and the topics are similar themselves, i.e. have similar word distributions.

The Hellinger distance between two distributions is another option that allows avoiding the shortcomings of the KL distance.

$$HD(p,q) = \frac{1}{\sqrt{2}} \sqrt{\sum_{1}^{T} (\sqrt{p_i} - \sqrt{q_i})^2}$$

The Hellinger distance varies from 0 to 1 and is defined for all values of p_i and q_i. A value of 1 means the distance is maximum and thus the distributions are very different. A value of 0 means the distributions are very similar. We can transform the Hellinger distance into a similarity measure by subtracting it from 1 such that a zero distance means a large similarity score and vice versa.

Lastly, we used the Manhattan distance between distributions p and q as defined below.

$$MD(p,q) = 2 \times (1 - \sum_{1}^{T} \min(p_i, q_i))$$

MD is symmetric, defined for any values of p and q, and ranges between 0 and 2. We can divide MD by 2 and subtract from 1 to transform it into a similarity measure.

4   Experimental Setup and Results

We present in this section results with the previously described methods on the Microsoft Research Paraphrase corpus (MSRP; [8]). The MSRP corpus is the largest publicly available annotated paraphrase corpus and has been used in most of the recent studies that addressed the problem of paraphrase identification. The corpus consists of 5,801 sentence pairs collected from newswire articles, 3,900 of which were labeled as paraphrases by human annotators. The whole set is divided into a training subset (4,076 sentences of which 2,753, or 67.5%, are true paraphrases), and a test subset (1,725 pairs of which 1,147, or 66.5%, are true paraphrases). The average number of words per sentence is 17.

A simple baseline for this corpus is the majority baseline, where all instances are labeled with the majority class label in the training corpus, which in MSRP is the positive class label. The baseline gives an accuracy and precision of 66.5% and perfect recall.

For the proposed methods, we first present results obtained using 300 dimensions for the LSA space, a standard value, and a similar number of topics for LDA. This number of dimensions has been empirically established by LSA researchers. Then, we vary the number of topics for the LDA model and observe changes in performance. Fewer topics usually means semantically less coherent topics as many words with different senses will be grouped under the same topic. More topics on the other hand would lead to somehow more coherent topics (this is an open research question, actually) but also sparser topic distributions for short texts as exemplified later.

We follow a training-testing methodology according to which we first train the proposed methods on a set of training data after which we use the learned models on testing data. In our case, we learn a threshold for the text-to-text similarity score above which a pair of sentences is deemed a paraphrase and any score below the threshold means the sentences are not paraphrases. We report performance of the various methods using accuracy (percentage of correct predictions), precision (the percentage of correct predictions out of the positive predictions), recall (percentage of correct predictions out of all true positives), F-measure (harmonic mean of precision and recall), and kappa statistics (a measure of agreement between our method's output and experts' labels while accounting for chance agreement).

We experimented with both word-to-word similarity measures and text-to-text similarity measures. The word-to-word similarity measures were expanded to work at sentence level using the two methods described next: greedy matching and optimal matching. For LDA, we used the word-to-word measure and text-to-text measures described in the previous section. For LSA, we use the cosine between two words' LSA vectors as a measure of word-to-word similarity. For LSA-based text-to-text similarity we first add up the word vectors for all the words in a text thus obtaining two vectors, one for each text, and then compute the cosine between these two text vectors.

Greedy Matching

In the greedy approach words from one sentence (usually the shorter sentence) are greedily matched, one by one, starting from the beginning of the sentence, with the most similar word from the other sentence. In case of duplicates, the order of the words in the two sentences was important such that the first occurrence matches with the first occurrence and so on. To be consistent across all methods presented here and for fairness of comparison across these methods, we require that words must be part of at most one pair.

The greedy method has the advantage of being simple. The obvious drawback of the greedy method is that it does not aim for a global maximum similarity score. The optimal method described next solves this issue.

Optimal Matching

The optimal method aims at finding the best overall word-to-word match, based only on the similarities between words. This is a well-known combinatorial optimization problem. The assignment problem is one of the fundamental combinatorial optimization problems and consists of finding a maximum weight matching in a weighted bipartite graph. Given a complete bipartite graph, $G = (S, T, E)$, with n worker vertices (S), n job vertices (T), and each edge $e_{s \in S, t \in T} \in E$ having a non-negative weight $w(s, t)$ indicating how qualified a worker is for a certain job, the task is to find a matching M from S to T with maximum weight. In case of different numbers of workers or jobs, dummy vertices could be used.

The assignment problem can be formulated as finding a permutation π for which $S_{OPT} = \sum_{i=1}^{n} w(s_i, t_{\pi(i)})$ is maximum. Such an assignment is called optimum assignment. An algorithm, the Kuhn-Munkres method [22, 23], has been proposed that can find a solution to it in polynomial time.

In our case, we model the semantic similarity problem as finding the optimum assignment between words in one text, T_1, and words in another text, T_2, where the fitness between words in the texts can be measured by any word-to-word semantic similarity function. That is, we are after a permutation π for which $S_{OPT} = \sum_{i=1}^{n} word\text{-}sim(v_i, w_{\pi(i)})$ is maximum where $word\text{-}sim$ can be any word-to-word similarity measure, and v and w are words from the texts T_1 and T_2, respectively. In our case, the word-sim are the word-to-word measures based on LDA and LSA.

Results

A summary of our experiments is shown in Table 1. These are results on MSRP test data obtained using the threshold for similarity that corresponds to the threshold learned from training data. The threshold that led to best accuracy on training data was selected. The threshold varied from method to method. The results obtained using the word-to-word similarity measures are labeled Greedy and Optimal in Table 1. The row labeled LSA shows results obtained when text-level LSA vectors were used, as explained earlier. The *Baseline* method indicates performance when labeling all instances with the dominant label of a true paraphrase. The rest of the rows in the table show results when the text-to-text similarity measures based on various distribution distances were used: IR (Information Radius), Hellinger, and Manhattan.

The LDA-Optimal offers competitive results. It provides best precision and kappa score. As noted from the table, the text-to-text similarity measures based on distribution distances perform close to chance. The problem seems to be rooted in the relative size of texts compared to the number of the topics in the LDA model. In the basic model, we used 300 topics (similar to the 300 dimensions used for LSA) for comparison purposes. The average sentence size in MSRP (after removing stopwords) is 10.3 for training data and 10.4 for testing data. That means that in a typical sentence

Table 1. Results on the MSRP test data

Method	Accuracy	Precision	Recall	F-Measure	Kappa
Baseline	66.55	66.53	1	79.90	0.22
LSA	**73.56**	75.34	**89.53**	**81.83**	34.61
LSA Greedy	72.86	75.50	87.61	81.11	33.89
LSA Optimal	73.04	76.72	85.35	80.80	35.95
LDA-IR	67.47	67.27	99.47	80.26	4.52
LDA-Hellinger	67.36	67.25	99.21	80.16	4.39
LDA-Manhattan	66.78	67.12	98.08	79.70	3.56
LDA-Greedy	73.04	76.07	86.74	81.05	35.01
LDA-Optimal	73.27	**77.05**	85.17	80.91	**36.74**

Table 2. Topic assignment for instance #23 in MSRP test data

Word	Topic	Word	Topic
senator	8	ashamed	1
Clinton	3	playing	7
ashamed	1	politics	8
playing	7	important	10
politics	8	issue	8
important	9	state	8
issue	8	budget	2
homeland	8	division	11
security	2	spokesman	1
funding	8	Andrew	3
		Rush	5

most of the 300 topics will not be assigned to any word leading to very similar topic distributions over the entire set of 300 topics. Even if the probability for topics that are not assigned to a word in a sentence is set to 0, the distance between two values of 0 is 0 which means the distributions are quite similar. To better illustrate this issue, we use the example below, which is instance #23 in MSRP test data.

Text A: *Senator Clinton should be ashamed of herself for playing politics with the important issue of homeland security funding, he said.*

Text B: *She should be ashamed of herself for playing politics with this important issue, said state budget division spokesman Andrew Rush.*

Table 2 shows the topic assignment for each of the non-stop words in the two sentences when using an LDA model of just 12 topics. We used this time 12 topics as it compares to the relative size of sentences in MSRP. We wanted to check whether using fewer topics somehow reduces the topic sparseness problem in short texts. Even in this case, because different words are assigned to the same topic, e.g. *senator*, *politics*, and *homeland* are all assigned to topic 8 which is about politics, there are some topics that do not occur in the sentence.

Figure 1 shows distributions of the 12 topics for the two sentences in the example above (left) and the two sentences in the example given in Introduction (right). From the figure, we notice that for the example on the right, more than half of the topics have identical probabilities in the two sentences (these are topics that are not assigned

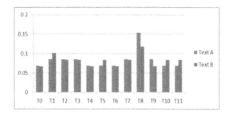
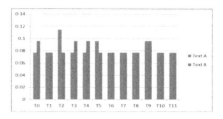

Fig. 1. Examples of topic distributions for the two sentences in instance #23 in MSRP test data (left) and the two sentences in the example given in Introduction (right)

Topic 2:	
number	0.014
money	0.012
system	0.010
business	0.009
information	0.009
special	0.009
set	0.009
job	0.009
amount	0.008
general	0.008

Topic 7:	
day	0.029
good	0.021
thought	0.0194
school	0.017
home	0.017
children	0.015
father	0.014
knew	0.013
told	0.0131
hard	0.011

Topic 8:	
states	0.020
world	0.019
united	0.015
government	0.013
american	0.012
state	0.012
war	0.011
power	0.009
president	0.008
groups	0.007

Fig. 2. Examples of topics and distributions overs words in three topics (top 10 words are shown for each topic)

Table 3. Results on MSRP test data with LDA-based methods for various number of topics

Method	Accuracy/Kappa (T=300)	Accuracy/Kappa (T=40)	Accuracy/Kappa (T=12)	Accuracy/Kappa (T=6)
LDA-IR	67.47/4.52	67.94/5.88	66.66/3.11	67.13/9.87
LDA-Hellinger	67.36/4.39	67.82/5.22	67.13/10.62	66.95/13.35
LDA-Manhattan	66.78/3.56	**68.23**/8.12	66.78/8.44	67.01/9.93
LDA-Greedy	73.04/35.01	72.63/34.72	71.94/26.20	71.71/32.36
LDA-Optimal	**73.27/36.74**	**73.50/36.77**	72.11/31.19	71.53/31.63

to any words). This topic sparseness issue leads to short distances between the topic distributions which in turn leads to very high similarity scores (low distance means the texts are close to each other semantically, i.e. high similarity). Figure 2 shows tops ten words in topics assigned to words in Table 2.

To further investigate the impact of number of topics on the behavior of the LDA-based measures, we experimented with LDA models that use 300, 40, 12, and 6 topics. The performance of these models on the MSRP test data is provided in Table 2. While the results slightly improve for the text-to-text similarity measures based on distribution distances the performance is still modest. The best results are obtained for T=40 topics.

5   Discussion, Conclusions, and Future Work

We presented in this paper our work with LDA-based semantic similarity measures. The conclusion of our investigation is that the word-to-word LDA-based measure (defined as the dot-product between the corresponding topic vectors) leads to competitive results. The proposed measure outperforms the algebraic method of LSA, in particular when combined with the Optimal matching method. As noted, LDA has the conceptual advantage of handling polysemy.

The distance-based similarity measures suffer from a data sparseness problem which could be alleviated if the number of topics used in the underlying Latent Dirichlet Allocation model is reduced to a number that is comparable or smaller than the average length of the texts for which a semantic similarity judgment is sought. We plan to further investigate and address the topic sparseness issue for the similarity measures based on distance among distributions. For instance, we would like to explore methods that try to optimize topic coherence, e.g. based on pointwise mutual information (PMI; [24]).

Another future line of work we would like to pursue is about combining LDA's topic distribution in a document with LSA-based word-to-word similarity. For instance, we would apply and use the LSA word-to-word similarity between two words only when an LDA model assigns the same topic to the two words, i.e. we would rely on the LSA score only when LDA indicates the two words have the same meaning as represented by an LDA topic.

Acknowledgments. This research was supported in part by Institute for Education Sciences under awards R305A100875. Any opinions, findings, and conclusions or recommendations expressed in this material are solely the authors' and do not necessarily reflect the views of the sponsoring agencies.

References

1. Ibrahim, A., Katz, B., Lin, J.: Extracting structural paraphrases from aligned monolingual corpora. In: Proceedings of the Second International Workshop on Paraphrasing (ACL 2003)
2. Iordanskaja, L., Kittredge, R., Polgere, A.: Lexical selection and paraphrase in a meaning-text generation model. In: Natural Language Generation in Artificial Intelligence and Computational Linguistics. Kluwer Academic (1991)
3. Graesser, A.C., Olney, A., Haynes, B.C., Chipman, P.: Autotutor: A cognitive system that simulates a tutor that facilitates learning through mixed-initiative dialogue. In: Cognitive Systems: Human Cognitive Models in Systems Design. Erlbaum, Mahwah (2005)
4. Rus, V., Graesser, A.C.: Deeper natural language processing for evaluating student answers in intelligent tutoring systems. In: Paper Presented at the Annual Meeting of the American Association of Artificial Intelligence (AAAI 2006), Boston, MA, July 16-20 (2006)
5. Rus, V., Nan, X., Shiva, S., Chen, Y.: Clustering of Defect Reports Using Graph Partitioning Algorithms. In: Proceedings of the 20th International Conference on Software and Knowledge Engineering, Boston, MA, July 2-4 (2009)
6. Fernando, S., Stevenson, M.: A semantic similarity approach to paraphrase detection. In: Proceedings of the Computational Linguistics UK, CLUK 2008 (2008)

7. Lintean, M., Rus, V.: Measuring Semantic Similarity in Short Texts through Greedy Pairing and Word Semantics. In: Proceedings of the 25th International Florida Artificial Intelligence Research Society Conference, Marco Island, FL (2012)
8. Dolan, B., Quirk, C., Brockett, C.: Unsupervised Construction of Large Paraphrase Corpora: Exploiting Massively Parallel News Sources. In: COLING 2004 (2004)
9. Blei, D.M., Ng, A.Y., Jordan, M.I.: Latent dirichlet allocation. The Journal of Machine Learning Research 3, 993–1022 (2003)
10. Landauer, T., McNamara, D.S., Dennis, S., Kintsch, W.: Handbook of Latent Semantic Analysis. Erlbaum, Mahwah (2007)
11. Miller, G.: Wordnet: a lexical database for English. Communications of the ACM 38(11), 39–41 (1995)
12. Pedersen, T., Patwardhan, S., Michelizzi, J.: WordNet: Similarity-Measuring the Relatedness of Concepts. In: The Proceedings of the Nineteenth National Conference on Artificial Intelligence (AAAI 2004), San Jose, CA (Intelligent Systems Demonstration), July 25-29, pp. 1024–1025 (2004)
13. Hirst, G., Stonge, D.: Lexical chains as representations of context for the detection and correction of malapropisms. In: Fellbaum, C. (ed.) WordNet: An Electronic Lexical Database. MIT Press (1998)
14. Banerjee, S., Pedersen, T.: Extended gloss overlaps as a measure of semantic relatedness. In: Proceedings of the Eighteenth International Joint Conference on Artificial Intelligence, pp. 805–810 (2003)
15. Patwardhan, S.: Incorporating dictionary and corpus information into a context vector measure of semantic relatedness. Master's thesis, Univ. of Minnesota, Duluth (2003)
16. Rus, V., Lintean, M., Graesser, A., McNamara, D.: Assessing Student Paraphrases Using Lexical Semantics and Word Weighting. In: Proceedings of the 14th International Conference on Artificial Intelligence in Education, Brighton, UK (2009)
17. Dagan, I., Glickman, O., Magnini, B.: The PASCAL Recognizing Textual Entailment Challenge. In: Proceedings of the Recognizing Textual Entailment Challenge Workshop (2005)
18. Lintean, M., Moldovan, C., Rus, V., McNamara, D.: The Role of Local and Global Weighting in Assessing the Semantic Similarity of Texts Using Latent Semantic Analysis. In: Proceedings of the 23rd International Florida Artificial Intelligence Research Society Conference, Daytona Beach, FL (2010)
19. Kozareva, Z., Montoyo, A.: Paraphrase Identification on the Basis of Supervised Machine Learning Techniques. In: Salakoski, T., Ginter, F., Pyysalo, S., Pahikkala, T. (eds.) FinTAL 2006. LNCS (LNAI), vol. 4139, pp. 524–533. Springer, Heidelberg (2006)
20. Celikyilmaz, A., Hakkani-Tür, D., Tur, G.: LDA Based Similarity Modeling for Question Answering. In: NAACL-HLT, Workshop on Semantic Search, Los Angeles, CA (June 2010)
21. Chen, X., Li, L., Xiao, H., Xu, G., Yang, Z., Kitsuregawa, M.: Recommending Related Microblogs: A Comparison between Topic and WordNet based Approaches. In: Proceedings of the 26th International Conference on Artificial Intelligence (2012)
22. Kuhn, H.W.: The Hungarian Method for the assignment problem. Naval Research Logistics Quarterly 2, 83–97 (1955)
23. Munkres, J.: Algorithms for the assignment and transportation problems. Journal of the Society for Industrial and Applied Mathematics 5(1), 32–38 (1957)
24. Newman, D., Lau, J.H., Grieser, K., Baldwin, T.: Automatic evaluation of topic coherence. In: Human Language Technologies: The Annual Conference of the North American Chapter of the Association for Computational Linguistics, Los Angeles, United States, pp. 100–108 (2010)

Evaluating the Premises and Results of Four Metaphor Identification Systems

Jonathan Dunn

Purdue University
West Lafayette, IN USA
jonathan.edwin.dunn@gmail.com

Abstract. This study first examines the implicit and explicit premises of four systems for identifying metaphoric utterances from unannotated input text. All four systems are then evaluated on a common data set in order to see which premises are most successful. The goal is to see if these systems can find metaphors in a corpus that is mostly non-metaphoric without over-identifying literal and humorous utterances as metaphors. Three of the systems are distributional semantic systems, including a source-target mapping method [1–4]; a word abstractness measurement method [5], [6,7]; and a semantic similarity measurement method [8,9]. The fourth is a knowledge-based system which uses a domain interaction method based on the SUMO ontology [10,11], implementing the hypothesis that metaphor is a product of the interactions among all of the concepts represented in an utterance [12,13].

1 Introduction

This study evaluates four different approaches to metaphor identification. First, each approach's premises and view of metaphor, whether explicit or implicit, are examined in order to understand the differing claims made about metaphor. Second, all four systems are evaluated on a single data set in order to compare their effectiveness. This is important because it helps us to understand which premises are valid (i.e., successful) and which are not (i.e., unsuccessful). Each approach posits certain properties of metaphors that can be used to distinguish metaphors from non-metaphors. The goal of this study is to determine if these properties are essential properties of metaphor or accidental properties that can distinguish metaphors from non-metaphors only in limited data sets.

Humor is used as a counterfactual to metaphor because it contains many of the same properties as metaphor (i.e., connections between different domains) but is interpreted in a very different way. In metaphor, the domains are seen as similar and the interpretation of the utterance involves synthesizing aspects of the two domains. In humor, however, the domains are seen as incongruous and the interpretation of the utterance focuses on dissimilarities between the two domains. For this reason, humor is a useful counterfactual for testing the precision of metaphor identification systems: do the properties posited to be unique to metaphor also show up in humor?

A. Gelbukh (Ed.): CICLing 2013, Part I, LNCS 7816, pp. 471–486, 2013.

This study uses a new data set to provide comparable evaluations of these four systems. The evaluation corpus consists of 25% metaphoric (500), 25% humorous (500), and 50% literal (2,000) utterances taken from the Corpus of Contemporary American English [14]. The evaluation corpus is organized into four top-level domains (ABSTRACT, MENTAL, SOCIAL, PHYSICAL) each of which is represented by instances of five different verbs. This organization ensures wide coverage and allows the results to be examined according to domain membership.

2   Premises of Metaphor Identification Systems

2.1   Possible Choices

The discussion of each metaphor identification system will highlight seven theoretical choices or assumptions which each must make, whether that choice is implicit or explicit. First, is metaphor based on conceptual source-target mappings; if so, are these mappings directly available in the linguistic utterance or are they mediated and thus not directly available? Second, is metaphor a binary or a gradient phenomenon? Third, is lexical meaning best discovered using distributional profiles or using human intuitions; are the two methodologies incompatible? Fourth, are lexical items assumed to point or refer to concepts in the human conceptual system? Fifth, are these concepts organized into domains and, if so, do domains behave differently? Sixth, is metaphor a property possessed by instances of lexical items, by grammatical relations or phrases, or by utterances as a whole? Seventh, do metaphors and non-metaphors belong to distinct populations or do both represent different tails in the distribution of a single population? These questions are discussed in reference to each of the systems as relevant.

2.2   Source-Target Mapping System

This section looks at a verb-noun clustering approach to identifying metaphors [1–4]. The system relies on the view that metaphor consists of a source and a target and that the two metaphorically mapped concepts are directly represented in the surface utterance (e.g., are present in the input sentence, so that no distinction is made between concepts and lexical items). Thus, since metaphor in this view consists of source-target mappings, the metaphor identification task consists in discovering whether these mappings are present or not present (a binary task: an utterance either is or is not metaphoric). The system moves from the linguistic utterance to the underlying conceptual mapping by assuming that the verb directly represents the source domain in the metaphoric mapping and that nouns (functioning as the subject and/or object of the verb) directly represent the target. This assumption is used to avoid the problem of determining which material in a metaphoric utterance is the "literal" material making up the source and which is the "metaphoric" material making up the target. This is a problem that must be faced under this view of metaphor because all that we see in the linguistic utterance is that some elements do not seem to match or go with other elements in literal language.

With this assumption in place, the system invokes the premise of distributional semantics, that the meaning of a lexical item is determined by (or at least described by) the patterns of its use as measured by the clustering of its surface arrangement in a large body of text. Thus, lexical items used in the same surface contexts have the same or similar meanings. We could perhaps posit a weaker distributional semantics premise, in which lexical items which occur in the same contexts have the same meaning and the same grammatical properties; other lexical items which have the same meaning are prevented from occurring in those contexts for syntactic or morphological reasons. The more similar the contexts, the more similar the meanings. The system combines this premise with the idea that there is a difference between the behavior of abstract and physical lexical items (requiring a sharp distinction between ABSTRACT and PHYSICAL domains). While physical lexical items cluster together (e.g., occur in the same contexts) according to their meaning, abstract lexical items cluster together according to their metaphoric association with particular source domains. In other words, these abstract lexical items derive their distributional properties from their metaphoric connection to particular source domains, so that lexical items which have very different meanings occur in the same contexts as a result of taking on the distributional properties of a single source domain.

This approach to metaphor identification is phrase-based, finding metaphors within grammatically related pairs (e.g., verb-object). This contrasts with word-level approaches (see the similarity and abstraction systems below) and utterance-level approaches (see the domain interaction system below). The focus on grammatical (e.g., syntactic) relations raises the problem of form-meaning mappings: are there metaphoric expressions in which the metaphoric mapping is represented by words that do not have a dependency relationship in the surface structure of the sentence? This lack of a dependency relationship can come about either because the concepts are not explicitly present in the linguistic utterance or because they are grammatically separated. The adoption of this operationalization of the source-target model brings with it the implicit premise that there is a one-to-one mapping between syntactic structure and semantic structure.

Further, because in Conceptual Metaphor Theory [15, 16] metaphoric sources have many different targets and metaphoric targets have many different sources, this clustering approach requires that these overlapping connections always correspond: if RELATIONSHIPS and ARGUMENTS can be WARS, then they must also both be able to be JOURNEYS. But what happens when the members of a cluster which share a few source domain mappings diverge strongly in their preference for other source domains? How does this affect the identified clusters, which are then used to generalize mappings from one metaphor to another? In other words, if A and B are clustered together because both map to C, and if A also maps with D but B does not, then a seed metaphor with an A-D mapping will incorrectly predict B-D mappings. This issue may cause false positives but will not cause false negatives.

Source-Target Mapping System: Methods. First, the source-target mapping system parses the linguistic data into grammatical relations using the RASP parser [17]. At the same time, verb and noun clusters are formed by looking at the grammatical contexts in which they occur in a larger corpus [18, 19]. Although statistical methods provide candidate clusters of verbs and nouns, humans intervene in the final selection of clusters. The final clusters are trained on the seed metaphoric utterances in order to learn what source-clusters (e.g., verbs) map metaphorically with what target-clusters (e.g., nouns). Finally, a selectional preference filter is used to eliminate false positives from the identifications; this filter will not reduce but may increase the number of false negatives, the rate of which is not reported in the original study.

2.3   Word Abstractness System

The second metaphor identification system [7] is based on the claim that metaphors occur in abstract contexts, so that metaphor identification requires a measure of abstractness for lexical items and their contexts. Before turning to the identification system itself, previous work by one of the principles needs to be considered [5]. Neuman's related work is in many ways the reverse of the source-target mapping system: rather than use clusters of similarly behaving lexical items to identify metaphoric mappings, Neuman uses clusters of similar metaphoric mappings to determine which lexical items have a similar meaning. In other words, Neuman argues that if we collect a large number of metaphoric expressions and determine which lexical items / concepts are involved in metaphoric mappings, then we can find the meanings of the words if we assume that lexical items / concepts which exist in mappings to the same source domain have the same meaning. Thus, it is the reverse of the source-target mapping system. Neuman's point is that distributional, bag-of-words semantics is too simplistic; a better system is a distributional, bag-of-relations system in which the focus is on semantic relations between concepts. Metaphor, he argues, is one such relation: "our basic thesis is that by analyzing metaphors in which our target term is embedded we may uncover its meaning" (2720).

Neuman's approach here depends upon the premise that metaphoric mappings are (1) mediated and (2) themselves as basic as or more basic than the concepts which form the source and target domains. Shannon [20] points out that cognitive approaches to metaphor assume that the source and target involved in a metaphoric mapping have a fixed and already existing set of properties, only some of which will be activated during the mapping. Shannon's claim is that metaphor itself is more basic than the concepts involved, which means that there is not a source and target directly available in the linguistic utterance. This is what Neuman calls a mediated mapping: while the metaphor is present in the utterance, the source and target are not linked to directly from the linguistic expression. The metaphor itself, existing on its own, provides that link first. The practical implication of a mediated mapping is that approaches to metaphor identification which require finding explicitly and overtly a source concept and target concept represented by lexical items in the linguistic utterance will miss a great many metaphors: those entities are not directly present in this view and so cannot be used for metaphor identification. The problem of mediated mappings is distinct from the problem of metaphors with mappings within a single domain [21] which, however, still raises problems for metaphor identification.

Both Neuman's related work and the domain interaction system below, which is based on Ontological Semantics [22], are explicit about the relationship between seen lexical items and the unseen concepts to which they refer. The idea is that natural language depends upon a human conceptual system of discrete, related concepts which can be modeled computationally using an ontology. Lexical items point or refer to concepts in this view; lexical items can point directly to a concept, they can point to a concept and specify or alter properties of that concept, and they can point to no concept but rather alter properties of the utterance as a whole [23]. Natural language meaning is also often under-specified in this view, so that concepts are present in the semantic structure of the utterance without being explicitly pointed to by lexical items. These relationships between lexical items and concepts affect all of the metaphor identification systems discussed in this paper, although they are only explicitly treated in the domain interaction system. At the same time, none of the systems here can deal with concepts that are not directly represented in the utterance by lexical items.

Word Abstractness System: Methods. In spite of Neuman's nuanced account of mediated metaphoric mappings, this metaphor identification system [7] does not rely on a principled view of metaphor: "Therefore we hypothesize that the degree of abstractness in a word's context is correlated with the likelihood that the word is used metaphorically" (680). The system focuses on the identification of metaphoric senses of a lexical item; certainly for some lexical items an abstract context will signal a metaphoric usage. But how well does this system transfer to new lexical items? And what about metaphors that occur in a non-abstract context: many metaphoric expressions describe physical scenes.

The word abstractness system relies, like the source-target mapping system, on a distinction between ABSTRACT and PHYSICAL domains, although the distinction is assumed to be gradient: lexical items are assigned a value that represents their relative abstractness. Metaphors are assumed to have a unique pattern of abstractness values, most likely patterns in which a few non-abstract lexical items occur in a highly abstract context. Unlike the source-target mapping system, this system can allow the implementation of a fuzzy or gradient threshold for metaphor identification.

The system first rates lexical items according to how abstract they are, on a scale from 0 to 1, with 1 being the most abstract. The approach to rating abstraction is taken from [6]; a list of rated lexical items is available from the authors. The system tags the words in the sentence with their parts of speech and finds the abstractness rating for each; if an abstractness rating is not available for a particular word form, the system attempts to find a match for its lemmatized form. For each sentence a feature vector is created that consists of five different combinations of abstractness ratings: (1) average of all non-proper nouns; (2) average of all proper nouns; (3) average of all verbs excluding target verb; (4)

average of all adjectives; (5) average of all adverbs. This vector is trained with a number of tokens of different verbs that are used metaphorically using a logistic regression learning algorithm. This is then applied to new instances of the same verbs as well as to new verbs.

2.4   Semantic Distance / Similarity System

This section looks at a metaphor identification system [8,9] that depends on the hypothesis that metaphoric material comes from a different origin (distribution) than non-metaphoric material. In other words, metaphor and non-metaphor are entirely separate, belonging to different populations with different properties. The system claims that metaphor can be identified by looking at semantic similarity measures within and between the metaphoric and non-metaphoric material in an utterance. Thus, literal and non-literal sentences or word usages are from two different categories and some mixture of properties will be able to determine which population or category a particular sentence or word usage belongs to.

The main property of non-literal language is that it does not exhibit semantic similarity with its context. The non-literal language does not fit, or exhibits a mismatch, with the semantic context in which it occurs. Thus, the task of metaphor identification is a matter of measuring semantic similarity. The distributional properties of lexical items are used as a representation of the meaning of the lexical items, so that lexical items which occur in the same contexts have the same meanings. As a result of this premise, the system claims that metaphors can be detected by finding unusual contexts; this is because semantic similarity or distance is measured using contexts in the first place (lexical items that do not frequently occur together will be measured as semantically dissimilar, so that this system detects infrequent co-occurrences). One result of this premise is that if a particular lexical item occurs often enough in a certain metaphoric mapping, this mapping will become part of the literal meaning of that lexical item: frequent metaphors will not be detected as metaphors. This is an interesting side-effect and may represent how speakers actually process metaphoric utterances. At the same time, the system seems to ignore the fact that unusual patterns (as measured using distributional semantic methods) have many possible sources, of which metaphor is only one. Humor, used as a counterfactual in the evaluation below, is another possible source.

Semantic Distance / Similarity System: Methods. The system adopts the distributional semantic premises and uses Normalized Google Distance [24] as the instrument for measuring semantic similarity or cohesion. Each sentence is represented using a feature vector with five different similarity measures: (1) the semantic similarity between the target expression and its context; (2) the average semantic similarity of the sentence as a whole; (3) the difference between the first and second measures; (4) a binary distinction between cases with a low or high difference between average and expression-specific semantic similarity; (5) the highest degree of similarity between the target expression and its context. A Bayes decision rule is used to determine which population the sentence is more likely to belong to, metaphoric or non-metaphoric, based on similarity of a sentence's feature vector with the feature vector of seed metaphors.

2.5   Domain Interaction System

The knowledge-based system, a domain interaction system called MIMIL (Measuring and Identifying Metaphor-in-Language), identifies metaphoric utterances using properties of all of the concepts pointed to by lexical items in the utterance. The system has two stages: first, determining what concepts are present in an utterance and what their properties are; second, using these properties to model metaphor. The first stage will be discussed below under methods.

The domain interaction system assumes, as discussed above involving Neuman's related work, that lexical items refer or point to concepts in the human conceptual system. These concepts have many properties and relations that connect them. Following Ontological Semantics [22], concepts are represented in part using two ontological properties: domain (for example, PHYSICAL vs. MENTAL), which is a product of the hierarchy of concepts; and event-status (for example, OBJECT vs. PROCESS / EVENT), which is independent of the hierarchy of concepts. The system assumes that every concept possesses these two properties and that these properties are sufficient for distinguishing metaphor from non-metaphor. In this way, the system takes concepts and their properties as identified by human intuitions (present in a knowledge-base) rather than as identified using distributional semantics.

The domain interaction system further assumes that metaphoricity is an utterance-level property that is not possessed by individual lexical items or individual grammatical relations. Thus, the system takes utterances as input (more specifically, the concepts referred to by the lexical items in the linguistic utterance). This means that all the concepts in the utterance, whether or not they are found in certain grammatical configurations, can interact metaphorically. The approach is somewhat similar to the source-target mapping system, except that it is formulated so that the source and target do not have to be identified as such. The advantage to this approach is that it does not assume a one-to-one mapping between syntactic structure and semantic structure and that it does not assume that nouns and verbs always represent the metaphoric target and source, respectively. A further advantage is that the system deals with concepts directly, rather than assuming that lexical items and concepts are equivalent. Finally, the domain interaction system covers both mediated (i.e., indirectly present) and unmediated metaphoric conceptual mappings, while the source-target mapping system covers only unmediated mappings (as discussed above).

On the other hand, these assumptions bring with them several weaknesses. The first is that the system removes all grammatical information from consideration: all concepts are assumed to interact equally with all other concepts. Even though there is not a one-to-one mapping between syntactic structure and semantic structure, there is a good deal of mapping between the two and the system ignores this. The second is that, while the system does not arbitrarily limit its scope to noun-verb relations (and thus includes more concepts in the utterance that are relevant for metaphor detection), it dilutes the influence of the relevant concepts by including irrelevant concepts as well. In other words, the source-target mapping system is limited because it takes a narrow approach to deciding what lexical items in the utterance are relevant for metaphor; but the domain interaction approach is limited because it takes a broad approach that avoids the issue altogether. The empirical question, to be tested below, is which simplifying assumption has fewer side-effects. Ultimately, both systems are implementations of the same underlying view of metaphor, that it involves the interaction between cognitive domains. They differ, however, in what properties of the interaction between domains are considered relevant to metaphor.

The domain interaction system differs from the semantic similarity / distance system because it assumes that metaphors and non-metaphors belong to different extremes of the same population. Thus, the system views metaphoricity as a continuous property which utterances possess to greater or lesser degrees. On one side of the distribution of this population are proto-typical metaphors, which speakers of a language would intuitively identify as metaphoric. On the other side of the distribution are proto-typical non-metaphors, which speakers of a language would intuitively identify as literal. In the middle, however, a majority of utterances have some amount of metaphoricity but are not clearly metaphoric or non-metaphoric (see [25] for further discussion). The implementation of the domain interaction system evaluated here is binary, so that utterances are taken to be metaphoric or non-metaphoric. However, like the word abstractness system and the semantic similarity / distance system, it can be converted into a gradient system that identifies different levels or degrees of metaphor (e.g., moderately metaphoric utterances vs. highly metaphoric utterances). The source-target mapping system is alone in not being easily converted into a gradient system (although it could be converted; for example, by manually assigning different weights to the seed metaphors used).

Domain Interaction System: Methods. The domain interaction system takes as input unrestricted and unannotated English text and uses existing resources to pre-process that text; the pre-processing constitutes the first stage discussed above which identifies the concepts referred to by the lexical items in the utterance. First, the system relies on OpenNLP [26] for tokenization, named entity recognition, and part of speech tagging. Second, the system relies on Morpha [27] for lemmatizing words. At this point, the lemmatized words are mapped to their WordNet synsets [28] using the part of speech tags to maintain a four-way distinction between nouns, verbs, adjectives, and adverbs. The system then maps the WordNet synsets onto concepts in the SUMO ontology [10] using the mappings provided [11]. This is done using the assumption that each lexical item is used in its default sense, so that no disambiguation takes place. Once the concepts present in the utterance have been identified in this manner, using the concepts present in the SUMO ontology, the system makes use of domain (ABSTRACT, PHYSICAL, SOCIAL, MENTAL) and event-status (PROCESS, STATE, OBJECT) properties of each concept present in the utterance. These are not present as such in the SUMO ontology, but were developed following

Ontological Semantics [22] as a knowledge-base specific to the domain interaction system.

The system claims that the interaction between the properties of the concepts referred to in the utterance can be used to identify metaphors. The implementation of the system evaluated here creates a feature vector using variables based on the properties of the concepts (discussed below).

3   Evaluation

This study replicates the methods of the systems in question in their most important details, adding new distinctions in order to test the explicit and implicit premises of the approaches. The unifying factor is the data set, which uses humorous utterances as a counterfactual for testing for the over-identification of metaphor.

First, the systems are evaluated using different classes: a three-way distinction between metaphor, humor, and literal language; a two-way distinction between metaphor and non-metaphor (a) with humor included in non-metaphor and (b) with humor excluded. These conditions allow us to test whether humor interferes with metaphor identification methods.

Second, the systems are evaluated using a four-way distinction between domains and without any distinction between domains. This allows us to test whether domain membership influences the behavior of metaphors (as revealed in the success rate of the identification systems).

3.1   Evaluation Methods for Source-Target Mapping System

The first part of evaluating the source-target mapping approach to metaphor identification was to cluster lexical items. The method for clustering verbs is described in [19]; [6] provide a resource of the most frequent 1,510 English verbs in the Gigaword corpus divided into 170 clusters. These clusters were used in the evaluation. The procedure used for clustering nouns in [4] is to include the frequency of grammatical relations (subject, object, indirect object), as annotated by the RASP parser, in a feature vector used to cluster nouns. In evaluating the source-target system, we took a different approach to obtaining noun clusters. Starting with 8,752 nouns examined by Iosif's SemSim system [29], we used a pairwise similarity matrix (measured using the Google-based Semantic Relatedness metric, as computed by Iosif) for the feature vector used for clustering nouns. The nouns were divided into 200 clusters using Weka's [30] implementation of the k means algorithm.

There are advantages and disadvantages to relying on the semantic relatedness metric rather than frequency of grammatical relations. On the one hand, the similarity measure is less sensitive to arbitrary patterns of object restrictions. In other words, many objects and indirect objects cannot occur with certain verbs, not because of their meaning but because of verb valency. This interferes with using grammatical relations as a substitute for meaning. The clusters put together using the similarity measure, however, will not all share the same valency but should have a related meaning. On the other hand, because the system detects similar combinations of a verb cluster and a noun cluster, valency is a salient property even though it is not directly related to meaning. Unlike the original system, no manual intervention was used in preparing the noun clusters. Finally, the evaluation did not need to filter out sentences with loose-valency verbs, those that accept a large variety of arguments, because the test corpus was designed around certain verbs chosen, in part, to avoid this property.

The search for metaphors was performed on the RASP-parsed version of the evaluation corpus; all verb-noun relations were included in the search. For each verb, 5 out of 25 metaphoric instances were used as seed cases, for a total of 105 seed metaphors. The seed metaphors were searched for across all verbs, not restricted to the verb they were taken from. Many of the seed metaphoric utterances contained multiple grammatically related clusters (e.g., verb-object) which were candidates for the metaphoric material in the utterance. No clear procedure was provided for choosing from among the candidate relations; in this evaluation we have erred on the side of inclusion by searching for all possible candidates. A total of 478 grammatical relations between clusters were identified in the 105 seed sentences; no manual intervention was used to trim this number down.

3.2   Evaluation Methods for Word Abstractness System

In replicating this study, we used the abstractness ratings from the authors. The corpus sentences were tagged using OpenNLP POSTagger [26] and all function words were removed. All words not found on the list of abstractness ratings (after reduced to their lemmatized form using Morpha [27]) were removed; empty slots in the feature vector (e.g., if there were no adjectives) were filled with a value of .5 for abstractness, following the original system. We started with the five attributes given by [7] and discussed above; we then augmented the feature vector with four additional variables: (6) average abstractness of all words, eliminating the grammatical distinction between them; (7) average abstractness of all words except for the target word; (8) the difference between the average abstractness of the sentence with the target word and without it; (9) the standard deviation of all the words. We tested these additional attributes because of the hypothesis that metaphors will cause mismatches between the total abstractness and the target word's abstractness.

3.3   Evaluation Methods for Semantic Distance / Similarity System

The evaluation of this approach tested a different distributional method for determining semantic similarity, Iosif's SemSim system [29]. There were two main reasons for not using the NGD [24] measure: (1) the test corpus had function words removed and other words reduced to their stems; the NGD results would not have taken this into account; (2) SemSim is more transparent in terms of its methodology and in terms of the corpus used. In this case, we used the American National Corpus (henceforth, OANC [31]), which consists of 14 million words taken from spoken and written contemporary American English. Thus, it has sources comparable to those used to create COCA, from which the test utterances were drawn (but COCA is not available to run SemSim on). The corpus was made comparable to the evaluation data by removing the most frequent functions words and running the Morpha analyzer to retrieve the lemmatized forms. SemSim's lexical classification system was then run on the entire OANC corpus for every word present in the evaluation data (H, the contextual window, was set at 2), creating an 8,690x8,690 matrix of similarity scores.

The pairwise similarity between words, comparable to NGD, was used to compute the 5 variables used in Sporleder and Li's system. To this we added four additional variables to test additional hypotheses: (6) the standard deviation of the similarity between the target word and the context; (7) the standard deviation of the similarity within the context. These were added to test the hypothesis that metaphor comes from a different source from the literal context, causing a mismatch in their similarity/distance. These caused two further variables to be included: (8) the difference between the standard deviations in similarity scores within the context and between target and context; and (9) the marker for negative differences in standard deviations that corresponds with variable (4) from the original study.

3.4   Evaluation Methods for Domain Interaction System

The domain interaction system has been implemented for the purposes of this study using a feature vector of variables created using the properties of the concepts referred to by lexical items in the utterance. The feature vector uses the following variables: (1) number of concepts in the utterance; (2-5) number of instances of each type of domain (ABSTRACT, PHYSICAL, SOCIAL, MENTAL); (6-8) number of instances of each type of event status (PROCESS, STATE, OBJECT); (9) number of instances of the domain with the highest number of instances; (10) number of instances of event-status with the highest number of instances; (11) sum of the individual domain variables minus (9); (12) sum of individual event-status variables minus (10); (13) number of domain types present at least once in the utterance; (14) number of event-status types present at least once in the utterance; (15) number of instances of the main domain divided by the number of concepts; (16) number of other domain instances divided by the number of concepts; (17) number of main event-status instances divided by the number of concepts; (18) number of other event-status instances divided by the number of concepts. This feature vector was evaluated using the same learning algorithms as the abstraction and similarity systems.

In creating this feature vector, four knowledge-bases (three existing and one new) were used: (1) the SUMO ontology; (2) WordNet synsets; (3) mappings between WordNet and SUMO; (4) domain and event-status properties of the SUMO concepts. The knowledge-bases used in the evaluation which are not available elsewhere can be found at http://www.jdunn.name.

4   Results

This section presents the results of the evaluations. Note that the different class comparisons (e.g., three-way vs. non-metaphor) will influence only the systems based on feature vectors. The "Joint" system takes variables from all the systems which use feature vectors (the abstractness, similarity, and domain interaction systems). To make the comparison as consistent as possible, evaluation of the semantic similarity / distance, word abstractness, joint, and domain interaction (MIMIL in the tables) systems was done, following [7], using Weka's implementation of the logistic regression learning algorithm. All instances were normalized before training and testing; the evaluations were performed using cross-validation (100 folds). The F-measures reported here are for metaphor classification only (i.e., precision for non-metaphor is not directly considered because this inflates the performance of the systems). This is done because some of the systems greatly over-identify literal utterances; however, because literal utterances dominate the evaluation data set, the over-identification of literal utterances would disproportionately raise the average F-measure for all classes in these systems. The feature vectors and other material used in the evaluation can be found at http://www.jdunn.name.

Table 1. Three-way distinction between metaphor, humor, and literal in all domains

System	True Pos.	False Pos.	True Neg.	False Neg.	F-Meas.
Similarity	1	0	2,482	504	0.004
Abstractness	1	2	2,482	505	0.004
Joint	67	44	2,446	444	0.215
MIMIL	133	382	2,437	63	0.374
Source-Tar.	113	461	2,038	300	0.229

As shown in Table 1, when tested on the three-way distinction between metaphor, humor, and literal utterances, the similarity and abstractness systems performed very poorly, essentially identifying no metaphors. The joint system performed worse than the domain interaction system, showing that the abstractness and similarity features reduce performance. As shown in later tests, the measurements of abstractness and semantic similarity, both at the word-level, simply do not distinguish between metaphor and non-metaphor in a realistic data set. The domain interaction and source-target mapping systems performed much better. Both systems identified a similar number of metaphors (133 and 113), but the domain interaction system had somewhat fewer false positives (382 vs. 461). More importantly, the source-target mapping system had a significantly higher number of false negatives (300 vs. 63). Using a higher number of seed metaphors would have lowered the source-target mapping system's false negative rate, but at the same time that would likely have raised the already high false positive rate.

Table 2. Two-way distinction with and without humor present in all domains

System	Data	True Pos.	False Pos.	True Neg.	False Neg.	F-Meas.
Similarity	+Humor	0	0	2,489	506	0.000
Abstractness	+Humor	1	3	2,486	505	0.004
Joint	+Humor	33	15	2,475	478	0.118
MIMIL	+Humor	90	31	2,469	425	0.283
Source-Tar.	+Humor	113	461	2,038	300	0.229
Similarity	-Humor	0	0	1,989	506	0.000
Abstractness	-Humor	2	5	1,984	504	0.008
Joint	-Humor	62	28	1,964	449	0.206
MIMIL	-Humor	125	46	1,954	390	0.364
Source-Tar.	-Humor	113	373	1,625	300	0.251

Table 2 shows that when tested using a two-way distinction that conflates literal and humorous utterances into a single non-metaphoric class (+Humor), the performance of MIMIL drops significantly, showing that humor is distinct from both metaphor and non-humor. The similarity and abstractness systems continue to perform very poorly; the joint system continues to perform more poorly than the domain interaction system on its own. One advantage of the source-target mapping system over the implementation of MIMIL evaluated here is that its identifications do not depend on the make up of the data set (only on the seed metaphors). Thus, its performance remains constant while MIMIL has more false negatives when humor and literal utterances are conflated into a single class. With humor removed altogether (-Humor), results similar to the three-way evaluation are achieved. Similarity and abstractness continue to perform poorly. MIMIL and the source-target mapping system identify a comparable number of metaphors (125 and 113). In this evaluation, however, MIMIL produces more false negatives (390 vs. 300) while the source-target mapping system produces more false positives (373 vs. 46).

As shown in Table 3, the source-target mapping system and the domain interaction system perform similarly within the ABSTRACT domain (the other systems are not shown here because their performance was too low). However, the performance of MIMIL is significantly less than on the data set as a whole (0.276 vs. 0.374 F-measure) while the source-target mapping system performs at the same level (0.239 vs. 0.229 F-measure). Within the MENTAL domain the source-target mapping system identifies more metaphors than MIMIL, but continues to have more false positives. MIMIL has more false negatives. Within the PHYSICAL domain, MIMIL greatly out-performs the source-target mapping system (0.629 vs. 0.268 F-measure). Further, within this domain both systems perform better than they did in any other domain. On the other hand, in the SOCIAL domain both systems perform more poorly than in any other domain. Here, also, the roles are reversed: the source-target mapping system significantly out performs MIMIL, which identifies almost no metaphors.

Table 3. Two-way distinction with humor present, by domain

System	Domain	True Pos.	False Pos.	True Neg.	False Neg.	F-Meas.
MIMIL	Abstract	25	16	609	115	0.276
Source-Tar.	Abstract	29	99	526	85	0.239
MIMIL	Mental	23	9	617	102	0.293
Source-Tar.	Mental	33	130	496	67	0.251
MIMIL	Physical	66	19	606	59	0.629
Source-Tar.	Physical	32	107	517	68	0.268
MIMIL	Social	4	1	623	121	0.062
Source-Tar.	Social	19	125	499	80	0.156

5   Conclusions

We can draw several interesting and useful conclusions from this evaluation. First, we see the importance of a justified theory underlying metaphor identification systems. Both the source-target mapping system and the domain interaction system (MIMIL) are concerned with explaining and justifying their choices; both greatly out-perform the systems which are not as firmly grounded in theory. Second, we see that domain membership has a significant influence on the performance of the systems. Third, we see that the two top systems have their best performance on different domains and classes.

This suggests that there are multiple types of metaphor and that each system is stronger at identifying one type over another type. In other words, some of the assumptions about metaphor discussed in the first part of this study are mutually exclusive, but others are not. For example, some metaphor identification systems assume that source-target mappings are explicitly present, some assume that they are indirectly present, and others assume that they are not relevant for metaphor identification. It is likely that some, but not all, metaphors have a mediated or unmediated mapping and that other metaphors have no mapping at all [13]. If this is the case, a synthesis of approaches to metaphor identification might allow stronger coverage overall. In other words, if there are multiple types of metaphor, and if the systems perform better on some types of metaphor as a result of their assumptions about metaphor, then a full-coverage metaphor identification system should include multiple synthesized methods.

For example, the source-target mapping system and the domain interaction system could be synthesized by ordering the application of the methods within a single system. The source-target mapping system could be run first and search for explicitly present mappings. This would miss many metaphors that do not have an explicit source-target mapping (e.g., metaphors whose mapping was mediated and thus not directly present, metaphors whose mapping was not present in a noun-verb relation, or metaphors which did not have an underlying conceptual mapping to begin with). The domain interaction system could then be run second in order to identify the metaphors which did not fall into the rather narrow scope of the source-target mapping system.

References

1. Shutova, E.: Models of metaphor in NLP. In: Hajiv, J., Carberry, S., Clark, S., Nivre, J. (eds.) Proceedings of ACL 2010, pp. 688–697. Association for Computational Linguistics, Stroudsburg (2010)
2. Shutova, E., Teufel, S.: Metaphor corpus annotated for source – target domain mappings. In: Calzolari, N., Choukri, K., Maegaard, B., Mariani, J., Odijk, J., Piperidis, S., Rosner, M., Tapias, D. (eds.) Proceedings of LREC 2010, pp. 3255–3261. European Language Resources Association, Paris (2010)
3. Shutova, E., Sun, L., Korhonen, A.: Metaphor identification using verb and noun clustering. In: Huang, C., Jurafsky, D. (eds.) Proceedings of COLING 2010, pp. 1002–1010. Tsinghua University Press, Beijing (2010)
4. Shutova, E., Teufel, S., Korhonen, A.: Statistical metaphor processing. Computational Linguistics 39 (2013) (forthcoming)
5. Neuman, Y., Nave, O.: Metaphor-based meaning excavation. Information Sciences 179, 2719–2728 (2009)
6. Turney, P., Littman, M.: Measuring praise and criticism: Inference of semantic orientation from association. ACM Transactions on Information Systems 21, 315–346 (2003)
7. Turney, P., Neuman, Y., Assaf, D., Cohen, Y.: Literal and metaphorical sense identification through concrete and abstract context. In: Barzilay, R., Johnson, M. (eds.) Proceedings of EMNLP 2011, pp. 680–690. Association for Computational Linguistics, Stroudsburg (2011)
8. Li, L., Sporleder, C.: Using Gaussian Mixture Models to detect figurative language in context. In: Kaplan, R., Burstein, J., Harper, M., Penn, G. (eds.) Proceedings of HLT-NAACL 2010, pp. 297–300. Association for Computational Linguistics, Stroudsburg (2010)
9. Sporleder, C., Li, L.: Contextual idiom detection without labelled data. In: Koehn, P., Mihalcea, R. (eds.) Proceedings of EMNLP 2009, pp. 315–323. Association for Computational Linguistics, Stroudsburg (2009)
10. Niles, I., Pease, A.: Towards a Standard Upper Ontology. In: Welty, C., Barry, C. (eds.) Proceedings of FOIS 2001, pp. 2–9. Association for Computational Linguistics, Stroudsburg (2001)
11. Niles, I., Pease, A.: Linking lexicons and ontologies: Mapping WordNet to the Suggested Upper Merged Ontology. In: Arabnia, H. (ed.) Proceedings of IEEE Intl. Conf. on Inf. and Knowl. Eng. (IKE 2003), pp. 412–416. IEEE Press, New York (2003)
12. Dunn, J.: Gradient semantic intuitions of metaphoric expressions. Metaphor and Symbol 26, 53–67 (2011)
13. Dunn, J.: How linguistic structure influences and helps to predict metaphoric meaning. Cognitive Linguistics 24, 33–66 (2013)
14. Davies, M.: The 385+ million word Corpus of Contemporary American English (1990–2008+): Design, architecture, and linguistic insights. International Journal of Corpus Linguistics 14, 159–190 (2009)
15. Lakoff, G., Johnson, M.: Metaphors We Live By. University of Chicago Press, Chicago (1980)
16. Lakoff, G., Johnson, M.: Philosophy in the Flesh: The embodied mind and its challenge to western thought. Basic Books, New York (1999)
17. Briscoe, E., Carroll, J., Watson, R.: The second release of the RASP system, pp. 77–80 (2006)

18. Sun, L., Korhonen, A.: Improving verb clustering with automatically acquired selectional preferences. In: Koehn, P., Mihalcea, R. (eds.) Proceedings of EMNLP 2009, pp. 638–647. Association for Computational Linguistics, Stroudsburg (2009)

19. Sun, L., Korhonen, A., Krymolowski, Y.: Verb Class Discovery from Rich Syntactic Data. In: Gelbukh, A. (ed.) CICLing 2008. LNCS, vol. 4919, pp. 16–27. Springer, Heidelberg (2008)

20. Shannon, B.: Metaphor: From fixedness and selection to differentiation and creation. Poetics Today 13, 659–685 (1992)

21. Barnden, J.: Metaphor and metonymy: Making their connections more slippery. Cognitive Linguistics 21, 1–34 (2010)

22. Nirenburg, S., Raskin, V.: Ontological Semantics. MIT Press, Cambridge (2004)

23. Raskin, V., Nirenburg, S.: An applied ontological semantic microtheory of adjective meaning for natural language processing. Machine Translation 13, 135–227 (1998)

24. Cilibrasi, R., Vitanyi, P.: The Google similarity distance. IEEE Transactions on Knowledge and Data Engineering 19, 370–383 (2007)

25. Gibbs, R.: Literal meaning and psychological theory. Cognitive Science 8, 275–304 (1984)

26. Apache: OpenNLP (2011), `http://opennlp.apache.org`

27. Guido, M., Carroll, J., Pearce, D.: Applied morphological processing of English. Natural Language Engineering 7, 207–223 (2001)

28. Princeton, U.: WordNet (2012), `http://wordnet.princeton.edu/`

29. Iosif, E., Potamianos, A.: SemSim: Resources for normalized semantic similarity computation using lexical networks. In: Calzolari, N., Choukri, K., Declerck, T., Doğan, M., Maegaard, B., Mariani, J., Odijk, J., Piperidis, S. (eds.) Proceedings of LREC 2012, pp. 3499–3504. European Language Resources Association, Paris (2012)

30. Witten, I., Frank, E.: Data Mining: Practical Machine Learning Tools and Techniques with Java Implementations. Morgan Kaufmann, San Francisco (2005)

31. Ide, N., Suderman, K.: The American National Corpus first release. In: Lino, M., Xavier, M., Ferreira, F., Costa, R., Silva, R. (eds.) Proceedings of LREC 2004, pp. 1681–1684. European Language Resources Association, Paris (2004)

Determining the Conceptual Space
of Metaphoric Expressions

David B. Bracewell, Marc T. Tomlinson, and Michael Mohler

Language Computer Corporation
Richardson, TX
{david,marc,michael}@languagecomputer.com

Abstract. We present a method of constructing the semantic signatures of target concepts expressed in metaphoric expressions as well as a method to determine the conceptual space of a metaphor using the constructed semantic signatures and a semantic expansion. We evaluate our methodology by focusing on metaphors where the target concept is *Governance*. Using the semantic signature constructed for this concept, we show that the conceptual spaces generated by our method are judged to be highly acceptable by humans.

1 Introduction

Metaphor is a pervasive literary mechanism allowing for the comparison of seemingly unrelated concepts. It has been widely studied in the linguistics literature [1,2,3,4] and more recently in computational linguistics [5]. The most prominent theory of metaphor is Lakoff's Contemporary Theory of Metaphor [6] in which concepts are mapped from an abstract *target* domain into a concrete *source* domain (e.g. "Time is Money" or "Love is War.") allowing for the target to be discussed and understood through the source.

As recent empirical studies have shown, metaphor is abundant in natural language occurring as often as every third sentence [7]. Because of its abundance and often unique usage, it is important to build a system that can recognize and understand metaphor in order to aid natural language processing applications such as authorship identification or semantic interpretation.

Automated methodologies for processing metaphor can be broken down into two main categories: recognition and interpretation. Interpretation of metaphor involves determining the intended, or literal, meaning of a metaphor [5]. The recognition of metaphor entails determining whether an expression is literal or figurative. Work on automated metaphor recognition dates back to the early 1990's with the work of Fass [8] which was based on detecting selectional preference violations while the more recent work of Shutova [9] accomplishes the same task using noun-verb clustering. An additional component of metaphor recognition is determining the conceptual metaphor for a given linguistic metaphor. CorMet [10] was one of the first systems to determine the source and target domain of a linguistic metaphor using domain-specific selectional preferences.

A. Gelbukh (Ed.): CICLing 2013, Part I, LNCS 7816, pp. 487–500, 2013.

In this paper, we examine an unsupervised method for determining the source and target concepts expressed in a metaphoric expression. The proposed method works by first building a semantic signature for a target concept which we define as a set of related WordNet senses. The senses are mined from automatically gathered Wikipedia pages that are highly relevant to the target concept. The conceptual space for a metaphoric expression is then generated using senses contained in the semantic signature for the target concept and the elements expressed in the text that correspond to the source concept. Candidates for the source concepts are taken from those words and phrases in the metaphoric expression that are not found in the semantic signature. The candidate source concepts are expanded and clustered, and then through a ranking process the final set of salient source concepts is determined.

We test the quality of the automatically generated semantic signature for the target domain of *Governance* which we define as anything that pertaining to the *application, enforcment*, and *setting* of rules. Governance can pertain to the actions of state or national governments or the actions of local organizational bodies, such as Parent Teacher Associations, school clubs, and sports teams. Using the automatically generated semantic signature for Governance, we then test the quality of the generated conceptual spaces using a subjective assessment. We demonstrate our methodologies using metaphoric expressions in English, Spanish, and Russian.

2   Related Work

Metaphor has been studied by researchers in many fields, including psychology, linguistics, sociology, anthropology, and computational linguistics. A number of theories of metaphor have been proposed, including the Contemporary Theory of Metaphor [6], the Conceptual Mapping Model [1], the Structure Mapping Model [11], and the Attribute Categorization Hypothesis [12]. Based on these theories, databases of metaphors, such as the Master Metaphor List (MML) [13] for English and the Hamburg Metaphor Database (HMD) [14] for French and German, have been constructed. The MML provides links between domains (source and target) and their conceptual mappings. It does not, however, define relations between the domains or mappings, e.g. how the source domains LIFE and PEOPLE are related. The HMD fuses EuroWordNet synsets with the MML source and target domains.

An active area of research in natural language processing has been the detection of figurative language [15,16,17,18]. One part of the more general task of detecting figurative language is the identification of metaphoric expressions in text [8,9,10]. Much of the early work on the identification of metaphor used hand-crafted world knowledge. The met* [8] system determined whether an expression was literal or figurative by detecting the violation of selectional preferences. Figurative expressions were then determined to be either metonymic, using hand-crafted patterns, or metaphoric, using a manually constructed database of analogies. The CorMet [10] system determined the source and target concepts of a metaphoric expression using domain-specific selectional preferences mined from Internet resources. More recent work has examined noun-verb clustering [9] which starts from a small seed set of one-word metaphors and results in clusters that represent source and target concepts connected via a metaphoric relation. These clusters are then used to annotate the metaphoricity of text.

Alongside the identification of metaphor is the interpretation of metaphor. One of the earliest systems to perform metaphor interpretation was the Metaphor Interpretation, Denotation, and Acquisition System (MIDAS) [19] which uses an iterative search method going from specific to more general concepts. After each iteration a mapping is generated from the more general to more specific metaphor. Other approaches have examined metaphoric interpretation as paraphrase generation. Shutova [18] examined ranking candidate paraphrases via a context-based probabilistic model. As with the early systems for metaphor identification, the KARMA [20] and the ATT-Meta [21] systems relied upon complex hand-coded world knowledge to determine valid inferences based on the source frame elements.

Also of relevance to the construction and use of semantic signatures is research on topic signatures. A topic signature is typically a set of related words which may have associated weights. Lin and Hovy [22] developed a method for the construction of topic signatures for automated summarization which they mined from a large corpus. Similarly, Harabagiu and Lacatusu [23] examined the use of topic signatures and enhanced topic signatures for multi-document summarization. In contrast, we examine in this paper the use of semantic signatures, which we define as a set of highly related and interlinked senses corresponding to a concept, in order to determine the elements of the target concept being expressed in a metaphoric expression.

3   Methodology

We propose a method for recognizing the conceptual representation of a metaphor by utilizing the semantic signature of the target concept along with a method for the conceptual expansion of the possible source concepts. We assume as input a metaphoric expression and generate the conceptual space that represents that expression. Based on Lakoff's Contemporary Theory of Metaphor, we define the representation of the conceptual space as a set of source and target concepts in which the concepts of the abstract target domain are mapped onto the concepts of the concrete source domain.

Previous work has shown that WordNets are a viable representation language for metaphor [24] and that ontologies, such as the Suggested Upper Merged Ontology (SUMO) [25], provide a good foundation for describing conceptual metaphors [1]. Based on this previous research, we use WordNet and SUMO to define the conceptual space of metaphors. WordNets are available in a number of languages, including English [26], Russian [27], and Spanish [28], and have been widely used in the evaluation of previous work. WordNet provides valuable semantic relations (e.g. hypernyms and hyponyms) with which concepts can be expanded, generalized, or specialized. Moreover, there exist mappings between WordNet and other ontologies, such as SUMO, and in most cases, there are interlingual links between WordNets of different languages.

In addition to WordNet and SUMO, we use Wikipedia as a source of semantic information. Wikipedia provides semantics in the form of intra-wiki links, category information, and interlingual links. As with WordNet, Wikipedia is available in a large number of languages, including the three we are using to evaluate our methodologies. As of this writing, the English version of Wikipedia includes over 4 million articles, and Spanish and Russian contain close to 1 million articles. Importantly, Wikipedia can provide information about concepts, such as named entities, that are not found in WordNet. These concepts can be linked back to WordNet through a variety of means, such as those discussed in several recent studies [29,30].

3.1   Creating the Semantic Signature for a Target Concept

When given a metaphoric expression, the first step in inducing its conceptual representation is to determine which elements of the expression relate to the source and which to the target. In the absence of any other information (e.g. selectional preference violations) which indicates the source and target elements of the expression, we can determine the target elements using the semantic signature of the target concept. Additionally, by defining a semantic signature for a target concept, we have a mechanism for automatically gathering and searching for metaphoric expressions whose conceptual representation involves the target in question.

We define a semantic signature as a set of highly related and interlinked (WordNet) senses corresponding to a particular domain with statistical reliability. For example, in the domain of Governance the concepts of "law," "government," and "administrator," along with their associated senses in WordNet, would be present in the signature. We generate the semantic signature using semantic knowledge encoded in the following resources: (1) the semantic network encoded in WordNet; (2) the semantic structure underlying Wikipedia; and (3) collocation statistics provided by the statistical analysis of a large corpora.

The target concept of a metaphor represents an abstract domain in which there exist a number of sub-concepts which describe different aspects of the target. We use Wikipedia to mine word senses for the semantic signature, as the world knowledge encoded in the text of Wikipedia and through its intra-wiki links and categories facilitates a significant exploration of the target concept. The construction of the semantic signature begins by utilizing the semantic markup in Wikipedia to collect articles that are highly related to the target concept. The process of collecting the relevant articles is similar to that of a traditional web crawler [31] or focused crawler [32]. The process, illustrated in Figure 1, begins by linking the target concept to one or more associated Wikipedia articles. This set of Wikipedia articles then becomes the "seed set" for the crawl. Linking to Wikipedia is done by searching for the target concept (and optionally content words making up the definition of the target concept) in the Wikipedia article

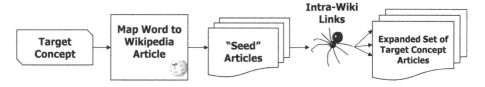

Fig. 1. Focused crawling of Wikipedia articles pertaining to the target concept using intra-wiki links

titles and redirects. In the case of Governance (the target concept with which we are evaluating the methodologies), there is a corresponding Wikipedia article with the same name.

The crawl process begins from the seed articles and spiders out using the intra-wiki links present in the articles, which typically link to other articles in the same domain. For example, the Wikipedia article on Governance contains links to articles like *Open source governance, Rule of law, Public management, Sovereign State*, and *Bureaucracy*. By initiating the crawl on these links, it becomes focused on the particular domain expressed in the seed articles. The crawling process continues until either no new articles are found or a predefined crawl depth (from the set of seed articles) has been reached. The algorithm is outlined in Figure 2. The result of the crawl is a set of Wikipedia articles whose domain is related to the target concept. From this set of articles, the semantic signature can be built by exploiting the semantic information provided by WordNet.

Process for finding articles associated with a target concept

Given a set of seed articles (S), and a maximum depth $(MaxDepth)$

```
1: set queue = { {sᵢ,0} : for each seed in S }
2: set concepts = {}
3: while(queue is not empty):
4:     set {article,depth} = queue.remove()
5:     foreach( link in intraWikiLinks(article) ):
6:     if( depth < MaxDepth ):
7:         concepts.add( link )
8:         queue.add( {link,depth+1} )
9: return concepts
```

Fig. 2. Process for finding articles associated with a target concept

The process of going from a set of target concept articles to a semantic signature is illustrated in Figure 3. The first step in the process of constructing the semantic signature from a set of target concept Wikipedia articles is to determine the possible word senses related to the target concept. The text of each of the gathered Wikipedia article is tokenized and each token associated with all of its possible WordNet senses (i.e. no word sense disambiguation is performed).

The word senses are then expanded using the lexical (e.g. derivationally related forms) and semantic relations (e.g. hypernym and hyponym) available in Word-Net. This results in a large number of word senses which may or may not be relevant to the target concept in part due to the polysemous nature of words.

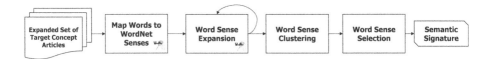

Fig. 3. Constructing the semantic signature of the target concept from Wikipedia articles pertaining to the target concept

Clustering is then performed to eliminate word senses that are not related to the target concept. Word sense clusters are built by first representing the word senses in the gathered articles as a graph $G = (V, E)$ where: V is a set of vertices defined for the word senses gathered from the Wikipedia articles; and E is a set of weighted edges defined between vertices based on WordNet similarity and word co-occurrence in a large, web-scale corpus. The weight, w, between two vertices is defined as:

$$w(v_1, v_2) = sim(sense_{v_1}, sense_{v_2}) * npmi(word_{v_1}, word_{v_2}) \tag{1}$$

where $sim(sense_{v_1}, sense_{v_2})$ is the Hirst and St-Onge [33] similarity metric which is linking terms via lexical chains in the WordNet ontology and $npmi(word_{v_1}, word_{v_2})$ is the normalized pointwise mutual information calculated as: $npmi(x, y) = (pmi(x, y))/(-log[p(x, y)])$ using the web or a large corpus. Edges are only added when the weight is greater than zero.

The graph is clustered to disambiguate the senses using the Chinese Whispers algorithm [34]. Chinese Whispers is a fast graph clustering algorithm capable of achieving a good clustering after only a few iterations. It was originally designed for word sense induction on large word graphs. Figure 4 shows the pseudocode for the version of the Chinese Whispers algorithm that we employed.

The clusters resulting from the Chinese Whispers algorithm contain semantically and topically similar word senses. The size of a cluster is in direct proportion to how central the concepts in that cluster are to the target. Given the clusters, the next step in the process uses the size of the cluster and a statistical analysis of large corpora to select the word senses that are significantly related to the target. First, "small" clusters (i.e. ones with less than three word senses) are removed, as they are unlikely to be related to the target concept. Second, word senses whose surface form appears in a standard stopword list (e.g. is, was, has) are removed. Finally, the remaining clusters are scored based on the average normalized pointwise mutual information (npmi) between the word senses in the cluster and the word senses in the set of phrases related to the target. Clusters with a score less than a threshold τ are removed. The set of target phrases used

Chinese Whispers

Given a set of vertices (V), an acceptance rate (α), and a maximum number of iterations $(maxItr)$

```
 1: foreach( v ∈ V):
 2:      set class(v) = i
 3:      set iteration = 0
 4: do:
 5:      shuffle(V)
 6:      foreach(v ∈ V):
 7:          set id = class(vₙ) of vₙ ∈ Neighbors(v) with max edge weight
 8:          if random() < α:
 9:              set class(v) = id
10:      set iteration = iteration +1
11: until converged or iteration > maxItr
```

Fig. 4. The Chinese Whispers graph clustering algorithm

- State   - Government   - Law
- Political   - System   - Country
- Power   - Economy   - United States

Fig. 5. A set of nine of the highest TF-IDF terms found for the Governance domain

in calculating the npmi are constructed from the gathered Wikipedia articles using TF-IDF (term frequency inverse document frequency), where TF is calculated within the gathered articles and IDF is calculated using the entire textual content of Wikipedia. A random sample of the top 20 target phrases determined for the concept of Governance is displayed in Figure 5. The set of word senses remaining after pruning clusters based on size and score is the set of concepts that make up the semantic signature.

4   Determining the Conceptual Space of a Linguistic Metaphor Using Semantic Signatures

We define the conceptual space of a linguistic metaphor as the combination of source and target concepts. The target concepts can be determined using the semantic signature. Identification of the source concepts can be done using semantic expansion, clustering, and selection techniques similar to those used for creating the semantic signature. The method for determining the conceptual space takes as input the linguistic metaphor and semantic signature of the target concept. Figure 6 shows an overview of the process for determining the source and target concepts for a metaphoric expression.

The process begins by mapping the words in the metaphoric expression to WordNet. Words and phrases are looked up in WordNet, and the four most fre-

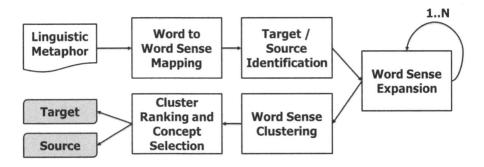

Fig. 6. Example of a generated conceptual space for a metaphoric expression

quent senses are assigned. In order to improve coverage and to capture entities and phrases not in WordNet, we also map phrases to Wikipedia articles. The phrases are mapped based on a statistical measure which takes into account the text of the article and intra-wiki links. The Wikipedia articles are then mapped to WordNet senses using the text in the categories of the article. For example, the organization "Bilderberg Group" is not present in WordNet, but is in Wikipedia and is under categories such as *Global trade and professional organizations, International business,* and *International non-governmental organizations.* From these categories we can determine that the Bilderberg Group relates to Word-Net senses such as "professional organization," "business," "international," and "nongovernmental organization."

The next step in the process identifies the target and possible source frame elements in the metaphoric expression. Target frame elements are those phrases which have a related WordNet sense in the semantic signature of the target concept. Those phrases that do not have a WordNet sense in the semantic signature are then treated as candidate source frame elements. The rest of the processing is performed separately over the source and target frame elements. The frame elements are then expanded using: (1) WordNet relations and (2) information from WordNet glosses. The senses are expanded using the lexical and semantic relations available from WordNet, including hypernymy, domain categories, and pertainymy. Source frame elements are also expanded with the content words found in the glosses associated with each of the noun and verb senses. The expansion results in a large set of possible concepts – some of which are not of interest for the metaphor expressed.

In order to determine the correct senses, clustering is performed using the same methodology as in the construction of the semantic signature. First, a graph is built from the senses with edge weights assigned based on similarity and co-occurrence. Then, the Chinese Whispers algorithm is used to cluster the graph. The clustering works to disambiguate the senses and to prioritize which senses are examined and incorporated into the conceptual space. Prioritization of the senses is done by ranking the clusters based on their size and the best scoring word sense in the cluster using:

$$rank(c) = size(c) \cdot \left(\frac{\sum_s score(s)}{|c|} \right) \tag{2}$$

where c is the cluster, s is a word sense in the cluster, and $|c|$ is the number of word senses in the cluster. The senses are scored using: (1) the degree distribution of the sense in the graph (more central word senses are given a higher weight); and (2) the minimum distance to the phrases appearing in the metaphoric expression calculated using Dijkstra's shortest path algorithm (weight concepts closer to the surface form higher). Formally, $score(s)$ is calculated as:

$$score(s) = \frac{degree(s) + minDistance(s, R)}{2} \tag{3}$$

where $degree(s)$ is degree distribution of s and $minDistance(s, R)$ returns the minimum distance in the graph between s and a phrase from the metaphoric expression.

Word Sense Selection Algorithm

Given a ranked set of clusters (C), an average cluster score (μ_c), and an average sense score (μ_s)

```
 1: set concepts={}
 2: set processed={}
 3: foreach(c ∈ C):
 4:     if( score(c) > μ_c and size(c) > 1:
 5:         foreach(s ∈ c):
 6:             if( score(s) > μ_s and snot ∈ processed:
 7:                 concepts.add( sumoConceptOf(s) )
 8:                 processed.add( s )
 9:                 processed.add( Neighbors(s) )
10: return concepts
```

Fig. 7. Algorithm for selecting the word senses to incorporate into the conceptual space

Clusters containing only one word sense or with a score less than the average cluster score (μ_c) are ignored. The remaining clusters and senses are then examined for incorporation into the conceptual space. The algorithm for selecting the concepts is shown in Figure 7. Senses from clusters that meet the selection criteria are candidates for incorporation into the conceptual space. Each sense is examined, with those in higher ranked clusters examined first. Senses are added as concepts when their score is greater than the average word sense score (μ_s). To decrease redundancy in the conceptual space neighbors of the added word sense in the word sense graph are ignored from future processing (depicted in the algorithm in Figure 7 as a processed set).

An example of a generated conceptual space is shown in Figure 8. In this example the target concept deals with politics and the identified sources relate

This political season promises a surplus of low blows.

TARGET:	POLITICAL SYMPATHIES , POLITICS
SOURCE:	AGENT, COMMITTING, DESTINATION, EXPRESSING, OVERMUCH, SEASON OF YEAR, SURPLUS, TIME DURATION, VIOLENT CONTEST

Fig. 8. Algorithm for selecting the word senses to incorporate into the conceptual space

to politics as a time (season), a violent contest (low blows), and an agent (the person dealing the low blows).

5   Experimentation

We have performed experiments to judge the quality of both the semantic signature and the conceptual space. We estimate the quality of our semantic signature by comparing our automatically generated signature for Governance with an expertly-created signature and by analyzing its coverage across a set of governance-related metaphors. However, for the evaluation of the conceptual space, we perform a subjective assessment as the acceptability of a metaphor and its mapping are heavily influenced by cultural and sub-cultural factors which make agreement on a single label difficult.

5.1   Semantic Signature of the Target Concept

We have evaluated the quality of our semantic signature using two experiments. In the first experiment, we measured the overlap between the automatically generated signature and one created by an expert. In the second experiment, we measured the percentage of human-annotated metaphoric expressions for which the semantic signature was able to find a target concept.

Manual creation of a semantic signature using WordNet is time intensive. Moreover, because of the fine-grained nature of WordNet senses, the agreement between two human-created signatures will not be perfect. An "expert-quality" semantic signature was created in English by a highly-trained linguist using WordNet. It took the linguist 56 hours to construct a 1,400 word-sense signature for Governance, while our automated approach took roughly half the time and resulted in a 4,000 word-sense signature. The overlap between the expert and automatically constructed semantic signatures was 88%. We have not yet made any attempt to determine the precision of the semantic signature.

The second experiment involved determining the number of governance-related metaphors for which the semantic signature could find a target concept. As a part of a larger effort, we employed two annotators per language (English, Russian, and Spanish) to annotate texts for governance-related metaphors. We currently have thousands of such annotations in each of the languages. For English, we found target concepts for 89% of 75 metaphors. For Spanish and Russian, we used the semantic signature constructed for English by mapping the words and phrases of the Spanish and Russian annotations into the equivalent English word senses using the available interlingual links. Using this translation, we found target concepts for 85% of 59 Spanish metaphoric expressions, while for Russian, we found target concepts in 72% of 64 metaphoric expressions.

5.2   Conceptual Space of Linguistic Metaphors

To evaluate the conceptual space, we selected a random sample of the metaphoric expressions identified by our annotators that were related to the target concept of Governance. We then generated the conceptual space for each of these metaphoric expressions. Annotators were asked to choose the three combinations of source and target concepts from our conceptual space represented the underyling conceptual metaphor. They were asked to rate the quality of these combinations using a scale from 1 to 10 with 1 being "completely wrong" and 10 being a "perfect match" for the conceptual metaphor. We evaluated our conceptual spaces using 14 English, 20 Russian, and 43 Spanish metaphoric expressions. Table 1 shows the average score for the first (#1), second (#2), and third (#3) ranked conceptual metaphors that were constructed by the annotators.

Table 1. The average score for the first (#1), second (#2), and third (#3) ranked conceptual metaphor chosen by the annotator

Language	Avg. Score for #1	Avg. Score for #2	Avg. Score for #3
English	8.14	6.50	4.29
Russian	9.00	7.90	6.65
Spanish	7.67	6.47	5.44

As can be seen from Table 1 the #1 scores yield an average greater than 7.5 on the 10 point scale for all three languages. This suggests that our algorithm is capable of generating correct conceptual spaces for the metaphoric expressions. This result is further supported by the fact that the degradation in score from the #1 chosen to the #3 chosen conceptual metaphor was only about 2 to 4 points, with the average scores still respectable for the #3 ranked conceptual metaphor.

6   Conclusion

We have presented a method for constructing the semantic signature of a target concept using WordNet, Wikipedia, and corpus statistics. We have shown that our semantic signature for the domain of Governance yields a high overlap with the semantic signature created by an expert. We have also presented a method for generating the conceptual space of a metaphoric expression using this semantic signature. Through experimentation, we found that the conceptual space was rated as highly acceptable by human judges.

One of the obstacles we faced during this process was the construction of language-specific semantic signatures. The English WordNet has 206,941 word senses which are combined into 117,659 synsets, while the Spanish and Russian WordNets that we obtained have significantly fewer word senses. Additionally, the version of the Russian WordNet we obtained did not contain any information regarding the semantic relations between synsets or the lexical relations between word senses. This lack of relation information forced us to translate senses using the interlingual links and perform expansion using the English WordNet. In the future, we will work to address these problems by using methodologies to automatically construct and enhance WordNets.

To augment our conceptual space generation, we plan on incorporating more resources such as FrameNet, Yago, and extended WordNet. In addition, we will explore enhanced techniques to improve the selection of concepts for the conceptual space. In particular, we will examine the use of co-ranking conceptual spaces based on metaphoric-preference (i.e. which targets and sources are most likely to appear in this language, genre, etc.) as discovered through a large repository of metaphoric expressions.

Acknowledgments. This research is supported by the Intelligence Advanced Research Projects Activity (IARPA) via Department of Defense US Army Research Laboratory contract number W911NF-12-C-0025. The U.S. Government is authorized to reproduce and distribute reprints for Governmental purposes notwithstanding any copyright annotation thereon. Disclaimer: The views and conclusions contained herein are those of the authors and should not be interpreted as necessarily representing the official policies or endorsements, either expressed or implied, of IARPA, DoD/ARL, or the U.S. Government.

References

1. Ahrens, K., Chung, S., Huang, C.: Conceptual metaphors: Ontology-based representation and corpora driven mapping principles. In: Proceedings of the ACL 2003 Workshop on Lexicon and Figurative Language, vol. 14, pp. 36–42. Association for Computational Linguistics (2003)
2. Wilks, Y.: Making preferences more active. Artificial Intelligence 11(3), 197–223 (1978)
3. Lakoff, G., Johnson, M.: Metaphors we live by, Chicago, London, vol. 111 (1980)
4. Tourangeau, R., Sternberg, R.: Understanding and appreciating metaphors. Cognition 11(3), 203–244 (1982)
5. Shutova, E.: Models of metaphor in nlp. In: Proceedings of the 48th Annual Meeting of the Association for Computational Linguistics, pp. 688–697. Association for Computational Linguistics (2010)
6. Lakoff, G., et al.: The contemporary theory of metaphor. Metaphor and Thought 2, 202–251 (1993)
7. Shutova, E., Teufel, S.: Metaphor corpus annotated for source-target domain mappings. In: Proceedings of LREC (2010)

8. Fass, D.: met*: a method for discriminating metonymy and metaphor by computer. Comput. Linguist. 17(1), 49–90 (1991)

9. Shutova, E., Sun, L., Korhonen, A.: Metaphor identification using verb and noun clustering. In: Proceedings of the 23rd International Conference on Computational Linguistics, COLING 2010, pp. 1002–1010. Association for Computational Linguistics, Stroudsburg (2010)

10. Mason, Z.J.: Cormet: a computational, corpus-based conventional metaphor extraction system. Comput. Linguist. 30(1), 23–44 (2004)

11. Wolff, P., Gentner, D.: Evidence for role-neutral initial processing of metaphors. Journal of Experimental Psychology: Learning, Memory, and Cognition 26(2), 529 (2000)

12. McGlone, M.: Conceptual metaphors and figurative language interpretation: Food for thought? Journal of Memory and Language 35(4), 544–565 (1996)

13. Lakoff, G.: Master Metaphor List. University of California (1994)

14. Eilts, C., Lönneker, B.: The hamburg metaphor database (2002)

15. Bogdanova, D.: A framework for figurative language detection based on sense differentiation. In: Proceedings of the ACL 2010 Student Research Workshop. ACLstudent 2010, pp. 67–72 (2010)

16. Li, L., Sporleder, C.: Using gaussian mixture models to detect figurative language in context. In: Human Language Technologies: The 2010 Annual Conference of the North American Chapter of the Association for Computational Linguistics, HLT 2010, pp. 297–300 (2010)

17. Peters, W., Wilks, Y.: Data-driven detection of figurative language use in electronic language resources. Metaphor and Symbol 18(3), 161–173 (2003)

18. Shutova, E.: Computational approaches to figurative language. PhD thesis, University of Cambridge (2011)

19. Martin, J.: A computational model of metaphor interpretation. Academic Press Professional, Inc. (1990)

20. Feldman, J., Narayanan, S.: Embodied meaning in a neural theory of language. Brain and Language 89(2), 385–392 (2004)

21. Barnden, J., Glasbey, S., Lee, M., Wallington, A.: Reasoning in metaphor understanding: The att-meta approach and system. In: Proceedings of the 19th International Conference on Computational Linguistics, vol. 2, pp. 1–5. Association for Computational Linguistics (2002)

22. Lin, C.Y., Hovy, E.: The automated acquisition of topic signatures for text summarization. In: Proceedings of the 18th Conference on Computational Linguistics, COLING 2000, vol. 1, pp. 495–501 (2000)

23. Harabagiu, S., Lacatusu, F.: Topic themes for multi-document summarization. In: Proceedings of the 28th Annual International ACM SIGIR Conference on Research and Development in Information Retrieval, pp. 202–209. ACM (2005)

24. Lönneker, B.: Is there a way to represent metaphors in wordnets?: insights from the hamburg metaphor database. In: Proceedings of the ACL 2003 Workshop on Lexicon and Figurative Language, LexFig 2003, vol. 14, pp. 18–27 (2003)

25. Pease, A., Niles, I., Li, J.: The suggested upper merged ontology: A large ontology for the semantic web and its applications. In: Working Notes of the AAAI 2002 Workshop on Ontologies and the Semantic Web, Edmonton, Canada, vol. 28 (2002)

26. Fellbaum, C.: Wordnet: An electronic lexical database (1998), WordNet is available from `http://www.cogsci.princeton.edu/wn`

27. Balkova, V., Sukhonogov, A., Yablonsky, S.: Russian wordnet. from uml-notation to internet/intranet database implementation. In: Proceedings of the Second International WordNet Conference (2004)

28. Atserias, J., Villarejo, L., Rigau, G., Agirre, E., Carroll, J., Magnini, B., Vossen, P.: The MEANING Multilingual Central Repository. In: Proceedings of the 2nd Global WordNet Conference (GWC), Brno, Czech Republic (January 2004)
29. Toral, A., Ferrández, O., Agirre, E., Munoz, R., Fakultatea, I., Donostia, B.: A study on linking wikipedia categories to wordnet synsets using text similarity. In: Proceedings of the 7th International Conference on Recent Advances in Natural Language Processing, pp. 449–454 (2009)
30. Niemann, E., Gurevych, I.: The people's web meets linguistic knowledge: automatic sense alignment of wikipedia and wordnet. In: Proceedings of the Ninth International Conference on Computational Semantics, IWCS 2011, pp. 205–214 (2011)
31. Heydon, A., Najork, M.: Mercator: A scalable, extensible web crawler. World Wide Web 2(4), 219–229 (1999)
32. Bracewell, D.B., Ren, F., Kuroiwa, S.: Mining News Sites to Create Special Domain News Collections. Computational Intelligence 4(1), 56–63 (2007)
33. Hirst, G., St-Onge, D.: Lexical chains as representations of context for the detection and correction of malapropisms. WordNet: An Electronic Lexical Database 305, 305–332 (1998)
34. Biemann, C.: Chinese whispers: an efficient graph clustering algorithm and its application to natural language processing problems. In: Proceedings of the First Workshop on Graph Based Methods for Natural Language Processing, pp. 73–80. Association for Computational Linguistics (2006)

What is being Measured in an Information Graphic?

Seniz Demir[1], Stephanie Elzer Schwartz[2], Richard Burns[3],
and Sandra Carberry[4]

[1] TUBITAK-BILGEM, Turkey
seniz.demir@tubitak.gov.tr
[2] Dept. of Computer Science, Millersville University, USA
stephanie.schwartz@millersville.edu
[3] Dept. of Computer Science, West Chester University, USA
rburns@wcupa.edu
[4] Dept. of Computer and Information Sciences, University of Delaware, USA
carberry@cis.udel.edu

Abstract. Information graphics (such as bar charts and line graphs)
are widely used in popular media. The majority of such non-pictorial
graphics have the purpose of communicating a high-level message which
is often not repeated in the text of the article. Thus, information graphics
together with the textual segments contribute to the overall purpose of an
article and cannot be ignored. Unfortunately, information graphics often
do not label the dependent axis with a full descriptor of what is being
measured. In order to realize the high-level message of an information
graphic in natural language, a referring expression for the dependent axis
must be generated. This task is complex in that the required referring
expression often must be constructed by extracting and melding pieces of
information from the textual content of the graphic. Our heuristic-based
solution to this problem has been shown to produce reasonable text for
simple bar charts. This paper presents the extensibility of that approach
to other kinds of graphics, in particular to grouped bar charts and line
graphs. We discuss the set of component texts contained in these two
kinds of graphics, how the methodology for simple bar charts can be
extended to these kinds, and the evaluation of the enhanced approach.

1 Introduction

Information graphics (such as simple bar charts and line graphs) are non-pictorial
graphics that depict entities and the relations between these entities. Some information graphics are constructed only to visualize raw data. However when
such graphics appear in newspapers and magazines, they generally have a communicative goal or message that is intended to be conveyed to the graph viewer.
The graphic designer makes deliberate choices in order to bring that message
out such as highlighting certain aspects of the graphic via annotations or the use
of different colors. For example, the bar chart in Figure 1 ostensibly conveys the
high-level message that "The mortgage program assets of the Chicago Federal

A. Gelbukh (Ed.): CICLing 2013, Part I, LNCS 7816, pp. 501–512, 2013.

Fig. 1. Graphic from Business Week

Home Loan Bank had a rising trend from 1998 to 2003". Clark [4] contends that language is not just text and utterances, but instead includes any deliberate signal (such as facial expressions) that is intended to convey a message. Thus under Clark's definition, an information graphic is a form of language.

Information graphics often appear as part of a multimodal document in popular media. Carberry et al. [3] contend that the high-level message of an information graphic is the primary contribution of the graphic to the article in which it appears. However, very often the intended message of a graphic is not repeated in the article's text. This is in contrast with scientific documents which generally describe what is graphed in their graphics. Moreover, the high-level message of a graphic often cannot be gleaned from the graphic's caption [6], as can be seen in Figure 1. Recognizing the intention of information graphics is an important step in fully comprehending a multimodal document. Thus, information graphics which contribute to the overall purpose of an article [9] cannot be ignored.

Several applications can greatly benefit from the realization of the graphic's high-level message such as the summarization of multimodal documents and the retrieval of graphics in a digital library. However, additional information should be extracted from the graphic if the high-level message is to be realized. One such piece of information is a descriptor for what is being measured in the graphic. Unfortunately, information graphics often do not label the dependent axis with a descriptor and in such cases the descriptor must be extracted from the text of the graphic. For example, the dependent axis label of the graphic in Figure 1 does not convey what is being measured in that graphic (i.e., "the dollar value of the Chicago Federal Home Loan Bank's mortgage program assets").

In our earlier work [5], we explored a corpus of simple bar charts to identify where to look for the pieces of a descriptor (which is referred to as the "measurement axis descriptor"). The insights gained from the study were used to develop a set of heuristics and augmentation rules that can be applied to a simple bar chart in order produce its measurement axis descriptor. The evaluation study showed that the approach generally produces reasonable descriptors as compared to several baselines. Although the heuristic-based approach is unique in that it showed that methodologies generally used in processing texts (such as finding patterns via corpus analysis) can successfully be applied to non-stereotypical forms of language, its extensibility to other kinds of graphics has not been previously validated.

Different kinds of graphics have different characteristics (such as the number of high-level messages, the type of underlying data – discrete versus continuous, and the distribution of component texts). Thus, one cannot assume that the heuristic-based solution would generate appropriate measurement axis descriptors for all kinds of graphics. This paper presents how a reasonable measurement axis descriptor for grouped bar charts and line graphs can be produced. The characteristics of these two kinds of graphics represent sufficient variability and hence the task of identifying descriptors for these kinds should be treated differently. We greatly drew from our earlier work on simple bar charts as we developed our generation approach. We first explored the component texts contained in these kinds of graphics and identified their similarities and differences from those of simple bar charts. We then investigated how the heuristic-based solution could be extended to grouped bar charts and line graphs and evaluated the generalizability of the overall methodology to more complex graphics.

The rest of this paper is organized as follows. Section 2 discusses related research and Section 3 describes the textual components of grouped bar charts and line graphs. Section 4 gives an overview of the heuristic-based approach and describes how this approach is extended to these kinds of graphics. Finally, Sections 5 presents the evaluation of the enhanced approach and Section 6 concludes the paper.

2   Related Work

The intention of information graphics has been studied by previous research. Kerpedjiev et al. [10] developed a method for automatically realizing intentions in graphics. The PostGraphe system [8] generates a multimodal report containing graphics with accompanying text once a data set is given along with the communicative intentions of the author. Elzer et al. [7] developed a Bayesian system that automatically recognizes the high-level message of simple bar charts. The intention recognition systems for more complex graphics were later developed. Wu et al. [13] implemented a system which recognizes the most likely high-level message in line graphs. Burns et al. [2] investigated the high-level messages that grouped bar charts can convey and developed a methodology for recognizing these messages.

Producing the measurement axis descriptor for a graphic is indeed a referring expression generation problem. Research has been studying the generation of referring expressions (determining a set of properties that would single out the target entity among other entities) for some years [11]. In recent years, it is considered as postprocessing in extractive multidocument summarization where the goal is to improve text coherence [1]. One prominent work in this area improves the coherence of a multidocument summary of newswire texts by regenerating referring expressions for the first and subsequent mentions of people names [12]. The referring expressions are extracted from input documents via a set of rewrite rules. Since our task is extracting a reasonable referring expression for the dependent axis from the graphic's text, it is similar to but more complex than this body of research. First, the referring expression often does not appear as a single unit and thus must be produced by extracting and combining pieces of information from the graphic's text. Second, even if it appears as a label on the dependent axis, it still needs to be augmented.

3   Textual Components of Information Graphics

Graphic designers often use text within and around an information graphic in order to present related information about the graphic. The set of component texts contained in graphics, individual or composite[1], are visually distinguished from one another by different fonts or styles, by blank lines, or by different directions and positions in the graphic. Despite the differences between the number of component texts present in a graph, an alignment or leveling of text contained in a graphic ("text levels") is observed. In this section, we describe the similarities and differences between the textual content of the three kinds of graphics that are considered in this work.

3.1   Simple Bar Charts

In our earlier study [5], a corpus of 107 bar charts (which we refer to as the "bar_chart_corpus") was collected from 11 different magazines (such as Business Week and Newsweek) and newspapers. These graphics were either individual bar charts or charts that appear along with other graphics (some of these graphs were of different kinds) in a composite. Seven text levels were observed in the collected corpus, but every text level did not appear in all graphics. Two of these text levels (Overall Caption and Overall Description) applied to composite graphs and referred to the entire collection of graphics in the composite whereas the remaining text levels applied to each individual graph:

- **Overall Caption:** The text that appears at the top of a composite graph (i.e., the lead caption on the composite) and serves as a caption for the whole set of individual graphs.
- **Overall Description:** The text that often appears under the Overall Caption in a composite but distinguished from it by a different font or a blank line. This text component is pertinent to all graphs in a composite and elaborates on them.
- **Caption:** The text that comprises the lead caption on a graphic.
- **Description:** The text that appears under the lead caption of a graphic.
- **Text In Graphic:** The only text that appears within the borders of a graphic.

---

[1] Composite graphics consist of multiple individual graphs.

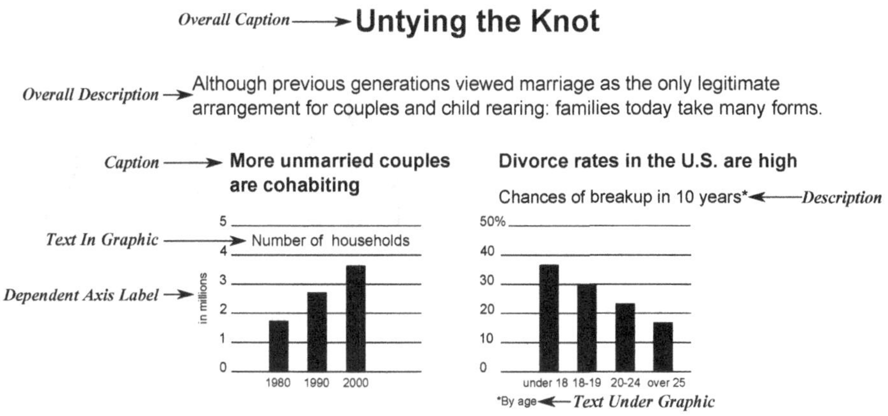
Overall Caption
Untying the Knot
Overall Description
Although previous generations viewed marriage as the only legitimate arrangement for couples and child rearing: families today take many forms.
Caption
More unmarried couples are cohabiting
Divorce rates in the U.S. are high
Chances of breakup in 10 years*
Description
Text In Graphic
Number of households
Dependent Axis Label
in millions
5
4
3
2
1
0
1980 1990 2000
50%
40
30
20
10
0
under 18 18-19 20-24 over 25
*By age
Text Under Graphic

Fig. 2. A composite graphic

- **Dependent Axis Label:** The text that explicitly labels the dependent axis of a graphic.
- **Text Under Graphic:** The text under a graphic that usually starts with a marker symbol (such as * or **) and is essentially a footnote[2].

For example, Figure 2 presents a composite graphic which contains all text levels recognized for simple bar charts. In the study, it was observed that almost all graphics have a Caption (99%) and the Description appeared in slightly more than half of the graphics (54.2%). The Text Under Graphic was the least observed text level (7.5%). Only 18.7% of the graphics had an explicitly labelled Dependent Axis Label. With the exception of one graphic, the Dependent Axis Label (if present) consisted solely of a phrase used to give the unit and/or scale of the values presented in the graphic (such as "in millions" in the left bar chart in Figure 2). These phrases, referred to as unit (e.g., "dollars") and scale (e.g., "billions") indicators, were also observed as part of other text levels.

3.2 Grouped Bar Charts

Grouped bar charts are a kind of information graphic which visually display quantifiable relationships of values with an additional grouping dimension. The grouping dimension makes these graphics more complex to comprehend than simple bar charts. We undertook a corpus analysis in order to explore the set of component texts contained in grouped bar charts. We collected a corpus of 120 grouped bar charts from popular media such as USA Today, The Economist, NewsWeek, and Time. The charts in the corpus (which we refer to as the "grouped_chart_corpus") are diverse in several ways, including the presentation

[2] Each Text Under Graphic has a referrer elsewhere that ends with the same marker and that referrer could be at any text level of the graphic. In the case of multiple such texts within the same graphic, each Text Under Graphic is differentiated with a different marker.

Arriving Late

The percentage of venture-capital financing in more-established companies is on the rise.

Venture-capital investments
■ Seed and early-stage companies
☐ Later-stage companies

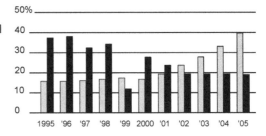

Fig. 3. Graphic from The Wall Street Journal

of the underlying data, the use of visual patterns, and the positioning of textual content. However, we recognized a leveling of component texts in grouped bar charts which is similar to that of simple bar charts. For instance, the text levels of grouped bar charts are also visually distinguishable from one another and not every text level presents in all grouped bar charts.

The seven text levels observed in simple bar charts were also recognized in grouped bar charts. In addition to these levels, another text level which we call the "Legend Descriptor" was observed in the corpus. The Legend Descriptor is the text that appears at the top of the labels of groupings in the chart. The Legend Descriptor is distinguished from the group labels and the preceding text (if exists) by a different font or a blank line. The Legend Descriptor and the group labels, surrounded with a box of solid lines in some cases, appear at different positions in grouped bar charts, including at the top of the graph and within the graph borders. For example, in the graph shown in Figure 3, the Legend Descriptor (i.e., "Venture-capital investments") and the group labels (i.e., "Seed and early-stage companies" and "Later-stage companies") appear at the left side of the dependent axis. The second column of Table 1 presents how often the recognized text levels appear in the grouped_chart_corpus. The Caption and the Text In Graphic are the most frequent and least frequent text levels in the corpus respectively.

Table 1. Text levels in grouped bar charts and line graphs

Text level	Frequency grouped bar charts	Frequency line graphs
Overall Caption	42 ~ 35%	24 ~ 20%
Overall Description	35 ~ 29.2%	20 ~ 16.7%
Caption	116 ~ 96.7%	115 ~ 95.8%
Description	94 ~ 78.3%	90 ~ 75%
Text In Graphic	5 ~ 4.17%	13 ~ 10.8%
Legend Descriptor	17 ~ 14.2%	
Dependent Axis Label	21 ~ 7.5%	12 ~ 10%
Text Under Graphic	8 ~ 6.7%	1 ~ 0.83%

Percent of adults who smoke in New York City

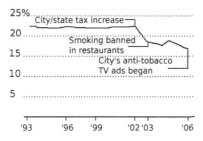

Fig. 4. Graphic from USA Today

3.3   Line Graphs

Line graphs are the preferred medium for conveying trends in quantitative data over an ordinal independent axis. For line graphs, we followed the same methodology for examining their component texts and how these texts are distributed around each graph. We first collected a corpus of 120 line graphs (which we refer to as the "line_graph_corpus") from newspapers (such as USA Today and Los Angeles Times) and magazines (such as BusinessWeek and Forbes). The seven text levels described in Section 3.1 were also recognized in this corpus. In addition, line graphs were observed to share a characteristic that was not seen in bar charts and grouped bar charts. In simple and grouped bar charts, the annotations are mainly for presenting values of specific bars. Although value annotations are also present in line graphs, the annotations at sample points of line graphs diverge substantially in nature. For example, Figure 4 presents a line graph from the corpus where the annotations are used for describing events occured within the time period depicted in the graph. As shown in the third column of Table 1, the Caption appears the most in the line_graph_corpus whereas the Text Under Graphic is the least observed text level.

4   Generating a Referring Expression for the Dependent Axis

We use the main aspects and overall methodology developed for simple bar charts in order to identify the measurement axis descriptor for grouped bar charts and line graphs. The methodology is based on the results of an analysis of a large corpus of simple bar charts (i.e., the bar_chart_corpus) and the identified measurement axis descriptors for these charts by two human annotators. The human annotators not only used the information residing within the component texts of a bar chart but also the graphic's article and their commonsense knowledge in order identify its measurement axis descriptor. There are a number of observations gained from this study, the most important of which are:

- The measurement axis descriptors often cannot be extracted as a whole from a single text level but rather formed by putting together pieces from component texts of the same graph or other graphs in the composite, or from the article's text.
- The measurement axis descriptor consists of a core noun phrase or wh-phrase that appears in one text level of the graphic, possibly augmented with words appearing elsewhere. For example, for the left bar chart in Figure 2, "number of households" is the core of the measurement axis descriptor "U.S.'s number of households".
- Even in cases where the dependent axis is explicitly labelled with a descriptor, it still has to be augmented with pieces of information extracted from other text levels.
- The textual components of a graphic form a hierarchy according to their placements and the core of the measurement axis descriptor generally appears in the lowest text level present in the graph. The ordering of the text levels in Table 1 (except the Legend Descriptor[3] and the Text Under Graphic) forms that hierarchy where the Dependent Axis Label and Overall Caption are at the bottom and top of it.
- If multiple sentences or sentence fragments are contained in a text level, the core of the measurement axis descriptor typically appears near the end of that level.
- Some cues are strong indicators of where to look for the core such as "Here is" and "Here are". The core generally follows such phrases and appears as a noun phrase in the same text level.

The methodology for generating a measurement axis descriptor consists of four steps. First, preprocessing deletes the unit/scale indicators and the ontological category of the bar labels (in cases where explicitly marked by the prepositions "by") from the text levels. Then, a set of heuristics is applied to the text levels in order to extract the core of the descriptor. Next, three kinds of augmentation rules are applied to the core in order to produce the measurement axis descriptor[4]. None of these augmentations might be applicable and in such cases the core forms the whole descriptor. Finally, postprocessing appends a phrase to indicate the unit of measurement to the front of the descriptor if it does not already contain the unit of measurement.

9 heuristics are developed for identifying the core. The heuristics are dependent on the parses of text levels and give preference to the text levels that are lower in the hierarchy. Heuristics are applied in order to the text levels (starting from the lowest text level) until the core is extracted from a level. The following are three sample heuristics:

- **Heuristic 2:** If Text In Graphic consists solely of a noun phrase, then that noun phrase is the core; otherwise, if Text In Graphic is a sentence, the noun phrase that is the subject of the sentence is the core.

---

[3] This text level is not observed in simple bar charts.

[4] In detail information about all heuristics and augmentation rules can be found in [5].

- **Heuristic 4:** If the current sentence at the text level consists solely of a wh-phrase followed by a colon (:) or a question mark (?), that wh-phrase is the core.
- **Heuristic 6:** If a fragment at the text level consists solely of a noun phrase, and the noun phrase is not a proper noun, that noun phrase is the core.

A sample augmentation rule used to fill out the descriptor is as follows:

- **Specialization of the noun phrase:** The core is augmented with a proper noun which specializes the descriptor to a specific entity; that proper noun is either the only proper noun present at any text level higher in the hierarchy than the level from which the core is extracted, or the only proper noun in the Overall Caption or the Caption.

For the graphic in Figure 1, Heuristic 2 identifies "mortgage program assets" in the Text In Graphic as the core. Since there is only one proper noun in the Description, the augmentation rule for specialization produces "Chicago Federal Home Loan Bank's mortgage program assets" as the augmented core. After adding a pre-fragment, the measurement axis descriptor becomes "The dollar value of Chicago Federal Home Loan Bank's mortgage program assets".

4.1   Measurement Axis Descriptor for Grouped Bar Charts

In grouped bar charts, we recognized all of the text levels observed in simple bar charts. We argue that the heuristics applicable to these text levels in simple bar charts can also be considered for the corresponding textual components of grouped bar charts. However, another text level, the Legend Descriptor, may appear in grouped bar charts. We therefore determined a new hierarchy of text levels for grouped bar charts which is the ordering shown Table 1 (except the Text Under Graphic). As in simple bar charts, the Overall Caption and the Dependent Axis Label are at the top and bottom of the hierarchy respectively. We also developed a new heuristic that is restricted to the Legend Descriptor:

- **Heuristic 10:** If Legend Descriptor consists solely of a noun phrase, then that noun phrase is the core; otherwise, if Legend Descriptor is a sentence, the noun phrase that is the subject of the sentence is the core.

Since the Legend Descriptor is the second lowest text level in the hierarchy, the heuristic specialized to this level is applied in between the heuristics that are restricted to the Dependent Axis Label (Heuristic 1) and the Text In Graphic (Heuristic 2). For example, for the graphic in Figure 3, our enhanced approach uses Heuristic 10 and a pre-fragment to construct "the percentage of venture-capital investments" as the referring expression for the dependent axis.

4.2   Measurement Axis Descriptor for Line Graphs

All text levels observed in line graphs are already covered by the heuristics developed for simple bar charts. Thus, rather than developing new heuristics specific to line graphs, we decided to use the heuristics of simple bar charts in order to produce the measurement axis descriptor for line graphs. For example, for the graphic in Figure 4, our approach runs Heuristic 6 to produce "the percent of adults who smoke in New York City" as the descriptor of that graph.

5 Evaluation

To evaluate our enhanced methodology on grouped bar charts and line graphs, we followed the same evaluation strategy that was carried out to evaluate the generated descriptors for simple bar charts [5]. First, a test corpus of graphics was collected and the produced measurement axis descriptors for these graphics were rated by two human evaluators. Each evaluator assigned a rating from 1 to 5 to each measurement axis descriptor: 1(very bad), 2 (poor, missing important information), 3 (good, contains the right information but is hard to understand), 4 (very good, interpreted as understandable but awkward), and 5 (excellent text). In cases where the evaluators assigned different ratings to the same graph, the lowest rating was recorded. Using the same scale, the evaluators also evaluated a number of baselines which used texts appearing at different text levels. For the baselines, if the evaluators differed in their ratings, the higher rating was recorded. This biased the evaluation toward better scores for the baselines in contrast to the identified measurement axis descriptors.

The evaluation of simple bar charts was performed on a corpus of 205 graphics [5]. The Dependent Axis Label, Text In Graphic, and Caption were used as baselines in the evaluation[5]. The evaluation score for the produced measurement axis descriptors was midway between good and very good and far better than the scores for the three baselines as shown in the second column of Table 2. For the evaluation of grouped bar charts, we collected a corpus of 120 graphics from popular media[6]. We used the Dependent Axis Label, Legend Descriptor, Text In Graphic, and Caption as baselines. The evaluation scores collected for the grouped bar charts are shown in the third column of Table 2. Similarly, we collected 120 line graphs from different newspapers and magazines in order to evaluate the generated descriptors for these graphs[7]. The Dependent Axis Label, Text In Graphic, and Caption were the baselines of the evaluation. The evaluation scores assigned by the same evaluators, who also rated the texts produced for grouped bar charts, are shown in the fourth column of Table 2.

The scores given in Table 2 show that our enhanced approach generates measurement axis descriptors which are midway between good and very good for grouped bar charts and line graphs. It is noteworthy to mention that the average score of the produced descriptors for grouped bar charts is higher than that of simple bar charts. Moreover, the average score assigned to the descriptors of line graphs is close to the score that simple bar charts received. In both grouped

---

[5] The Description was not used as a baseline in the evaluation since that text level is most often full sentences and thus would generally produce very poor results.

[6] These graphics are different from those contained in the grouped_chart_corpus.

[7] None of these graphics are contained in the line_graph_corpus.

Table 2. Evaluation results of simple bar charts, grouped bar charts, and line graphs

Evaluated Text	Avg. Score simple bar charts	Avg. Score grouped bar charts	Avg. Score line graphs
Measurement Axis Descriptor	**3.574**	**3.867**	**3.467**
Dependent Axis Label	1.475	1.367	1.075
Legend Descriptor		1.383	
Text In Graphic	1.757	1.142	1.25
Caption	1.876	2.683	2.733

bar charts and line graphs, the evaluation score of the produced descriptors is far better than the scores for the baselines. Our analysis also showed that the text level from where the core of the measurement axis descriptor is extracted varies among grouped bar charts and line graphs (shown in Table 3) and that the three kinds of augmentation are used in both kinds of graphics. Although the average score that the Caption received in grouped bar charts and line graphs is significantly higher than the scores for other baselines, there are a number of cases where this text is not enough on its own. For example, for the graphic in Figure 3, the Caption "Arriving Late" does not convey what is being measured in the graphic. Thus, both evaluators assigned the lowest rating to this text level in the evaluation. On the other hand, the produced descriptor for the same graph "The percentage of venture-capital investments" received the highest score of 5 from both evaluators.

Table 3. The use of heuristics in grouped bar charts and line graphs

Corpus	H_1	H_2	H_3	H_4	H_5	H_6	H_7	H_8	H_9	H_10
Grouped bar charts	11	5	1	4	13	64	4	5	0	13
Line graphs	5	9	0	1	16	66	7	8	8	0

If a text level presents in a graphic, our enhanced generation approach uses the same heuristics applicable to that level no matter in which kind of graphic it appears. However, it was unclear that this approach would work well in different kinds of graphics before the evaluations. But, the collected scores demonstrated that the heuristics applicable to a particular text level work reasonably well in different kinds of graphics. This can be seen from the fact that the average scores of the measurement axis descriptor for different kinds of graphics are close to each other in the rating scale and higher than the scores for the baselines.

6   Conclusion

This paper presents how a heuristic-based approach for generating a referring expression for the dependent axis (i.e., what is being measured in the graphic) of simple bar charts can be extended to different kinds of graphics, in particular to grouped bar charts and line graphs. We identified the textual components of grouped bar charts and line graphs and described the similarities and differences of these components from the texts of simple bar charts. We presented how our existing approach can be enhanced to cover the observed text levels in these kinds of graphics. An evaluation study showed that our approach generally produces reasonable referents for the dependent axis of grouped bar charts and line graphs. In the future, we will address the task of generating such referents for the dependent axis of all kinds of information graphics.

References

1. Belz, A., Kow, E., Viethen, J., Gatt, A.: The grec challenge 2008: Overview and evaluation results. In: The Proceedings of the 5th International Natural Language Generation Conference (2008)
2. Burns, R., Carberry, S., Elzer, S., Chester, D.: Automatically Recognizing Intended Messages in Grouped Bar Charts. In: Cox, P., Plimmer, B., Rodgers, P. (eds.) Diagrams 2012. LNCS, vol. 7352, pp. 8–22. Springer, Heidelberg (2012)
3. Carberry, S., Elzer, S., Demir, S.: Information graphics: An untapped resource for digital libraries. In: The Proceedings of the ACM Special Interest Group on Information Retrieval Conference, pp. 581–588 (2006)
4. Clark, H.: Using Language. Cambridge University Press (1996)
5. Demir, S., Carberry, S., Elzer, S.: Issues in realizing the overall message of a bar chart. In: Recent Advances in Natural Language Processing, vol. 5, pp. 311–320. John Benjamins (2007)
6. Elzer, S., Carberry, S., Chester, D., Demir, S., Green, N., Zukerman, I., Trnka, K.: Exploring and exploiting the limited utility of captions in recognizing intention in information graphics. In: The Proceedings of the Annual Meeting on Association for Computational Linguistics, pp. 223–230 (2005)
7. Elzer, S., Carberry, S., Zukerman, I., Chester, D., Green, N., Demir, S.: A probabilistic framework for recognizing intention in information graphics. In: The Proceedings of the International Joint Conference on Artificial Intelligence, pp. 1042–1047 (2005)
8. Fasciano, M., Lapalme, G.: Intentions in the coordinated generation of graphics and text from tabular data. Knowledge and Information Systems 2(3), 310–339 (2000)
9. Grosz, B., Sidner, C.: Attention, intentions, and the structure of discourse. Computational Linguistics 12(3), 175–204 (1986)
10. Kerpedjiev, S., Green, N., Moore, J., Roth, S.: Saying it in graphics: from intentions to visualizations. In: The Proceedings of the Symposium on Information Visualization, pp. 97–101 (1998)
11. Krahmer, E., van Erk, S., Verleg, A.: Graph-based generation of referring expressions. Computational Linguistics 29(1), 53–72 (2003)
12. Nenkova, A., McKeown, K.: References to named entities: a corpus study. In: The Proceedings of the Conference of the North American Chapter of the Association for Computational Linguistics on Human Language Technology, pp. 70–72 (2003)
13. Wu, P., Carberry, S., Elzer, S., Chester, D.: Recognizing the Intended Message of Line Graphs. In: Goel, A.K., Jamnik, M., Narayanan, N.H. (eds.) Diagrams 2010. LNCS, vol. 6170, pp. 220–234. Springer, Heidelberg (2010)

Comparing Discourse Tree Structures

Elena Mitocariu[1], Daniel Alexandru Anechitei[1], and Dan Cristea[1,2]

[1] "Al.I.Cuza" University of Iasi, Faculty of Computer Science
16, General Berthelot St., 700483 – Iasi, Romania
[2] Romanian Academy, Institute for Computer Science
2, T. Codrescu St., 700481 – Iasi Romania
{elena.mitocariu,daniel.anechitei,dcristea}@info.uaic.ro

Abstract. The existing discourse parsing systems make use of different theories to put at the basis of processes of building discourse trees. Many of them use Recall, Precision and F-measure to compare discourse tree structures. These measures can be used only on topologically identical structures. However, there are known cases when two different tree structures of the same text can express the same discourse interpretation, or something very similar. In these cases Precision, Recall and F-measures are not so conclusive. In this paper, we propose three new scores for comparing discourse trees. These scores take into consideration more and more constraints. As basic elements of building the discourse structure we use those embraced by two discourse theories: Rhetorical Structure Theory (RST) and Veins Theory, both using binary trees augmented with nuclearity notation. We will ignore the second notation used in RST – the name of relations. The first score takes into account the coverage of inner nodes. The second score complements the first score with the nuclearity of the relation. The third score computes Precisions, Recall and F-measures on the vein expressions of the elementary discourse units. We show that these measures reveal comparable scores there where the differences in structure are not doubled by differences in interpretation.

Keywords: discourse parser, Rhetorical Structure Theory, Veins Theory, evaluation, discourse tree structure.

1 Introduction

Discourse parsing systems have an important role in many NLP applications such as question-answering, summarization or deep understanding of texts. Driven by one or another of the discourse theories, the majority of parsers produce tree structures. However, although correctly revealing relations between parts of the discourse, the topology of trees sometimes evidence differences which are not anchored in true discourse facts. To give just one example, a narration (sequence of elementary discourse units of equal salience) could be represented by binary trees in many ways. This paper proposes new measures of evaluating discourse trees which shadows idiosyncrasies of notation, while bringing forth differences which are supported by discourse phenomena.

A. Gelbukh (Ed.): CICLing 2013, Part I, LNCS 7816, pp. 513–522, 2013.

We propose three scores for comparing the discourse trees and we exemplify them from the point of view of Rhetorical Structure Theory and Veins Theory. The first score takes into account only the coverage of the nodes. The second score takes into consideration the nuclearity of the relations. The third score computes Precisions, Recall and F-measures on vein expressions. These metrics can be applied to either RST or VT annotated trees.

Among the theories dealing with the organization of the discourse, Rhetorical Structure Theory (RST) is well accepted and, when the discourse structure should be seen in correlation with referentiality, Veins Theory (VT) proposes to look at the discourse structure from another angle. Both use binary tree[1] representations, in which leafs are elementary discourse units (*edus*), such as clauses or short sentences, and internal nodes represent larger text spans. RST uses a labeling function that attaches relation names and nuclearities to its inner nodes, while VT ignores the names of relations. Because of this simplified representation, one can say that VT is included in RST. For instance, on an annotated RST corpus one can calculate head the vein expressions, which are of primary importance only in VT. However, the conclusions that VT develops go in a different direction then those set by RST.

To compare discourse structures, the PARSEVAL measures [1], based on Precision, Recall and F-measure, are used. The main discontent is that these measures cannot be used to account for structures which, although different, have rather similar interpretations.

In the following, we will consider binary discourse trees on which parent nodes specify the nuclearities of the left and right daughters. For example, in Fig. 1 the node labeled 1 is a nucleus and the node labeled 2 is a satellite.

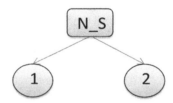

Fig. 1. The discourse tree representation

The automatic identification of discourse structure should include, as a preliminary step, a method for the segmentation of the discourse in *edus*. Discourse tree structures are the result of discourse parsing systems and many of them are developed based on the RST terminology. Since Marcu's first attempt at developing a rule-based discourse parser [2] on the base of discourse markers, several algorithms for discourse parsing have been proposed, both statistical and rule-based [3]. The SPADE algorithm doesn't use cue words for identification of rhetorical relations [4], using instead

---

[1] A binary tree is a tree in which every node has at most two children. The binary trees used to represent discourse structures are always strict (proper) binary trees (each parent node has exactly two children).

syntactic and lexical fetures. It is a sentence-level discourse parser which exploits a set of lexico-gramatical dominance relations corresponding to the syntactic representation of sentences. In [5] is described an algorithm that scores and labels relations between spans, where the discourse parse trees are the result of a supervised leaning. A probabilistic parser for parsing the discourse structure of dialogue is presented in [6].

In the following, we make first a brief presentation of the two theories we have used to exemplify the structure of discourse: Rhetorical Structure Theory – in section 2, and Veins Theory – in section 3, by insisting only on notions linked to representational issues. In section 4 we introduce and discuss the scores and in section 5 we give some conclusions.

2   Rhetorical Structure Theory

Rhetorical Structure Theory developed by Thompson and Mann is one of the most used theories among those dealing with the structure of discourse. It is used in discourse analysis and text generation [7]. Like in the majority of discourse theories, RST segments the text in *edu*s, which could be clauses or sentences expressing elementary events, situations, statements. In RST, the spans of the text have a different importance in the whole discourse. Nuclei are the most important spans in the text and are compulsory for the coherence of the discourse. When they are *edu*s, their deletion makes the text incoherent. When they are larger than *edu*s, at least parts of them should be kept if a simplification is intended. Unlike nuclei, satellites only give additional information. They complement the statements/situations/events communicated by the nuclei and, if deleted from the text, the overall understability will not be harmed, even if some details will be missing.

Thus, the discourse includes nucleus-satellite relations and multinuclearity relations [8]. In a nucleus-satellite (hypotactic) relation, the whole span includes exactly one nucleus and any number of satellites. In a multinuclear (paratactic) relation, all subspans have the same importance and are considered nuclear. Although, the RST trees are not necessarily binary, equivalent transformations can be imagined that make them binary. For instance, a 3-arguments paratactic relation can be transformed into 2 binary paratactic relations with the same name.

3   Veins Theory

Taking from RST the notion of nucleus and satellite, but ignoring the name of the relations, VT attaches two expressions to each node of a binary discourse tree: the head and the vein.

The head expression is a list of the most salient *edu*s that the node covers and is calculated *bottom-up*, as follows:

— if the node is a leaf, the head expression is its label;
— else, the head expression is the concatenation of the head expressions of its nuclear children.

By definition, the vein expression of a node n is that list of elementary discourse units which are of primary importance for understanding the span the node n covers in the context of the whole discourse. It is calculated *top-down*, as follows:

— the vein expression of the root node is its head expression;
— otherwise, if the node is a nucleus and his father has the vein expression v, then:

  • if the node has a left satellite sibling with head h, then its vein expression is *seq(mark(h), v)*;
  • else: v;

— finally, if the node is a satellite with the head h and the vein of his father is v, then:

  • if it is a left son, its vein expression is *seq (h,v)*;
  • else its vein expression is *(seq(h,simpl(v))).*

The *seq*, *simpl*, *mark*, are function which can be described thus:

— *seq* returns the concatenation of two sequences of nodes, considering the frontier ordering, from left to right;
— *mark* returns the same symbols as in its argument, but marked in some way (for example between parentheses or primed);
— *simpl* eliminates all marked symbols from its argument [9].

4   The Scores

In discourse representation theories, it is notorious that a certain vision on a discourse tree structure can be mirrored by different tree topologies, which not necessarily express significant differences in the interpretation. As such, because representing discourse structures by trees has become a standard in many theories (SDRT, RST, VT, etc.), it would be useful to have a measure by which to compare these structures and show to what extend they mean the same thing or not. In the technology of discourse parsing, this issue is particularly important for at least three reasons. First, a corpus of gold files should be prepared by hand, using the competencies of human annotators. And, because the appreciation of a discourse structure is to such a large extend subjective, we would like to know the opinions of more experts on the issue. Then, whenever corpora are built in NLP and more annotators are involved, their outcomes should be compared for agreement. Second, evaluation of a discourse parser is usually done by comparing the parser's output against this gold representation contributed by humans. Finally, modifying certain parameters of the parser could yield trees which are slightly different, while most of the rest of the structure could be shared. But how much identical are these trees? Where in the structure of the tree, differences start to be significant? To what extend some differences can be ignored, therefore characterizing stable decisions of the parser?

In this section we will propose a number of scores that are intended to offer better comparisons between discourse tree structures.

There are many possibilities to represent by binary trees a discourse structure. The number of different binary trees that can be drawn with n terminal nodes is given by the Catalan numbers [10]:

$$\#\text{trees} = \frac{(2*n-2)!}{n!(n-1)!}.$$ (1)

For example, in case of a discourse with 4 *edu*s, formula (1) above indicates 5 different binary trees, which are depicted in (Fig. 2).

4.1   The Overlapping Score

In [3] it is shown that a discourse tree can be formally represented as a set of tuples of the form $R_k[i,m,j]$, denoting a relation and its two arguments, as text spans: the left one spanning between *edu*s i and m, and the right one spanning between *edu*s $m+1$ and j. For example, in the tree of Fig. 2(c) the tuples are : $R_1[1,1,4]$, $R_2[2,2,4]$ and $R_3[3,3,4]$. The tuple representing the root relation will cover all leafs.

The Overlapping Score (OS) is taking into consideration only the coverage of the nodes. The representation with tuples can be used for that. By enduring an abuse of language, we will say that two tuples overlap if their starting and ending positions are identical ($i_1=i_2$ and $j_1=j_2$). With this convention, the score will be computed by dividing the number of overlapping tuples of a discourse tree structure of the same text, to the total numbers of tuples (number of leafs -1) as presented in equation (2):

$$OS = \frac{\#overlapping\ tuples}{\#total\ tuples}.$$ (2)

At least one pair of tuples are overlapping, the ones describing the root relation and which should cover the whole range of nodes. Two different tree structures with the same text segmentation in clauses will have the same number of leafs.

To exemplify OS, we will compute it for some pairs of trees in the family of trees of (Fig. 2) (imagining that one represents a gold file and the second is contributed by a discourse parser). For example if we consider the discourse trees from Fig. 2(a) and Fig. 2(b), the number of overlapping tuples is two: $R_3[1,1,2]$ overlaps with $R_2[1,1,2]$ (covering nodes: 1 and 2) and $R_1[1,3,4]$ is identical with $R_1[1,2,4]$ (covering nodes 1, 2, 3, 4). The total number of tuples for this example is 3. The remaining pair of tuples ($R_2[1,2,3]$ and $R_3[3,3,4]$ are not overlapping because $i_1 \neq i_2$ and also $j_1 \neq j_2$ Applying equation (2) we obtain OS = 2/3.

If we analyze the trees from Fig. 2(a) and Fig. 2(c), the number of overlapping tuples is 1 and the OS will be 1/3. Thus we can conclude that the discourse tree structure from Fig. 2(a) is more similar to the discourse tree structure of Fig. 2(b) than to that in Fig. 2(c). If OS equals 1 it means that the two discourse trees structures are identical from the point of view of the topology.

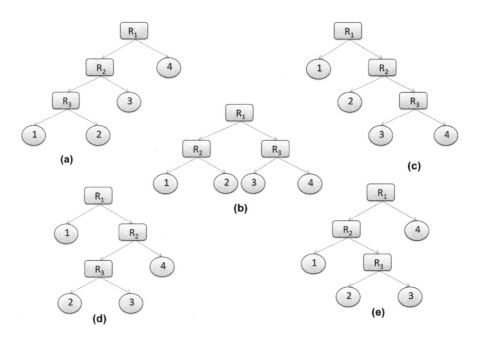

Fig. 2. Different representations as binary trees for a text with 4 *edus*

4.2   The Nuclearity Score

As known, in both RST and VT, the nuclearity is very important. Based on OS, we can develop a Nuclearity Score (NS). This will take into consideration, besides the overlapping spans under nodes, as given by the OS, the nuclearity of the relations for the corresponding overlapping tuples.

NS is calculated by multiplying OS with the ratio between the number of overlapping tuples with identical nuclearity (let's call these nuclearity-overlapping tuples) and the total number of overlapping tuples.

$$\text{NS} = \text{OS} * \frac{\#nuclearity-overlapping\ tuples}{\#overlaping\ tuples}. \tag{3}$$

After applying formula (2) in (3) and simplification with #overlapping tuples, formula (4) is obtained:

$$\text{NS} = \frac{\#nuclearity-overlapping\ tuples}{total\ \#\ tuples}. \tag{4}$$

The difference between Fig. 2 and Fig. 3 is that the last one includes also marking for the nuclearity of the parrent nodes. To exemplify how NS is computed, let's consider the discourse trees of Fig. 3(b) and Fig. 3(c): there are two pairs of overlapping tuples: (N_S[1,2,4] – N_S[1,1,4]) and (N_S[3,3,4] – S_N[3,3,4]), out of which only one pair is also a nuclearity-overlapping tuple: (N_S[1,2,4] – N_S[1,1,4]). Thus, formula (4) will give NS = 1/3.

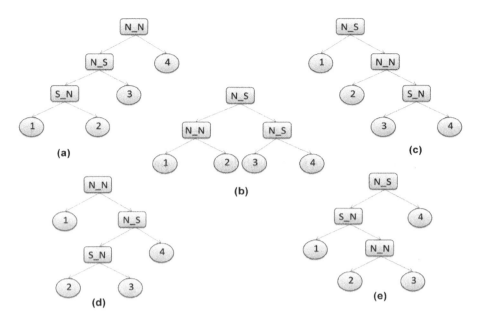

Fig. 3. Examples of tree representations including nuclearity markings

Like in the previous case, if the score is 1, it means that both trees are identically in terms of the topology but also the nuclearity. The way the two scores have been defined makes that the Nuclearity Score be less then or equal with the Overlapping Score for the same pair of trees, showing that NS is more restrictive than OS. When NS is significantly lower than OS, this will highlight differences in nuclearity. Let's note also that nuclearity is relevant only for sub-structures that overlap.

An indicator like NS is important in applications like summarization, where nuclei are relevant. As an observation, considering the fact that the scores depend on each other, this means that if OS is nil then NS will be also nil.

4.3   The Vein Scores

The third proposal we make for comparing discourse structures takes into consideration Precisions, Recall and F-measures on vein expressions. Vein expressions can be computed only on discourse trees that display the nuclearity of nodes. As shown in Sections 2 and 3 above this can be done on both RST and VT annotated trees and the computation proceeds as was presented in Section 3 above. VT uses the vein expressions to measure the coherence of discourse segments as well as of the global discourse [8]. As such, Vein Scores, in addition to the other two scores defined earlier, place centrally the possibility to compare discourses on the ground of coherence, in the measure in which this is implicitly incorporated in tree structure. The formulas are as follows:

$$\text{VS-Precision} = \frac{\sum_{i=1}^{N} \frac{IL_i}{T_i}}{N}. \tag{5}$$

$$\text{VS-Recall} = \frac{\sum_{i=1}^{N} \frac{IL_i}{G_i}}{N}. \tag{6}$$

$$\text{VS-FMeasure} = \frac{2 * \text{VS-Precision} * \text{VS-Recall}}{\text{VS-Precision} + \text{VS-Recall}}. \tag{7}$$

IL_i - represents the number of identical labels from the vein expression of the *edu i* in the test tree and the gold tree;

T_i - represents the total number of labels of the vein expression for *edu i* in the test tree;

G_i - represents the total number of labels of the vein expression for *edu i* in the gold tree;

N - represents the total number of *edu*s.

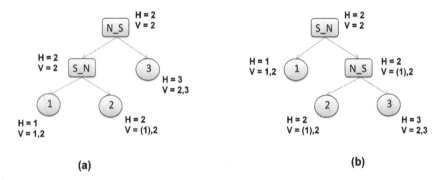

(a)   **(b)**

Fig. 4. Two discourse tree representations including the head and vein expressions for a text with three elementary discourse units

Let us consider the two trees in Fig. 4, with three *edu*s in which the head (H) and vein (V) expressions are noted. Suppose that the discourse tree of Fig. 4(a) is a test tree and the tree of Fig. 4(b) is the gold tree. Applying formulas (5), (6) and (7) we obtain:

$$\text{VS-Precision} = \frac{\sum_{i=1}^{N} \frac{IL_i}{T_i}}{N} = \frac{\frac{2}{2} + \frac{2}{2} + \frac{2}{2}}{3} = \frac{3}{3} = 1$$

$$\text{VS-Recall} = \frac{\sum_{i=1}^{N} \frac{IL_i}{G_i}}{N} = \frac{\frac{2}{2} + \frac{2}{2} + \frac{2}{2}}{3} = \frac{3}{3} = 1$$

$$\text{VS-FMeasure} = \frac{2 * \text{VS-Precision} * \text{VS-Recall}}{\text{VS-Precision} + \text{VS-Recall}} = \frac{2 * 1 * 1}{1 + 1} = 1$$

Although the trees from Fig. 4 are different in terms of structure and nuclearity, it can be seen that the vein expressions of the *edus* of both trees are identical, which means that they convey the same meaning.

Let us consider another example:

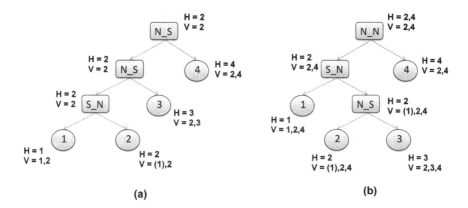

Fig. 5. Two discourse tree representations including the head and vein expressions for a text with four elementary discourse units

Applying the same formulas (5), (6) and (7) we obtain the following results:

$$\text{VS-Precision} = \frac{\sum_{i=1}^{N} \frac{IL_i}{T_i}}{N} = \frac{\frac{2}{2} + \frac{2}{2} + \frac{2}{2} + \frac{2}{2}}{4} = \frac{4}{4} = 1$$

$$\text{VS-Recall} = \frac{\sum_{i=1}^{N} \frac{IL_i}{G_i}}{N} = \frac{\frac{2}{3} + \frac{2}{3} + \frac{2}{3} + \frac{2}{2}}{4} = \frac{3}{4} = 0.75$$

$$\text{VS-FMeasure} = \frac{2 * \text{VS-Precision} * \text{VS-Recall}}{\text{VS-Precision} + \text{VS-Recall}} = \frac{2 * 1 * \frac{3}{4}}{1 + \frac{3}{4}} = \frac{6}{7} \approx 0.857$$

From Fig. 5 it can be seen that the two trees are totally different in terms of topology and nuclearity, but two of the 4 *edus* display the same vein expressions. Correspondingly, the scores computed above reflect this resemblance.

If the first two scores are relevant for comparing discourse tree structures, this one is more relevant for summarization applications. Thus we develop a method of comparing discourse trees based on three scores beginning from the discourse structure to the summarization application, considering both Rhetorical Structure Theory and Veins Theory.

5   Conclusion

We have proposed three new scores for comparing discourse tree structures. The first one measures the identical coverings among all substructures of the discourse structure, the second one factors this number with identical nuclearities and the third one compares vein expressions of the terminal nodes. The idea behind these scores is to consider less distant tree structures that could manifest differences in topology if these differences are not relevant for the interpretation of the discourse. Also, our measures make possible to evaluate test structures produced by a parser even against gold structures annotated with a different theory than the one the parser relies on. Moreover, different parsers can now be evaluated and compared better, because not relevant idiosyncrasies of discourse structures pass now unnoticed by the scores.

The three measures connect well with different interests that could motivate the comparison of structures. One of the applications in which they are most useful is summarization, where the nuclearity of the units has a great importance in recognizing the main ideas.

References

1. Roark, B., Harper, M., Charniak, E., Dorr, B., Johnson, M., Kahne, J.G., Liuf, Y., Ostendorf, M., Hale, J., Krasnyanskaya, A., Lease, M., Shafran, I., Snover, M., Stewart, R., Yung, L.: SParseval: Evaluation metrics for parsing speech. In: Proceedings of LREC (2006)
2. Marcu., D.: The theory and Practice of Discourse Parsing and Summarization. MIT press (2000)
3. Soricut, R., Marcu, D.: Sentence Level Discourse Parsing using Syntactic and Lexical Information. In: Proceedings of the Human Language Technology and North American Association for Computational Linguistics Conference, Edmonton, pp. 149–156 (2003)
4. Hernault, H., Prendinger, H., duVerle, D., Ishizuka, M.: HILDA: A Discourse Parser Using Support Vector Machine Classification. Dialogue and Discourse, pp. 1–33 (2010)
5. Reitter, D.: Simple signals for complex rhetorics: On rhetorical analysis with rich-features support vector models. LDV-Forum. GLDV-Journal for Computational Linguistics and Language Technology, 38–52 (2003)
6. Baldridge, J., Lascarides, A.: Probabilistic head-driven parsing for discourse structure. In: Proceedings of the Ninth Conference on Computational Natural Language Learning, Ann Arbor, Michigan, pp. 96–103 (2005)
7. Mann, W.C., Thompson, S.A.: Rhetorical Structure Theory: Toward a functional theory of text organization. Text, 243–281 (1988)
8. Taboada, M., Mann, W.C.: Rhetorical Structure Theory: looking back and moving ahead. Discourse Studies, 423–459 (2006)
9. Cristea, D., Ide, N., Romary, L.: Veins theory: A model of global discourse cohesion and coherence. In: Proceedings of the 17th International Conference on Computational Linguistics, Montreal, pp. 281–285 (1998)
10. Davis, T.: Catalan Numbers, http://www.geometer.org/mathcircles

Assessment of Different Workflow Strategies for Annotating Discourse Relations: A Case Study with HDRB

Himanshu Sharma[1], Praveen Dakwale[1], Dipti M. Sharma[1],
Rashmi Prasad[2], and Aravind Joshi[3]

[1] LTRC,IIIT-Hyderabad
[2] University of Wisconsin-Milwaukee
[3] University of Pennsylvania
himanshu_s@students.iiit.ac.in,
{dakwale.praveen,diptims}gmail.com,
prasadr@uwm.edu,
joshi@minus.seas.upenn.edu

Abstract. In this paper we present our experiments with different annotation workflows for annotating discourse relations in the Hindi Discourse Relation Bank(HDRB). In view of the growing interest in the development of discourse data-banks based on the PDTB framework and the complexities associated with the discourse annotation, it is important to study and analyze approaches and practices followed in the annotation process. The ultimate goal is to find an optimal balance between accurate description of discourse relations and maximal inter-rater reliability. We address the question of the choice of annotation work-flow for discourse and how it affects the consistency and hence the quality of annotation. We conduct multiple annotation experiments using different work-flow strategies, and evaluate their impact on inter-annotator agreement. Our results show that the choice of annotation work-flow has a significant effect on the annotation load and the comprehension of discourse relations for annotators, as is reflected in the inter-annotator agreement results.

1 Introduction

In recent years, due to the growing interest in discourse analysis in NLP, there is an increasing requirement for corpora annotated with discourse information, for research across different languages. A significant number of corpora are being developed for various languages based on different frameworks for discourse analysis. The most widely used among these is the Penn Discourse Treebank(PDTB)[1] for English which is based on Discourse Lexicalized Tree Adjoining Grammar (DLTAG). This framework is based on the notion that connectives are discourse-level predicates which project predicate-argument structure on par with verbs or abstract objects at sentence level [2]. PDTB involves annotation of mainly two

A. Gelbukh (Ed.): CICLing 2013, Part I, LNCS 7816, pp. 523–532, 2013.

types of discourse relations: Explicit relations, which are annotated when the relation is realized by a connective from syntactically pre-defined categories; and Implicit relations, which are annotated between adjacent pairs of sentences, not related by any of the "Explicit" connectives. Along with these relations, cases where no appropriate Implicit connective can be inserted between the adjacent sentence pairs, are annotated as Alternative Lexicalization(AltLex), Entity relation(EntRel) and No Relation(NoRel).

Following PDTB, development of various discourse annotated corpora for different languages have been initiated based on the similar framework as that of PDTB such as,[3] for Chinese, [4] for Turkish,[5] for Arabic and [6] for Hindi. Besides these discourse annotation for Czech is another important work [7], in which Explicit discourse relations were annotated similar to PDTB, however it is projected as a new annotation layer above the existing layers in the Prague Dependency Treebank.

However, in order to develop better quality corpora, various problems associated with the consistent annotation of discourse relations, need to be handled efficiently. Well defined guidelines, manuals and annotation frameworks are developed in order to reduce these inconsistencies. Various evaluation studies of these corpora such as [8] show low consistencies in terms of inter-annotator agreement. These inconsistencies arise due to many different reasons such as inherent complexity in the identification of discourse relation, identification of multiple elements comprising the relation like connective, arguments etc. and the annotation of larger text spans as arguments.

PDTB based discourse annotation involves identifying different types of discourse relations (Explicit, Implicit, Altlex, Entrel, NoRel). Annotation works related to different languages adopted different annotation work-flows. Some of which adapted simultaneous annotation of above mentioned types of relation, while others adopt different type of relations to be annotated in separate steps. Choice of annotation work-flow can affect the accuracy in discourse annotation as well as the inter-annotator reliability. Thus one needs to device an optimal strategy which achieves better accuracy and consistency. In this paper, we attempt to investigate the effects that different annotation work-flow models and related strategies have on annotation quality. For this purpose, we conduct multiple inter-annotator agreement experiments over the Hindi Discourse relation Bank(HDRB)[6].

2   PDTB and Hindi Discourse Relation Bank

Hindi Discourse relation Bank is an ongoing project aimed at the development of a corpus annotated with discourse relations[6],which mostly follows the framework laid by PDTB. As discussed in section (1), this framework involves annotation of mainly two types of discourse relations : Explicit and Implicit.

The corpus taken for HDRB is a subset of the corpus on which Hindi syntactic dependency annotation is being independently done[9]. All the source corpus texts are taken from Hindi News data. It comprises news articles from several domains, such as politics, sports, films, etc. The current size of the HDRB annotated corpus is approximately 75k words.

Although HDRB follows the approach of PDTB, HDRB 1.0 guidelines depart from PDTB work-flow in some points such as semantic labeling convention (for arguments, as compared to that of syntactically driven labeling in PDTB), subtype level annotation, uniform treatment of pragmatic relations etc. as discussed in [6]. But, as concluded in the evaluation studies of [8], these modifications lead to greater difficulty in annotation, hence they propose a rollback to the approach adopted by PDTB on some points. Also, unlike PDTB, the current annotation on HDRB does not involve annotation of attribution and supplements.

Apart from the differences stated above, an important divergence that HDRB has from PDTB, is that of annotation work-flow. In the next section, we discuss in detail about the different work-flows adopted by annotation projects in different languages pertaining to different requirements, and the advantages and disadvantages associated with them. In Section (5), we provide the results of an inter-annotator study over HDRB comparing the agreement in each work-flow.

3   Background and Motivation

Discourse annotation projects in different languages have adopted mainly two different work-flows for annotation, which involve either separate or simultaneous annotation of different types of discourse relations. In this section, we discuss these two work-flows in detail.

– **Workflow 1 :** *Separate annotation of different types of relations.* PDTB adhered to separate annotation of Explicit and Implicit Connectives [1], i.e. in PDTB, for the first phase, Explicit connectives were annotated throughout the corpus, and the second phase followed the annotation of Implicit relations throughout the same corpus. They adopted this work-flow because Implicit relations are complex and in many cases, are ambiguous as compared to Explicit relations. Separating the annotation of two relation types in two separate steps not only reduces the annotation load, but as stated in [3], making use of explicit relation as a prototype also helps pin them down. Moreover, since PDTB was the first project to apply the proposed framework, they attempted to first undertake the relatively easier task of annotating Explicit relations and later moved on to the more ambiguous and difficult task of identifying Implicit relations. It is because of the incremental development of PDTB that the categories of EntRel and AltLex were devised after the annotation of Implicit connectives, when the annotation revealed that annotators were not able to insert a connective in many cases [10]. Following PDTB, other discourse annotation projects such as [4],[5] etc. also adopted the same work-flow.

– **Workflow 2 :** *Simultaneous annotation of all types of relations.* Some of the discourse annotation projects diverge from the work-flow adopted by

PDTB. For ex [3] adopt the work-flow involving simultaneous annotation of Explicit and Implicit relations, they argue that if the number of Implicit relations is very large as compared to the Explicit relations, then it may not be worthwhile to set up separate tasks to cover a smaller set of data. Similarly, HDRB also considers the same work-flow based on the hypothesis that simultaneous annotation is necessary to ensure that annotators were taking the full context of the discourse into account during the annotation process, as is stated in [8]. However, later they conclude that in order to reduce the load on annotators and hence to maintain the consistency in the annotation, work-flow should be split into two steps as adopted by PDTB initially.

In view of the above discussed advantages and disadvantages of separate and simultaneous annotation, we attempt to empirically determine and compare the quality and consistency obtained by using different work-flows and hence determine the better of these approaches in terms of inter-annotator agreement. In this paper, we describe the experiments that were conducted over different work-flows for HDRB annotation and discuss the qualitative and quantitative analysis of the results.

– **Workflow 3** : *Text-wise independent annotation of all types of relation.* Along with the two work-flows as described above, we also consider a third work-flow as a midway between the above work-flows. It involves separate annotation of Explicit and Implicit connective, but at the level of individual discourses or texts rather than annotation of one type of relation throughout the corpus in a single step. That is, first step involves annotation of Explicit connective in an individual text(article) immediately followed by the annotation of Implicit relations in the same text, and also Altlex, Entrel and NoRel wherever required. Since, in HDRB the corpus taken for the annotation consists of individual articles taken from news data, each news article is almost a coherent single discourse/text. It would be intuitive to consider such a work-flow because it divides the annotation process in separate steps, hence reducing the load on annotators, while at the same time it maintains the attention of the annotators locally to a single text, and hence the overall context of the text is retained in annotator's cognition.

4   Experimental Set Up

We conducted experiments to calculate the inter-annotator agreement over the three work-flows as described in Section (3.1). For these experiments, 6 annotators were grouped in 3 separate groups i.e. 2 annotators in per group. Each group was instructed to follow one of the three work-flows. After the annotation, we calculated the inter-annotator agreement based on different parameters(described in detail in section 3.2) for the pair of annotators in each group. Each annotator was provided with the same corpus consisting of 34 data files, each of which have the text discussing a particular news item. The overall size of the experimental corpus was 12k words.

All the annotators were native speakers of Hindi, they were provided with the same set of guidelines(but different work-flow instructions), and have no previous experience in discourse annotation. Thus all the other parameters of the experiments except work-flow remained constant. We briefly describe the three different annotation work-flows below:

4.1 Annotation Work-Flows

— Group 1 : Annotators in Group-1 were instructed to first annotate all the Explicit connectives throughout the corpus. For this, they were provided with a pre-defined list of Explicit connectives, and were also asked to identify the connectives which belong to certain grammatical categories, as is mentioned in the HDRB guidelines. In the second step, they were instructed to annotate Implicit relations between pairs of adjacent sentences. In the later steps, they identified and annotated the cases for AltLex, Entrel or NoRel, wherever they found it difficult to insert an Implicit connective, respectively.

— Group 2 : In Group-2, the annotators were instructed to conduct simultaneous annotation of all types of relations i.e Explicit, Implicit, AltLex, Entrel and NoRel. They were instructed to annotate Explicit connectives as defined by the guidelines, whenever encountered in the text. At the same time, they were asked to infer the Implicit Relations between pairs of adjacent sentences which were not already related by an explicit connective. In addition, they were directed to identify and annotate AltLex, Entrel or NoRel whenever an Implicit relation could not be inferred.

— Group 3 : Group-3 of annotators were instructed to annotate the relations in individual phases within the same discourse/text. In other words, they annotated all the Explicit connectives pertaining to a file in the first step, next, Implicit Relations between pairs of adjacent sentences in the same file were annotated. Similarly, AltLex, EntRel and NoRel were annotated respectively over the same file in different phases. The same cycle of phases, as is described above, was carried out for each data file in the corpus. The annotation by this group was similar to Group-1. But, unlike Group-1, where the annotation was done simultaneously on complete dataset, this group carried out each phase one by one on a single text and then moved to the next one.

4.2 Inter-annotator Agreement Measures

Agreement over Connective Identification. Identification of connectives is the first step for annotation of Explicit relations. Since a particular lexical element may or may not be a discourse connective all the time (due to different syntactic roles), the annotation task is to identify whether a that lexical element is a discourse connective or not.

We report the number of connectives identified by each annotator and the total number of common connectives identified. In case of Implicit Relations, annotators were asked to first infer the discourse sense between two arguments and then choose a connective which possibly represents this discourse sense. Hence, the agreement is calculated over Implicit connectives inserted by the annotators at the same position in the text.

Agreement over Connective Arguments. Each discourse relation has two arguments, hence, similar to the earlier studies [11], two metrics were considered while calculating inter-annotator agreement over these discourse arguments.

- Metric 1 : For Arg1 and Arg2, agreement is recorded separately, considering them as two separate tokens. The value is 1, when both annotators make identical textual selection of a single token (any individual argument) and 0 otherwise. Thus, the number of instances in this is twice the number of relations in the discourse (since there are two arguments in a relation).

- Metric 2 : For any relation, agreement-value is 1 only when a pair of annotators make identical textual selections for both tokens (both Arg1 and Arg2) and 0 if either one or both don't match.

Agreement over Sense Identification. As discourse annotation in PDTB framework involves annotation of sense for the identified discourse relation from a predefined sense hierarchy, the agreement over the discourse sense should also be calculated. PDTB, HDRB, Chinese and Turkish Discourse Treebanks report the agreement over sense identification using either percentage agreement or F-score, while Arabic Discourse Treebank Experiments[5] report kappa and alpha reliability statistics over sense agreement. Since, sense for each relation could be annotated from a pre-defined set of senses along with a certain hierarchy, kappa and alpha statistics are better measures for determining agreement over sense. The Sense of a relation can only be calculated in the case of Explicit, Implicit and AltLex relations.

We use Fleiss's Kappa [12] statistics for sense agreement in Explicit and Implicit for each group. Fleiss's Kappa is defined as :

$$\kappa = \frac{Pr(a) - Pr(e)}{1 - Pr(e)} \tag{1}$$

The factor (1 - Pr(e)) gives the degree of agreement that is attainable above chance, and, (Pr(a) - Pr(e)) gives the degree of agreement actually achieved above chance.

We report sense agreement for three level of sense identification as follows:

- Class level sense : Contingency, Comparison, Expansion, Temporal.
- Sub-Type level sense : 27 sub-type level senses as defined by [6].

5   Experimental Results and Discussion

In view of the discussion in section (4.2), we calculate agreement for each group over connectives, arguments and senses :

5.1   Inter-annotator Agreement Results

Table(1) shows the agreement over connectives for each group in terms of connectives identified by each annotator and the total number of connectives on which both the annotators agree.

Table 1. Inter Annotator Agreements over Explicit,Implicit Connectives

	Group1(Separate)	Group2(Simultaneous)	Group3(Textwise)
Explicit			
Connectives identified by Ann1	169	192	202
Connectives identified by Ann2	221	215	203
Matching connectives	189	180	175
Implicit			
Connectives identified by Ann1	217	236	356
Connectives identified by Ann2	235	232	368
Matching connectives	157	193	321

Table(2) shows the agreement over arguments for each discourse relation in terms of Measure-1 and Measure-2 as stated in section(4.2)

Table 2. Inter Annotator Agreements over Arguments

	Group1(Separate)	Group2(Simultaneous)	Group3(Text-wise)
Explicit			
Argument agreement (Measure 1)	0.52	0.71	0.831
Argument agreement (Measure 2)	0.40	0.63	0.708
Implicit			
Argument agreement (Measure 1)	0.51	0.73	0.783
Argument agreement (Measure 2)	0.49	0.72	0.613

Table(3) shows inter-annotator agreement for Class-level as well as Subtype-level(Full sense) in terms of Kappa measure as stated in section (4.2)

Table 3. Inter Annotator Agreements over discourse Sense

	Group1(Separate)	Group2(Simultaneous)	Group3(Text-wise)
Explicit			
Class level Sense agreement (Kappa)	0.74	0.857	0.966
Subtype level Sense agreement (Kappa)	0.62	0.780	0.779
Implicit			
Class level Sense agreement (Kappa)	0.61	0.77	0.230
Subtype level Sense agreement (Kappa)	0.49	0.76	0.240

5.2   Discussion and Error Analysis

As observed from our results, inter-annotator agreement values vary over all the parameters across the three groups. This suggests that the choice of annotation work-flow affects the inter-annotator agreement, as we proposed initially.

The agreement values for connectives and arguments in case of both Explicit and Implicit relations for Group-3 (Text-wise separate annotation) are comparatively higher than other groups. Also, Group-3 has higher sense agreement in case of Explicit relations. As we proposed in Section (3), dividing the annotation process in steps, while keeping the discourse context in the annotators' cognition, could help increase the consistency in annotation. The higher consistency observed for Group-3 supports this assumption.

However, in case of Implicit annotation, the kappa agreement value over sense annotation for Group-3 is lower than the other two groups. This observation is in contrast with the agreement values over connectives and arguments, where Group-3 has higher agreement. This suggests possibility of other factors affecting the sense identification, such as the annotators' previous experience in linguistic annotation. As stated earlier, none of the annotators had any previous experience in discourse annotation. But Group-3 had experience in syntactic (dependency) annotation. Thus, there is a possiblity of this annotation experience affecting or helping the annotators in identfying the discourse relations. However, this analysis requires further experiments with annotators having different annotation experience.

Group-1 has lower agreement values for most of the parameters. This suggests that separate annotation of different types of relations decreases the overall inter-annotator agreement. The possible reason behind this could be the discontinuity in the text comprehension which leads to different understanding/interpretation of the discourse relation by different annotators.

6   Conclusion and Future Work

In this paper we presented our experiments with HDRB, to compare inter-annotator agreement over Explicit and Implicit relations with three annotation work-flows, i.e. *simultaneous*, *separate* and *text-wise* annotation. The results show that the choice of annotation work-flow affects all the elements in discourse annotation; specifically connectives, arguments and sense identification. The most significant observation from our experiments is the higher agreement in *text-wise* annotation work-flow as compared to the other two work-flows. This increased consistency in annotation can be attributed to the low annotation load along with the availability of the text context in the annotators' mind while following this work-flow.

However, we observe that some of the results such as lower agreement for sense annotation could not be explained only on the basis of choice of the work-flow. This suggests that there is a possibility of other factors affecting the process, such as the linguistic background or annotation experience of the annotator. Further studies and experiments are required to analyze these parameters along with the choice of annotation strategies, which we plan to conduct as a future work. Nevertheless, the varying agreement values in our current set of experiments correctly verify our initial assumption about the effects of choice of the work-flow over the consistency in annotation.

References

1. Prasad, R., Dinesh, N., Lee, A., Miltsakaki, E., Robaldo, L., Joshi, A.K., Webber, B.L.: The Penn Discourse TreeBank 2.0. In: LREC (2008)
2. Webber, B., Stone, M., Joshi, A., Knott, A.: Anaphora and Discourse Structure. Computational Linguistics 29, 545–587 (2003)
3. Yuping, Z., Nianwen, X.: PDTB-style discourse annotation of Chinese text. In: Proceedings of the 50th Annual Meeting of the Association for Computational Linguistics, ACL 2012, pp. 69–77 (2012)
4. Zeyrek, D., Demirşahin, I., Sevdik-Çalli, A., Balaban, H.Ö., Yalçinkaya, İ., Turan, Ü.D.: The Annotation Scheme of the Turkish Discourse Bank and an evaluation of inconsistent annotations. In: Proceedings of the Fourth Linguistic Annotation Workshop, LAW IV 2010, pp. 282–289 (2010)
5. Al-Saif, A., Markert, K.: The Leeds Arabic Discourse Treebank: Annotating Discourse Connectives for Arabic. In: LREC (2010)
6. Oza, U., Prasad, R., Kolachina, S., Sharma, D.M., Joshi, A.: The Hindi Discourse Relation Bank. In: Proceedings of the Third Linguistic Annotation Workshop, ACL-IJCNLP 2009, pp. 158–161 (2009)
7. Mladová, L., Zikánová, Š., Hajičová, E.: From Sentence to Discourse: Building an Annotation Scheme for Discourse based on Prague Dependency Treebank. In: Proceedings of Language Resources and Evaluation, LREC (2008)
8. Kolachina, S., Prasad, R., Sharma, D.M., Joshi, A.: Evaluation of Discourse Relation Annotation in the Hindi Discourse Relation Bank. In: Proceedings of the Eight International Conference on Language Resources and Evaluation, LREC 2012 (2012)
9. Begum, R., Husain, S., Dhwaj, A., Sharma, D.M., Bai, L., Sangal, R.: Dependency Annotation Scheme for Indian Languages. In: Proceedings of the Third International Joint Conference on Natural Language Processing, IJCNLP (2008)

10. Prasad, R., Joshi, A., Webber, B.: Realization of Discourse Relations by Other Means: Alternative Lexicalizations. In: Coling 2010: Posters, Coling 2010 Organizing Committee, pp. 1023–1031 (2010)
11. Miltsakaki, E., Prasad, R., Joshi, A., Webber, B.: Annotating Discourse Connectives And Their Arguments. In: Proceedings of the HLT/NAACL Workshop on Frontiers in Corpus Annotation (2004)
12. Fleiss, J.: Measuring nominal scale agreement among many raters. Psychological Bulletin (1971)

Building a Discourse Parser for Informal Mathematical Discourse in the Context of a Controlled Natural Language

Raúl Ernesto Gutiérrez de Piñerez Reyes and Juan Francisco Díaz Frias

EISC, Universidad del Valle
{raul.gutierrez,juanfco.diaz}@correounivalle.edu.co
http://www.univalle.edu.co/

Abstract. The lack of specific data sets makes difficult the discourse parsing for Informal Mathematical Discourse (IMD). In this paper, we propose a data driven approach to identify arguments and connectives in an IMD structure within the context of Controlled Natural Language (CNL). Our approach follows a low-level discourse parsing under Peen Discourse TreeBank (PDTB) guidelines. Three classifiers have been trained: one that identifies the `Arg2`, other that locates the relative position of `Arg1` and a third that identifies the (`Arg1` and `Arg2`) arguments of each connective. These classifiers are instances of Support Vector Machines (SVMs), fed from an own Mathematical TreeBank. Finally, our approach defines an End-to-End discourse parser into IMD, whose results will be used to classify of informal deductive proofs via the low level discourse in IMD processing.

Keywords: Discourse parser, Support Vector Machines, Informal Discourse Mathematical, Controlled Natural Language, Connectives, Arguments.

1 Introduction

Informal Mathematical Discourse (IMD) is characterized by a mixture of natural language and symbolic expressions in the context of textbooks, publications in mathematics, and mathematical proof. With regard to linguistic aspects, processing informal mathematical discourse is influenced by the notational context, the imprecision of informal language and the mixture of both: symbolic and natural language [20]. For computational systems, it is expensive to formalize mathematical texts, these may contain some complex and rich linguistic features of natural language such as the use of anaphoric pronouns and references, rephrasing and distributive vs. collective readings, etc. Additionally, the mathematical proofs may contain reasoning gaps, being hard to fill by using automated theorem proving (ATP). Therefore, this wide gap between informal argumentation and formal mathematics can be reduced by including controlled natural languages, given that they allow the elimination of processing ambiguity and reducing the complexity within IMD. In this same address, in [11] the narrative

A. Gelbukh (Ed.): CICLing 2013, Part I, LNCS 7816, pp. 533–544, 2013.

structures of mathematical texts are annotated by using graphs. Logical precedents of mathematical relations are also annotated between two chunks of texts parallel to the reasoning structure. In this sense, in [19,20], the IMD processing and the corpus linguistics are articulated, furthermore the linguistic meaning, the dialogs and the tutoring task are annotated. In order to build the corpus, a series of dialog sessions we carried out applying three theorems of the set theory, using a *Wizard-of-Oz* experiment. A corpus of 22 samples was collected at the end of the dialog sessions. Our approach can be though of, as being complementary to the set theory in the sense that this approach analyses linguistic phenomena by using lexical-syntactic annotations for processing IMD. During this research we have employed a Mathematical Treebank [8], a syntactically annotated corpus with deductive informal proofs, intended to serve as a linguistic resource in information extraction tasks within the CNL context. Our approach treats automatic discourse processing of paraphrased mathematical proofs by using a data driven approach so as to identify arguments of explicit IMD connectives in the CNL context. In this paper, we will focus on the segmentation and labeling of discourse arguments (**Arg1** and **Arg2**) which are propositions within a mathematical proof. In our CNL microstucture, these propositions have been thought of as paraphrased logical expressions and are part of the abstract objects [1] defined in the Penn Discourse TreeBank (PDTB) [15]. In addition, we built the Mathematical TreeBank (M-TreeBank), a corpus manually annotated with syntactic structures, and the Mathematical Discourse Treebank (MD-TreeBank) following PDTB guidelines. We annotate M-TreeBank and MD-TreeBank from a pure corpus. Our pure corpus is composed of a standard set of 150 proofs in which their sentences are well-formed within the CNL context. The MD-TreeBank was annotated with the mathematical relations (connectives) and its corresponding arguments (propositions), as well as the semantic classes of each connective and attributions. According to the type of the sentences presented in the CNL macrostructure, we defined the attributions and we annotated the type of informal deductive argumentation (proof models).

This paper is organized as follows: in §2 we present an introduction on CNL; Then, in §3 we explain the Mathematical TreeBank as our corpus of gold-standard parse trees; next, in §4 we describe argument annotation in PDTB on the one hand, and we explain in detail argument annotation in our Mathematical Discourse TreeBank on the other hand; in §5 we present both, the pipeline architecture for discourse segmentation and several feature sets used for classification; in §6 we describe the experimental setup, along with an evaluation and results; finally, we draw our conclusions in §7.

2   Controlled Natural Language within IMD

Controlled Natural Languages within IMD are subsets of natural languages whose grammars and dictionaries have been restricted to reduce or eliminate both ambiguity and complexity [3,9]. We developed a controlled natural language for mathematics, which is a defined subset of Spanish with restricted grammar and dictionary. We distinguished between *macrostructure* (argumentation text structure) and *microstructure* (grammatical structure in a sentence). Our CNL macrostructure of an argumentation is structured by a set of sentences such as: *assumptions, statements, justifications and deductions*. We found 15 proof models characterized by those starting with one assumption and those starting with two assumptions. The proof remainder, consists of statements, justifications, and deductions. In that order, assumptions are always introduced by an assumption trigger or clue word (e.g., *sea [let], suponga [suppose], and suponga que [suppose that]*). A statement always starts with connectives such as: *entonces [then], por lo tanto [therefore], de modo que [so that], en particular [in particular]*, among others. Justifications always start with connectives, for instance: *porque [because], debido a que [due to]*. Deductions always start with words *por lo tanto [therefore]* and *en consecuencia [in consequence]*; both of them followed by the clue word *se deduce que [deduce that]*. In the CNL microstructure of an argumentation, the use of nouns, adjectives, verbs, articles, determiners, connectives (*entonces, por lo tanto, porque, o [or], etc*) and the morphological information, operates in the same way as in natural Spanish discourse. In the CNL microstruture, it is also explicit the use of subordinate and coordinated clauses, as well as the use of anaphoric pronouns, also references are replaced by justifications. Our approach presents and extremely constrained CNL microstructure for mathematics in which complex linguistic phenomena is avoided.

3   The Mathematical TreeBank

The Mathematical TreeBank (M-TreeBank) is a corpus manually annotated with syntactic structures for supporting study and the analysis of IMD phenomena in a low-level discourse. The M-TreeBank consists of about 748 sentence/tree pairs in spanish that are included within 150 mathematical argumentations of a pure corpus. The pure corpus is composed of a standard set of 150 proofs in which their sentences are well-formed within the CNL context. The pure corpus is a transformation of an original corpus; defined after a previous experiment in which the students were required to demonstrate two theroems of set theory: (1) si $A \subseteq B$ entonces $U \subseteq \overline{A} \cup B$; (2) $P(A \cap B) = P(A) \cap P(B)$. The annotation scheme basically follows the Penn TreeBank II (PTB) scheme [14], human-annotated. Based on PTB, we used the clausal structure, as well as most of the labels and based on the UAM Spanish TreeBank scheme [16] we adapted the annotation structure of the connectives on the one hand, and the annotation scheme for morphological information on the other. The set of morphological features was adapted from Freeling 2.2.[1] The treebank trees have a syntactic structure of constituents similar to that of PTB. These trees contain paraphrased logical expressions along with connective annotations, morphological information, indefinite and definite descriptions annotation. The Mathematical TreeBank is

---

[1] http://nlp.lsi.upc.edu/freeling/

used for extraction tasks as an important resource in lexical, mathematical and syntactic features characterization.

4   Discursive Annotation into IMD

In this section, we present an overview on PDTB and we describe in detail the Mathematical Discourse TreeBank (MD-TreeBank) as a discoursive corpus defined on the annotation protocol of Penn Discourse TreeBank [15].

4.1   The Penn Discourse Treebank (PDTB)

The Penn Discourse TreeBank (PDTB) [15] is a large corpus containing one million words from the Wall Street Journal [14], annotated with discourse relations and its corresponding arguments. Discourse relations refer to discourse connectives operating as those predicates that takes two text spans as their arguments. The text span, which is syntactically bound to the connective, is called Arg2, while the other argument is called Arg1. In PDTB there are two types of relations; one established by *explicit* connectives and the other, established by *implicit* connectives. Discourse relations are explictly obtained from well-defined syntactic classes (subordinating conjunctions, coordinating conjunctions, and adverbs), or from *alternative lexicalizations* (AltLex), meaning other expressions that cannot be classified as connectives. In PDTB, the position of Arg1 w.r.t. the discourse connective shows that 60.9% of the explicit relations are same sentence (SS); 39.1% are precedent sentence (PS), and 0% are following sentence (FS). In PS, the position of Arg1 in previous, adjacent sentence (IPS) is 30.1% and in previous, non adjacent sentence (NAPS) is 9.0%.

4.2   Discursive Annotation of a Pure Corpus

The Mathematical Discourse TreeBank (MD-TreeBank) is a starting point of discursive annotation into IMD based on our CNL macrostructure. It consists in a discursive corpus over the pure corpus. Our MD-TreeBank was annotated with 150 proofs and it was defined in the annotation protocol of Peen Discourse TreeBank (PDTB). As in PDTB, MD-TreeBank is syntactically supported by M-TreeBank. In MD-TreeBank, we only followed explicit relations and we focused on the types and subtypes of class levels (*Temporal, Contingency, Comparison, Expansion*). We defined nine classes specifying both, the functionality and the deductive character of connectives. Among the defined classes we have: *alternative-conjunctive, alternative-disyunctive, cause-justification, cause-reason, cause-transitive, conclusion-result, conditional-hypothetical, instantiation*, and *restatement-equivalence*. In addition, we focused on attributions of arguments, specifically, the property *type* in which we identified six types of attributions according to CNL macrostructure. The attribution type labels defined are: AFIRM: Statements, COND: Conditionals, SUP: Assumptions, EXP: Explanations, DED: Deductions, JUS: Justifications. In MD-TreeBank, the discourse arguments (Arg1 and Arg2) were annotated as paraphrased logic propositions. These propositions kept a deductive order, sharing a lineal order among sentences into mathematical proof. Each proof is annotated with a sequence of connectives following the structure of the CNL. However, notice that this annotation is only linguistic and that its logical representation is not intended for annotating. As operating in PDTB, the argument in the clause that is syntactically bound to the connective is called Arg2; the other one is called Arg1. For example, the causal relation in (3) belongs to the type "Pragmatic Cause" with the subtype label "Justification" in which the porque connective indicates that Arg1 is expressing a claim and Arg2 is providing jusification for this claim.

(3) Entonces *el elemento x pertenece al conjunto complemento de A o al conjunto A* [*the element x belows to the complement set A or to set A*] porque [*because*] **la unión de el conjunto complemento de A con el conjunto A es igual al conjunto universal U** [*The union of complement set A with set A is equal to universal set U*].

In MD-TreeBank the statistics of the position of Arg1 w.r.t. the discourse connective is shown in Table 1. In MD-TreeBank, the position of Arg1 w.r.t. the discourse connective shows that 55% of the explicit relations are same sentence (SS); 45% are precedent sentence (PS). In PS, the position Arg1 in previous, adjacent sentence (IPS) is 20% and previous, non adjacent sentence (NAPS) is 25%.

Table 1. Statistics of position Arg1 in MD-TreeBank

Position	
Arg1, in same sentence (SS) as connective	55%
Arg1, in previous, adjacent sentence (IPS)	20%
Arg1, in previous, non-adjacent sentence (NAPS)	25%

5   IMD Segmentation

The Identification and labeling of IMD arguments (propositions) are modeled as supervised classification tasks. In this section, we present a discourse parser that, given an input argumentation, it automatically extracts the discourse arguments (Arg1 and Arg2) linked to each connective of the proof. We develop the argument segmentation task as a set of steps in which a step output feeds the input of the next step, based on Support Vector Machines (SVM) [17]. We use SVMs which are a set of maximum-margin classifiers that minimize the classification error and maximize the geometric margin. SVM's have been used in many practical tasks of natural language processing. In this work, they are used for identifying arguments Arg1 and Arg2 given a mathematical proof. As explained in [12,7], our IMD parser labels the Arg1 and Arg2 of every discourse connective of a mathematical proof. It is done in three steps: (1) identifying the Arg2, (2) identifying the locations of Arg1 as SS or PS, and (3) identifying the Arg1 and labeling of

`Arg1` and `Arg2` spans. Figure 1, shows the pipeline architecture used for such a process. First (1), we identify the `Arg2` and then we extract the `Arg2` span by using a rule-based algorithm; next (2), we identify the relative position of `Arg1` as same sentence (SS) or precedent sentence (PS). Then the relative position is propagated to an argument extractor and the `Arg1` is labeled correspondingly; finally (3), `Arg1` is identified given (`Arg2`,SS,PS) spans, and arguments (`Arg1` and `Arg2`) are extracted. A preprocessing phase must be carried out before the segmentation and labeling of the arguments from the input text (mathematical proof). First, we take the input text and we segment it in sentences; second, each sentence must be tagged; third we use a statistical parser model is used (Bikel's parser [2]) for obtaining the parse tree for each sentence. Next, we use a rule-based algorithm for building a features vector for each proposition found in each parse tree. Each proposition is assigned a numerical transformed feature vector which will serve as a dataset of SVMs (Figure 1 shows the training data features in bold line and the test data features in dot line).

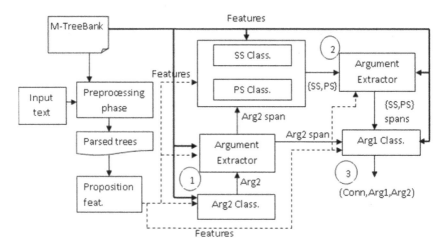

Fig. 1. System pipeline for the argument parser

5.1   SVMs for IMD Segmentation

Support Vector Machines (SVM) as defined in [17] are a kind of large-margin classifiers which are used to model binary classifiers. According to Manning [13], an SVM is a vector-space-based machine learning method, where the goal is to find a decision boundary between two classes that is maximally far from any point in the training data. Using SVMs we classify and label the connectives and their corresponding propositions (`Arg1` and `Arg2`) within a mathematical proof. In this work, we define three classifiers sequentially trained by using manually annotated proofs taken from the M-Treebank corpus (see Figure 1). In our IMD parsing, we implement the `Arg1` and `Arg2` identification as a task of binary classification by using SVMs. We extract feature vectors from M-Treebank for training each classifier where each feature vector have a fixed length. All feature vectors of our SVMs are representations of mathematical propositions within input text. Each proposition will have four assigned feature vectors, indexed by each one of the features: (1) a vector of 0,1's indicating if a proposition has a feature or not; (2) a vector of natural numbers (tf), indicating how many times a feature appears in a proposition; (3) a vector of real numbers (idf) indicating the relation between the frequency of a feature in a proposition with respect to the total number of propositions in the corpus, and (4) a vector of weights that corresponds to the scalar multiplication of vectors tf and idf.

Features. We define a set of lexical, syntactical and mathematical features, containing the necessary information for identifying and labeling CNL propositions within a mathematical proof. In contrast, to some approaches ([18] and [5]), that select pairwise the best heads of Arg1 and Arg2, our methodology is based on both the extraction and identification of argument spans (Arg1 and Arg2), as done in [7]. For identifying Arg2, we use the following features: the connective string (CONN) and its syntactic category (POS). The contextual features of the connectives were defined as the three previous words (3p) and the three next words (3n), the sentence number (NoS) in which the connective is found, and the Arg2 span. For identifying the relative position of Arg1 as SS or PS, the features used are: the connective string (CONN), its syntactic category (POS), sentence number (NoS) in which the connective is found, and the Arg2 span. For identifying Arg1 span we use the following features: the connective string (CONN), its syntactic category (POS), sentence number (NoS) in which the connective is found, the Arg2 span, and the SS and PS spans. Finally, built from a proposition, we define mathematical features as DeD-chunks. These chunks represent the indefinite and definite descriptions [21] and paraphrased set operations.

5.2 The Arguments Extraction Process

Once Arg1 and Arg2 arguments are classified, the result is propagated to the arguments extractor, which extracts the Arg1 and Arg2 spans accordingly. The arguments extractor has been designed based on the syntactic rules of our CNL. As in [4,12], we found some syntactic relations: coordinating conjunctions, conditionals, justifications, etc. For example, in Figure 2(a) the coordinating conjunctions divides the {Arg1,Arg2} top node into two argument spans ("CONN-CC" is the connective POS of o [or], y [and], etc). In Figure 2(b) the conditionals also divides the {Arg1,Arg2} top node into two argument spans. As an example, if "si-entonces [if then]" (connectives of Arg1 and Arg2 down nodes) are embedded within Arg2; then, "En particular [In particular]" is the connective of this Arg2 top node. Justifications trigger syntactic relations as shown in Figure 2(c), where "Entonces [Then]" can be identified as the connective of Arg1 top node and "debido a que [due to]" can be identified as the connective of Arg2 down node. Notice that, in the same sentence (SS), we have: Arg1 coming before Arg2 (see Figure 2(a)-(b)), Arg2 embedded within Arg1 (see Figure 2(c)).

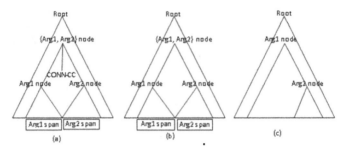

Fig. 2. Syntactic relations of `Arg1` and `Arg2` subtree nodes into the parsing trees

For instance, the following rule extracts `Arg1` and `Arg2` of a same sentence.

if the connective C is *porque [because]* | *debido a que [due to]* | *dado que [given that]* **and** C's parent node is SBAR and C's grandparent node left is CONN-AFIR **then** set C's grandparent right sibling as `Arg1` node and set C's parent right sibling as `Arg2` node

According to this rule, C's grandparent right sibling and C's parent right sibling are S sentences. In this rule, our arguments extractor labels these S sentences with two possibilities: `Arg1` node or `Arg2` node. In our CNL, all arguments' parent nodes are S sentences which consists of a noun phrase (NP) and a verb phrase (VP). Our arguments extractor was designed in order to identify the `Arg1` and `Arg2` subtree nodes within the sentence parse tree for all explicit connectives. Finally, when the `Arg1` was classified as PS, we located it in the `Arg1` in the immediately previous sentence of the connective (IPS) and in some non-adjacent previous sentence of the connective (NAPS).

6   Experiments and Results

In our experiments we employed three datasets. For each one of the experiments, we obtained 328 test proofs and 150 training proofs. A first dataset was built for measuring the performance of `Arg2` classifier. We obtained 1158 training vectors defining 50% negative examples and 50% positive examples. Similarly, we determined a second dataset for training the `Arg1` argument position classifier. We obtained 965 training vectors; 491 of them being SS negative examples and 474 of them are PS positive examples. A third dataset was defined in order to identify `Arg1` given the `Arg2` argument. We obtained 2123 training vectors: 1158 positive examples from `Arg2` and 965 negative examples from `Arg1`. We used the SVM-Light [10], an implementation of Support Vector Machine (SVM) in C for all classification tasks. We also used ROC (Receiver Operating Characteristic) [6] to illustrate the performance of binary classifiers and to organize the datasets.

6.1   Evaluation Methodology

Our results are presented using precision (P) as well as recall (R) and F_1 measures. All our extraction experiments are based on standard parse trees taken

Table 2. Results for the `Arg2` classifier

Features	P	R	F_1
CONN+NoS	0.37	0.29	0.56
CONN+NoS+3p+3n	0.49	0.62	0.48
CONN+NoS+POS+3p+3n	0.75	0.79	0.75

Table 3. Results for the `Arg2` argument extractor

Model-Features	P	R	F_1
M1=CONN+NoS+POS+Arg2-s	0.96	0.96	0.96
M2=CONN+NoS+POS+Arg2-s+mf	0.98	0.98	0.98

from our M-TreeBank. We take advantage of the annotation protocols from our MD-TreBank and we follow these guidelines to extract the mathematical propositions of a proof within the CNL context. We performed a 10-fold cross-validation, since we ensure that 150 proofs can be partitioned into 10 equal size proofs; that is, we define 15 demonstration models at the 150 proofs. Observe that for training of classifiers, we define three feature models (M1,M2,M3). With purpose of measuring the performance of `Arg2` classifier, we carried out the experiments under three settings referring to the connective and their contextual features (see Figure 2). At first, we use the connective string (CONN) and the sentence number (NoS) in which the connective is found; then, we use the connective string with its three previous words (3p) and three next words (3n); finally, we added the connective POS to previous features set. In Table 2, we report the results of the `Arg2` classifier. As it was expected, the usefulness of integrating the contextual and syntactic information into the connective was not enough. To improve this result, we added the `Arg2` span (Arg2-s) and the mathematical features (mf) so as to evaluate the performance of the `Arg2` extractor. The Table 3 illustrates the results of identifying the `Arg2` subtree nodes. As expected, the inclusion of Arg2-s feature increased 21% the accuracy of the classifier. This increment represents the difference between the F_1 latter of 0.75 (third column of Table 2) and F_1 of 0.96 (third column of M1 in Table 3). Moreover, an increment of 0.02% on the accuracy (F_1 score) of the arguments extractor is shown in Table 3. We believe that the increment in M2 model is due to the incorporation of mathematical features. Specifically, each proposition has on average two types of mathematical features. We define six types of mathematical features such as "DeD", "InD", "NPU", "NPI", "NPC", "NPP". These mathematical features represent definite and indefinite descriptions as well as set operations from propositions. Then, we evaluate the performance of the `Arg1` extractor into two steps. First of all, we train the arguments extractor for separate; for each one SS and PS we use the M1 model (see Table 4); second, we trained the arguments extractor for the SS and PS at once, by using M2 model. The results of identifying the `Arg1` and `Arg2` subtree nodes for the SS case, as well as the PS case are illustrated in 4. Notice that SS F_1 and PS F_1 on M2 model are the

Table 4. Results for the `Arg1` argument extractor as SS and PS

Model-Features	SS F_1	PS F_1
M1=CONN+NoS+POS+Arg2-s	0.91	0.95
M2=CONN+NoS+POS+Arg2-s+mf	0.98	0.98

Table 5. Overall results for the `Arg1` and `Arg2` classification

Model-Features	Arg1 F_1	Arg2 F_1
M2=CONN+NoS+POS+Arg2-s+mf	0.97	0.97
M3=CONN+NoS+POS+Arg2-s+mf+SS-s+PS-s	0.98	0.98

same. This appreciation is due to that the syntactic relations between (`Arg1` and `Arg2`) have the same CNL microstructure. The increments of SS F_1 and PS F_1 between M1 model and M2 model (Table 4) can be explained by including the mf features. These latter are contextual features in the propositional level rather than the connective level. The `Arg1` classifier was evaluated following the pipeline architecture. We trained this classifier under two models. We first evaluated the performance of the classifier using M2 model features. Second, we evaluated the performance of the classifier added to M2 model, the SS and PS spans (SS-s and PS-s). As it was expected, in Table 5, we cam observe that the `Arg1` F_1 and `Arg2` F_1 are the same for the two models. There is a small error propagation (1%) between the `Arg1` argument extractor (SS F_1 of 0.98 and PS F_1 of 0.98) over M2 and the `Arg1` classifier (`Arg1` F_1 of 0.97) over M2. In the second row of Table 5 it is shown the increment of 1%; for both `Arg1` F_1 and `Arg2` F_1 over M3 model. We believe that adding of SS-s and PS-s features to M2 model eliminates small error margin, and improves the accuracy of the classifier.

7   Conclusions

In this work, we presented an algorithm that performs IMD parsing within CNL context; in which and end-to-end system is implemented. We considered that CNL microstructure and macrostructure were adapted to PTB and PDTB guidelines within the IMD context. Both argument spans and mathematical features were included in order to increment the accuracy of classifiers and the result were propagated to the arguments extractor. When the argument spans and the mathematical features as contextual features of arguments were used, the results increased the precision of the classifier. For example, in the classification of `Arg2`, the inclusion of `Arg2` span increased 21% the accuracy of classifier (see Tables 2 and 3). Similarly, the inclusion of mathematical features (mf) into M2 model (Table 3) and into M2 model (Table 4) increased their accuracy 2% and 3.5%, respectively. With respect to overall results of `Arg1` classifier (Table 5), we can observe a small increment when adding of SS (SS-s) and PS (PS-s) spans. In future works, we must consider adding more theorems of set theory. We also conclude that these latter results were obtained because all the propositions have the same CNL microstructure. All these results will be used for classification of informal deductive proofs using low-level discourse in IMD processing.

References

1. Asher, N.: Reference to Abstract Objects in Discourse. Kluwer Academic Publishers (1993)
2. Bikel, D.: Design of a Multilingual, Parallel Processing Statistical Parsing Engine. In: 2nd International Conference on Human Language Technology Research HLT 2002, pp. 178–182. Morgan Kaufmann Publishers Inc., San Francisco (2002)
3. Cramer, M., Fisseni, B., Koepke, P., Kühlwein, D., Schröder, B., Veldman, J.: The Naproche Project Controlled Natural Language Proof Checking of Mathematical Texts. In: Fuchs, N.E. (ed.) CNL 2009. LNCS, vol. 5972, pp. 170–186. Springer, Heidelberg (2010)
4. Dinesh, N., Lee, A., Miltsakaki, E., Prasad, R., Joshi, A., Webber, B.: Attribution and the (Non-)Alignment of Syntactic and Discourse Arguments of Connectives. In: CorpusAnno 2005 Proceedings of the Workshop on Frontiers in Corpus Annotations II: Pie in the Sky, pp. 29–36. Association for Computational Linguistics (ACM), Stroudsburg (2005)
5. Elwell, R., Baldridge, J.: Discourse Connective Argument Identification with Connective Specific Rankers. In: ICSC 2008 Proceedings of the 2008 IEEE International Conference on Semantic Computing, pp. 198–205. IEEE Computer Society, Washington (2008)
6. Fawcet, T.: An Introduction to ROC Analysis. Pattern Recognition Letters- Specialissue: ROC Analysis in Pattern, 861–874 (2006)
7. Ghosh, S., Johansson, R., Riccardi, G., Tonelli, S.: Shallow Discourse Parsing with Conditional Random Fields. In: 5th International Joint Conference on Natural Language Processing, Chiang Mai, Thailand, pp. 1071–1079 (2011)
8. Gutiérrez de Piñerez, R.E., Díaz, J.F.: Preprocessing of Informal Mathematical Discourse in Context of Controlled Natural Language. In: Proceedings of the 21st ACM International Conference on Information and Knowledge Management (CIKM 2012). Association for Computing Machinery, ACM (2012)
9. Humayoun, M., Raffalli, C.: MathAbs: A Representational Language for Mathematics. In: Proceedings of 8th International Conference on Frontiers of Information Technology, Islamabad, Pakistan, p. 37 (2010)
10. Joachims, T.: Making large-Scale SVM Learning Practical. In: Schlkopf, B., Burges, C., Smola, A. (eds.) Advances in Kernel Methods - Support Vector Learning. MIT Press (1999)
11. Kamareddine, F., Maarek, M., Retel, K., Wells, J.B.: Narrative Structure of Mathematical Texts. In: Kauers, M., Kerber, M., Miner, R., Windsteiger, W. (eds.) MKM/CALCULEMUS 2007. LNCS (LNAI), vol. 4573, pp. 296–312. Springer, Heidelberg (2007)
12. Lin, Z., Ng, H.T., Kan, M.: A PDTB-Styled End-to-End Discourse Parser. The Computing Research Repository 1011 (2011)
13. Manning, C.D., Raghavan, P., Schtze, H.: Introduction to Information Retrieval. Cambridge University Press, New York (2008)
14. Marcus, M., Santorini, B., Ann Marcinkiewicz, A.: Building a Large Annotated Corpus of English: the Penn Treebank. Computational Linguistics 19(2), 313–330 (1993)

15. Prasad, R., Dinesh, N., Lee, A., Miltsakaki, E., Robaldo, L., Joshi, A., Webber, B.: The Penn Discourse TreeBank 2.0. In: Proceedings of the 6th International Conference on Languages Resources and Evaluations (LREC 2008), Marrakech, Marocco (2008)
16. Ruesga, S.L., Sandoval, S.L., Len, L.F.: Spanish Treebank: Specifications version 5. Universidad Autnoma de Madrid (1999)
17. Vapnik, V.N.: The Nature of Statistical Learning Theory. Springer, New York (1995)
18. Wellner, B., Pustejovsky, J.: Automatically Identifying the Arguments of Discourse Connectives. In: Proceedings of the 2007 Joint Conference on Empirical Methods in Natural Language Processing and Computational Natural Language Learning (EMNLP-CoNLL), pp. 92–101. Association for Computational Linguistics, Prague (2007)
19. Wolska, M., Vo, B.Q., Tsovaltzi, D., Kruijff-Korbayov, I., Karagjosova, E., Horacek, H., Fiedler, A., Benzmller, C.: Annotated Corpus of Tutorial Dialogs on Mathematical Theorem Proving. In: Proceedings of 4th International Conference on Language Resources and Evaluation, Lisbon, Portugal, pp. 1007–1010 (2004)
20. Wolska, M.: A Language Engineering Architecture for Processing Informal Mathematical Discourse. In: Towards Digital Mathematics Library, Birmingham, United Kingdom, pp. 131–136. Masaryk University (2008)
21. Zinn, C.: Understanding Informal Mathematical Discourse. Ph.D. thesis. Universitat Erlangen-Nürnberg Institut für Informatik (2004)

Discriminative Learning
of First-Order Weighted Abduction
from Partial Discource Explanations

Kazeto Yamamoto, Naoya Inoue, Yotaro Watanabe,
Naoaki Okazaki, and Kentaro Inui

Tohoku University

Abstract. Abduction is inference to the best explanation. Abduction has long
been studied in a wide range of contexts and is widely used for modeling artificial
intelligence systems, such as diagnostic systems and plan recognition systems.
Recent advances in the techniques of automatic world knowledge acquisition and
inference technique warrant applying abduction with large knowledge bases to
real-life problems. However, less attention has been paid to how to automatically
learn score functions, which rank candidate explanations in order of their plausibility. In this paper, we propose a novel approach for learning the score function
of first-order logic-based weighted abduction [1] in a supervised manner. Because
the manual annotation of abductive explanations (i.e. a set of literals that explains
observations) is a time-consuming task in many cases, we propose a framework
to learn the score function from partially annotated abductive explanations (i.e. a
subset of those literals). More specifically, we assume that we apply abduction to
a specific task, where a subset of the best explanation is associated with output
labels, and the rest are regarded as hidden variables. We then formulate the learning problem as a task of discriminative structured learning with hidden variables.
Our experiments show that our framework successfully reduces the loss in each
iteration on a plan recognition dataset.

1 Introduction

Making the implicit information (e.g. corefference relations, agent's plans, etc.) in sentences explicit is an important technique for various tasks in natural language processing. We are trying to construct discource understanding frameworks using abduction as
the framework of making the implicit explicit.

Abduction is inference to the best explanation. Applying abduction to discource understanding was studied in 1980s and 1990s. The most important study of those is *Interpretation as Abduction* (IA) [1] by Hobbs et al. They showed the process of natural
language interpretation can reasonably be described as abductive inference. For example, let us carry out interpretation on "John went to the bank. He got a loan." Treating
world knowledge as background knowledge, target sentence to interpret as observation,
IA applies abduction and outputs the best explanation as the result of interpretation. The
interpretation of the example sentence with abduction is showed in Figure 1. From the
result of interpretation, we can retrieve such implicit information as the goal of "went
to the bank" is "got a loan" and that a coreference relation exists between "John" and
"He".

A. Gelbukh (Ed.): CICLing 2013, Part I, LNCS 7816, pp. 545–558, 2013.

While the lack of world knowledge resources hampered applying abduction to real-life problems
in the 1980s and 1990s, a number of techniques for acquiring
world knowledge resources have
been developed in the last decade
[2,3,4,5,6, etc.]. In addition, the
development of an efficient inference technique of abduction warrant the application of abduction
with large knowledge bases to

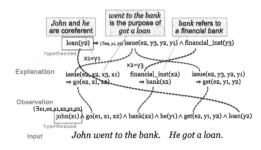

Fig. 1. An example of discource interpretation with abduction

real-life problems [7]. Consequently, several researchers have started applying abduction to real-life problems exploiting large knowledge bases. For instance, inspired by
[1], [8] propose an abduction-based natural language processing framework using forty
thousands axioms extracted from the popular ontological resources, WordNet [2] and
FrameNet [3]. They evaluate their approach on the real-life natural language processing
task of textual entailment recognition [9].

However, less attention has been paid to how to automatically learn score functions, which rank candidate explanations in order of their plausibility (henceforth, we
call it the *cost function*). To apply abductive inference to a wide range of tasks, this
non-trivial issue needs to be addressed because the criterion of plausibility is highly
task-dependent, as discussed in Section 2.2. A notable exception is a series of studies
[10,11,12], which emulate abduction in the probabilistic deductive inference framework, Markov Logic Networks (MLNs) [13]. MLN-based approaches can exploit several choices of weight learning methods originally developed for MLNs [14,15, etc.].
However, MLN-based abduction has severe problems when they are applied to discource processing which we will discuss in Section 5.

In this paper, we propose a novel supervised approach for learning the cost function of first-order logic-based abduction. This is a framework to learn the cost function
from subsets of explanations (henceforth, we call it *partial abductive explanations*).
More specifically, we assume that we apply abduction to a specific task, where a subset
of the best explanation is associated with output labels, and the rest are regarded as
hidden variables. We then formulate the learning problem as the task of discriminative
structured learning with hidden variables. As the cost function, we use the parametrized
non-linear cost function proposed by [1].

This paper is organized as follows. We first give a brief review of abduction and the
cost function of Hobbs et al.'s Weighted Abduction (Section 2). We then show our approach for learning the parametrized cost function of Weighted Abduction (Section 3).
We start with the simple case where complete abductive explanations are given (Section 3.3), and then describe a method for learning with partial explanations (Section
3.4). Finally, we demonstrate how our learning algorithm works with a dataset of plan
recognition (Section 4).

2   Background

2.1   Abduction

Abduction is inference to the best explanation. Formally, logical abduction is defined as follows:

- **Given:** Background knowledge B, and observations O, where both B and O are sets of first-order logical formulas.
- **Find:** A *hypothesis (explanation)* H such that $H \cup B \models O$, $H \cup B \not\models \perp$, where H is a set of first-order logical formulas[1]. We say that p is *hypothesized* if $H \cup B \models p$, and that p is *explained* if $(\exists q)\ q \rightarrow p \in B$ and $H \cup B \models q$.

Typically, several hypotheses H explaining O exist. We call each of them a *candidate hypothesis*, and each literal in a hypothesis an *elemental hypothesis*. The goal of abduction is to find the best hypothesis among candidate hypotheses by a specific evaluation measure. In this paper, we formulate abduction as the task of finding the minimum-cost hypothesis $\hat{H}$ among a set $\mathcal{H}$ of candidate hypotheses. Formally, we find $\hat{H} = \arg\min_{H \in \mathcal{H}} c(H)$, where c is a function $\mathcal{H} \rightarrow \mathbb{R}$, which is called the *cost function*. We call the best hypothesis $\hat{H}$ the *solution hypothesis*. In the literature, several kinds of cost functions have been proposed, including cost-based and probability-based [16,17,1,18,12, etc.].

2.2   Weighted Abduction

Hobbs et al. [1] propose a cost function that can evaluate two types of plausibility of hypotheses simultaneously: *correctness* and *informativeness*. Correctness represents how reliable the contents of information are. Informativeness is how specific the information is. Hobbs et al. parametrized the cost function in a way that one can construct a cost function that favors more specific and thus more informative explanations, or less specific but more reliable explanations in terms of a specific task by altering the parameters. The resulting framework is called *Weighted Abduction*.

   In principle, the cost function gives a penalty for assuming specific and unreliable information but rewards for inferring the same information from different observations. To the best of our knowledge, Hobbs et al.'s Weighted Abduction is the only framework that considers the appropriateness of a hypothesis' specificity. Hobbs et al. exploit this cost function for text understanding where the key idea is that interpreting sentences is to find the lowest-cost abductive explanation[2] to the logical forms of the sentences in a agreement with a correctness-informativeness tradeoff. However, they do not elaborate on how to give the parameters of cost function. Ovchinnikova et al.[8,19] proposed the methods of determining the parameters of cost function statistically from annotated corpora, though, there are no existing methods for training the parameters discriminatively.

---

[1] Throughout the paper, $\models$ and $\perp$ represent logical entailment and logical contradiction respectively.

[2] Hobbs did not mention cases where there are multiple lowest-cost explanations. So we assume those cases do not occur in this paper.

Let us describe the cost function in more formal way. Following [1], we use the following representations for background knowledge, observations, and hypothesis throughout the paper:

- **Background Knowledge** B: a set of first-order logical Horn clause whose literals in its body are assigned positive real-valued *weights*. We use a notation p^w to indicate "a literal p has the weight w" (e.g. $p(x)^{0.6} \wedge q(x)^{0.6} \Rightarrow r(x)$). We use a vector θ to represent weights on each literal, where each element refers to one weight on a literal in a specific axiom.
- **Observations** O: an existentially quantified conjunction of literals. Each literal has a positive real-valued cost. We use a notation $p^{\$c}$ to denote "a literal p has the cost c," and $c(p)$ to denote "the cost of the literal p" (e.g. $p(x)^{\$10} \wedge q(x)^{\$10}$).
- **Hypothesis** H: an existentially quantified conjunction of literals. Each literal also has a positive real-valued cost (e.g. $r(x)^{\$10} \wedge s(x)^{\$10}$). We define *unification* as an operation that merges two literals to one literal with the smaller cost, assuming that the arguments of the two literals are the same. For example, given the hypothesis $p(x)^{\$30} \wedge p(y)^{\$10} \wedge q(y)^{\$10}$, the unification of $p(x)^{\$30}$ and $p(y)^{\$10}$ yields another hypothesis $p(x)^{\$10} \wedge q(x)^{\$10}$, assuming $x = y$. The unification operation is applicable to any hypotheses, unless $H \cup B \not\models \perp$. We say that $p(x)$ is *unified* with $p(y)$ if $x = y \wedge c_\theta(p(x)) \geq c_\theta(p(y))$.[3]

Given a weight vector θ, the cost function of H is defined as the sum of all the costs of elemental hypotheses in H:

$$c_\theta(H) = \sum_{h \in P_H} c_\theta(h)$$

$$= \sum_{h \in P_H} \left[\prod_{i \in chain(h)} \theta_i \right] c(obs(h)), \qquad (1)$$

where P_H is a set of elemental hypotheses that are not explained nor unified, $chain(h)$ is a set of indices to a literal in axioms that are used for hypothesizing h, and $obs(h)$ is an observed literal that is back-chained on to hypothesize h. Henceforth, we refer to a weight vector θ as the *parameter* of cost functions.

3   Discriminative Weight Learning of Weighted Abduction

In this section, we propose a method to learn the parameters of cost function in Weighted Abduction by recasting the parameter estimation problem as an online discriminative leaning problem with hidden variables.

The idea is four-fold:

1. We train the cost function with only partially specified gold abductive explanations which we represent as a partial set of the required literals (*gold partial explanations*).

---

[3] If $cp(x) = cp(y)$, we regard that either $p(x)$ or $p(y)$ is unified.

2. We automatically infer complete correct abductive explanations from gold partial explanations by abductive inference.
3. We optimize the parameters of the cost function by minimizing the loss function where the loss is given by the difference of the costs of the minimal-cost hypothesis and the complete correct abductive explanations.
4. We employ feed-forward neural networks to calculate the gradient of each parameter.

In the rest of this section, we first formalize explanation in Weighted Abduction with directed acyclic graphs (Section 3.1), and we then describe the outline of our learning method (Section 3.2) and elaborate on our learning framework in the simple case where complete abductive explanations are given (Section 3.3). We then describe a method for learning the parameters from partial abductive explanations (Section 3.4). Finally, we describe how to update the parameters through error back-propagation in FFNNs (Section 3.5).

3.1 Preliminaries

In this paper, we express the hypotheses of Weighted Abduction as directed acyclic graphs (DAG) following Charniak [20]. Namely, we regard each literal in the hypothesis as a node of DAG and each of relation between literals as an edge of DAG. We call these graphs *proof graph* and use a notation $G_{O,B,H}$ to denote the proof graph made from the observation O, the background knowledge B and the hypothesis H.

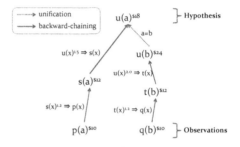

Fig. 2. An example proof tree

We define following two types of edge in proof graphs:

- **Backward-Chaining:** Given the tail node's literal $p(x)^{\$c_1}$ and the head node's literal $q(x)^{\$c_2}$, this relation indicates that $q(x) \cup B \models p(x)$. Namely, $q(x)$ is hypothesized with $p(x)$. Then, the cost of head node's literal is caluclated by multiplication of the cost of tail node's literal and the weight of background knowledge (e.g. $c_2 = c_1 w$, where $q(x)^w \Rightarrow p(x)$).
- **Unification:** Given the tail node's literal $p(x)$ and the head node's literal $p(y)$, this relation indicate that $p(x)$ and $p(y)$ are unified and $x = y$.

Between the tail node and the head node of each edge in a proof graph, the relation that the head node's literal explain the tail node's literal exists. Thus, the set of literals of leaf nodes in proof graphs corresponds to P_H and the set of literals of root nodes in proof graphs corresponds to O.

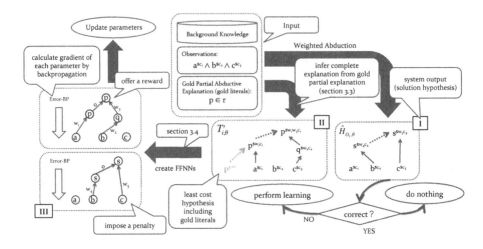

Fig. 3. Outline of proposed parameter learning method

We show an example proof graph in Figure 2. This is the proof graph made from the following background knowledge, observation and hypothesis:

$$B = \{ \forall x \, (s(x)^{1.2} \Rightarrow p(x)), \forall x (s(x)^{1.2} \Rightarrow q(x)),$$
$$\forall x \, (u(x)^{1.5} \Rightarrow s(x)), \forall x (u(x)^{2.0} \Rightarrow t(x)) \}, \tag{2}$$
$$O = \exists x \, (p(a)^{\$10} \wedge q(b)^{\$10}) \tag{3}$$
$$H = \exists x \, (u(a)^{\$18} \wedge u(b)^{\$24} \wedge s(a)^{\$12} \wedge t(b)^{\$12} \wedge a = b) \tag{4}$$

The cost of a hypothesis is calculated with Equation 1. Therefore, the cost of this hypothesis is calculated as $c(H) = \sum_{h \in P_H} c(h) = \$18$.

3.2 Outline of Our Method

In this section, we describe the outline of our learning method. The overall framework is illustrated in Figure 3.

First, we assume each training example to be a pair (O_i, τ_i), where O_i is an observation and τ_i is a gold partial explanation. A gold partial explanation is a set of literals that must be included in the correct abductive explanation T_i for the input observation O_i, i.e. $T_i \cup B \models O_i$ and $\tau_i \subseteq T_i$.

Next, we consider the online version of parameter learning [4]. For each cycle, given (O_i, τ_i), we perform Weighted Abduction for the observations O_i and background knowledge B with parameters θ, and get the solution hypothesis $\hat{H}_{O_i, \theta}$ ([I] in Figure 3). If $\hat{H}_{O_i, \theta}$ does not include τ_i (i.e. $\hat{H}_{O_i, \theta}$ is an incorrect prediction), we update the parameters so that $\hat{H}_{O_i, \theta}$ includes τ_i. In order to do so, we first infer a complete abductive explanation $\hat{T}_{i, \theta}$ from the gold partial explanation τ_i ([II]). We then update parameters θ by imposing

---

[4] The batch version can also be considered by accumulating the gradients for each cycle before updating the weights.

a penalty to the wrong solution hypothesis $\hat{H}_{O_i,\theta}$ and offering a reward to the inferred correct complete abductive explanation $\hat{T}_{i,\theta}$. To compute these updates, we translate $\hat{H}_{O_i,\theta}$ and $\hat{T}_{i,\theta}$ to feed-forward neural networks and perform backpropagation on them ([III]).

In this paper, we assume that there is enough knowledge to infer the correct explanation in each problem (we call this *the knowledge completeness assumption*). If this assumption were not satisfied, which means that the correct explanation is not included in the candidate hypotheses, then we could not infer the correct explanation irrespectively of parameters. In the following discussion, we do not consider a case of knowledge base shortage.

3.3 Learning from Complete Abductive Explanations

Let us first assume that we have a set of training examples labeled with a complete abductive explanation. Namely, we consider a training dataset $D = \{(O_1, T_1), (O_2, T_2), ..., (O_n, T_n)\}$, where O_i is an observation and T_i is the gold (correct) complete abductive explanation for O_i, i.e. $T_i \cup B \models O_i$.

For each labeled example (O_i, T_i), the solution hypothesis $\hat{H}_i$ is obtained by:

$$\hat{H}_i = \arg\min_{H \in \mathcal{H}_i} c_\theta(H) \tag{5}$$

We consider that a solution hypothesis $\hat{H}_i$ is correct if $\hat{H}_i = T_i$. Now we consider a loss function that calculates how far the current solution hypothesis is from the gold explanation, analogously to standard learning algorithms. If the current solution hypothesis is correct, the loss is zero. If $\hat{H}_i \neq T_i$, on the other hand, we consider the loss as given by the following loss function:

$$E(O_i, \theta, T_i) = \begin{cases} \frac{1}{2}\left(\frac{c_\theta(T_i)-c_\theta(\hat{H}_i)}{c_\theta(T_i)+c_\theta(\hat{H}_i)} + m\right)^2 + \lambda\theta \cdot \theta & (\hat{H}_i \neq T_i) \\ 0 & (\hat{H}_i = T_i) \end{cases}, \tag{6}$$

where $\lambda\theta \cdot \theta$ is a regularization term and m is a margin. Our goal is to learn the cost function c_θ that has minimal prediction errors. This goal is accomplished by learning parameters θ^* which minimize the total loss as below:

$$\theta^* = \arg\min_\theta \sum_{(O,T) \in D} E(O, \theta, T) \tag{7}$$

We describe how to minimize the loss in Section 3.5.

Note that we use the ratio of cost functions $\frac{1}{2}\left(\frac{c_\theta(T_i)-c_\theta(\hat{H}_i)}{c_\theta(T_i)+c_\theta(\hat{H}_i)} + m\right)^2$ as the loss function, instead of $\frac{1}{2}\left(c_\theta(T_i) - c_\theta(\hat{H}_i)\right)^2$. In the following, we shortly justify the use of the ratio of cost function. Let us suppose that we employ $\frac{1}{2}\left(c_\theta(T_i) - c_\theta(\hat{H}_i)\right)^2$ as the loss function. Then, we can minimize the loss function by minimizing the weight terms that appear in both $c_\theta(T_i)$ and $c_\theta(H_i)$, namely the weights assigned to axioms that are used in both T_i and H_i. For instance, given $O_i = \{p(a)^{\$c}\}$, $B = \{q(x)^{w_0} \Rightarrow p(x), s(x)^{w_1} \Rightarrow q(x), t(x)^{w_2} \Rightarrow$

$q(x)$}, $T_i = \{s(a)^{\$w_0w_1c}\}$, $H_i = \{t(a)^{\$w_0w_2c}\}$, we can minimize the value of loss function by minimizing the value of w_0. As a result, the learning procedure just decreases w_0 as much as possible to minimize the loss function. This prevents our framework from learning a meaningful cost function, because the minimization of weights does not imply that we can infer the gold hypothesis as the solution hypothesis. To avoid this problem, we employ the ratio of cost functions.

3.4 Learning from Partial Abductive Explanations

In the above, we assumed that each training example has a complete abductive explanation. However, this assumption is not realistic in many cases because it is usually prohibitively costly for human annotators to give a complete abductive explanation for each given input. This leads us to consider representing a training example as a pair of observation O_i and gold partial explanation τ_i, which is a partial set of literals that must be included in the explanation of O_i. In the case of Figure 3, we assumed that the correct hypothesis for the given observation is partially specified by the literal $p \in \tau$.

This way of simplification is essential in real-life tasks. In plan recognition, for example, it is not an easy job for human annotators to give a complete explanation to an input sequence of observed events, but they can tell whether it is a shopping story or a robbing story much more easily, which can be indicated by a small set of gold literals.

Now, our goal is to learn the cost function from partial explanations $D = \{(O_1, \tau_1), (O_2, \tau_2), ..., (O_n, \tau_n)\}$. Regarding whether each gold literal is included in the solution hypothesis $\hat{H}_{O_i,\theta}$ and the structure of the proof graph $G_{O_i,B,\hat{H}_{O_i,\theta}}$ as hidden states, this task can be seen as discriminative structure learning with hidden states. The issue is how to infer the complete correct explanation $\acute{T}_{i,\theta}$ from a given incomplete set τ_i of gold literals. Fortunately, this can be done straightforwardly by adding the gold literals τ_i to the observation O_i:

$$O_i^+ = O_i \cup \{t^{\$\infty} \mid t \in \tau_i\}, \tag{8}$$

where each gold literal is assigned an infinitive cost. Then, the solution hypothesis $\hat{H}_{O_i^+,\theta}$ is equivalent to the complete correct explanation $\acute{T}_{i,\theta}$ if the following conditions are satisfied:

- A hypothesis including τ_i exists in the candidate hypotheses for O_i (the knowledge completeness assumption).
- $\hat{H}_{O_i^+,\theta}$ has no backward chaining from $t \in \tau_i$.

Figure 3 ([II]) illustrates a simple case, where $\hat{H}_{O_i^+,\theta}$ is inferred by adding the gold literal p to the observation. Since this added literal p is assigned an infinitive cost, it is strongly motivated to derive an explanation including that p, resulting in obtaining the correct explanation $\acute{T}_{i,\theta}$.

When these conditions are satisfied, because each t has a huge cost, the system selects as the solution hypothesis $\hat{H}_{O_i^+,\theta}$ the hypothesis in which most literals in τ_i unify with other literals. Then, assuming the existence of a hypothesis including τ_i in the candidate hypotheses for O_i, there is the hypothesis in which each of the literals in τ_i unifies to a literal in the candidate hypotheses for O_i^+, and it is selected as solution hypothesis

$\hat{H}_{O_i^+,\theta}$. Because the cost of t must be 0 when it is unified with an other literal included in O_i or hypothesized from O_i, the cost of $\hat{H}_{O_i^+,\theta}$ is equal to cost of $\hat{T}_{i,\theta}$. So $\hat{H}_{O_i^+,\theta}$ must be equal to $\hat{T}_{i,\theta}$.

It should be note that we can check whether candidate hypotheses satisfy the above-mentioned conditions by checking the cost of the solution hypothesis, because any non-unified $t^\$\infty$ will result in a huge cost.

3.5 Updating Parameters with FFNNs

To update parameters, we want to compute the gradient of the loss function for each parameter. However, since the cost function and the loss function are both nonlinear to their parameters, their gradients cannot be computed straightforwardly.

To solve this problem, we propose employing feed-forward neural networks (FFNNs). An FFNN is a directed acyclic graph where the output of each node j is given by:

$$z_j = h(a_j), \qquad (9)$$

$$a_j = \sum_{i \in \{i | e_{i \to j} \in \mathbb{B}\}} z_i \times \mathbf{w}_{i \to j}, \quad (10)$$

where z_i denotes the output of node i, a_i denotes the degree of activation of node i, $h(a)$ is an activation function, $e_{i \to j}$ denotes a directed edge from node i to node j, and $\mathbf{w}_{i \to j}$ denotes the weight of $e_{i \to j}$.

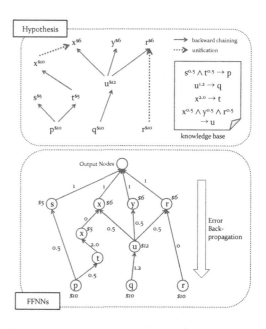

Fig. 4. Example of transforming hypotheses into FFNNs

Then, we express the cost function of H with a FFNN. This is achieved by applying the following convertion to $G_{O,B,H}$:

1. The cost of each literal in $G_{O,B,H}$ is the output of the node in the corresponding FFNN.
2. Each backward-chaining edge in $G_{O,B,H}$ is an edge with weight w in the FFNN where w denote the weight of the background knowledge of the corresponding backward-chaining edge.
3. Each unification edge in $G_{O,B,H}$ is an edge with weight 0 in the FFNN.
4. The activation function of each layer in FFNNs is $h(a) = a$.
5. An *output node*) is added to the FFNN, making new edges with weight 1 between output node and each node that corresponds to each literal in P_H (i.e. leef nodes in the proof graph).

Then, the value of the output node is equal to the cost function $c_\theta(H)$ in Weighted Abduction.

We show that the cost function of Weighted Abduction is converted into equivalent FFNNs as shown in Figure 4. This indicates the FFNN can express the cost function of Weighted Abduction. Therefore, we are able to apply various techniques in FFNNs to learning parameters of Weighted Abduction. Namely, gradients of the loss function can be caluclated easily by using the backpropagation technique of FFNNs.

Moreover, FFNNs are flexible framework and can express various functions by changing the activation functions or the network's structure. Thus, this idea can be apply to not only Weighted Abduction but other various frameworks of abduction.

3.6 Procedures of Parameter Learning

The overall learning procedure is given in Algorithm 1. First, the solution hypothesis is inferred from observation O, and if it does not include gold literals τ, it is treated as a negative example H^- (Line 3-6). Next, the positive example H^+ is inferred from observation O^+ (Line 7,8). The loss is then calculated from the costs of H^+ and H^- (Line 9) H^+ and H^- is converted into FFNNs (Line 10). The gradient of the loss function for each nonzero cost literal is assigned to the corresponding node in the FFNN (Line 11-15). The gradients of the loss function for costs of the other literals are calculated by applying standard backpropagation to the converted FFNNs (Line 16). Updating the parameters is performed with these gradients (Line 17). The parameters are trained iteratively until the learning converges.

Algorithm 1. Parameter learning

1: **Input:** $B, \theta, \mathbb{D}$
2: **repeat**
3:   **for all** $(O, \tau) \in \mathbb{D}$ **do**
4:     $\hat{H} \leftarrow Inference(O, \theta)$
5:     **if** $\tau \not\subseteq \hat{H}$ **then**
6:       $H^- \leftarrow \hat{H}$
7:       $O^+ = O \cup \{t^{\$\infty} \mid t \in \tau\}$
8:       $H^+ \leftarrow Inference(O^+, \theta)$
9:       $E_{O,\theta} \leftarrow LossFunction(H^+, H^-)$
10:      $N \leftarrow MakeFFNN(H^+, H^-)$
11:      **for all** $h \in P_{H^+} \cup P_{H^-}$ **do**
12:        **if** $c(h) > 0$ **then**
13:          assign gradient $\frac{\partial E_{O,\theta}}{\partial c(h)}$
14:        **end if**
15:      **end for**
16:      do backpropagation
17:      $\theta \leftarrow UpdateWeights(\theta, \nabla E_{O,\theta})$
18:    **end if**
19: **end for**
20: **until** convergence
21: **Output:** θ

---

4   Evaluation

We evaluate the proposed learning procedure on the dataset of plan recognition. In this experiment, we address the following questions: (i) does our leaning procedure actually decrease prediction errors? (ii) are models trained by our learning procedure robust to unseen data? To answer these questions, we evaluate prediction performance on a plan recognition dataset in the two settings: a closed test (i.e., the same dataset is used for both training and testing) and an open test (i.e., two distinct datasets are used for training and testing). In order to obtain the lowest-cost hypotheses, we used the Integer Linear Programming-based abductive reasoner proposed by [7].

4.1 Dataset

We used [21]'s story understanding dataset, which is widely used for evaluation of abductive plan recognition systems [10,18,12]. In this dataset, we need to abductively infer the top-level plans of characters from actions which are represented by the logical forms. For example, given *"Bill went to the liquor-store. He pointed a gun at the owner,"* plan recognition systems need to infer *Bill*'s plan. The dataset consists of development set and a test set, each of which includes 25 plan recognition problems. The dataset contains on average 12.6 literals in observed logical forms. The background knowledge base contains of 107 Horn clauses.

In our evaluation, we introduced two types of axioms in addition to the original 107 axioms. First, to make the predicates representing top-level plans (e.g. shopping, robbing) disjoint, we generated 73 disjointness axioms (e.g. $robbing(x) \Rightarrow \neg shopping(x)$). Note that it is still possible to infer multiple top-level plans for one problem, because we are able to hypothesize $robbing(x) \wedge shopping(y)$. Second, we generated axioms of superplan-subplans relations (e.g. $going\_by\_plane(x) \Rightarrow going\_by\_vehicle(x)$). In total, we used 220 background axioms for our evaluation.

For evaluating the prediction performance of our system, we focused on how well the system infers top-level plans, and their subparts (i.e. subplans, role-fillers), following [12]. More specifically, we use precision (ratio of inferred literals that are correct), recall (ratio of correct literals that are inferred by the system), and F-measure (harmonic mean of precision and recall), because the gold data often has multiple top-level plan predicates.

4.2 Experimental Setting

We applied weight regularization in order to prevent overfitting to the training set. The hyperparameter for regularization λ was set to 0.1. For parameter updating, we employed the annealing approach; $\theta_{new} = \theta - \eta_0 k^i \nabla E_\theta$ where η_0 (initial learning rate) was set to 0.0001, k (annealing parameter) was set to 0.95 and i is the number of iterations. The hyperparameters were selected based on performances on the development set. All weights were initialized to 0.0.

4.3 Results and Discussion

At first, we report results of the closed test where the development set was used for both training and testing. Figure 5 shows the values of the loss function at each iteration on the development set. The curve indicates that our learning procedure successfully reduces values of the loss function at each iteration. The reason for the fluctuation in values is thought to be the existence of hidden variables.

In the open test, we trained our model on the development set and then tested on the test set. Figure 6 shows plots of values of the three measures (i.e. Precision, Recall and F-measure) on the test set at each iteration. Although the values are also fluctuate as with the closed test, performance rises in terms of all measures compared to the performances at iteration zero (i.e. initial values). The results suggest that the learning procedure is robust to unseen data.

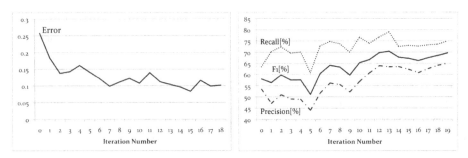

Fig. 5. Loss function values (closed test)   **Fig. 6.** Open test results

Singla and Mooney [12] report that the MLN-based approach achieve 72.10 F-measure on the same test set, which is slightly better than our results. However, our experimental setting and Singla's are different on various point such as framework of abduction (i.e. Weighted Abduction vs. MLN-based abduction), method of parameter learning (i.e. FFNNs vs. MLNs), method of parameter initialization (i.e. constant value vs. manually tuning). Therefore, it is unable to compare usefulness of these frameworks.

It has taken about half an hour to perform training for each iteration. Most of the time was spent in obtaining solution hypotheses using ILP-based abductive inference.

5   Related Work

As mentioned in Section 1, abduction has been extensively studied in a wide range of contexts. However, less attention has been paid to how to automatically learn cost functions. In the field of Statistical Relational Learning, some researchers [10,12,11, etc.] employ Markov Logic Networks [13] to emulate abductive inference. MLNs provide well-studied software packages of inference and learning.

However, MLN-based approaches require special procedures to convert abduction problems into deduction problems because of the deductive nature of MLNs. The pioneering work of MLN-based abduction [10] converts background axioms into MLN logical formulae by (i) reversing implication and (ii) constructing axioms representing mutual exclusiveness of explanation (e.g. the set of background knowledge axioms $\{p_1 \rightarrow q, p_2 \rightarrow q, p_3 \rightarrow q\}$ is converted into the following MLN formulae: $q \rightarrow p_1 \vee p_2 \vee p_3$, $q \rightarrow \neg p_1 \vee \neg p_2$, $q \rightarrow \neg p_1 \vee \neg p_3$ etc.). As the readers can imagine, MLN-based approach suffers from the inefficiency of inference due to the increase of converted axioms. Therefore, learning would not scale to larger problems due to the severe overhead [22]. [12] report that their MLN-based abduction models cannot be trained in larger dataset.

Moreover, when MLN-based approaches are applied to abduction-based discourse processing, a critical problem arises. MLN-based approaches represent a hypothesis as a truth assignment to ground atoms in the Herbrand base of background knowledge, while our framework represents a hypothesis as a set of first-order literals or equalities of logical variables. This means that a hypothesis generated by MLN-based approaches loses the first-order information in the input text. As shown in Section 1, each logical variable in the observation corresponds to a mention in the discourse; thus losing this information would be a serious drawback in discourse processing. For example, suppose that MLN-based approaches produce the hypothesis $president(A)$, $male(A)$, $doctor(B)$, $male(B)$ (A and B are constants) to the observation $\exists p, m_1, d, m_2 \{president(p) \wedge male(m_1) \wedge doctor(d) \wedge male(m_2)\}$. Then, we can interpret this hypothesis as two types of first-order logical forms: $president(p) \wedge male(m_1) \wedge doctor(d) \wedge male(m_2) \wedge p = m_1 \wedge d = m_2$, or $president(p) \wedge male(m_1) \wedge doctor(d) \wedge male(m_2) \wedge p = m_2 \wedge d = m_1$. This means that we cannot decide which discourse mentions are identified as coreferential in the hypothesis generated by MLN-based approaches. Some previous work [23,24] represent coreference relations by introducing special predicates that describe two logical variables are equal, but they use MLNs to create a classifier (i.e. binary log-linear classification model that utilizes a number of features) rather than reasoner. Therefore, it is a non-trivial issue to use these coreference representations with logical inference aimed at complicated commonsense reasoning, which is our goal in abudction-based discourse processing.

6   Conclusion

We have proposed a supervised approach for learning the cost function of Weighted Abduction. We formulated the learning procedure in the framework of structured learning with hidden variables. Our approach enables us to learn the non-linear cost function from partial abductive explanations, which is the typical situation in real-life tasks because constructing complete abductive explanations is usually a cost-consuming task. To the best of our knowledge, this is the first work to address the issue of automatic parameter learning of the cost function of Weighted Abduction, which can evaluate both the correctness and informativeness of explanations. In our evaluation, we found that our learning procedure can reduce the value of loss function in each iteration, and learned weights are also robust to unseen dataset.

Our future work includes large-scale evaluation of our learning procedure. We plan to evaluate our procedure on the popular natural language processing tasks, coference resolution with a massive set of axioms extracted from several language resources (e.g. WordNet [2]). It is also a problem that it takes long time to training weights. This problem will be critical in training on a large data set. We will address this problem by improving of abductive reasoner and optimization methods. As discussed in [1], coreference relation correponds to the unification of two logical variables. We therefore plan to incorporate a term that represents the cost of variable unification in the cost function of Weighted Abduction.

Acknowledgments. This work is partially supported by Grant-in-Aid for JSPS Fellows (22-9719), Grant-in-Aid for Scientific Research (23700157, 23240018), and PRESTO, JST.

References

1. Hobbs, J.R., Stickel, M., Martin, P., Edwards, D.: Interpretation as abduction. Artificial Intelligence 63, 69–142 (1993)
2. Fellbaum, C. (ed.): WordNet: an electronic lexical database. MIT Press (1998)

3. Ruppenhofer, J., Ellsworth, M., Petruck, M., Johnson, C., Scheffczyk, J.: FrameNet II: Extended Theory and Practice. Technical report, Berkeley, USA (2010)
4. Chambers, N., Jurafsky, D.: Unsupervised Learning of Narrative Schemas and their Participants. In: ACL, pp. 602–610 (2009)
5. Schoenmackers, S., Davis, J., Etzioni, O., Weld, D.: Learning First-order Horn Clauses from Web Text. In: EMNLP, pp. 1088–1098 (2010)
6. Hovy, D., Zhang, C., Hovy, E., Penas, A.: Unsupervised discovery of domain-specific knowledge from text. In: ACL, pp. 1466–1475 (2011)
7. Inoue, N., Inui, K.: ILP-Based Reasoning for Weighted Abduction. In: AAAI Workshop on Plan, Activity and Intent Recognition (2011)
8. Ovchinnikova, E., Montazeri, N., Alexandrov, T., Hobbs, J.R., McCord, M., Mulkar-Mehta, R.: Abductive Reasoning with a Large Knowledge Base for Discourse Processing. In: IWCS, Oxford, UK, pp. 225–234 (2011)
9. Dagan, I., Dolan, B., Magnini, B., Roth, D.: Recognizing textual entailment: Rational, evaluation and approaches - Erratum. NLE 16, 105 (2010)
10. Kate, R.J., Mooney, R.J.: Probabilistic Abduction using Markov Logic Networks. In: PAIRS (2009)
11. Blythe, J., Hobbs, J.R., Domingos, P., Kate, R.J., Mooney, R.J.: Implementing Weighted Abduction in Markov Logic. In: IWCS, Oxford, UK, pp. 55–64 (2011)
12. Singla, P., Domingos, P.: Abductive Markov Logic for Plan Recognition. In: AAAI, pp. 1069–1075 (2011)
13. Richardson, M., Domingos, P.: Markov logic networks. In: ML, pp. 107–136 (2006)
14. Huynh, T.N., Mooney, R.J.: Max-Margin Weight Learning for Markov Logic Networks. In: Proceedings of the International Workshop on Statistical Relational Learning, SRL 2009 (2009)
15. Lowd, D., Domingos, P.: Efficient Weight Learning for Markov Logic Networks. In: Kok, J.N., Koronacki, J., Lopez de Mantaras, R., Matwin, S., Mladenič, D., Skowron, A. (eds.) PKDD 2007. LNCS (LNAI), vol. 4702, pp. 200–211. Springer, Heidelberg (2007)
16. Charniak, E., Goldman, R.P.: A Probabilistic Model of Plan Recognition. In: AAAI, pp. 160–165 (1991)
17. Poole, D.: Probabilistic Horn abduction and Bayesian networks. Artificial Intelligence 64 (1), 81–129 (1993)
18. Raghavan, S., Mooney, R.J.: Bayesian Abductive Logic Programs. In: STARAI, pp. 82–87 (2010)
19. Ovchinnikova, E.: Integration of World Knowledge for Natural Language Understanding. Atlantis Press (2012)
20. Charniak, E., Shimony, S.E.: Probabilistic semantics for cost based abduction. In: AAAI, pp. 106–111 (1990)
21. Ng, H.T., Mooney, R.J.: Abductive Plan Recognition and Diagnosis: A Comprehensive Empirical Evaluation. In: KR, pp. 499–508 (1992)
22. Inoue, N., Inui, K.: Large-scale Cost-based Abduction in Full-fledged First-order Logic with Cutting Plane Inference. In: Proceedings of the 12th European Conference on Logics in Artificial Intelligence (2012) (to appear)
23. Poon, H., Domingos, P.: Joint unsupervised coreference resolution with markov logic. In: Proceedings of EMNLP, pp. 650–659 (2008)
24. Song, Y., Jiang, J., Zhao, W.X., Li, S., Wang, H.: Joint learning for coreference resolution with markov logic. In: Proceedings of the 2012 Joint Conference on Empirical Methods in Natural Language Processing and Computational Natural Language Learning, pp. 1245–1254. ACL (2012)

Facilitating the Analysis of Discourse Phenomena in an Interoperable NLP Platform

Riza Theresa Batista-Navarro*, Georgios Kontonatsios*, Claudiu Mihăilă*,
Paul Thompson*, Rafal Rak, Raheel Nawaz,
Ioannis Korkontzelos, and Sophia Ananiadou

The National Centre for Text Mining,
The University of Manchester,
131 Princess Street, Manchester M1 7DN, UK
{rbatistanavarro,gkontonatsios,cmihaila,pthompson,
rrak,rnawaz,ikorkontzelos,sananiadou}@cs.man.ac.uk

Abstract. The analysis of discourse phenomena is essential in many natural language processing (NLP) applications. The growing diversity of available corpora and NLP tools brings a multitude of representation formats. In order to alleviate the problem of incompatible formats when constructing complex text mining pipelines, the Unstructured Information Management Architecture (UIMA) provides a standard means of communication between tools and resources. U-Compare, a text mining workflow construction platform based on UIMA, further enhances interoperability through a shared system of data types, allowing free combination of compliant components into workflows. Although U-Compare and its type system already support syntactic and semantic analyses, support for the analysis of discourse phenomena was previously lacking. In response, we have extended the U-Compare type system with new discourse-level types. We illustrate processing and visualisation of discourse information in U-Compare by providing several new deserialisation components for corpora containing discourse annotations. The new U-Compare is downloadable from http://nactem.ac.uk/ucompare.

Keywords: UIMA, interoperabilty, U-Compare, discourse, causality, coreference, meta-knowledge.

1 Introduction

One of the most important outcomes of recent research into biomedical text mining is the large number of newly created, manually annotated corpora. Such corpora were originally designed for training and evaluating systems that perform specific tasks. However, such resources can additionally provide support for other tasks. Data reuse is both highly demanded and occurs frequently, as it saves important amounts of human effort, time and money. For instance, the

---

* The authors have contributed equally to the development of this work and production of the manuscript.

A. Gelbukh (Ed.): CICLing 2013, Part I, LNCS 7816, pp. 559–571, 2013.
© Springer-Verlag Berlin Heidelberg 2013

GENIA corpus [1], which initially contained only named entity annotations, has subsequently been extended by different researchers and groups to include event annotations and meta-knowledge information [2].

Corpora containing more complex types of semantic information, including discourse phenomena, have begun to appear recently. *Discourse* is defined as a coherent sequence of textual zones (e.g., clauses and sentences). Different zones contribute different information to the discourse, and are connected logically by *discourse relations*, which are characterised by labels such as "causal", "temporal" and "conditional". In turn, these allow more complex knowledge about the facts mentioned in the discourse to be inferred. Discourse relations can be either explicit or implicit, depending on whether or not they are represented in text using overt *discourse connectives* (or *triggers*). Discourse annotations cover several types of information that attempt to explain how the structure of text leads to a coherent discourse. They include:

- classification of previously mentioned textual zones and structured events (e.g., biological processes) according to their general interpretation (e.g., background knowledge, hypotheses, observations, etc.) in the discourse
- determining which textual zones are linked together in terms of discourse relations (e.g., causality) and how these links should be characterised
- resolving coreference within or across sentences.

Understanding discourse-level information is important for several tasks, such as automatic summarisation and question answering [3, 4]. Due to the complex nature of discourse phenomena, their annotation is extremely time-consuming and hence there is an urgent need to make these annotated corpora available to the community in a format that is readily usable.

As mentioned above, annotated corpora constitute an important resource for the development of NLP tools, e.g., tokenisation, part-of-speech tagging and syntactic parsing. Individually, these tools do not usually address a complete NLP task, but when combined together into workflows, they become much more powerful. For example, each of the above-mentioned tool types is a prerequisite for the next: part-of-speech tagging must be preceded by tokenisation, whilst tokens with parts-of-speech are needed as input to syntactic parsers. Discourse analysis, a more complex task, requires a range of linguistic pre-processing of input texts, and hence a pipeline of tools must be run prior to executing a discourse analysis tool.

In order to allow such multi-step pipelines to be built with the minimum effort, it is desirable to combine the different NLP tools in a straightforward manner. Unfortunately, for many existing tools, such combination is not trivial due to the fact that tools are implemented in different programming languages, and have differing input and output formats. Evaluation of tools against gold standard corpora may also be problematic if the annotation format is different from the output of the tool. Thus, creating and evaluating text processing pipelines can require a considerable amount of effort from developers, in order to allow different resources (i.e., tools and corpora) to "communicate". The lack of ease with which resources can be reused can be a major obstacle to the rapid creation of complex NLP applications.

The Unstructured Information Management Architecture (UIMA) [5] is a framework that offers a solution to the problem of interoperability of NLP resources by providing a standardised mechanism by which resources communicate. However, UIMA does not impose any recommended set of types, and developers are left to define their own types.

U-Compare is a UIMA-based, text mining workflow construction platform that enables interoperability of NLP components by defining a *sharable* type system. This covers a wide range of annotation types that are common to many NLP components. The type system was designed such that there are several hierarchical levels which can be extended according to the individual requirements of different components, yet it is possible for components to be compatible at higher levels of the hierarchy. A further feature of U-Compare is its graphical user interface, via which users can rapidly create NLP pipelines using a simple drag-and-drop mechanism, with no requirement for programming skills. A key functionality of U-Compare is its evaluation and comparison mechanism that allows users to identify optimal NLP pipelines [6]. Users can also visualise agreements and disagreements between different pipelines directly in the text.

U-Compare contains the world's largest repository of type-system compatible modules [7]. Users can freely combine these components into pipelines that form syntactic (e.g., part-of-speech tagging, dependency parsing) or semantic (e.g., named entity recognition, event extraction) aggregate applications. However, despite this large library of components, U-Compare did not previously support the analysis of discourse phenomena. In this paper, we describe our work to extend U-Compare in order to facilitate the construction of NLP pipelines that annotate discourse phenomena in free text. The contributions of this paper can be summarised as follows: (a) expansion of the U-Compare type system to model discourse phenomena, (b) facilitation of the visualisation of discourse annotations, and (c) implementation of data deserialisers (readers) for various corpora containing discourse phenomena annotation, which can be used within UIMA processing pipelines and within U-Compare in particular. As a consequence, U-Compare provides a convenient environment in which to test and evaluate discourse analysis components. Given the wide range of available lower-level processing tools available in U-Compare, most tools that produce the types of information required as input to discourse components are present in the U-Compare library. Annotations produced by the new discourse components are visualised automatically, and different pipelines can be easily evaluated against various discourse phenomena corpora.

2   Discourse Phemomena and Corpora

In this section, we discuss in greater detail the three specific types of discourse phenomena that have been the focus of our extensions to U-Compare, i.e., causality, coreference and characterisation/interpretation of discourse segments. We subsequently explain the characteristics of the annotated corpora that have been made interoperable with other U-Compare components.

2.1 Causality in the Biomedical Domain

Statements regarding causal associations have long been studied in general language, mostly as part of more complex tasks. Comparatively little work has been carried out in the biomedical domain, although causal associations between biological entities, events and processes are central to most claims of interest [8]. Despite this, a unified theory of causality has not yet emerged, be it in general or specialised language. Many tasks, such as question answering and automatic summarisation, require the extraction of information that spans across several sentences, together with the recognition of relations that exist across sentence boundaries, in order to achieve high levels of performance. Take, for instance, the case in example 1, where the trigger *Therefore* signals a justification between the two sentences: because "a normal response [...] glutamic acid residues", the authors believe that the "regulation of PmrB [...] some amino acids".

(1) In the case of PmrB, a normal response to mild acid pH requires not only a periplasmic histidine but also several glutamic acid residues.
Therefore, regulation of PmrB activity may involve protonation of one or more of these amino acids.

Not all causality relations are as obvious as the previous one, where the trigger is explicit and is typically used to denote causality. In example 2, there is an implicit discourse causal association between the first half of the sentence, "This medium lacked [...] 10 mM MgCl2", and the latter half, "which represses [...] PmrA-activated genes". This is due to the fact that, generally, bacterial gene expression could be affected by specific properties of growth media, such as pH and concentration of metals.

(2) This medium lacked $Fe3+$ or $Al3+$, the only known PmrB ligands (Wosten et al., 2000), and contained 10 mM MgCl2, which represses expression of PmrA-activated genes (Soncini and Groisman, 1996; Kox et al., 2000).

Amongst the large number of corpora that have been developed for biomedical text mining, several include the annotation of statements regarding causal associations, such as GENIA [1] and GREC [9]. However, these corpora do not include an exhaustive coverage of causal statements and the granularity of the annotation of such statements is rather limited. Other corpora, e.g., BioCause [10] and BioDRB [11], contain annotations of causality and other discourse relations in biomedical journal articles, and here we focus on these.

2.2 Coreference

Coreference is a phenomenon involving linguistic expressions referring to a unique referent [12]. It is often associated with the phenomenon of *anaphora*, which is characterised by an expression (*anaphor*) whose interpretation depends on an entity previously mentioned in the same discourse (*antecedent*). Coreference and anaphora are extensively used in both spoken and written discourse, and are central to discourse theories such as the Centering Theory [13], a theory of local discourse coherence.

The task of grouping all coreferring mentions in text into respective coreference chains is known as *coreference resolution*. *Anaphora resolution*, on the other hand, is the process of determining the antecedent of an anaphor. Whilst the output of anaphora resolution is a set of anaphor-antecedent pairs, that of coreference resolution is a set of coreference chains which can be treated as equivalence classes. Despite this difference, an overlap between them may be observed in several cases. Often, an anaphor and its antecedent are coreferential (i.e., they have the same referent) and may be placed in the same coreference chain, as in the following example:

(3) John gave his wife a necklace for her birthday. She thanked him for it.

Example 3 contains the following coreference chains: {*John, his, him*}, {*his wife, her, She*} and {*necklace, it*}. The anaphor and antecedent in each of the anaphoric pairs {*his, John*}, {*her, his wife*}, {*She, his wife*}, {*him, John*} and {*it, necklace*} fall within the same chain. In some scenarios, however, they do not fall within the same chain, such as in the following example:

(4) Peter received his paycheque yesterday but John didn't get one.

Example 4 includes the anaphoric pairs {*his, Peter*} and {*one, his paycheque*}. Whilst *his* and *Peter* refer to the same person and thus belong to the same coreference chain, *one* and *his paycheque* do not corefer since the paycheque that John is expecting is a different entity from Peter's paycheque. The coreference chains in this example, therefore, are: {*Peter, his*}, {*his paycheque* (Peter's)}, {*yesterday*}, {*John*} and {*one* (John's paycheque)}.

The phenomenon of coreference has diverse representation formats: in some corpora, coreferring mentions are represented as coreference chains; in most of them, however, they are linked in the same way as an anaphor and its antecedent. In the 2005 Automatic Content Extraction (ACE'05) corpus [14], coreferring mentions pertaining to the same entity/relation/event are grouped together, forming a coreference chain. In contrast, in GENIA MedCo [15], HANAPIN [16] and the BioNLP Shared Task [17] corpora, a pair of mentions referring to the same entity are linked like an anaphor and its antecedent are.

Both the coreference and anaphora resolution tasks are non-trivial and are still considered unsolved, especially in scientific text. Providing NLP researchers with access to tools and resources for these tasks is hence desirable for the development and improvement of coreference and anaphora resolution systems.

2.3   Meta-knowledge

There have been several efforts to annotate and automatically recognise aspects of the characterisation or interpretation of textual discourse segments.

Collectively, these different aspects have been defined as *meta-knowledge* [2]. Recognition of meta-knowledge is important for tasks such as isolating new knowledge claims in research papers [18], maintaining models of biomedical processes [19] or curating biomedical databases [20]. The following sentences expressing similar information illustrate different aspects of meta-knowledge:

(5) It is *known* that leukotriene B4 (LTB4) affects the expression of the proto-oncogenes c-jun and c-fos.

(6) We *examined* whether leukotriene B4 (LTB4) affects the expression of the proto-oncogenes c-jun and c-fos.

(7) *Previous studies* have shown that leukotriene B4 (LTB4) does *not* affect the expression of the proto-oncogenes c-jun and c-fos.

(8) These results *suggest* that leukotriene B4 (LTB4) *partially* affects the expression of the proto-oncogenes c-jun and c-fos.

The italicised words and expressions highlight how the interpretations of these sentences differ in both subtle and more significant ways. In sentence (5), the word *known* indicates that generally accepted background knowledge is being expressed, whilst in sentence (6), the word *examined* denotes the very different interpretation that the truth value of the information described is under investigation. In sentence (7), a finding coming from a previous study is reported (*Previous studies*), and the presence of the word *not* indicates that this is a negated finding. In sentence (8), an analysis of results is presented, as denoted by *suggests*, whilst the word *partially* indicates that the reaction expressed by the sentence occurs at lower intensity than would be expected by default.

Most previous work on meta-knowledge annotation and recognition has been concerned with annotation of scientific texts at the sentence level with a single aspect of meta-knowledge, usually negation/speculation detection [21] or classification according to rhetorical function or general information content, using categories such as background knowledge, methods, hypotheses, experimental observations, etc. Different schemes and classifiers (both rule and machine learning-based) have been created for both abstracts [22] and full papers [23].

A small number of annotation schemes have considered more than one aspect or dimension of meta-knowledge. For example, the CoreSC annotation scheme [24] augments general information content categories with additional attributes, such as *New* and *Old*, to denote current or previous work. In order to account for the fact that longer sentences may contain several types of information, another corpus [25] annotated sentence segments with five different aspects of meta-knowledge. A related effort [2] also annotates five meta-knowledge dimensions, i.e., knowledge type, certainty level, manner, polarity and knowledge source, but at the level of *events* rather than segments. Events are structured representations of pieces of biomedical knowledge, formed of multiple discontinuous spans of text. They usually consist of an *event trigger*, i.e., a word or phrase that characterises the event, together with its arguments, e.g., theme, agent, cause, location, etc. Like segments, there may be multiple events in a single sentence. Annotation of meta-knowledge at the event level can be used to train event extraction systems that allow constraints to be specified not only in terms of event types/participants, but also in terms of meta-knowledge interpretations of individual events [26].

3   Related Work

Text mining platforms alleviate the requirement for programming and technical skills and thus minimise the effort required to conduct *in silico* experiments. Text mining frameworks that support the construction of workflows define a common, standardised and well-defined architecture for developing pipelined applications and are becoming highly popular and widely applicable. Of the numerous text mining workflow construction platforms currently available [27–30], we focus on those that enable interoperability of resources and allow the representation of discourse phenomena.

GATE [28] is a workflow construction framework that has been used for the development of various text mining applications (e.g., automatic summarisation, cross-lingual mining and ontology building). It supports the representation and processing of the discourse phenomenon of coreference. However, GATE mainly focusses on assisting in specific programming tasks (i.e., large collection of software libraries) rather than interoperability of text mining resources. It also lacks a common, hierarchical data type system. In contrast, U-Compare comes with a library of heterogenous components (i.e., native components written in various languages or web services) which can communicate with each other under one common type system.

Heart of Gold [29] is an XML-based architecture primarily aimed at the development of text analysis pipelines. It focusses on the interoperability of components, allowing tools developed using different programming languages to be combined. However, it does not support the processing nor representation of discourse phenomena.

Specifically focussed on the analysis of medical records, cTAKES [27] was developed as another UIMA-based workflow construction platform. As in the case of U-Compare, cTAKES defines UIMA types, but it supports the representation and analysis of only two discourse phenomena, namely coreference and negation.

Argo [30] is a web-based, UIMA-compliant workflow construction platform that supports the U-Compare type system. The two platforms are therefore fully compatible (i.e., components and workflows are interchangable between U-Compare and Argo). Currently, Argo does not support the analysis of discourse phenomena. However, due to the compatibility of the two platforms, the type system extensions described here are also applicable to Argo.

4   Discourse Phenomena in U-Compare

In this section, we discuss how the U-Compare type system has been extended to allow the representation of the discourse phenomena that we are considering, namely, causality, coreference and meta-knowledge.

4.1   Type System Extensions

The U-Compare type system can be divided into three hierarchies, namely syntactic (e.g., part-of-speech tags), document (e.g., title, author) and semantic (e.g., named entities) levels. We have introduced a new sub-hierarchy of semantic annotations that allows us to model discourse phenomena. The extended U-Compare type system is depicted in Figure 1. As can be observed from the figure, the newly created *Discourse phenomena hierarchy* has been added within the *Semantic hierarchy*. In this new hierarchy, `DiscoursePhenomenon` is the top level discourse data type, from which various annotations descend. These include three classes of annotations: one for events (`EventAnnotation`), one for coreference (`CoreferenceAnnotation`) and one for discourse relations (`DiscourseAnnotation`).

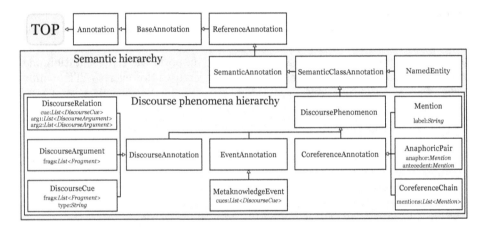

Fig. 1. The extended U-Compare type system

A discourse annotation is modelled using two text-bound data types, whilst a third type links the previous two. `DiscourseCue` marks discourse connectives that can either cover spans of text, in the case of explicit triggers, or have a length of zero, in the case of implicit triggers. In the latter case, the trigger that best fits the situation (if provided by annotators) is displayed in the annotation panel of U-Compare. `DiscourseArgument` represents a span of text that plays a role in the discourse relation. Both types are modelled using `Fragments` in order to allow the representation of discontinous annotations. `DiscourseRelation` combines the previous types to create a discourse relation. It contains three lists, one for discourse connectives and two for each of the arguments of the relation.

Using this abstract hierarchy of data types, it becomes possible to directly model a variety of discourse relations, such as causality, temporality or conditionality. Moreover, relations from different resources become interoperable and can be seamlessly combined and processed by existing text analysis components in U-Compare. Figure 2 illustrates a causal relation from the BioCause corpus displayed in the U-Compare GUI. As can be noticed, the two arguments and the trigger are underlined. Moreover, there are two arrows connecting the trigger with each of the arguments. The arrows also display the role of each argument in the relation.

This acid pH-promoted increase appears to be specific to a subset of PhoP-activated genes (our unpublished results) that includes pmrD because expression of the PhoP-regulated
                                                Evidence   Effect
slyA gene and the PhoP-independent corA gene was not affected by the pH of the medium.

Fig. 2. BioCause Causality

In the case of coreference, we have introduced three data types under `Coref-erenceAnnotation`. `Mention` represents any expression annotated in text, whilst `AnaphoricPair` consists of two `Mention` objects corresponding to an anaphor and its antecedent. In Figure 3, for example, sample coreference annotations in the HANAPIN corpus [16] are represented in the GUI as three `AnaphoricPair` objects. The last data type is `CoreferenceChain`, which consists of a list of `Mentions` referring to the same entity.

A new cyclic diamine, 1,5-diazacyclohenicosane (1), was isolated from
                 Coref   Coref
samples of the marine sponge Mycale sp. collected at Lamu Island (Kenya).
                 Coref

Fig. 3. HANAPIN Coreference

`CoreferenceChain` was used, for instance, in representing the grouping of the underlined mentions in Figure 4, which all belong to the same coreference chain in the ACE'05 corpus. The newly added data types allow for the representation of different types of coreference annotations found in various corpora, enabling U-Compare to interpret them regardless of whether they come in the form of coreference chains or anaphoric pairs.

You know, Senator, inadvertently, you gave support in this book to one of
      Coref   Coref   Coref
my pet political themes, which you probably don't agree with. On page 256
           Coref
you say "As the year went by" -- this is when you were in the Senate.
                Coref

Fig. 4. ACE'05 Coreference

The extensions to the U-Compare type system for meta-knowledge are geared towards its representation at the event level. This information is encoded by extending the existing `Event` annotation type with a new type `MetaKnowledgeEvent`, which includes attributes to store the values of all five meta-knowledge dimensions. In addition, the `DiscourseCue` type (previously introduced in the context of causality) is used to annotate different types of meta-knowledge (e.g., negation or certainty level cues). There is a *type* attribute, which allows the dimension that the cue represents (e.g., certainty level) to be encoded.

The representation of event-based meta-knowledge annotation in U-Compare is illustrated in Figure 5. Two event triggers are underlined, i.e., "activation" and "increases". Each of the two events has different meta-knowledge information associated with it. The word "partial" shows the intensity (manner cue) of the "activation" event. For the "increases" event, the word "suggests" gives information about both the knowledge type (KT) and certainty level (CL) of the event: it shows that the event is the result of an analysis based on the previous "activation" event and that there is a lack of complete certainty about the event on the part of the author.

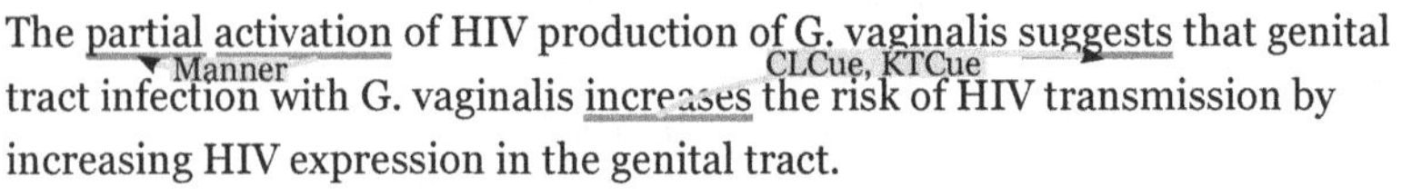

The partial activation of HIV production of G. vaginalis suggests that genital
     ▼ Manner   CLCue, KTCue
tract infection with G. vaginalis increases the risk of HIV transmission by
increasing HIV expression in the genital tract.

Fig. 5. Meta-knowledge

4.2   From Resource Readers to Semantic Labellers

We have made a number of tools available through U-Compare that automatically read various discourse annotated corpora, create UIMA annotations that include the newly introduced types described above, and visualise these annotations in U-Compare. The new corpus readers are capable of handling several corpora annotated with the discourse phenomena previously mentioned. These include the GENIA Event Corpus [2], BioCause [10], BioDRB [11], ACE'05 (Coreference) [14], BioNLP Shared Tasks [17], GENIA MedCo [15] and HANAPIN [16]. The above resources are interoperable with U-Compare's existing processing components. They can be combined into pipelines using a drag-and-drop graphical user interface, without any requirement for programming skills. The only prerequisite is that the input types of a component are compliant with the output types of preceding components.

We present a use case in which heterogeneous components have been combined with the aim of determining the semantic labels of otherwise semantically empty expressions (e.g., pronouns) in the GENIA MedCo corpus.

The created pipeline consists of the following components:

1. Collection Reader: GENIA MedCo Coreference Reader
2. Syntactic components: OSCAR Tokeniser [6]
3. Semantic components: ABNER NLPBA [31] and OSCAR Maximum Entropy Markov Model [6] Named Entity Recognisers (NERs)

Figure 6 illustrates an example document from the corpus annotated by this pipeline. With the incorporation of NERs into the pipeline, "NAC" and "DC", for example, have been identified as chemical molecules (CM), allowing users to infer that the pronouns *its* and *their* also refer to chemical molecules. If it is desired that coreferential expressions inherit the semantic types predicted by the NERs, our new Coreference Annotation Merger component can be optionally included in the pipeline to propagate labels to the coreferential expressions. Such propagation of semantic labels will allow users to subsequently extract relations or events involving coreferential expressions [32].

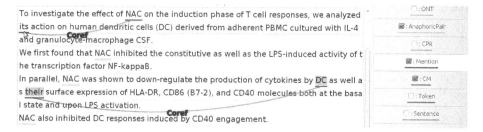

To investigate the effect of NAC on the induction phase of T cell responses, we analyzed its action on human dendritic cells (DC) derived from adherent PBMC cultured with IL-4 and granulocyte-macrophage CSF.
We first found that NAC inhibited the constitutive as well as the LPS-induced activity of the transcription factor NF-kappaB.
In parallel, NAC was shown to down-regulate the production of cytokines by DC as well as their surface expression of HLA-DR, CD86 (B7-2), and CD40 molecules both at the basal state and upon LPS activation.
NAC also inhibited DC responses induced by CD40 engagement.

Fig. 6. Sample document with results of our semantic labelling pipeline

5 Conclusions and Future Work

In this paper, we have presented extensions that were implemented on top of the existing workflow construction platform U-Compare to allow the processing and visualisation of discourse phenomena. We have introduced a new hierarchy of annotation types, now integrated into the type system of U-Compare, that models various discourse phenomena, including causality, coreference and meta-knowledge. Furthermore, we have implemented readers for several annotated corpora that can read and map the annotations to U-Compare's type system and display complex relations (e.g., coreference chains and causal relations) for easy interpretation. Our goal is to enhance the interoperability of such resources so that they can be seamlessly incorporated into text mining pipelines that process and analyse discourse phenomena. Moreover, building on the interoperability of U-Compare, we have shown in practice how discourse annotated corpora can become compatible with U-Compare's processing tools. As a case study, we have designed a pipeline that automatically assigns semantic labels to otherwise semantically empty expressions (e.g., pronouns) in the GENIA MedCo corpus.

It should be noted that automatically produced annotations cannot be included directly in gold standard annotated corpora. As post-processing work, curators need to manually remove erroneous tags. As future work, we plan to integrate tools that aim to reduce the human effort in producing or enriching gold-standard annotated corpora into U-Compare. Such tools [33] are incrementally trained on annotations produced by curators in order to correctly predict unseen annotations in text.

Acknowledgement. The work described in this paper has been funded by the MetaNet4U project (ICT PSP Programme, Grant Agreement: No 270893) and the JISC-funded "Integrated Social History Environment for Research (ISHER)-Digging into Social Unrest" project, which is part of the "Digging into Data Challenge."

References

1. Kim, J.D., Ohta, T., Tsujii, J.: Corpus annotation for mining biomedical events from literature. BMC Bioinformatics 9, 10 (2008)
2. Thompson, P., Nawaz, R., McNaught, J., Ananiadou, S.: Enriching a biomedical event corpus with meta-knowledge annotation. BMC Bioinformatics 12, 393 (2011)
3. Marcu, D.: The Theory and Practice of Discourse Parsing and Summarization. MIT Press, Cambridge (2000)
4. Sun, M., Chai, J.Y.: Discourse processing for context question answering based on linguistic knowledge. Knowledge-Based Systems 20, 511–526 (2007)
5. Ferrucci, D., Lally, A.: UIMA: an architectural approach to unstructured information processing in the corporate research environment. Natural Language Engineering 10, 327–348 (2004)
6. Kolluru, B., Hawizy, L., Murray-Rust, P., Tsujii, J., Ananiadou, S.: Using Workflows to Explore and Optimise Named Entity Recognition for Chemistry. PLoS ONE 6, e20181 (2011)
7. Kano, Y., Baumgartner Jr., W.A., McCrochon, L., Ananiadou, S., Cohen, K.B., Hunter, L., Tsujii, J.: U-Compare: share and compare text mining tools with UIMA. Bioinfomatics 25, 1997–1998 (2009)
8. Kleinberg, S., Hripcsak, G.: A review of causal inference for biomedical informatics. Journal of Biomedical Informatics 44, 1102–1112 (2011)
9. Thompson, P., Iqbal, S., McNaught, J., Ananiadou, S.: Construction of an annotated corpus to support biomedical information extraction. BMC Bioinformatics 10, 349 (2009)
10. Mihăilă, C., Ohta, T., Pyysalo, S., Ananiadou, S.: BioCause: Annotating and analysing causality in the biomedical domain. BMC Bioinformatics 14, 2 (2013)
11. Prasad, R., McRoy, S., Frid, N., Joshi, A., Yu, H.: The Biomedical Discourse Relation Bank. BMC Bioinformatics 12, 188 (2011)
12. Jurafsky, D., Martin, J.H.: Speech and Language Processing, 2nd edn. Prentice Hall Series in Artificial Intelligence. Prentice Hall (2008)
13. Grosz, B.J., Weinstein, S., Joshi, A.K.: Centering: A Framework for Modeling the Local Coherence of Discourse. Comp. Ling. 21, 203–225 (1995)
14. Walker, C.: ACE 2005 Multilingual Training Corpus (2006)
15. Su, J., Yang, X., Hong, H., Tateisi, Y., Tsujii, J.: Coreference Resolution in Biomedical Texts: a Machine Learning Approach. In: Ashburner, M., Leser, U., Rebholz-Schuhmann, D. (eds.) Ontologies and Text Mining for Life Sciences: Current Status and Future Perspectives. Dagstuhl Seminar Proceedings, vol. 08131 (2008)
16. Batista-Navarro, R.T.B., Ananiadou, S.: Building a coreference-annotated corpus from the domain of biochemistry. In: Proceedings of BioNLP 2011, pp. 83–91 (2011)
17. Stenetorp, P., Topić, G., Pyysalo, S., Ohta, T., Kim, J.D., Tsujii, J.: BioNLP Shared Task 2011: Supporting Resources. In: Proceedings of the BioNLP Shared Task 2011 Workshop, pp. 112–120. ACL (2011)
18. Sandor, A., de Waard, A.: Identifying Claimed Knowledge Updates in Biomedical Research Articles. In: Proceedings of the Workshop on Detecting Structure in Scholarly Discourse (DSSD), pp. 7–10 (2012)
19. Oda, K., Kim, J.D., Ohta, T., Okanohara, D., Matsuzaki, T., Tateisi, Y., Tsujii, J.: New challenges for text mining: mapping between text and manually curated pathways. BMC Bioinformatics 9, S5 (2008)
20. Yeh, A., Hirschman, L., Morgan, A.: Evaluation of text data mining for database curation: lessons learned from the KDD Challenge Cup. Bioinformatics 19, 331–339 (2003)

21. Medlock, B., Briscoe, T.: Weakly supervised learning for hedge classification in scientific literature. In: Proceedings of ACL, pp. 992–999 (2007)
22. McKnight, L., Srinivasan, P.: Categorization of sentence types in medical abstracts. In: Proceedings of the AMIA Annual Symposium, pp. 440–444 (2003)
23. Mizuta, Y., Korhonen, A., Mullen, T., Collier, N.: Zone analysis in biology articles as a basis for information extraction. Int. J. Med. Inf. 75, 468–487 (2006)
24. Liakata, M., Teufel, S., Siddharthan, A., Batchelor, C.: Corpora for the conceptualisation and zoning of scientific papers. In: Proceedings of LREC, pp. 2054–2061 (2010)
25. Wilbur, W.J., Rzhetsky, A., Shatkay, H.: New directions in biomedical text annotation: definitions, guidelines and corpus construction. BMC Bioinformatics 7, 356 (2006)
26. Miwa, M., Thompson, P., McNaught, J., Kell, D., Ananiadou, S.: Extracting semantically enriched events from biomedical literature. BMC Bioinformatics 13, 108 (2012)
27. Savova, G., Masanz, J., Ogren, P., Zheng, J., Sohn, S., Kipper-Schuler, K., Chute, C.: Mayo clinical Text Analysis and Knowledge Extraction System (cTAKES): architecture, component evaluation and applications. Journal of the American Medical Informatics Association 17, 507–513 (2010)
28. Cunningham, H., Hanbury, A., Rüger, S.: Scaling Up High-Value Retrieval to Medium-Volume Data. In: Cunningham, H., Hanbury, A., Rüger, S. (eds.) IRFC 2010. LNCS, vol. 6107, pp. 1–5. Springer, Heidelberg (2010)
29. Schäfer, U.: Middleware for creating and combining multi-dimensional NLP markup. In: Proceedings of the 5th Workshop on NLP and XML: Multi-Dimensional Markup in Natural Language Processing, pp. 81–84. ACL (2006)
30. Rak, R., Rowley, A., Black, W., Ananiadou, S.: Argo: an integrative, interactive, text mining-based workbench supporting curation. Database: The Journal of Biological Databases and Curation 2012 (2012)
31. Settles, B.: Biomedical named entity recognition using conditional random fields and rich feature sets. In: Proceedings of the International Joint Workshop on NLP in Biomedicine and its Applications, pp. 104–107. ACL (2004)
32. Gabbard, R., Freedman, M., Weischedel, R.: Coreference for Learning to Extract Relations: Yes Virginia, Coreference Matters. In: Proceedings of the 49th Annual Meeting of the Association for Computational Linguistics: Human Language Technologies, pp. 288–293. Association for Computational Linguistics, Portland (2011)
33. Tsuruoka, Y., Tsujii, J., Ananiadou, S.: Accelerating the annotation of sparse named entities by dynamic sentence selection. BMC Bioinformatics 9, S8 (2008)

Author Index